World War II
Harbor Defenses of San Diego

Commander (Ret.) H.R. (Bart) Everett, USN

Coast Defense Study Group Press
McLean, Virginia USA

PLEASE DIRECT ANY COMMENTS OR CORRECTIONS TO THE PUBLISHER - INFO@CDSG.ORG

IBSN 978-0-9748167-5-3 (Hardcover B&W)
LIBRARY OF CONGRESS CATALOG CARD NUMBER 2021938050

First Edition: August 2021
Printed in the USA by Ingram Spark

Front Cover Image: Drawing of Battery Ashburn's 16-inch gun in its casemate at Fort Rosecrans, CA.
Drawn by Wendy Kitchens

Title Page Image: San Diego, CA 1930 USGS Map

Rear Cover Image: USACE Plan for Battery Ashburn (BCN 126) 1942

Book Illustrations by Lee Guidry and Lindsay Seligman

Publisher Info: Coast Defense Study Group Press is a division of Coast Defense Study Group (www.cdsg.org), publishes books of historical interest, especially seacoast fortifications. Under the CDSG Press label, our organization has published:

- *Notes on Seacoast Fortification Construction* by Col. Eben E. Winslow, 1920, 428 pp. 1994 reprint HC with drawings
- *Seacoast Artillery Weapons Technical Manual (TM) 9-210* by U.S. War Dept. 1944, 202 pp. 1995 reprint PB
- *The Service of Coast Artillery* by F. Hines & F. Ward, 1910, 736 pp. 1997 reprint HC
- *Permanent Fortifications & Sea-Coast Defences* by U.S. Congress, 1862, 544 pp. 1998 reprint HC
- *American Coast Artillery Matériel* Ordnance Dept. Doc#2042 by U.S. War Dept., 1922, 528 pp., 2001 Reprint HC
- *American Seacoast Defenses: A Reference Guide* (3rd Edition) by Mark A. Berhow, (2015) 732 pp. HC
- *The Endicott & Taft Board Reports,* reprint of original reports of 1886 and 1905 by U.S. Congress, 525 pp. 2007 HC
- *Artillerists and Engineers: The Beginnings of US Fortifications 1794-1815* by Col. Wade, U.S. Army, 226 pp. 2011 PB

CDSG Press is interested in new titles, especially those dealing with fortifications, please contact Terry McGovern at 703/538-5403 or at tcmcgovern@att. net if you have a title that you are seeking to have published. Visit www.cdsg.org/press.

1700 Oak Lane
McLean, Virginia 22101 USA
tcmcgovern@att.net

Table of Contents

FOREWORD

I was sitting at my desk one day when I received a call from the head of the Physical Security Office, informing me there was a "busload of tourists" at the laboratory gate, seeking directions to "see the batteries." There was a fairly simple response: Drive south a couple of miles to Cabrillo National Monument and just before you arrive at the entrance station look off to your right and you will see the back entrance to Battery Ashburn. You can't park there, but you can see it as it flashes past. There are no interpretive signs or photo spots, just a massive concrete edifice with the name and date somehow incised precisely into the stone, a gate-like affair, and a small door leading… somewhere. That was pretty much the extent of opportunities for the public to see anything of the coastal defense installations on Point Loma.

At the monument itself, there was a reasonable amount of material about the numerous military fortifications sprinkled liberally around the less than half-dozen square miles of the southern half of the Point Loma peninsula. But "seeing" them, visiting them, was out of the question.

Intrigued by the security officer's description of "busload," I made the five-minute walk out to the security gate and—sure enough!—there was a large tour bus idling there, filled with thirty or forty "tourists." The gentleman in charge explained this was a group whose title I've forgotten, but it was something like (descriptively, if not necessarily accurately) the West Coast Defense Installation Enthusiasts. They had previously visited the Presidio in San Francisco, seeing what there was to see related to their interests, and now were in San Diego for a similar purpose. I asked casually how the Bay Area experience was, to a chorus of negatives about vandalism and graffiti. I suspect it must have been one of those days when my responsibility as the public face and voice of a major Navy research and engineering laboratory outweighed concerns about security and restricted access. I excused myself and walked inside the closest building to find the physical security officer, and proposed a plan. He headed off to call the guards at the several security posts we would encounter, while I re-boarded the bus and announced, "Have I got a deal for you!"

The individual who reads this book with some attention will understand that Battery Ashburn, mentioned above, is just about the only coast-defense installation visible to the general public on Point Loma. The majority of the dozens of batteries, searchlight stations, and radar sites described in painstaking detail in this book are on federal government property occupied by the Naval Submarine Base and the eight-decades-old Navy laboratory today called Naval Information Warfare Center Pacific. And those installations are out of sight, around corners, down hills, and behind restricted-access gates guarded by security officers who will not let anyone through who does not have the appropriate security badge. Or individuals (one or two generally) accompanied by someone with an escort badge. It happens I had one of those badges, and with the support of that security officer I intended to take not one or two visitors, but thirty or forty, and a tour bus, to see what practically no one ever gets to see—the amazing wealth of military history hidden around those corners and down those hills.

I showed them rusted rail tracks that once carried massive shells to the two eight-inch guns of Battery Strong, located on top of the hill behind the laboratory's Shipboard Antenna Model Range; the vegetation-shrouded lower entrance to Strong just off Woodward Road; Battery Woodward itself, at which stop I cautioned photographers to avoid pointing their cameras at an installation now housing a highly classified project; the sharp-relief indentations in a World War II fortification's concrete wall from sandbags long since disintegrated into dust at the Radar Range.

On a number of occasions during the tour, I got what I'd hoped for: expressions nearly identical to those on my children's faces the morning of December 25th.

My description earlier about security and "no one allowed beyond this point" was a purposeful expression related to the value of the book you now hold in your hands. If you live in San Diego, or travel there someday, and drive out to Point Loma asking, "Can I see the batteries, please?" you will get that statement about drive south and look for Battery Ashburn. That's about it. Yes, at Cabrillo Monument's Visitor Center there are wonderful resources to learn about the military fortifications of Point Loma. Opportunities to see the real things? Not many. And reasonably so, because without that rigidly restricted access, those fortifications almost certainly would embody the "vandalism and graffiti" of the Presidio.

In spite of that, through the incredible efforts of Bart Everett, you are about to embark on a nearly first-hand tour of those batteries, command posts, base-end stations, searchlight shelters. And while I will not be narrating this tour (which I did poorly, by the way), your experience will not be limited to dull gray pages of prose attempting to substitute for physical presence. Rather you will be treated to hundreds of enlightening photographs, precision maps and on-site hand-drawn representations, intricately detailed schematics, and the voices of the long-dead battery gun crews, all relating in an up-close-and-personal manner anything and everything one could possibly want to know about the harbor-defense installations of Point Loma.

Developed over decades, prior to and during the early stages of World War II, those installations stood in anxious anticipation of a formidable force of enemy warships appearing over the western horizon, deliberately positioned to defend a strategically vital port that is home to one of the U.S. Navy's two largest complexes, as well as (and not coincidentally) a metropolis billing itself as America's Finest City.

With this volume, Bart Everett demonstrates his usual complete technical mastery of his subject, coupled with his customary ability to write easily understandable prose. As always, he shows an uncanny ability and willingness to research a subject to its very roots, and then dig under them to see if there is anything else. His intriguing side excursions (I've never heard of incendiary balloons! Have you?) include the same fascinating technical details and pertinent illustrations.

Employing masterful blending of technical specifications, first-hand accounts from oral histories, and meaningful juxtaposition of world and resultant local events, he provides thorough explanations of a bewildering array of electro-mechanical devices related to warfare and the structures that housed them.

Most importantly, based on my initiating this discussion by mentioning a tour, he takes the reader through the weed-choked gullies and rugged canyons and enlarged sinkholes and collapsed tunnels of Point Loma on an expedition of discovery, carefully cutting back vegetation to reveal a decaying wood and metal structure unvisited and in fact unseen since the late 1940s. Artfully interspersing photos of himself and his co-discoverers with those of World War II military personnel at the same location, he then proceeds to recount precisely what was here, its exact physical dimensions and the equipment required for its operation, describing those operations in extraordinary detail.

In addition to the tour I conducted for the "busload of tourists," I used that escort badge on a couple of occasions to take Howard Overton, the Cabrillo Monument chief ranger at the time, to visit sites like Battery North /Battery Gillespie to which, despite his position, he had no access. Overton's knowledge of the subject, amply exemplified by the multitude of references to his work in this book, was vastly superior to mine. On the other hand, on several occasions I saw that morning-of-December-25th expression on his face.

I expect those military history enthusiasts who study this book will have the same facial expression many times.

Tom LaPuzza
Retired Public Affairs Officer
Point Loma Navy Laboratory

PREFACE

In 1986, I was transferred as an Engineering Duty Officer from the Naval Sea Systems Command in Washington, DC, to the Naval Ocean Systems Center (NOSC), now Naval Information Warfare Center Pacific (NIWC PAC), on the Point Loma peninsula in San Diego, CA. My orders were to recruit and train a hand picked group of engineers to address identified Navy needs in the emerging field of robotics. Our fledgling unmanned systems laboratory at NOSC was located on Woodward Road, overlooking the scenic Pacific Ocean in the northwest corner of what was once Fort Rosecrans. Amazingly, this area had changed very little since WWII, having been locked up on Navy property, hence untouched by the curious public. I kept looking over my shoulder as the years ticked by for fear some Congressman would move us out to El Centro, then sell off this ocean-view real estate to reduce the national debt.

Due to its strategic location, the Point Loma peninsula quite naturally has a rich and extended history of military significance associated with Fort Rosecrans and the Harbor Defenses of San Diego (HDSD). A number of World War I and II fortifications along both sides of the point are still extant, generally contained within three principal locations: the Cabrillo National Monument, the Naval Information Warfare Systems Center Pacific (NIWC Pacific),[1] and the Naval Submarine Base. Public access to the two military sites, however, is by necessity restricted, resulting in much less awareness of the surviving infrastructure and its storied role in history. At the same time, however, this restricted access has helped preserve much of the historic Fort Rosecrans legacy.

The Naval Information Warfare Center Pacific (NIWC PAC) and some of its predecessor organizations such as NOSC have been involved in various aspects of robotics and unmanned systems development since the early 1960s. I became interested in the Center's WWII history in 1989, when testing a teleoperated HMMWV-based robotic vehicle in our newly designated Robotic Test Area, which is situated in what used to be the northwest corner of Fort Rosecrans. A wheel of this unmanned vehicle broke through a dirt-encrusted section of plywood covering a skylight for an extensive underground tunnel-and-bunker network behind a 5-inch WWII coast-artillery battery. I was forever intrigued by the wartime artifacts this incident revealed.

It was to be 17 years before I would learn the names and roles of all the historical HDSD fortifications, partly because there were less capable internet searches at the time, and my primary focus was by necessity elsewhere. As intriguing to me as the discovery of this underground complex was the fact that no one at the Center seemed to know anything at all about these underground bunkers and their connecting tunnels. Even today there are many conflicting reports in the written records, and far fewer veterans around to remember first hand. Each year that goes by further reduces the number of surviving HDSD structures due to erosion, decay, and new construction, as most military facilities generally do not have sufficient budget for maintenance of outdated structures.

Fortunately, an extensive survey of the historic resources located within the Cabrillo National Monument area was compiled for the National Park Service in April 2000, which is available online, and there are several amplifying publications. In addition, at least three cultural resource surveys have been commissioned to catalogue former Fort Rosecrans structures on Navy property. All these valuable documents provided significant input to the compilation of this book.

---

1   The Space and Naval Warfare Systems Center Pacific (SSC Pacific), from which I retired in 2016, was renamed the Naval Information Warfare Center Pacific (NIWC Pacific) in February 2019.

My growing interest in the Center's extensive WWII heritage was fueled by the surprising number of well-preserved above-ground coast-defense structures that were still visibly extant throughout this area. Some of these provided an opportunistic variety of readily available test sites for evaluating our mobile robotic systems. I progressively "rediscovered" more and more historical remains over my 30 years at the Center, with each such revelation contributing more awareness of the overall picture. The fact that we were repurposing much of this historic infrastructure to train modern-day Warfighters in the use of military robots would have been a rather amazing prospect in the eyes of the troops that constructed and manned these facilities.

Who in that bygone era could have possibly imagined robots would one day autonomously explore these same bunkers, building accurate maps of their floorplans while annotating any detected items of tactical interest, then reconnecting with their human handlers to upload the results? But in the course of my book research, much to my own amazement, I "rediscovered" something else: there had been almost as many military robots deployed world-wide during WWII as we have today: air, land, and sea! [2] And all now mostly forgotten, much like the tunnels, bunkers, searchlight silos, fire-control radars, and gun batteries of Fort Rosecrans...

While the Center has conducted a few site surveys to document these historic artifacts, the resulting reports shed little light on how things worked back then from a technical standpoint. As an engineer, I was very interested in such, and as a semi-romantic, I hated to see all this history disappear. So I began photographing everything as it currently stood, poured over the historic site surveys, collected all the period photos that could be found, and wrote this book describing it all.

A number of people helped in this quest, particularly Tom Lapuzza, our Public Affairs Officer, who enthusiastically sanctioned my desire to further investigate and document this long-forgotten wartime infrastructure. To start things off, Tom provided a copy of a 1982 Flower and Roth cultural-resource survey, along with a couple of others, which collectively served as a preliminary roadmap of where to start looking during my subsequent weekend explorations.

Tom also introduced me to Terry DiMattio, Superintendent of the Cabrillo National Monument on the southern tip of Point Loma, who turned me over to Bob Munson, the park historian. Bob pointed me to a National Park Service publication compiled and edited by Chief Ranger Howard Overton, entitled *The 19th Coast Artillery and Fort Rosecrans: Remembrances*. This invaluable manuscript chronicled Overton's oral-history interviews with 28 surviving members of the 19th Coast Artillery Regiment, all of whom had been stationed at Fort Rosecrans during the war. Bob also allowed me to scan the monument's extensive collection of Fort Rosecrans photos for inclusion in the book, and introduced me to Navy Region Southwest archeologist Dr. Andy Yatsko, who further endorsed my efforts in an encouraging e-mail.

I am also indebted to retired Navy Commander and Coast Defense Study Group (CDSG) member Al Grobmeier for our many discussions, in which he shared his extensive knowledge of Fort Rosecrans, and for letting me scan his photographs and slides. It was through Al that I learned about the CDSG and consequently became a member myself, which opened the door to a wealth of historic documentation. Similarly, CDSC member Lee Guidry was of invaluable assistance in accompanying me on several weekend visits to selected HDSD sites, letting me scan his historical photograph collection, and providing copies of his meticulous field sketches created while assisting with the Center's 1992 Keniston cultural-resource survey.

---

2   See Everett, H.R., *Unmanned Systems of World Wars I and II*, MIT Press, 2015.

Thanks also to Randy Peacock of our Facilities Department for providing full access to the Center's historical records and photographs, and to my son Todd, who accompanied me for safety purposes on numerous weekend field trips. And last but by no means least, thanks to my good friend Jeff Bowen, our Facilities Outside Plant Engineer, who has an amazing knowledge of the location, purpose, and condition of every structure and utility connection on Center property. Jeff clued me to numerous abandoned WWII sites, plus several that had been repurposed, which I then attempted to identify in their original context. He had to do this after hours or on weekends, of course, which further increased my appreciation.

In summary, many factors contributed to the level of detail found in this book:

1. While working at the Center for 30 years, I was surrounded by a truly amazing collection of WWII infrastructure, much of which was in surprisingly good condition.
2. I received a lot of support from several knowledgeable individuals and organizations.
3. I was given access to the Center's cultural-resource surveys, which proved extremely useful.
4. I was graciously allowed to scan the historic photo collections from both the Navy and Cabrillo National Monument archives.
5. I received additional photos and drawings, plus considerable subject-matter expertise, from Coast Defense Study Group members Mark Berhow, Lee Guidry, and Al Grobmeier.
6. I had learned vacuum-tube theory building mobile robots in high school!

In response to political issues outside our control, the name of our Center has changed eight times in 80 years, to include three times while I was there. This was unfortunate from a marketing standpoint, as my attempts to cultivate awareness of our robotics group were reset to square one each time. In contrast, the Naval Research Lab on the east coast has retained its name and reputation since 1923, hence almost everyone in the scientific community knows who they are. But more importantly from the reader's perspective, these continuing name changes may introduce confusion in the following chapters, which I hope to mitigate with the chronological listing provided in the Appendix.

1
Background

A military presence has been a major focal point of San Diego's development since the earliest European ship sailed into the nation's most southwesterly port. With the arrival of sailing for Spain in 1542, the port--then named San Miguel--took on life as a strategic location for shelter and defense. Surveyed and mapped by Sebastián Vizcaíno in 1602, and given the permanent name of San Diego de Alcalá, the port's virtues were once again noted. Lacking any apparent mineral wealth, the local Kumeyaay natives remained little disturbed until 1769, when Spanish military detachments arrived by sea and land. They accompanied missionaries under Father Junípero Serra for the purpose of founding Mission San Diego de Alcalá, first in Upper California.

San Diego remained sparsely occupied during the Spanish period, constructing in 1774 a Presidio—fortress with soldiers and families—on the hill overlooking San Diego Bay. They built Fort San Joaquin (Guijarros) in 1797 on today's Ballast Point, guarding the entrance to the harbor. A very few battles took place—most against British and American smugglers—well into the Mexican period (1821-1846). The US-Mexican War brought an all-out attack by the American Pacific Squadron with troops marching south under John Fremont and those attacking from the east under Stephen Watts Kearny at the Battle of San Pasqual. The outnumbered Californios were forced to surrender on December 6, 1846. The American military, including army, navy and marines; Mormon Battalion volunteers; and multi-ethnic residents of the Pueblo of San Diego, always peaceful, lived under a fairly amicable truce until the Treaty of Guadalupe Hidalgo was signed on February 2, 1848, giving the area to the United States. Some barracks were built and the harbor modified, but the major activities in the port included fishing, the arrival of visitors, and, with the discovery of gold near Sacramento, a stream of ships stopping for provisions on their way north to San Francisco.

The early American period—1850 to World War I—saw some token military efforts during the Civil War and a shore detail kept a constant lookout. With the arrival of Alonzo Horton in 1868, San Diego's focus shifted from Old Town to a new development flanking today's Broadway and referred to as Horton's Addition or New Town. The completion of the transcontinental railroad and subsequent lines to Southern California and San Diego set off an economic boom of the 1880s soon followed by the national depression labeled as the Panic of 1893.

Although far removed on the West Coast, San Diego felt repercussions of the Cuban Revolution of 1897. The city revived some long-forgotten plans to fortify the entrance to San Diego Harbor while the monitor *Monadnock* guarded the harbors of San Diego and San Pedro. A torpedo placement was planned for the seaward side of Ballast Point for firing wire-controlled torpedoes and a remote-controlled system of detonated mines. In the meantime, San Diego decided to hold a midwinter carnival on February 22, 1897, with a fancy-dress ball and Rear Admiral Charles Beardsley, Retiring Commander U.S. Navy, Pacific Squadron as guest of honor along with a large portion of the Pacific Fleet.

In 1898, when war was actually declared against Spain, concerned citizens held a parade honoring the resolution and burned a Spanish flag in front of the San Diego Union office. Five hundred men volunteered to defend the city in case of attack and held a celebration when Admiral Dewey captured Manila. By 1898 Company D, Third U.S. Artillery had arrived to occupy the new fortifications on Ballast Point. Troops were moved from the San Diego Barracks to the new Army installation on Point Loma in 1903. It was named Fort Rosecrans after Civil War hero General William S. Rosecrans.

William Kettner, later Congressman Kettner, arrived in San Diego in 1907. Theodore Roosevelt sent the Great White Fleet of sixteen front-line battleships to pay a visit in 1908; Kettner became an immediate promoter of the US Navy as an economic benefit to San Diego. Soon after a Naval Quarantine Station and a Coal Station were built to serve the navy in the harbor. Plans were begun as early as 1909 to host an exposition honoring completion of the Panama Canal proposed to open in 1914. City Park—renamed Balboa Park—was the chosen site. It proceeded according to plan, but not without controversy. Considering the attendance in 1915 of both Theodore and Franklin Roosevelt, promoters of San Diego as a prominent naval base, the fair was an economic boon to the port. Plans were interrupted slightly by the outbreak of the Mexican Revolution in 1910 with some military and civilian exchanges during the Battle of Tijuana in 1911.

When the Panama-California Exposition opened in 1915, not only was completion of the canal celebrated, the fair became a popular stopping place for those in the military and those wanting to join the service. Marine Corps Colonel Joseph H. Pendleton became an enthusiastic supporter of San Diego during the Exposition. San Diego's role became even more important with the outbreak of World War I and the founding of a Naval Training Station and Naval Hospital in Balboa Park. Tent encampments were necessary until more permanent Navy facilities could be built following the war. The US Army established Camp Kearny on the mesa north of Mission Valley in 1917. It was completed in August and occupied until 1920. The navy also secured permanent facilities on Point Loma and in Balboa Park as San Diego congressmen and city officials favored these moves. In 1922, San Diego was designated as headquarters for the Eleventh Naval District.

Franklin D. Roosevelt, long a promoter of San Diego as a naval base, became president in 1932 as the national economy reached an all-time low. By 1934, the Depression reached its full effect in San Diego, but a number of government measures encouraged recovery. Prohibition of liquor—the eighteenth amendment— was repealed in 1933 and a second exposition was planned for Balboa Park in 1935. The Navy spent $1.4 million and the Army $1.8 million on construction projects. Major Rueben H. Fleet of Consolidated Aircraft Corporation of Buffalo, New York, made a far-reaching decision when he moved his plant with 800 employees and $9 million in orders to San Diego. After examining several locations, Fleet found San Diego to have everything he needed—a good airport, a publicly-owned waterfront, an excellent harbor, a city large enough to furnish labor and materials, and a proper climate for test flying and materials. Consolidated Aircraft began operating in San Diego in October 1935.

The federal census of 1940 showed that San Diego's population was nearing 300,000. With the outbreak of World War II, the rate of San Diego's growth increased tremendously. Local aircraft plants attracted workers from other states while all military establishments were expanded and new facilities acquired. San Diego's climate was ideal for the year-round training of US Army, Navy, and Marine Corps recruits. The Navy represented more than $2.5 million in monthly payrolls and expenditures. Linda Vista, known in its early stages as Defense Housing Project No. 4002, was a $14 million project covering an area of 1,459 acres overlooking Mission Valley. Sponsored by the National Housing Authority, ground was broken on October 31, 1941, and 3,000 houses were constructed in 200 days. In May, 1,846 more units were built on large lots with paved streets and four schools were located in the immediate vicinity. Camp Callan for US Army artillerymen occupied a five-mile stretch of land along Torrey Pines Mesa. An amphibious base was developed by the Navy on Coronado Strand along with Brown Field on Otay Mesa; Ream Field, in Imperial Beach; and Miramar Naval Air Station. The Marines acquired more than 123,000 acres of historic Rancho Santa Margarita near Oceanside to build Camp Pendleton—the world's largest military base. The Marines also set up Camp Elliott on Kearny Mesa.

With defense precautions in San Diego dictating that thousands of street lamps be blacked out, headlights be partially covered with black hoods, camouflage installed on defense installations, and "victory gardens" planted to supplement scarce vegetables, the city stood at alert. Consumers did without new cars and new appliances while simple needs such as sugar, butter, meat, coffee and nylon stockings were rationed. Local Japanese residents were removed to internment camps and restrictions were imposed on travel. San Diego remained in a state of readiness until the end of the war in August 1945 and has continued as one of the nation's strongest military cities until the present time.

The high concentration of military resources in the San Diego region during World War II drove the need to invest in strong harbor defenses and placed San Diego high on the priority list for the construction and manning of these defenses. In a forthcoming chapter, the threats from Japan are detailed and given San Diego's West Coast location, also drove the number and density of San Diego's harbor defenses. These temporary and permanent defenses built upon San Diego's earlier harbor defenses of the Endicott-Taft Period. The following chapter details these defensive works from 1890 to 1915.

Background Chapter by Iris H.W. Engstrand, PhD

2

The Endicott-Taft Period

US coast-defense upgrades following the Civil War are generally categorized into two overlapping time frames: the Endicott Period (1890-1910), which includes the Spanish-American War, and the Taft period (1905-1915), which includes the start of WWI. Additional upgrades and improvements at Fort Rosecrans naturally took place prior to and during WWII. Each of these timeframes will be generally addressed in the following sections, with more detailed treatment by subject area (gun batteries, searchlights, fire control, etc.) in subsequent chapters.

Fort Rosecrans

On 26 February 1852, President Millard Fillmore signed an Executive Order creating the 1,300-acre Point Loma Military Reservation in San Diego, CA (Thompson, 1991). The US Army took possession of this Reservation and nearby Ballast Point in 1870 (Keniston, 1996), but it was not occupied until February 1898 (Carey, 2005), and named Fort Rosecrans by general order on 22 July of the following year (Berhow, 2020). As reported by Thompson (1991):

> "When the 30th Company, Coast Artillery, arrived from duty in the Philippines in 1901, it occupied both San Diego Barracks and Fort Rosecrans. Major Anthony W. Vodges, Artillery Corps, commanded both posts. A month later, the 115th Company, Coast Artillery, was formed. For the next two years, one company was stationed at Fort Rosecrans and one at the Barracks."

Figure 2-1 shows the Fort Rosecrans Military Reservation on the Point Loma peninsula as of 28 March 1934, with three other Government installations adjoining or within the property:

- The Naval Fuel Depot on the north end, which included the Quarantine Station operated by the US Treasury Department.

- Two lighthouses, one on Ballast Point and the other at the southern tip of Point Loma, operated by the US Coast Guard.

- The Cabrillo National Monument, operated by the US Department of the Interior, on a half-acre of land surrounding the old Spanish lighthouse.

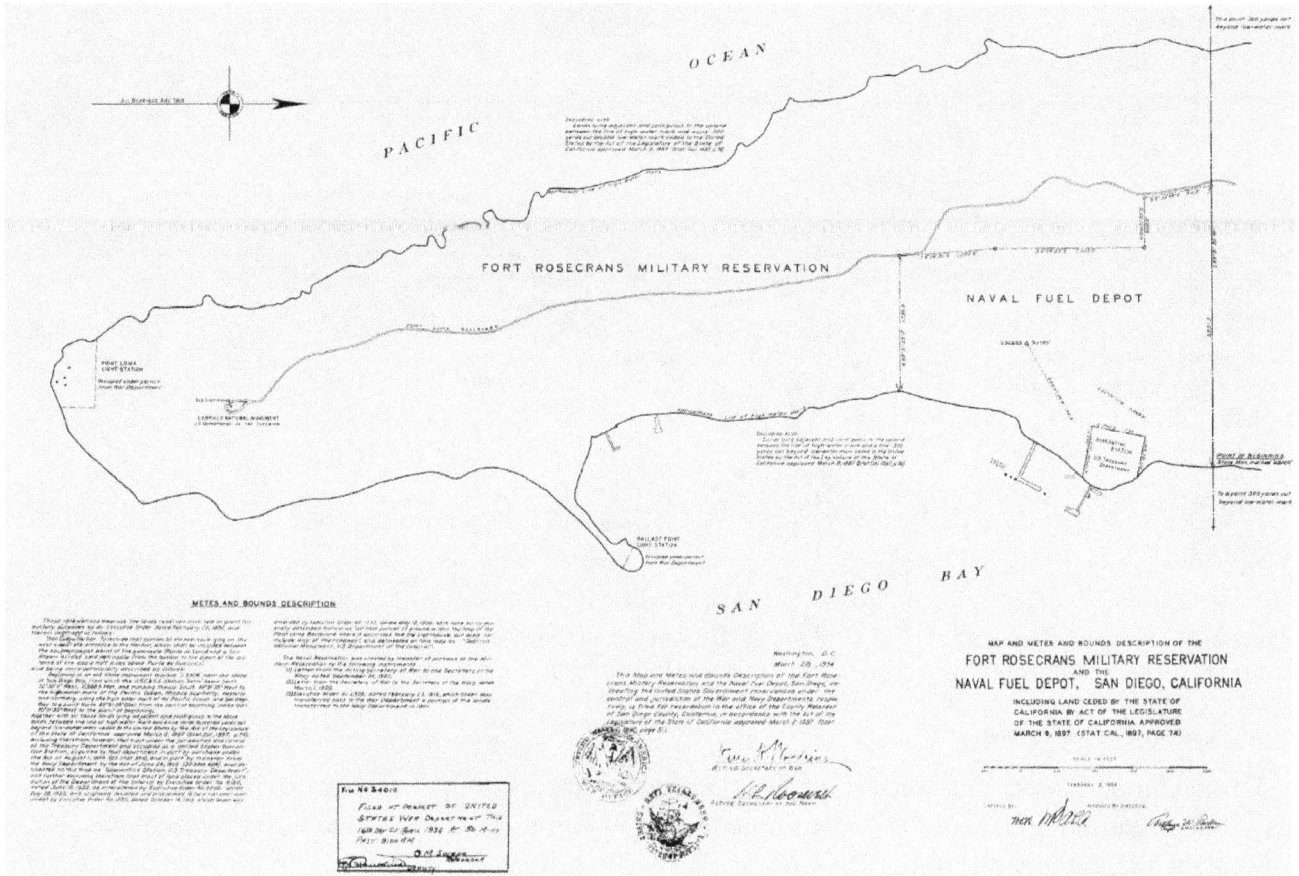

Figure 2-1. This 2 February 1934 map shows the Fort Rosecrans Military Reservation on the southern tip of Point Loma, with the Naval Fuel Depot in the northeast corner at center right (courtesy SSC Pacific).

Endicott Period (1890 – 1910)

On 3 March 1885, a formal "Board of Fortifications or Other Defenses" was appointed by then President Grover Cleveland to review the perceived deterioration of US coastal fortifications, headed by Secretary of War William C. Endicott (Thompson, 1991). The Endicott Board subsequently recommended a comprehensive national upgrade of harbor-defense capabilities in December that same year, to include the following significant innovations (Duchesneau, 2006):

- Controlled minefields to protect harbor entrances, command detonated from shore-based stations.
- Groups of 12-inch breech-loading mortars that could rain down a spread of projectiles onto the more vulnerable decks of enemy vessels.
- Large-bore disappearing rifles hidden behind earth-covered parapets.

These concealed defense upgrades provided significantly improved effectiveness with increased protection against sea-based attacks from hostile navies (Duchesneau, 2006).

The locations of Endicott-period Batteries McGrath, Calef-Wilkeson, and Fetterman are depicted at lower left in the 29 June 1921 map presented in Figure 2-2, with Ballast Point out of image to the bottom. Taft-period Battery White in Powerhouse Canyon is at upper left, as further discussed later. The mining casemate is not called out in the map legend at left, but is marked by an arrow from its tactical symbol (a circle inside a square). The L-shaped wharf for the mine-planter boats is shown at bottom center, with the engineer's boathouse labeled as "12".

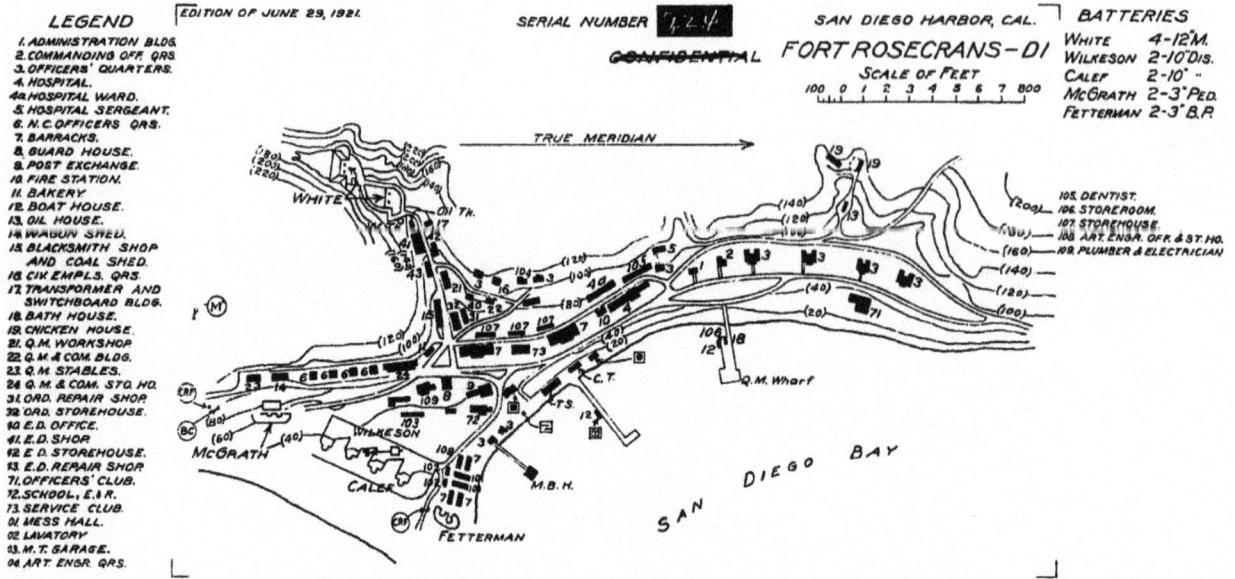

Figure 2-2. This "Fort Rosecrans - D1" (Detail Map 1) US Army map shows the location of the various buildings and gun batteries as of 29 June 1921, with Battery Whistler (four 12-inch mortars) out of image to the left. Note the L-shaped mine-planter wharf at bottom center and the quartermasters wharf to its right, and the coincidence rangefinder (CRF) position for Battery Fetterman left of bottom center.

In the United States, controlled minefields played an important harbor-defense role from the start of the Endicott Period until about midway through WWII (Figure 2-3). The fundamental principles of this concept, which originated during the American Civil War, were developed over a 20-year period at the US Army's Engineer School at Willets Point, NY, under the direction of Commandant Colonel Henry L. Abbott (Berhow, 2004). The key advantage of a controlled minefield was the ability for friendly ships to enter and leave port without having to open and close underwater nets that sealed off the harbor entrance.

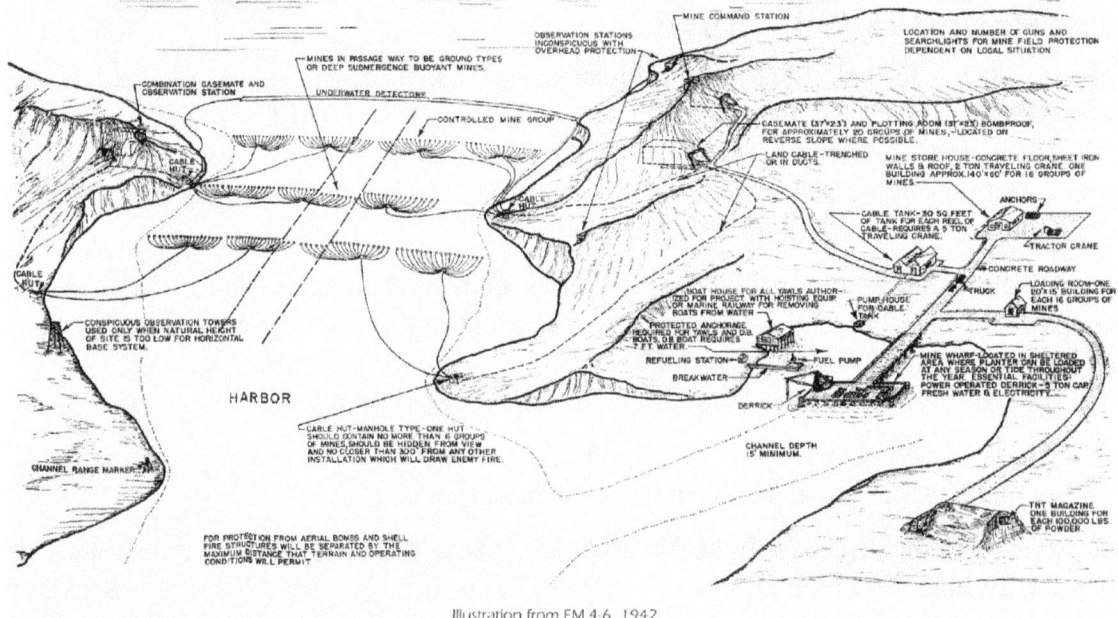

Illustration from FM 4-6, 1942

Figure 2-3. The controlled-minefield defense involved electrically actuated submerged mines that could be detonated on command from the mining casemate (just right of top center), based on the observed and plotted positions of enemy vessels (US Army illustration adapted from FM 4-6, 1942).

The mine-planting concept is illustrated in Figure 2-4a. An adjustable mooring rope attached to a 2,500-pound anchor was used to hold the buoyant mine at an appropriate depth so it would be beneath the hull of an enemy ship passing overhead. A single-conductor detonating cable from each mine in the planted group was attached to an underwater distribution box, which in turn was connected via a multi-conductor cable to the mining casemate on shore.[1] When an enemy vessel was observed to pass over the known location of a submerged mine, that particular mine was electrically detonated from a control panel in the mining casemate.[2]

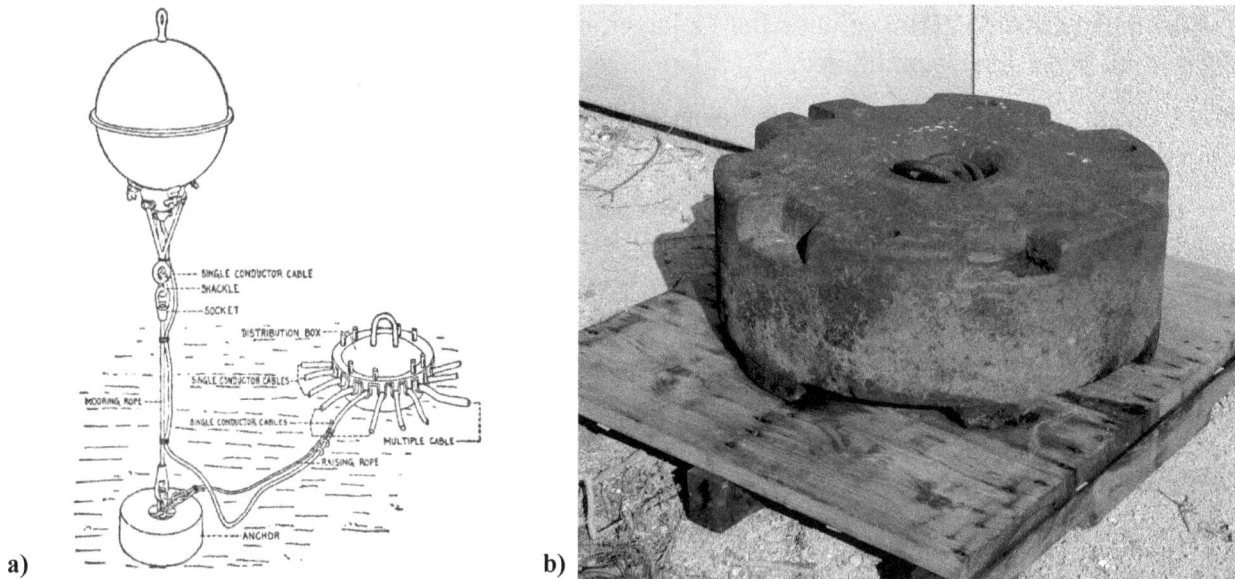

Figure 2-4. a) Anchor and cable-distribution box for controlled mines of the type used in the San Diego harbor entrance during the Spanish American War and WWI (adapted from Berhow, 2004). **b)** This 2,500-pound mine anchor was found on Center property just west of Woodward Road in November 2007.

The San Diego harbor entrance was mined during the Spanish American War but not in WWII, with two sets of submarine nets installed at the harbor entrance during the latter conflict. Bogart (2008) claims no controlled minefields were planted anywhere in the US during WWI. Our later discussion of Battery Meed, however, shows Major J.C. Gilmore (1913) referenced a "present minefield located outside the breakwater" in his 29 September 1913 memo to the Commanding General, Western Department. WWI started less than a year later on 28 July 1914.

The mine anchor shown in Figure 2-4b above was discovered in heavy brush just west of Woodward Road near Battery Gillespie. This 2,500-pound anchor presumably was brought over from the east side of the point to secure a guy wire for improvised-antenna testing at the U.S. Navy Electronics Laboratory during the 1950-to-1960 timeframe. Bogart (2008), reports a post-war experimental minefield was planted off San Diego in 1947:

> "The purpose of the experimental field was to develop a mine capable of detecting and destroying a midget submarine. The experience of WW II being that, except for the U-47's penetration of Scapa Flow, all other undersea attacks on harbors were carried out by midget submarines."

---

1   In the 1880s, the largest insulated cable that could be manufactured in quantity contained just seven conductors (Berhow, 2004).

2   For additional details, see the "Controlled Mines in US Army Seacoast Defenses" section in *American Seacoast Defenses* (Berhow, 2004).

Mining Casemate

According to a construction report prepared by Major E.G.B. Davis (1896), Corps of Engineers, the first mining casemate at Fort Rosecrans was located between Guns 3 and 4 at Battery Wilkeson (see Figure 2-7 on the next page and Figure 2-14 in the next section):

> "The mining casemate has been placed in the traverse between the two guns, in the location recommended by the Board of Engineers January 16, 1895. The amount of vibration that the instruments will receive from the shock of discharge of the guns is an uncertain quantity."

This was apparently an interim solution until a larger and more appropriate facility could be constructed later, as funds at the time were limited.

Work on this second mining casemate began in early 1897 (Hathaway et al., 1987). As reported by Thompson (1991):

> "In 1897, 1st. Lt. Meyler supervised the construction of such a casemate located one-eighth of a mile north of Battery Wilkeson. A concrete cable tank and a mine storehouse soon followed."

This new mining casemate,[3] circled in Figure 2-5 below, became operational in the spring of 1898 (Hathaway et al., 1987). The locations of the associated cable tank (C.T.) and mine (torpedo)[4] storehouse (T.S.), completed in June 1900, are seen just inland from the mine-planting wharf (up arrow).[5]

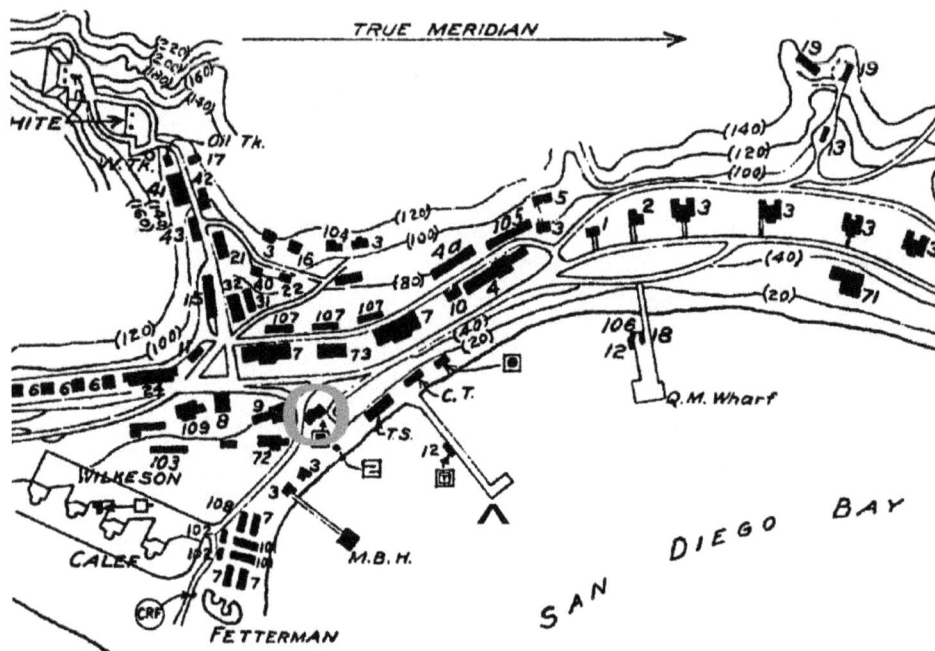

Figure 2-5. The new mining casemate (circled) is north of Battery Wilkeson (lower left quadrant) in this enlarged section of the 29 June 1921 map presented earlier in Figure 2, with the L-shaped mine-planter wharf to the right (up arrow). Note the engineer's boathouse (callout 12) midway down the wharf, and the cable tank (C.T.) and torpedo storehouse (T.S.) along the shoreline above the wharf.

In April 1898, Captain Meyler received orders to mine the San Diego harbor entrance,[6] with instructions to secure the help of a 'corps of 120 patriotic citizens' to plant the mines (Thompson, 1991). Eighty

---

3　The mining casemate was designated Building 167 on what is now Naval Submarine Base Point Loma.

4　The Army initially used the term "torpedo" for what was later called a "mine" following the introduction of "automobile torpedoes" that could be launched from shore, and later from submarines (Everett, 2015).

5　This 29 June 1921 map actually shows the new cable tank and torpedo storehouse that were completed in 1910 (not the earlier 1900 structures, which presumably would have also been located near the wharf).

6　First Lieutenant James J. Meyler was promoted to Captain in July 1898 (Thompson, 1991).

concerned citizens subsequently volunteered, supervised by five newly arrived enlisted engineers trained in mine laying,[7] with the necessary cable, explosives, and supplies shipped in by steamer from San Francisco (Thompson, 1991). The mining casemate became operational in 1898 (Hathaway et al., 1987); fifteen mines were planted in the harbor entrance channel in that May and recovered in September (Meyler, 1901). Plan and section views of the mine storehouse are seen in Figure 2-6.

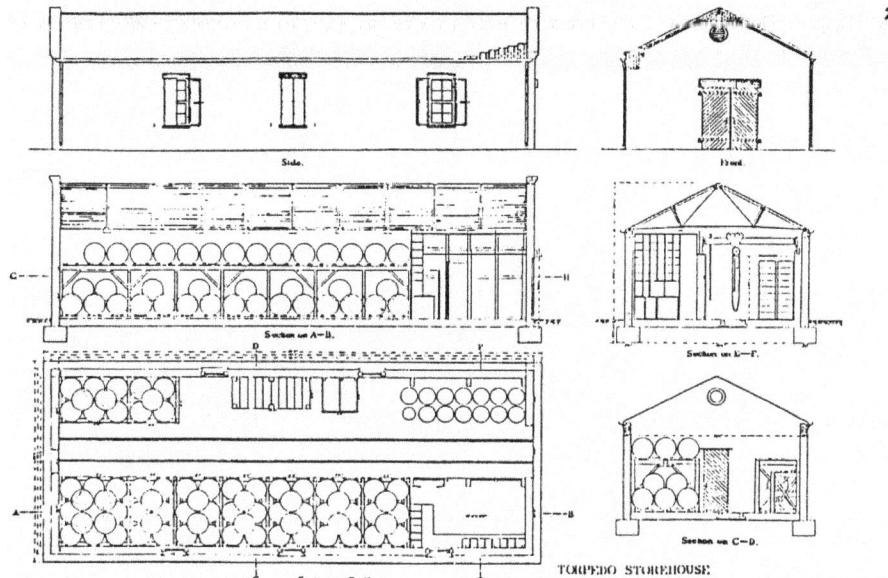

Figure 2-6. Plan and section views for the mine (torpedo) storehouse, which was located along the shoreline near the L-shaped mine-planting wharf at the U.S. Quarantine Station, La Playa (see again Figure 2-2).

It soon became clear the new mining casemate was too small, however, as the generator had to be placed in a wooden shed above ground, thus prompting plans for an immediate expansion (Hathaway et. al., 1987). Plan and section views for these envisioned engine and battery-room additions, dated 6 December 1898, are presented in Figure 2-7.

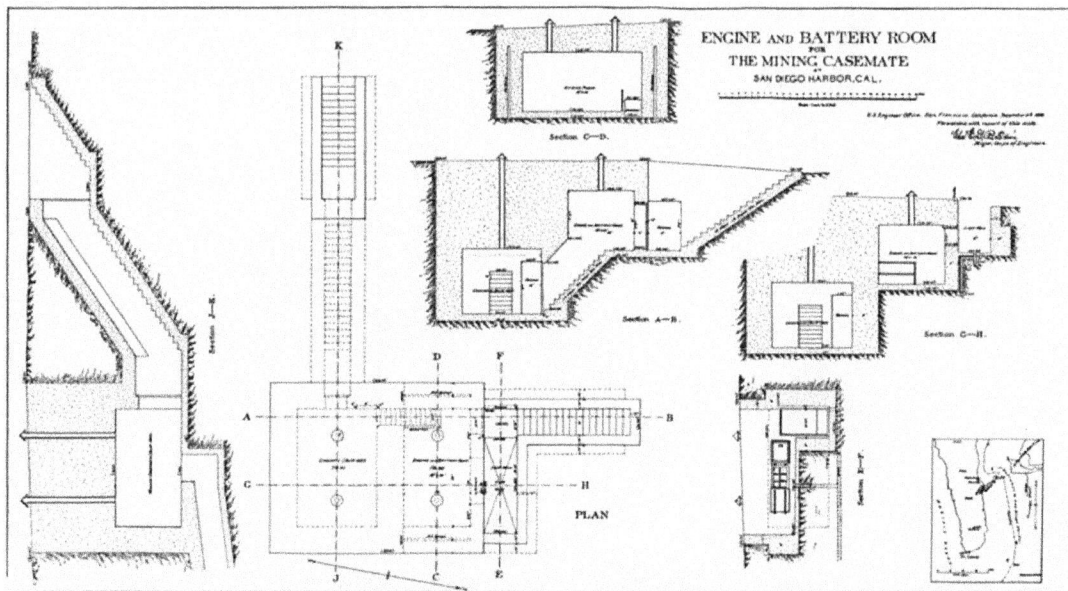

Figure 2-7. Plan and section views for the proposed engine- and battery-room additions to the new mining casemate from a 6 December 1898 status report prepared by Major E.G.B. Davis, Corps of Engineers, San Francisco. A new set of access stairs is seen on the north side of the plan view and in Section A-B.

---

7   These five engineers arrived at Fort Rosecrans on 11 May 1898 (Thompson, 1991).

The cable-gallery trough beneath the original casemate shown earlier in the lower-left-corner Section View J-K (see again Figure 2-7), which led east to San Diego Bay, is seen in more detail in Figure 2-8 below. As explained by Hathaway, et. al. (1987):

> "A concrete tunnel built into the casemate floor, known as a cable gallery, provided a conduit for electrical cables that led from the facility to mines in the harbor. The gallery was 2 feet wide, 3 feet high, 69 feet long, and extended eastward to just beyond the original shoreline, at which point its opening was 1 foot below the mean-high-water line. From that point, a concrete trough extended another 60 feet into the bay, where it terminated 2 feet below the mean-low-water line."

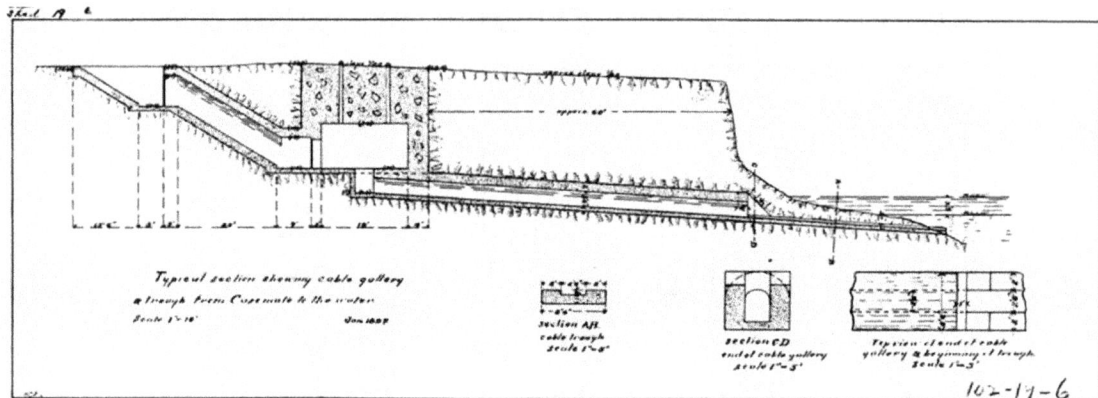

Figure 2-8. Section view dated January 1897 showing the flooded cable trough (center) leading from the cable gallery beneath the mining casemate (white rectangle left of center) out to the water at far right.

The 1897 mining casemate built by Captain Meyler's volunteers is depicted by the rectangular space in the lower-left corner of the "Section J-K" diagram shown in Figure 2-9, which also shows the planned additions above and to the right. A new engine and battery room was to be built at this upper level (top center), with a new set of access stairs leading down from the north. The 5-foot-wide "Recess" to the right of the engine-battery-room addition was for a planned light well,[8] as seen later in the plan view of Figure 2-10.

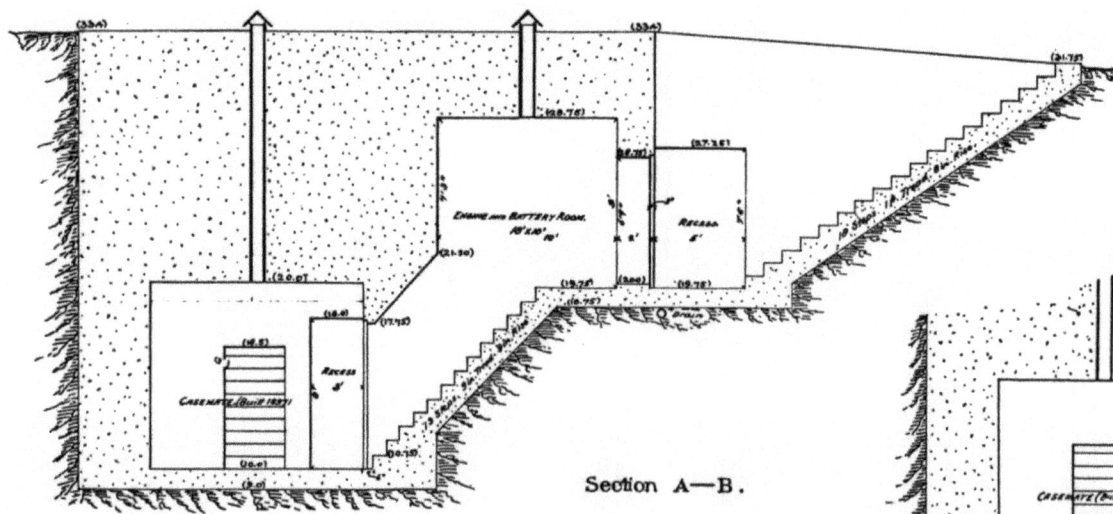

Figure 2-9. The 1897 mining casemate is depicted by the rectangle in the lower-left quadrant of "Section A-B" from Figure 2-7, enlarged for clarity, with the original access stairs on the west side. The proposed Engine and Battery Room addition (top center) included a new set of access stairs on the north side.

---

8   A light well or air shaft is an unroofed external space provided within the volume of a large building to allow light and air to reach what would otherwise be a dark or unventilated area (Wikipedia, 2019).

Figure 2-10, also enlarged from Figure 2-7, presents a plan view of these new additions, with the original access stairs at upper left, and the northern access stairs leading down to the engine and battery room in the upper-right quadrant. A narrow horizontal window on the north wall of the engine room, which looked into the adjacent light well, could be folded down for ventilation. The third and final set of stairs seen at top center continues down to the lower-level mining-casemate Control Room at image left (see also Figure 2-9 above).

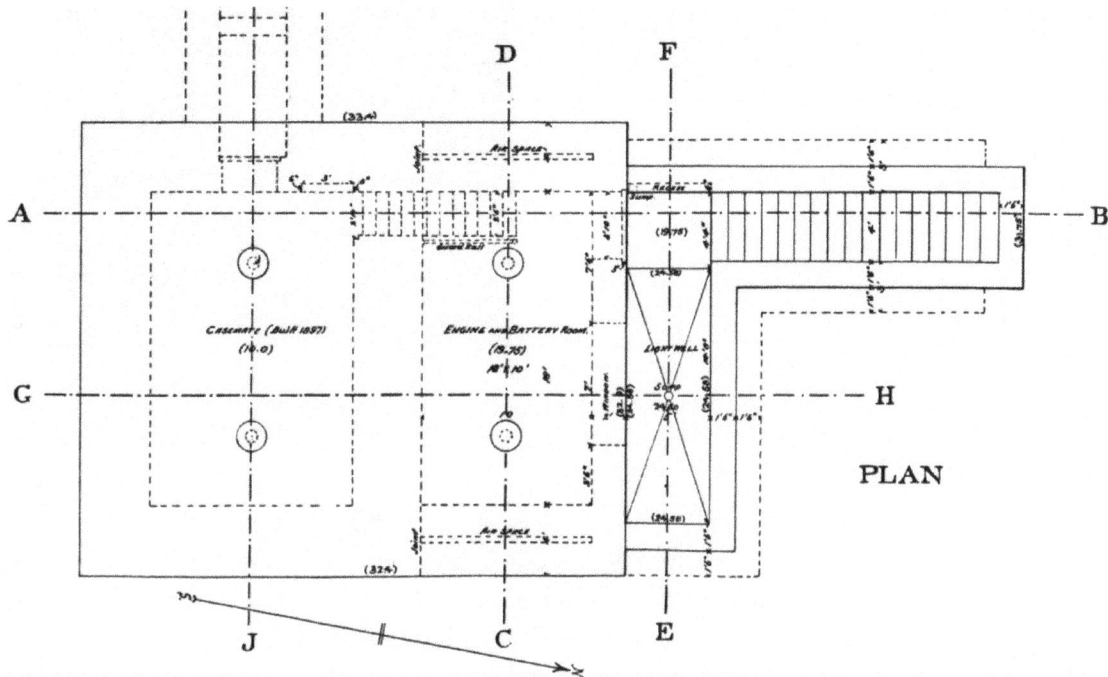

Figure 2-10. The 1897 mining casemate is represented by the large rectangle on the left. The new additions just to its right include the combined engine and battery room in the middle and the light well on the north side.

Construction on the new engine room began in March and was completed in June 1899, with its storage battery installed a year later. A February 1905 inspection report, however, found the mining casemate, cable tank, and mine storehouse to still be inadequate (Hathaway et. al., 1987):

> "As a result of these deficiencies, a board of officers recommended on October 7, 1905, that sleeping and battery rooms be added to the casemate, and that additional facilities including a mine planting wharf, storehouse, cable tank, searchlights, loading room, and fire control station be constructed."

Work began accordingly in 1909 on both a new sleeping room and a dedicated battery room on the north side of the light well seen earlier in Figure 2-10, which were completed in January of the following year (Hathaway et. al., 1987):

> "The sleeping room measured 9 feet x 16 feet and the battery room 10 feet x 16 feet. Both were 11 feet in height. The walls were 2 feet thick and the structure was covered over with approximately 4 feet of earth... A frame roof was built over the light well and a 60-foot tunnel was constructed to connect the basement of Building 158 with the western entrance."

The location of this twice-enlarged mining casemate is seen in the bottom-right corner of Figure 2-11a below. The associated mine-planter wharf was located about 800 feet north of Battery Wilkeson, per Form 3 of the Report of Completed Works, corrected to 15 September 1925. The engineer's boathouse (structure 12 on the 1921 map shown earlier in Figure 2-2) is at the midpoint of the wharf on the southeast side. The mines and their associated cables were transported by rail cart from the loading room to the end of the 362-foot wharf, then loaded aboard a mine planter using a 5-ton pillar crane (Figure 2-11b).

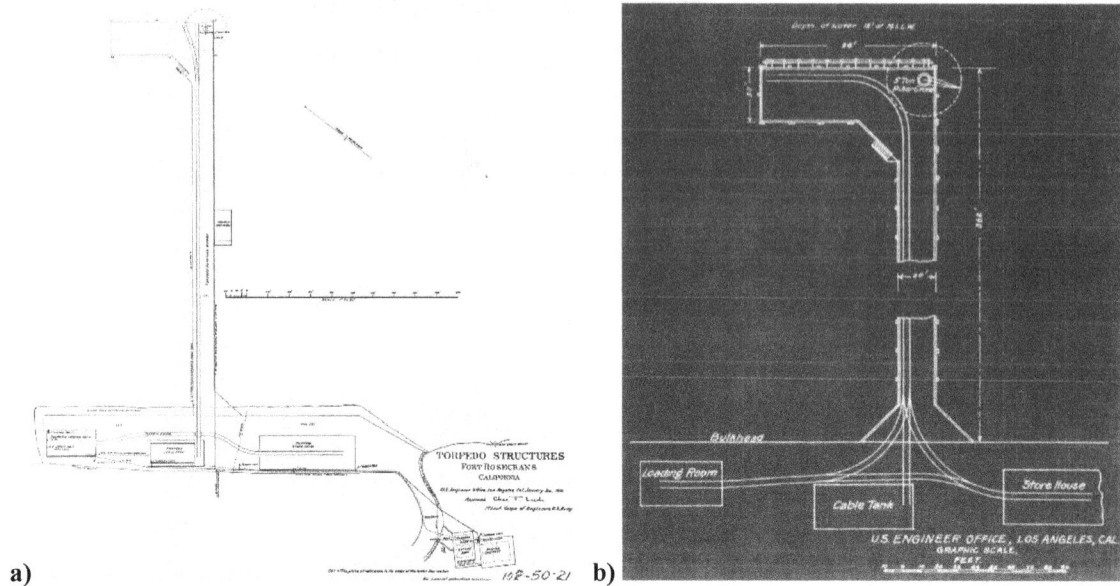

Figure 2-11. a) The 20 January 1911 "Torpedo Structures" drawing shows the mine planting wharf in relation to the mining casemate (lower-right corner), which was removed in the late 1980s to enable new construction. **b)** This undated blueprint of the wharf more clearly identifies the adjacent support buildings.

The undated photo presented in Figure 2-12 shows the remote detonation of a controlled mine during a live-fire exercise at Fort Rosecrans, presumably sometime prior to WWI, with the target sled just to the left of the blast. The mine-planter boat off to the right had apparently just made a turn to starboard. Close inspection of the original image shows a couple of rowboats some distance behind the target sled, three navigational markers for the harbor-entrance channel, and a relatively undeveloped Point Loma peninsula running south-to north in the background.

Figure 2-12. A controlled-mine explosion in the entrance to San Diego Bay during a live-fire exercise, looking west towards Point Loma in background right (undated US Army photo courtesy Cabrillo National Monument). Note the triangular towed-target sled just left of the blast.

Battery Wilkeson

The first Endicott Period battery to be completed in San Diego was Battery Wilkeson, named for Brevet Lieutenant Colonel Bayard Wilkeson, who was killed at Gettysburg on 1 July 1863 (Ruhge, 2019d). According to the Report of Completed Works, corrected to 15 September 1925, construction commenced on 21 January 1897 and was completed on 31 June 1899, with a total cost of $217,262. Overlooking the harbor entrance just north of Ballast Point (see again Figure 2-2), the new reinforced-concrete structure was armed with four 10-inch rifles on *M1896* disappearing carriages (Figure 2-13).

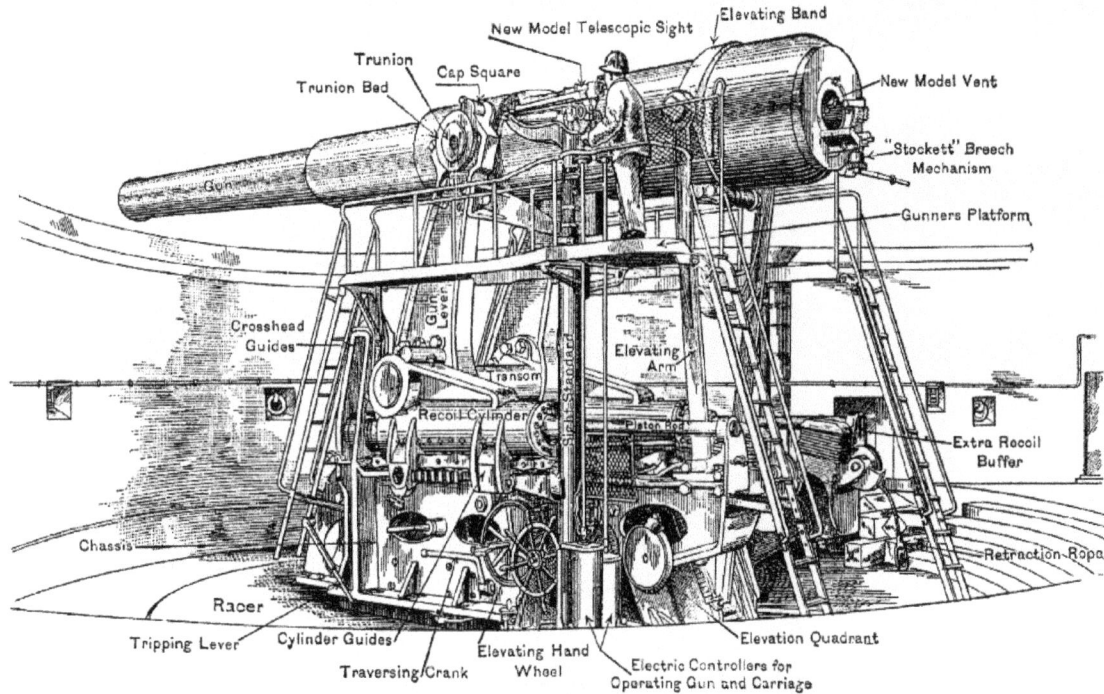

Figure 2-13. Drawing of a limited-fire disappearing-carriage *M1896* mounting a 12-inch gun (adapted from Berhow, 2004), similar to the 10-inch guns at Battery Wilkeson.

Battery Wilkeson was built in stages, starting with emplacements 2, 3 and 4, which were transferred on 22 March 1898; emplacement 1 was transferred two years later on 15 February 1900. As seen in Figure 2-14, Guns 3 and 4 on the northern end had different angles of fire than Guns 1 and 2 on the southern end. This shifted angle of fire was originally specified for Gun 4 only (Davis, 1896):[9]

> "The left flank gun is to have an angle of fire of 150 degrees, 90 degrees to the left and 60 degrees to the right of the normal to the crest line. The adjacent gun is to have an angle of fire of 120 degrees, 60 degrees on each side of the normal."

In 1915, Battery Wilkeson was organizationally divided into two separate batteries (Figure 2-14): Battery Wilkeson, with former Guns 1 and 2, and Battery Calef (named for Colonel John H. Calef), with former Guns 3 and 4 (Stanton, 2019). This pair of two-gun batteries was later merged back into a single battery called Calef-Wilkeson in 1919 (RCW, 1925).

---

9   Major E.G.B. Davis, Corps of Engineers, San Francisco, is quoting in his status report to Brigadier General John M. Wilson, Chief of Engineers, US Army, Washington DC, an earlier undated report he received from a Lieutenant Deakyne, who is not identified.

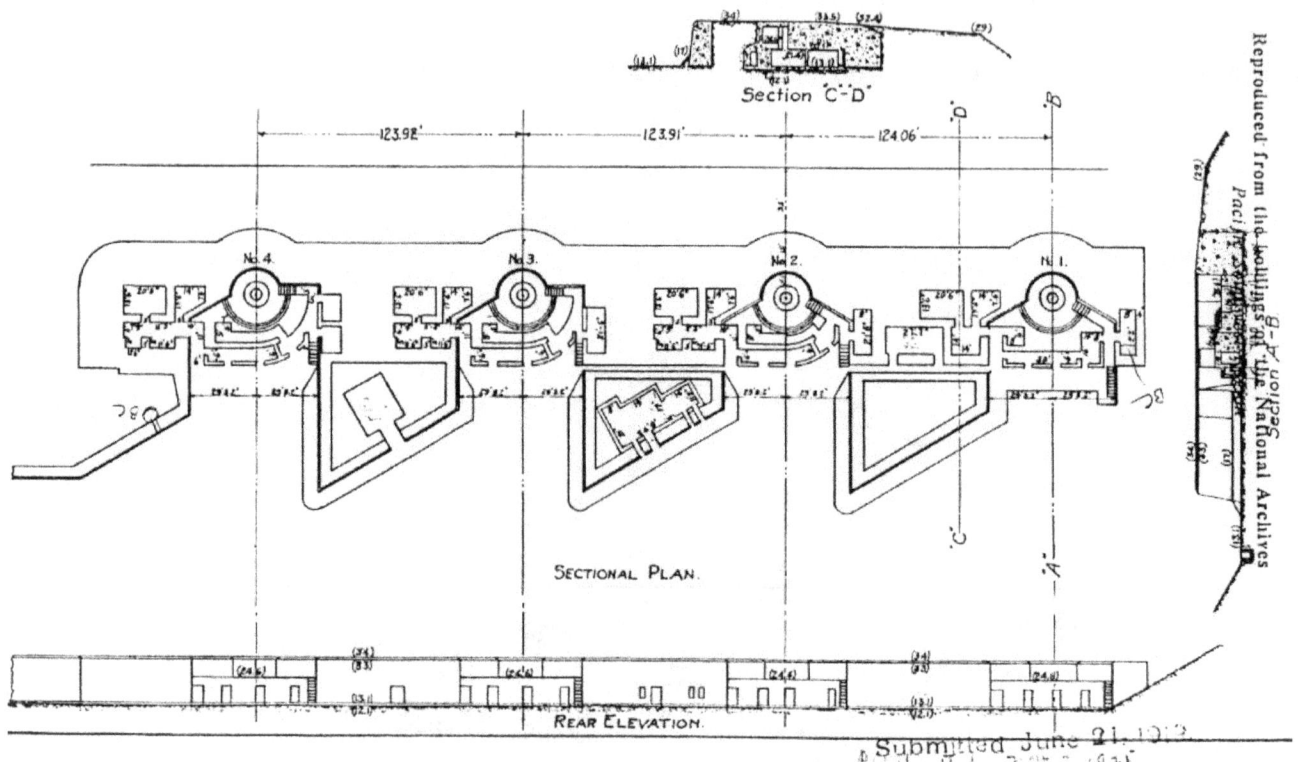

Figure 2-14. Plan view of Batteries Wilkeson and Calef, corrected to 27 March 1919 (NARA). Emplacements 2, 3, and 4 were transferred on 22 March 1898, with emplacement 1 delayed until 15 February 1900. Note the power room off the traverse behind Guns 2 and 3, and the original mining casemate behind Guns 3 and 4.

This symmetrical field of fire initially proposed for the "adjacent" Gun 3 (like that shown for Guns 1 and 2) was changed some 9 months later, however, as Davis (1897) dutifully reported in a follow-up report:

> "The plan of the second emplacement was modified by Department indorsement by arranging the crest in the same way as the crest of the flank emplacement,[10] so as to give a fire to the left in prolongation of the battery front, this provision being considered desirable in order to meet the contingency of the flank gun being disabled.

Emplacements 2, 3, and 4 were transferred to the Coast Artillery on 22 March 1898, with emplacement 1 transferred some 2 years later on 15 February 1900. Equipped with a depression position finder (DPF),[11] the battery commander station (Figure 2-15) was some distance away, dug into a high knoll just below the old Spanish lighthouse on the southern tip of Point Loma (CHA, 1985).

---

10 Davis was apparently using a left-to-right numbering scheme in his report, rather than the traditional numbering scheme used in the later RCW drawing shown in Figure 2-14. Plan view of Batteries Wilkeson and Calef, corrected to 27 March 1919. Emplacements 2, 3, and 4 were transferred on 22 March 1898, with emplacement 1 delayed until 15 February 1900. Note the power room off the traverse behind Guns 2 and 3, and the original mining casemate behind Guns 3 and 4..

11 Further discussion and photos of the Depression Position Finder will be presented in Chapter 7 (Fire Control).

Sheet 2.

BATTERY COMMANDERS STATION
FOR
FORT ROSECRANS.
SAN DIEGO HARBOR, CAL.

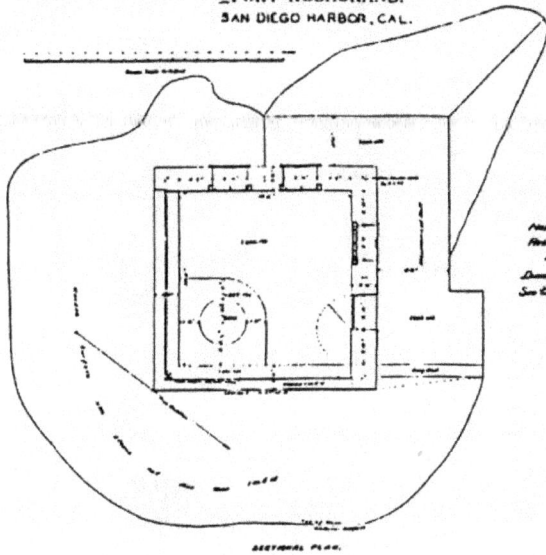

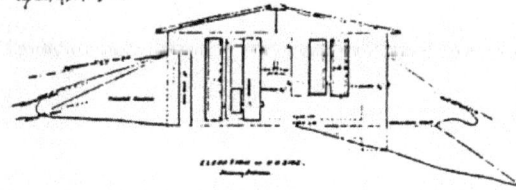

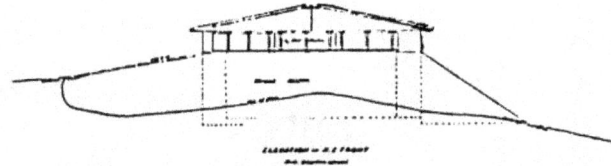

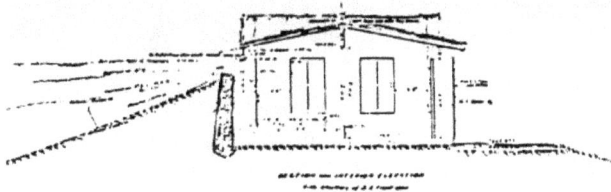

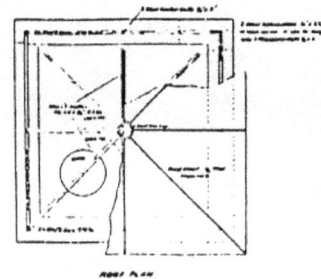

Figure 2-15. Plan and section views of the battery commander station for Batteries Wilkeson and Calef, located south of the battery near the old Spanish lighthouse on what is now on Cabrillo National Monument property (US Army diagram adapted from CHA, 1985).

Figure 2-16 shows plan and section views of the power room in the traverse between Guns 2 and 3 (see again Figure 2-14). There was a single 25-kilowatt motor-generator set on the left and two 25-kilowatt AC generator sets in the center, with their two radiators on the right. The battery was also connected to commercial power. After the fort's Central Power House was constructed in the ravine below Battery White in 1905, the electrical equipment in the Battery Wilkeson traverse was replaced by a three-phase 1100-volt AC generator and a 125-volt DC generator, both manufactured by Westinghouse.[12]

---

12 Data from Form 5 of the "Power for Substation at Battery Calef" Report of Completed Works, corrected to 27 March 1919.

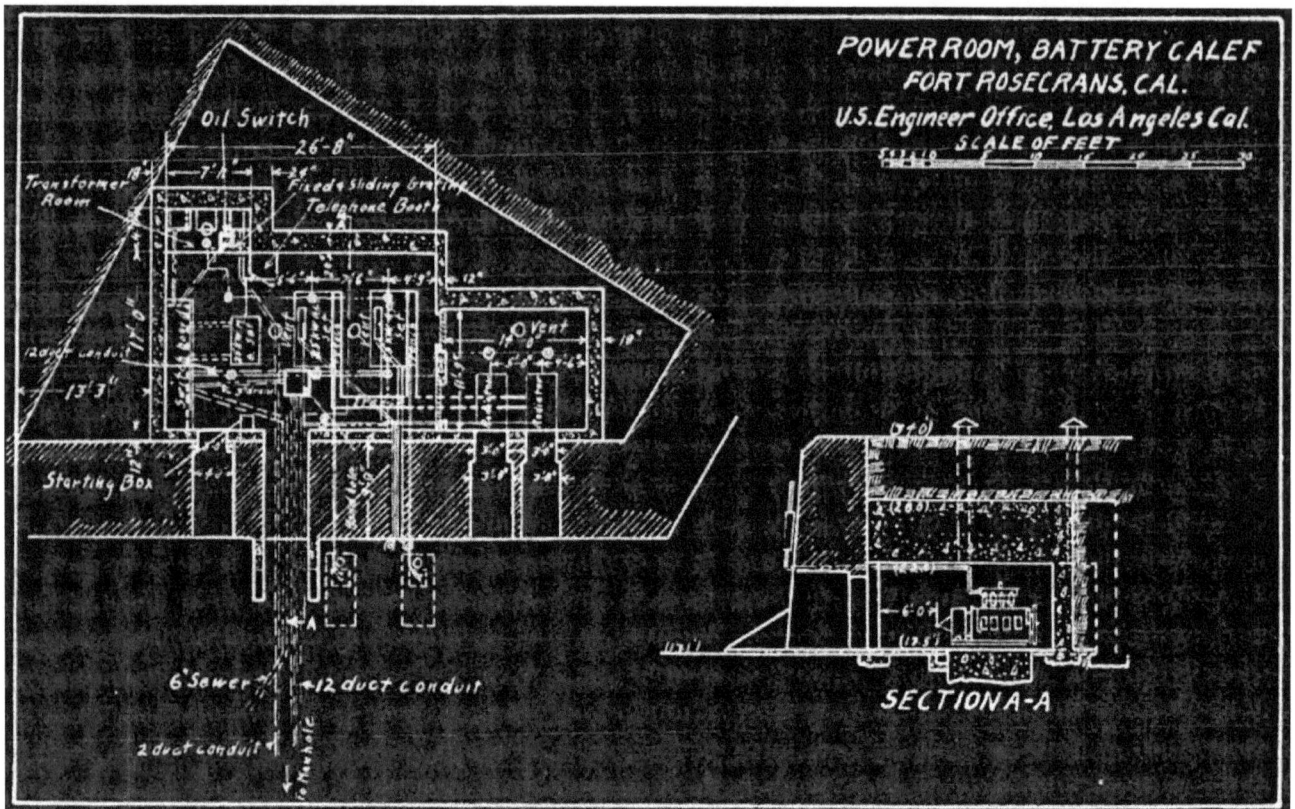

Figure 2-16. The plan view (left) in this undated blueprint of the power room at Battery Wilkeson shows it was originally equipped with (from left to right) a single 25-kilowatt motor-generator set, two 25-kilowatt AC generator sets, and two radiators (NARA).

Battery Wilkeson was proof fired in November 1899 (Ruhge, 2019d). The 20 January 1903 photo in Figure 2-17 shows the four 10-inch disappearing-rifle emplacements, looking northeast along Ballast Point towards the old Coast Guard lighthouse, with North Island in the background. The Gun 1 position (bottom right), which was transferred 2 years later than Guns 2 through 4, looks decidedly newer, with its 10-inch rifle covered by a tarp. According to Glaze (2017), the construction crane seen at upper left behind the Gun 4 position was involved in mounting the 3-inch guns at Battery Fetterman.[13]

---

13 Construction began on Battery Fetterman in May 1898 and was completed on 13 February 1903. (See again 1921 map showing battery locations presented earlier in Figure 2-2. This "Fort Rosecrans - D1" (Detail Map 1) US Army map shows the location of the various buildings and gun batteries as of 29 June 1921, with Battery Whistler (four 12-inch mortars) out of image to the left. Note the L-shaped mine-planter wharf at bottom center and the quartermasters wharf to its right, and the coincidence rangefinder (CRF) position for Battery Fetterman left of bottom center.)

Battery WILKESON :
    Fort Roscerans. Calif.
              Jany 20/03

Figure 2-17. The four 10-inch disappearing-rifle emplacements of Battery Wilkeson are seen looking northeast along Ballast Point towards the old Coast Guard lighthouse on 20 January 1903 (US Army photo courtesy SSC Pacific). Note the construction crane at Battery Fetterman just beyond Gun 4.

Figure 2-18 provides a close-up view of the Gun 1 emplacement in Battery Wilkeson during a live-fire target practice in 1916. The newly introduced 10-inch disappearing rifle was counterweighted to facilitate being raised just before firing, after which it sank down out of sight, where it could more easily be reloaded (Duchesneau, 2006). When situated at a reasonable elevation behind protective earthworks, such an arrangement provided a distinct tactical advantage over earlier designs.

Figure 2-18. Target practice with a 10-inch disappearing rifle at Battery Wilkeson, circa 1916 (US Army photo courtesy Cabrillo National Monument). A soldier operating a coincidence rangefinder is seen to the right of Gun 1, with the San Diego jetty extending south from North Island in the upper-right corner.

A close-up view of a 10-inch disappearing rifle is presented in Figure 2-19. The procedures for transporting ammunition from the Battery Wilkeson magazines on the lower level to the guns above were explained by Davis (1896):

> "In serving ammunition, the powder will be carried by hand from the magazine to the corridor and there placed on the ammunition truck. The shot and shell will be carried on trolleys from the shot and shell room to the truck. The truck is raised on the lift to the loading platform and then moved to the gun. Two cranes for each gun will raise the ammunition when the lifts are disabled."

Figure 2-19. Close-up view of a breach-loading 10-inch disappearing rifle, location and date unknown (US Army photo courtesy Al Grobmeier). Note artilleryman using the sighting telescope in upper-left corner.

The below excerpt is from a 4 July 1979 letter from Hollis T. Gillespie to Ranger Brett Jones at Cabrillo National Monument (Overton, 1993): [14]

> "I am enclosing two pictures of #2 gun of Battery Wilkeson in firing position. The wooden platform was used as a place for the breech crew to stand while firing a sub-caliber 1-lb. projectile for target practice. The barrel for the sub-caliber shell was fitted inside the 10-in. bore of the gun and allowed for a less expensive and less dangerous method of target practice, while giving the range crew experience at tracking a target."

The 10-inch disappearing rifles were ordered salvaged in the fall of 1942 (Thompson, 1991) and the battery was officially deactivated the following year (Ruhge, 2019d).

Figure 2-20a shows Battery Calef-Wilkeson, looking north towards San Diego Bay on 13 April 1968, with the Gun 1 emplacement in the foreground. The wooden structure atop the upper level (left of bottom center) of this southernmost emplacement served as the local battery commander station when the battery was active (see "BC" annotation in the plan view of Figure 2-14). The second wooden structure seen precariously situated along the shoreline in the bottom-right corner was perhaps an observation station with an azimuth instrument for some later nearby battery.[15] Figure 2-20b shows emplacements 3 and 4 looking east.

---

14 Hollis Gillespie, Battery B, was assigned to Guns 1 and 2 at Battery Calef-Wilkeson just prior to WWII, then served on Battery McGrath when Calef-Wilkeson was declared obsolete (Overton, 1993).

15 This structure would have been destroyed by the muzzle blast from Gun 1 at Battery Wilkeson.

Figure 2-20. a) Battery Calef-Wilkeson is seen looking north towards San Diego Bay on 13 April 1968, with the old Coast Guard lighthouse facility on Ballast Point in upper-right corner. **b)** The former Battery Calef seen looking east towards the entrance to San Diego Bay on 13 April 1968 (photos courtesy Al Grobmeier).

Referring again to Figure 2-14, Gun emplacements 1 and 2 at Battery Calef-Wilkeson had a different lower-level layout from that of Guns 3 and 4, and even each other. The extended stairway to the right of a straight rear wall with three doors indicates Figure 2-21 below shows the rear of the Gun 1 emplacement, which has been repurposed by the Navy for storage and training. The local battery commander station previously seen on the top level in Figure 2-20a above would have been located behind the chain link fence in the upper-right corner of this photo.

Figure 2-21. This rear view of the battery shows the lower-level doorways of the Gun 1 emplacement, which has been repurposed by the Navy as seen on 30 October 2006 (photo courtesy Jeff Bowen). Note radio antenna at upper right, the former location of the battery commander station seen earlier in Figure 2-20a.

Battery Fetterman

Named for 2nd Lieutenant George Fetterman of the 3rd US Artillery (Ruhge, 2019), Battery Fetterman was equipped with two 3-inch rapid-fire *M1898* guns on balanced-pillar mounts. Construction began in 1898 at the former site of the old Spanish Fort Guijarros, just to the north of Battery Wilkeson near Ballast Point. Plan and section views from the Report of Completed Works are presented in Figure 2-22. Construction was completed on 13 February 1903, with transfer to the Coast Artillery on 30 April at a cost of $8,865.

As reported by Captain James J. Meyler (US Engineer's Office, Los Angeles) in a status report to Brigadier General John M. Wilson (US Army Chief of Engineers, Washington, DC) (Meyler, 1899a):

"The site selected, on a piece of new made-level ground with a top reference of about 8 feet above mean low water, seems well adapted to these small-calibre guns. It is about one acre in extent, and has been made during the past year from material excavated on the site of the 10-inch gun battery. It will enable the guns of the battery to thoroughly sweep the mine fields as well as the land approaches on either side of the harbor entrance."

The "10-inch gun battery" cited above was of course Battery Calef-Wilkeson.

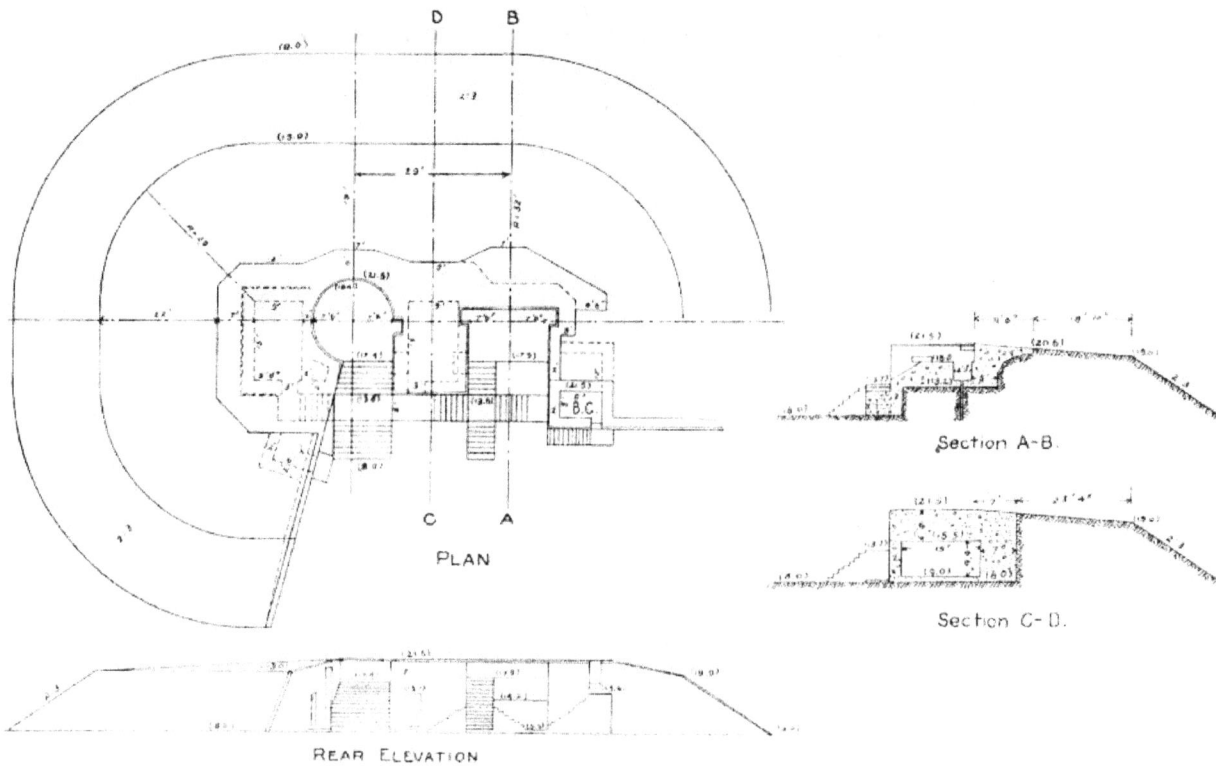

Figure 2-22. Plan and section views of Battery Fetterman from the Report of Completed Works, corrected to 15 September 1925, by which time the two 3-inch guns had been removed.

A 10 October 1903 entry detailing the defenses of San Diego in an engineer's journal prepared by Major J.H. Willard provide the following data for Battery Fetterman:

- 2 3-inch R.F. guns, Model 1898; Nos. 95 and 96, Driggs-Seabury Co.

- 2 3-inch balanced-pillar mounts, Model 1898; Nos. 95 and 96, Driggs-Seabury Co.

Following WWI, Battery Fetterman's 3-inch guns and carriages were approved for removal on 27 March 1920 (Willard, undated). The carriages were ordered scrapped on 26 May that same year and the guns transferred to Watervliet Arsenal 6 weeks later on 9 July (Stanton & Thayer, 2019b). Interestingly, Form 2 of a Report of Completed Works for a "C.R.F. Station, Battery Fetterman," corrected to 1 December 1921 (Figure 2-23), indicates this coincidence-range-finder station 50 feet south of the battery was transferred in 19 May 1921, well after the battery was disarmed. The "Type of Observing Instrument" for this station was recorded as "None received."

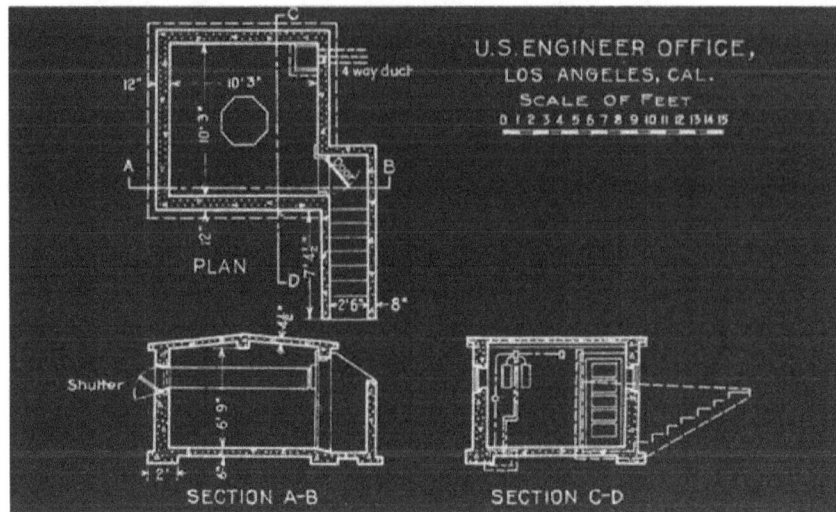

Figure 2-23. Plan and section views for the "C. R. F. Station, Battery Fetterman" from Form 2 of the Report of Completed Works, corrected to 1 December 1921 (NARA). This station 50 feet south of Fetterman may have been contemplated for use by nearby Battery Calef-Wilkeson (Figure 2-24).

Figure 2-24 below shows the abandoned emplacements of Battery Fetterman in the turn of Rosecrans Street at photo center on 27 December 1937, just to the right of Battery Calef-Wilkeson, with Ballast Point extending northeast at upper right. The "C. R. F. Station, Battery Fetterman" discussed above may possibly be the rectangular structure south of Fetterman near the Gun 4 emplacement of Battery Calef-Wilkeson (up arrow). The distance appears to be greater than 50 feet however, and it does not correlate well with the CRF location shown earlier in Figure 2-15. A far better perspective is seen later in Figure 2-52.

Figure 2-24. Battery Fetterman (image center) is just right of Battery Calef-Wilkeson (lower left quadrant) in this 27 December 1937 photo, with Ballast Point extending to the northeast at upper right. Note what may be the C.R.F station (up arrow) south of Fetterman (US Navy photo LSF 84-4-67 courtesy SSC Pacific).

On 21 August 1939, the request for authority to demolish the abandoned Battery Fetterman was granted by Headquarters Ninth Corps Area, Presidio of San Francisco. The battery was destroyed in 1940 and later replaced by an Anti-Motor-Torpedo-Boat (AMTB) battery of two 90-millimter guns at the same site (AMTB Fetterman), as will be further discussed in Chapter 6.

Battery Meed

Battery James Meed was named after US Army Captain James Meed, who was killed fighting the British in the War of 1812. This battery was built at Fort Pio Pico (Figure 2-25), a subpost of Fort Rosecrans constructed in 1901 on the Zuninga Shoal Tract, North Island Military Reservation, Coronado, CA (Meyler, 1900; Ruhlen, 1959; Roberts, 2016). The Report of Completed Works for "Battery Pio Pico," corrected to 27 March 1919, indicates construction started in September 1901, was completed in February 1902, with the battery transferred on 10 May that same year at a final cost of $10,000. Additional support structures included a cook house, a watchman's house, and a wharf.

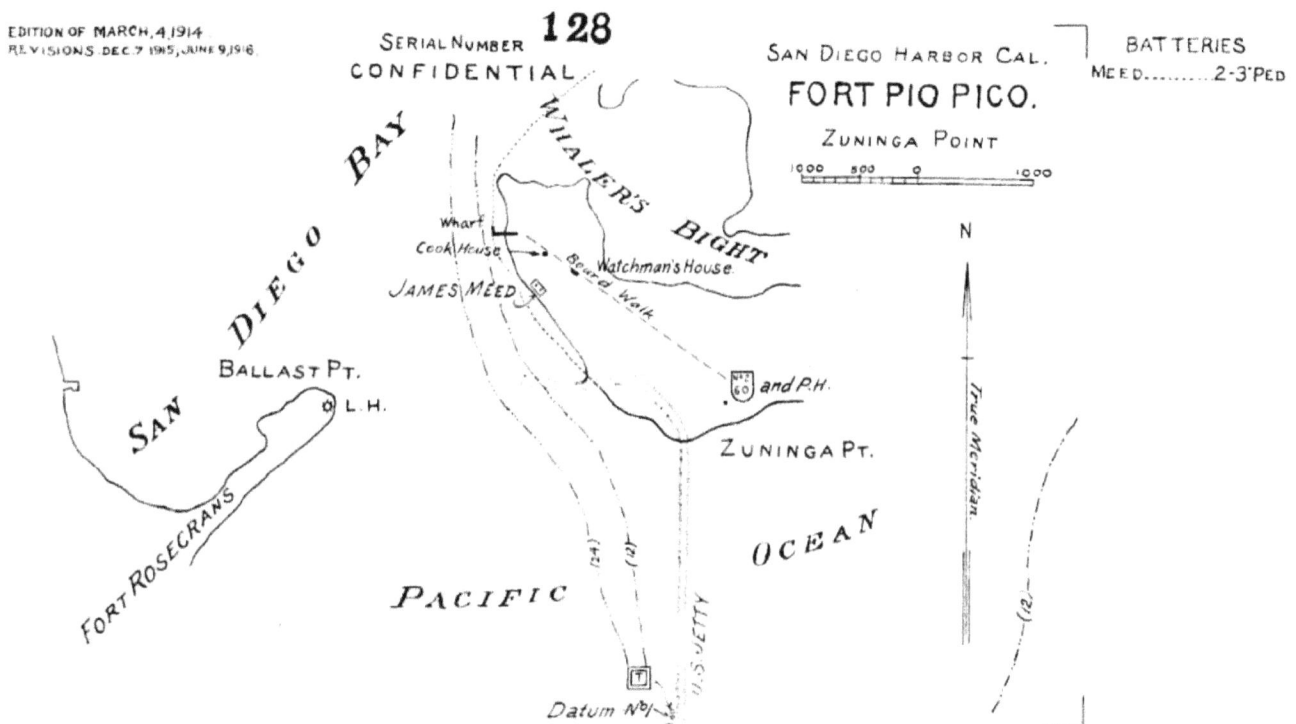

Figure 2-25. Battery James Meed (image center), armed with a pair of pedestal-mount 3-inch guns, was constructed at Fort Pio Pico on North Island, across the harbor entrance from Ballast Point. Troops from Fort Rosecrans were shuttled back and forth daily by boat via the wharf just north of the battery.

A FortWiki page entitled "Fort Pio Pico" indicates the battery was accepted for service in 1902, but without its armament (Stanton & Taylor, 2019c):

> "The carriages arrived on 26 November 1909 and the guns arrived sometime after that. By 13 April 1910 the battery was fully armed."

Roberts (2016), on the other hand, reports two 3-inch 15-pound model 1903 pedestal-mount guns were installed in 1904, but provides no references. This claim appears negated, however, by a 31 January 1907 Report of Completed Batteries (RCN) for Battery Meed, Fort Pio Pico, submitted on 8 January 1908 by Amos A. Fries, Corps of Engineers, Los Angeles, CA, which specifically stated:

> "No guns or carriages on hand. Carriages have been ordered from Watertown Arsenal."

Another FortWiki page entitled "Battery Meed" provides the following unreferenced account (Stanton & Thayer, 2019d):

> "The battery was accepted in 1902 without guns or carriages. The *M1903* carriages were mounted and turned over to the engineers on 26 November 1909 without guns. The guns arrived and were mounted by 13 April 1910."

Form 1 from the Report of Completed Works for "Battery Pio Pico," corrected to 27 March 1919, unfortunately provides no further clarifying information regarding armament disposition other than "Guns removed to Battery McGrath January 17, 1919." As seen in Figure 2-26, the Form 7 plan and section views of the battery have a similar hand-written annotation.

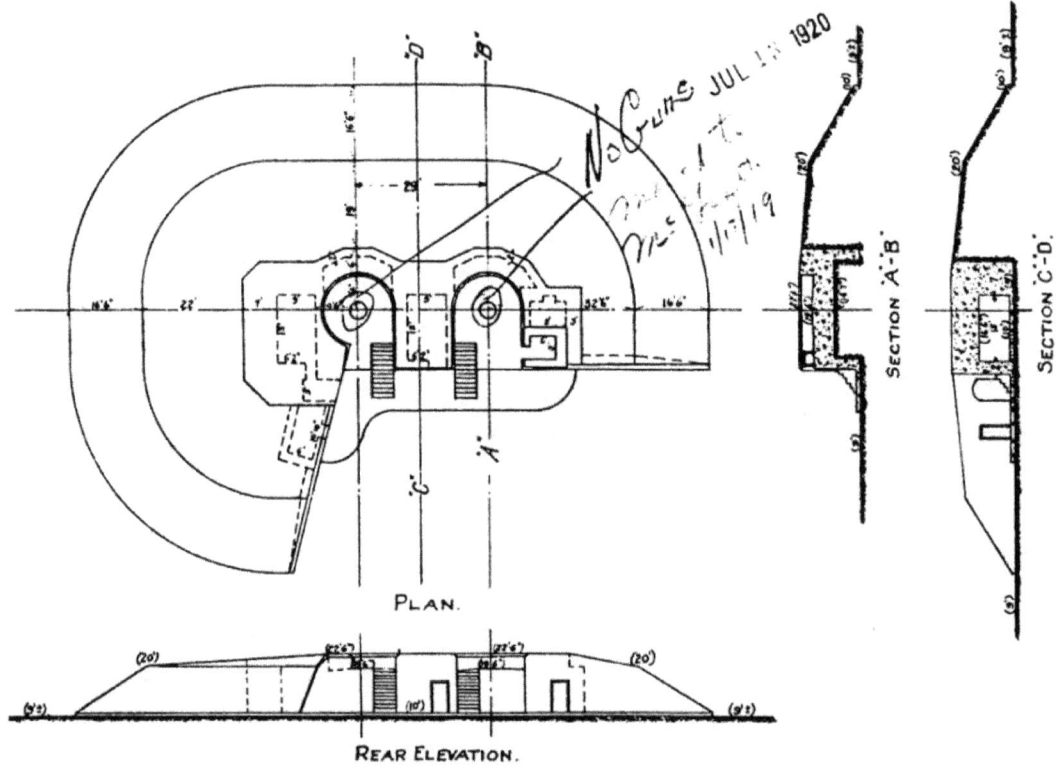

Figure 2-26. Plan and section views of Battery James Meed at Fort Pio Pico, reproduced from the Report of Completed Works corrected to 27 March 1919 (NARA). The hand-written inscription indicates the two 3-inch guns were moved to Battery McGrath on 17 January 1919.

The guns were mounted on the upper level of this two-story battery (Figure 2-27), with the shell and powder brought up by hand from the magazines on the lower level below, since there were no ammunition hoists installed (Stanton & Thayer, 2019d). As earlier pointed out by Captain James J. Meyler (1900), Corps of Engineers, Los Angeles:

> "The guns of the 15-pdr. Battery will be able to thoroughly sweep the mine fields, as well as the land approaches on either side of the harbor entrance."

A complementary role was assigned to Battery Fetterman, also equipped with two 3-inch guns, on the opposite side of the harbor entrance.

Figure 2-27. View of the Gun 1 emplacement at Battery Meed, looking south during target practice at Fort Pio Pico in 1911 (George Ruhlen Jr. photo courtesy Al Grobmeier). These two 3-inch guns were later transferred to Battery McGrath.

The interesting statement below, included by Captain Meyler in his August 1900 memorandum to Brigadier General John M. Wilson, paints a decidedly different picture of Coronado as we know it today (Meyler, 1900):

"As the Zuninga Shoal Tract adjoins the hunting grounds of the Coronado Beach property, it is considered desirable to use steel instead of wooden doors, and, on completion of the work, to build a barbed wire fence along the boundary line between the two properties."

On 2 September 1913, Major William G. Davis, Coast Artillery Corps, of Headquarters, Coast Defenses of San Diego, Fort Rosecrans, CA, sent a memo to the District Engineer, Los Angeles, CA, expressing concern over beach erosion at Fort Pio Pico (Davis, 1913):

"It has come under the observation of this office that the wing-jetty built last winter for the protection of Battery Meed is not of sufficient height. During recent storms high seas have washed over the jetty and encroached upon the beach in front of this battery. It is requested that steps be taken before the opening of the stormy season to have the height of the jetty increased."

This was rather short notice, however, in that the winter rainy season in San Diego typically starts in November. Major J.C. Gilmore, Coast Artillery Corps, responded to this memorandum on 10 September 1913, requesting the estimated cost of the desired work.

On 29 September of that same month, an estimate to increase 350 feet of the jetty to a height of 9 feet above mean lower low water was received in the amount of $9,935.00. Major Gilmore (1913) subsequently pointed out the following concern to the Commanding General, Western Department, which had previously been expressed by the Pacific Coast Artillery District Commander on 19 April 1913:[16]

"The position of Battery Meed was originally selected to cover the mine-fields of that day, located even further inside the channel than the battery itself. The present mine field, however, is located outside the break-water, and a glance at the accompanying map will show that Battery Meed covers it only at its extreme ranges and then only imperfectly."

---

16 A.G.O. 2032794 – Hq. West. Dep. 292-E.

Gilmore (1913) subsequently recommended that Battery Meed instead be relocated across the harbor entrance to Billy Goat Point on Point Loma. Battery Meed was damaged in a storm a few months later in 1914, however, after which its two 3-inch guns were removed for installation in Battery McGrath at Fort Rosecrans in 1919 (Ruhlen, 1959).[17] Fort Po Pico was subsequently abandoned by the Army, then demolished in 1922 (NAF, 2019).

A slightly different account, however, is provided by Linder (2019):

"Although Fort Pio Pico was damaged by a storm in 1914, stalwart defenders of Coronado continued to serve at their posts on Zuniga Point throughout World War I. Coronado's guns were ultimately dismounted and transferred to Point Loma in 1918 when it was discovered that beach erosion had further undermined Battery Meed."

The disconnect in dates is explained in the following excerpt from Thompson (1991):

"Lt. Col. W.F. Hase inspected the Fort Rosecrans batteries early in 1918 and discovered that the sea was undermining the 3-inch Battery Meed at Fort Pio Pico. His recommendation that the guns be dismounted and the battery abandoned was approved. Since the 5-inch guns at Battery McGrath had gone to war, the Engineers modified its emplacements and mounted Meed's guns in their stead in May 1919" (Hase, 1918a, 1918b).

In summary, it seems the guns from Battery Meed were transferred to Battery McGrath in 1918, but not installed until 1919.

The entire Fort Pio Pico area was eventually turned over to the Navy under Executive Order 7215, dated 26 October 1935 (Stanton & Thayer, 2019c). The Army's Rockwell Field, established on North Island in 1917, was also transferred to the Navy at this time, and later became Naval Air Station North Island. Ruhge (2019b) reports the original Fort Pio Pico location was eradicated in 1941 when the ship channel was widened.

Taft Period (1905 - 1915)

Following the Spanish-American War, a new fortifications board was appointed by President Theodore Roosevelt in 1905. Headed by Secretary of War William Howard Taft, the Taft Board reviewed the Endicott recommendations, and subsequently accelerated the installation of many projected features that had yet to be developed or installed (Lewis, 1979). These included more searchlights for nighttime illumination and a new triangulation system of fire control that was superior to the elementary sighting instruments attached to the individual guns, as will be extensively discussed in Chapters 7 and 8.

Another significant outcome was the introduction of 12-inch mortars to expand the range and field of fire of defensive batteries. A WWI-era engineer's journal entry by Major J.W. Willard, U.S. Army Corps of Engineers, under the subheading "Fort Rosecrans" reads as follows (Willard, undated):

"ADDITIONAL: Under date of June 11, 1914, the Chief of Staff approved memorandum of the Chief of Coast Artillery, June 8, 1914, (81892/14) recommending inclusion in estimates for 1916 construction of mortar batteries at Fort Rosecrans. Under date of May 18, 1914, the Chief of Staff approved memorandum of the Chief of Coast Artillery, May 18, 1914, (94312/5) recommending the convening of a board of officers to consider plans for mortar batteries at Fort Rosecrans. The board so convened recommended 2 batteries, one of 8 mortars in Powerhouse Canyon, Fort Rosecrans, and one of 4 mortars on the U.S. Naval Reservation adjoining the Fort Rosecrans Reservation." [18]

---

17 As previously noted, Battery McGrath's 5-inch guns had been sent overseas during WWI.

18 See also the Fort Rosecrans map presented earlier in Figure 2-2. This "Fort Rosecrans - D1" (Detail Map 1) US Army map shows the location of the various buildings and gun batteries as of 29 June 1921, with Battery Whistler (four 12-inch mortars) out of image to the left. Note the L-shaped mine-planter wharf at bottom center and the quartermasters wharf to its right, and the coincidence rangefinder (CRF) position for Battery Fetterman left of bottom center..

The four-mortar battery west corner of the Naval Reservation emerged as Battery Whistler, while the proposed eight-mortar battery in Powerhouse Canyon was reduced to four mortars in what became Battery John White.

Battery Whistler

Named for Civil War Colonel Joseph N.G. Whistler, Battery Whistler was located near the northern boundary of Fort Rosecrans on the east side of the main road running north/south along the top of the Point Loma ridge, now Catalina Boulevard (Figure 2-28). According to the Report of Completed Works, construction commenced on 15 March 1916 and was completed on 19 August 1919, with transfer to the Coast Artillery on that date at a final cost of $10,000. This installation complemented a similar capability simultaneously under construction in nearby Powerhouse Canyon, Battery John White, as discussed in the next section. Both these batteries had an all-round field of fire.

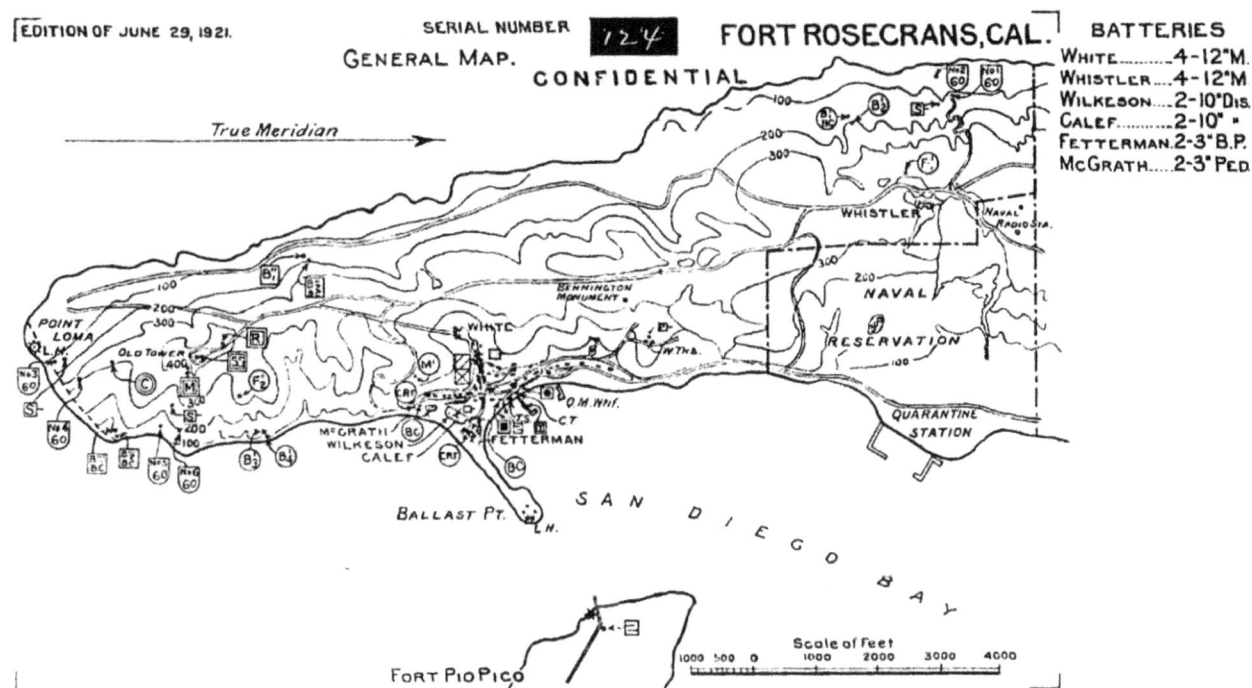

Figure 2-28. Battery Whistler was located west of the Naval Reservation on the northern portion of Fort Rosecrans, just east of what is now Catalina Boulevard at image right (NARA).

By 31 December 1917, the battery had been armed with two 12-inch *M1890* mortars mounted on *M1896* carriages in each of its gun pits. As shown in Figure 2-29, powder and shells were stored in earth-covered concrete magazines along the front of the battery. Targeting solutions calculated in the centrally located plotting room at the rear of the battery were conveyed to data booths in each of the gun pits. The facility was serviced by commercial power and the central Fort Rosecrans powerplant, with a backup pair of 25-kilowatt AC generators in the engine room behind Pit B.

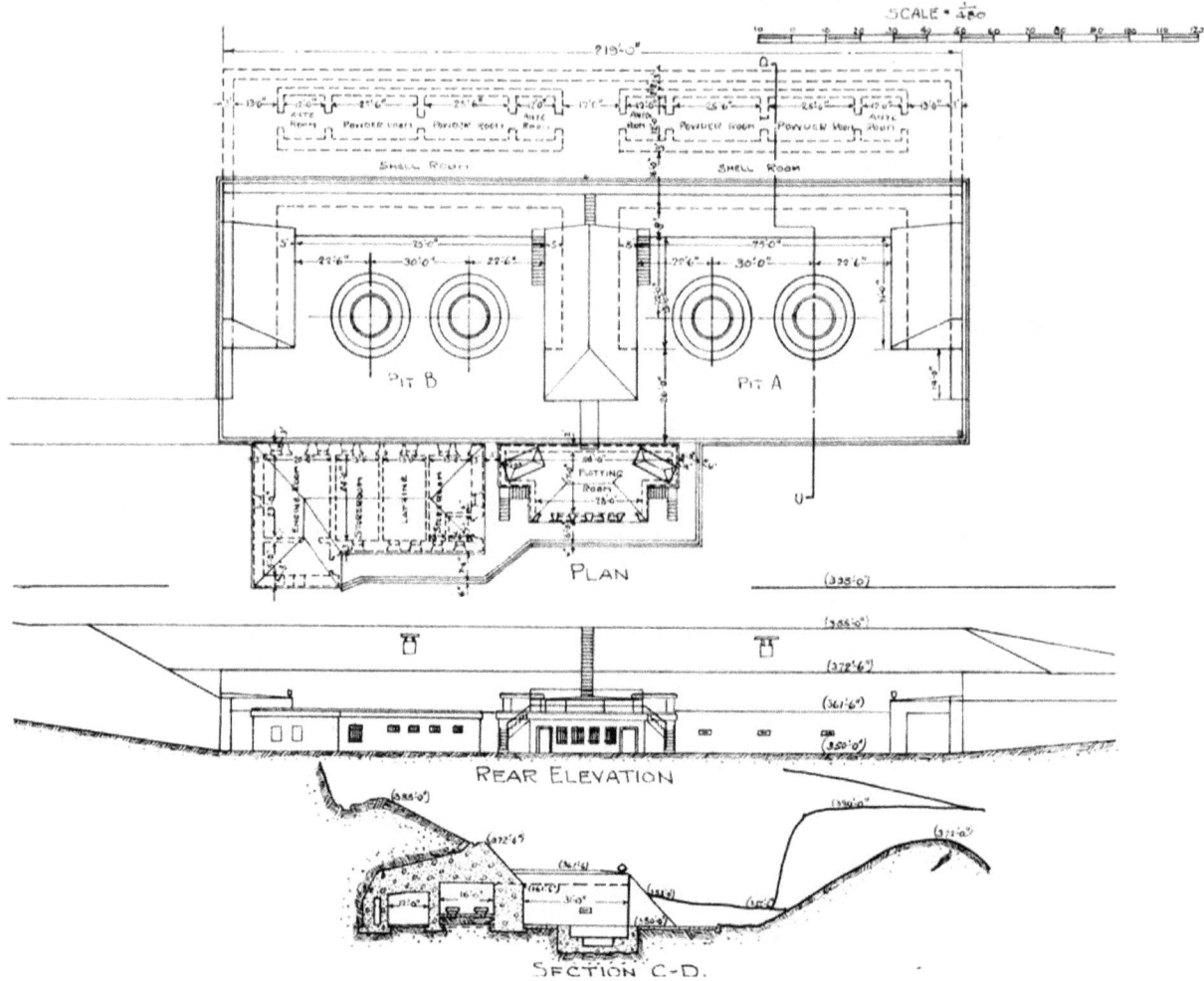

Figure 2-29. Plan and section views of Battery Whistler from Form 7 of the Report of Completed Works, corrected to 27 March 1919 (NARA). The powder and shell rooms were beneath the sloped front section, while the engine room for the two backup generators is at lower left behind Pit B in the plan view. Note also the Plotting Room at bottom center in the plan view.

Figure 2-30 shows horse-drawn graders smoothing the terrain surrounding the ongoing construction on 25 October 1916, with Catalina Boulevard seen winding southward in background right towards the Cabrillo National Monument. Additional fill dirt has yet to be added to the front side of the battery at far right, which would eventually rise to the top of the two concrete 18-by-18-inch vent shafts for the shell and powder rooms seen in the upper-right corner. The topside barracks area would be constructed on the east side of Catalina Boulevard just south of this battery during WWII.

Figure 2-30. Battery Whistler is seen looking south during construction on 25 October 1916, with Catalina Boulevard in background right (US Army photo courtesy Cabrillo National Monument). The four 12-inch mortars would finally be mounted on 31 December 1917. Note the two vents in the upper-right corner.

The undated photo in Figure 2-31 below shows the Navy Radio and Sound Laboratory at right center, looking west towards the Pacific Ocean in the background. Gun pits A and B of Battery Whistler are seen south (left) of the Lab, on the east side of Catalina Boulevard. The upper WWII barracks southeast of Battery Whistler have not yet been constructed. Note the city of San Diego water tank at the bend in Catalina Boulevard at upper right, just north of the Fort Rosecrans fence line, which runs east-west down to the cliffs overlooking the shoreline.

Figure 2-31. Aerial view of Battery Whistler (left of center) looking west, undated, but prior to construction of the WWII barracks to the south (photo courtesy SSC Pacific). The city water tank at upper right on Catalina Boulevard marks the northern boundary of the fort, with the fence line running east-west.

The aerial photo presented in Figure 2-32, also undated, shows the upper WWII barracks on the far side of Battery Whistler, looking southeast towards the entrance to San Diego Bay. Note the access road for Battery Whistler in the upper-right corner coming down from the barracks area, and the protective earthwork now surrounding the concrete air vents on the west side of the battery. The above-ground concrete structure at photo center behind Gun Pits A and B housed the engine room for the two back-up generators on its south end, with two storerooms plus a latrine in the center, and the plotting room on the north end (see again Figure 2-29).

Figure 2-32. Undated WWII-era photo of Battery Whistler, looking southeast towards the upper WWII barracks area, with the entrance to San Diego Bay seen in background left. The origin of the two Civil-War-era sling carts in the foreground and their role at this location are unknown (US Navy photo courtesy SSC Pacific).[19]

Battery Whistler was declared obsolete in 1942, some 23 years after its completion. Following the end of WWII, Dr. Waldo Lyon, then head of the Subsurface Studies Branch in the Research Division of the Navy Electronics Lab, acquired the abandoned facility in 1948. According to the *Navy Newstand* website (ASL, 2019), construction reportedly began on the new "Submarine Research Facility" atop this structure in 1952. Figure 2-33, however, shows the new overhead gantry crane for this lab already installed in August 1949 (see also Figure 2-34).

---

19 These sling carts allowed Civil War artillerymen to quickly remove and roll away a damaged gun barrel, then replace it with a new one in the field (LaPuzza, 2020). They were at some point refurbished by Waldo Lyon and other members of the Arctic Submarine Lab, and years later donated by the Center to Fort Pulaski National Monument in Savannah, GA.

Figure 2-33. The four 12-inch mortars at Battery Whistler were excessed in 1942 and scrapped on 18 December 1943. This August 1949 aerial photo shows the newly installed overhead gantry crane for the Center's Submarine Research Facility (US Navy photo courtesy SSC Pacific).

The undated (probably mid-1950s) photo in Figure 2-34 shows the finished Submarine Research Facility, looking west towards the Pacific Ocean. The original inscription attached to this photo reads:

"Advanced radiographic and spectrographic studies were enhanced by use of ingenious equipment at this NEL Submarine Research Facility, which possessed the only betatron on the West Coast." [20]

As previously mentioned, the Submarine Research Facility would be renamed the "Arctic Submarine Lab" in 1969, with Dr. Waldo Lyon as its first Director (ASL, 2019).

---

20 Betatron – an early type of particle accelerator for producing high-energy electron beams.

Figure 2-34. The "Submarine Research Facility" (Building 371) would become the "Arctic Submarine Lab" in 1969 (US Navy photo courtesy SSC Pacific). Note the Fort Rosecrans topside chapel on the far side of Catalina Boulevard at upper left, and the San Pedro radio tower under construction just left of top center.

Dr. Waldo Lyon is probably best known for guiding the nuclear submarine *USS Nautilus (SSN-571)* on its historic transpolar journey under the northern polar icepack in August 1958, using a new technique known as inertial navigation (Figure 2-35a). The *Nautilus* departed Honolulu on 23 July 1958, crossed under the North Pole at 2315 EDT on 3 August, and surfaced 4 days later off Greenland (DoA, 1959). Then Rear Admiral Hyman Rickover, the legendary father of the US Navy's nuclear power program, is seen standing with Dr. Lyon on the deck of the *Nautilus* in Figure 2-35b.

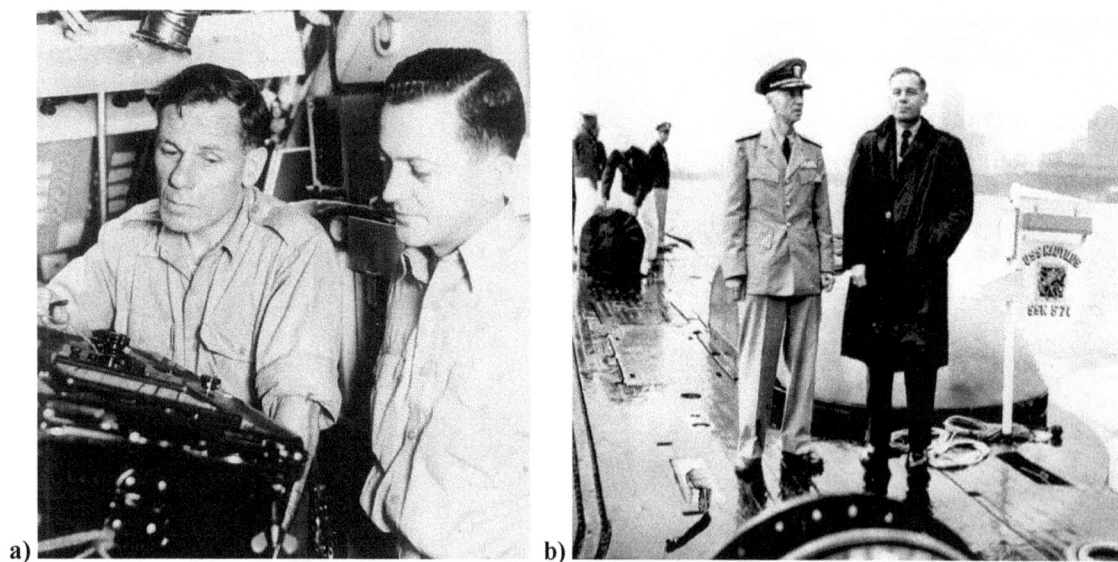

Figure 2-35. a) Dr. Lyon (left) focuses on the sonar console with *Nautilus* skipper Commander W.R. Anderson while operating beneath the ice. **b)** Rear Admiral Hyman Rickover stands with Dr. Lyon on the deck of the world's first nuclear submarine, the *USS Nautilus* (US Navy photos courtesy SSC Pacific).

Figure 2-36a shows Captain John Phelps, Commanding Officer of the U.S. Navy Electronics Lab in San Diego, CA, greeting Dr. Lyon in front of Building 33 upon his return from the historic transpolar expedition of *USS Nautilus* in 1958. Given the obvious strategic advantages associated with under-ice navigation, the Arctic Submarine Experimental Pool and Cryogenic Facilities were completed at the Battery Whistler site the following year (ASL, 2019). Figure 2-36b shows Virginia Lyon accepting the Distinguished Federal Service Award from President John F. Kennedy on behalf of her husband in 1962.

a)   b)

Figure 2-36. a) Dr. Lyon is welcomed on return from his 1958 voyage by Captain John Phelps, Commanding Officer, U.S. Navy Electronics Lab. **b)** Virginia Lyon accepts the Distinguished Federal Service Award from President Kennedy on behalf of her husband in 1962. (US Navy photos courtesy SSC Pacific).

On 4 June 1944, the destroyer escort *USS Pillsbury (DE-133)* of US Navy *Task Force 22.3* had captured the German *U-505*, a Kriegsmarine *Type IXc* long-range U-boat, off the west coast of Africa (Green, 2003). Figure 2-37 shows a *Pillsbury* boarding party securing a tow line to the submarine, which would soon yield valuable intelligence vital to the war effort (NNS, 2004). In 1954, this submarine was transported via the St. Lawrence Seaway and through the Great Lakes and put on display at the Museum of Science and Industry in Chicago, IL.

Figure 2-37. A boarding party from the destroyer escort *USS Pillsbury (DE-133)* secures a tow line to the *U-505*, a *Type IXc* German U-boat captured by the US Navy on 4 June 1944 (US Navy photo courtesy SSC Pacific).

Removed shortly after the 1944 capture, the submarine's periscope was later shipped to the former Battery Whistler area at the Navy Electronics Laboratory in San Diego. The undated concept drawing shown in Figure 2-38 illustrates the installation of this repurposed periscope in the cold-water tank at the Submarine Research Facility (Building 371). As more recently reported in the *Los Angeles Times* (Perry, 2002):

"Building 371 contained a 250,000-gallon pool of seawater that could be chilled to minus 50 degrees Fahrenheit to simulate arctic conditions. The periscope allowed scientists in a space beneath the pool to study the performance of equipment above the waterline and to give prospective submarine skippers the simulated experience of 'breaking through' the ice."

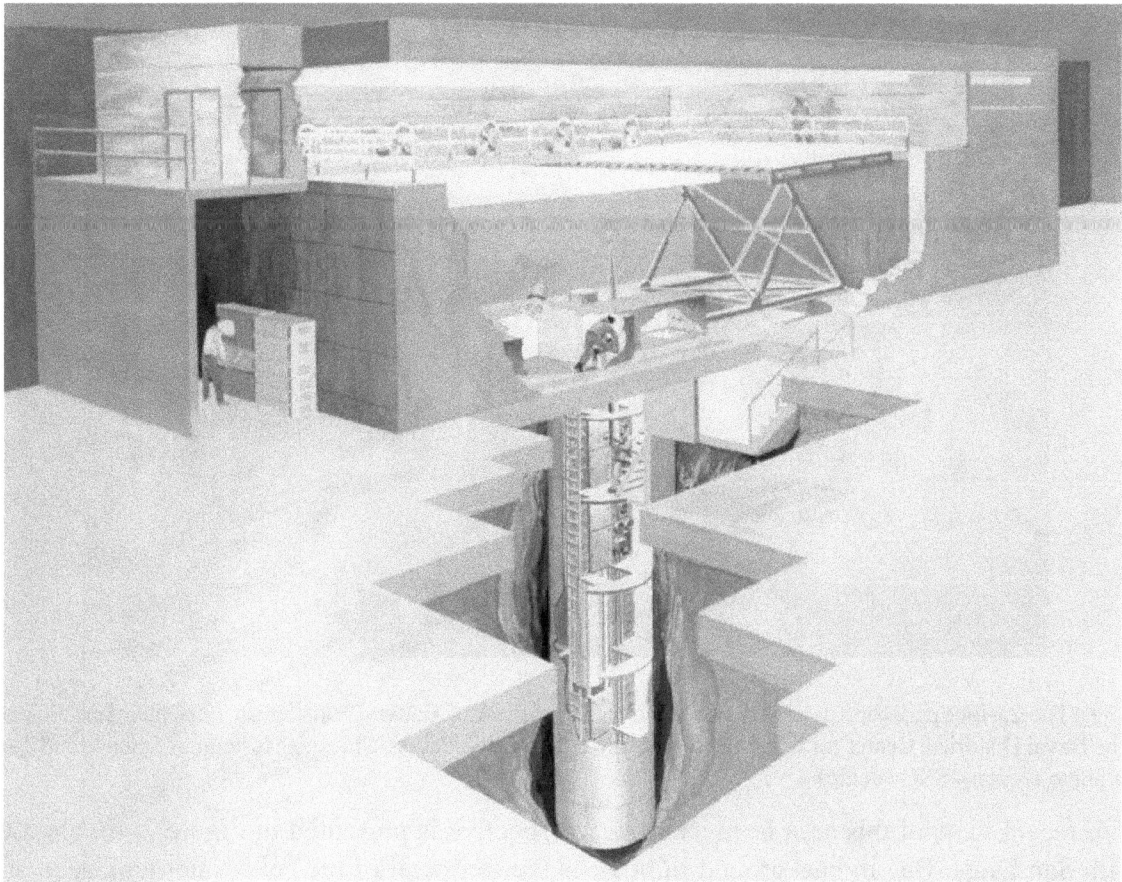

Figure 2-38. Undated artist concept of the German *U-505* periscope installation beneath the cold-water tank in Building 371 at the Submarine Research Facility, which was renamed the Arctic Submarine Laboratory in 1969 (drawing courtesy SSC Pacific).

No longer required following the end of the Cold War, Building 371 at the former Battery Whistler site on Point Loma was eventually demolished in 2002 (Perry, 2002). Figure 2-39 shows the 25-foot periscope assembly being removed from the bottom of what had once been the cold-water tank in Pit A, its eastern wall now destroyed, looking north on 12 September of that year. Navy Region Southwest transferred ownership of the periscope to the Naval Historical Center in Washington, DC,[21] which subsequently arranged for the artifact to be loaned to the Museum of Science and Industry in Chicago, where it was added to the *U-505* exhibit.

---

21 The Naval Historical Center was renamed the Naval History and Heritage Command in 2008.

Figure 2-39. The 25-foot periscope assembly was removed from Pit A of Battery Whistler on 12 September 2002 and transferred to the Naval Historical Center for indefinite loan to the *U-505* exhibit at the Chicago Museum of Science and Technology (US Navy photo courtesy SSC Pacific).

A more recent view of this area from a similar perspective is presented in Figure 2-40, circa October 2007, with San Diego Bay in background right. The western wall of the cold-water tank seen at far left in Figure 2-39 above has now also been demolished, and the periscope shaft shown earlier in Figure 2-27 filled in and sealed with concrete. The small concrete room at background center, identical to one on the south end of the battery, is for some reason not identified in the Report of Completed Works drawing presented earlier in Figure 2-29.

Figure 2-40. Remains of mortar Pit A of Battery Whistler, looking north in October 2007, with the keyhole-shaped concrete patch covering the former location of the *U-505* periscope at image center (see again original battery layout in Figure 2-29.)

Battery John White

Named for Colonel John Vassar White, U.S. Army, Battery White was intended to further beef up the Harbor Defenses of San Diego. A proposed drawing of this installation layout, which featured four 12-inch *M1890MI* mortars on *M1896MI* carriages in two separate pits, was submitted to the Secretary of the Army for approval in November 1915. Construction commenced west of Battery Wilkeson in Power-house Canyon ravine on 8 November 1915 (see again Figure 2-2) and was completed the next year with the layout shown in Figure 2-41.

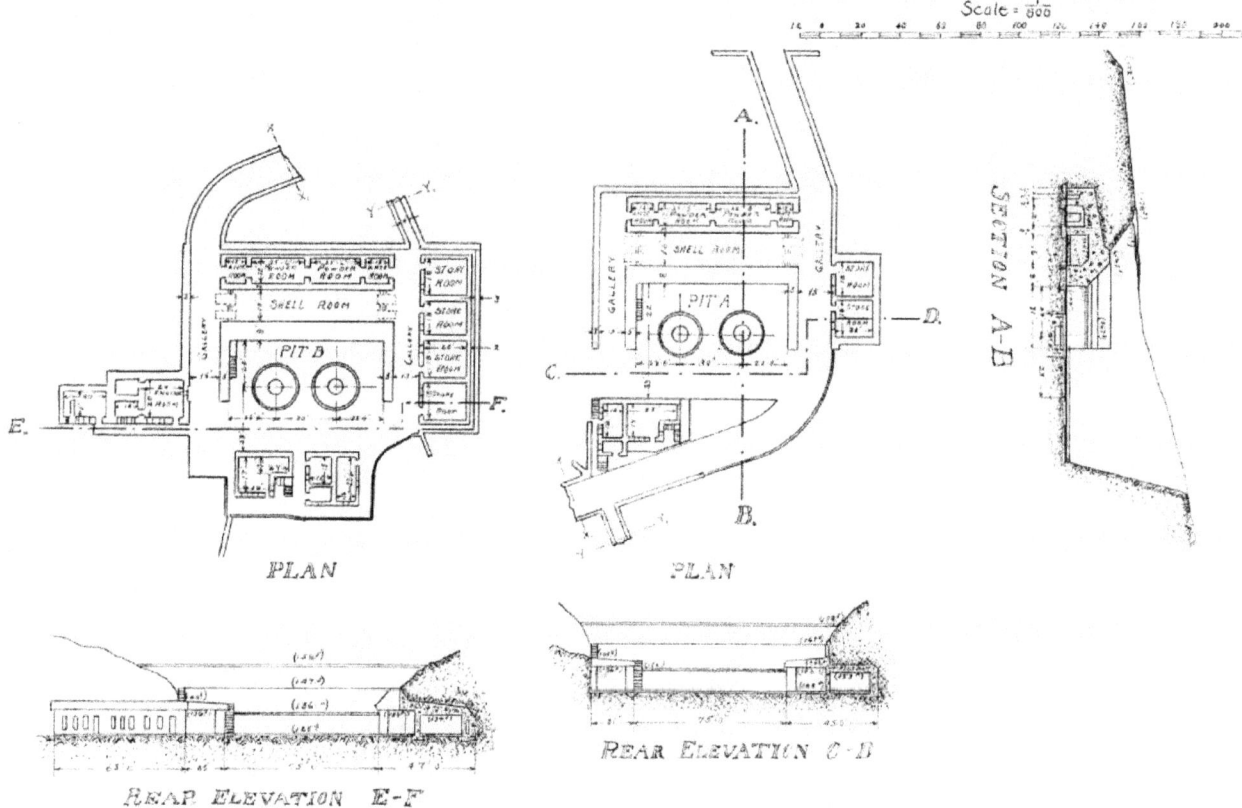

Figure 2-41. Plan and section views from Form 7 of the RCW for the 12-inch mortar Battery John White, undated (NARA). The trunnion elevation of Gun Pit A on the right was 149.70 feet, while that of Gun Pit B on the left was 129.65 feet. The power house is seen at the bottom of the plan view for Pit A at image center.

The battery was connected to commercial electrical power, with a pair of back-up 25-kilowatt genera-tors in a local powerhouse between the gun pits. As delivery of the four 12-inch mortars and carriages was delayed, the guns were not mounted until 31 December 1917, then proof fired a year later (Ruhge, 2016). Battery White was finally transferred to the 3rd Coast Artillery Regiment on 19 August 1919 at a cost of $144,200. The aerial photo presented in Figure 2-42 below shows Pit A and Pit B in Powerhouse Canyon with their mortars mounted in November 1922.

Figure 2-42. Guns 1 and 2 in Pit A are seen in the lower-left quadrant, with Guns 3 and 4 in Pit B (right of center), looking north circa November 1922 (US Army photo courtesy Al Grobmeier). Note the tunnel entrance at lower left in front of Pit A in the Powerhouse Canyon ravine.

Figure 2-43 below provides a later view of Battery White, looking west in the mid to late 1930s. The associated back-up power house is seen inside the curve of the ramp just right of the Gun 2 emplacement in Pit A (upper-left corner). The three wooden buildings in the lower-right corner are identified as the "E.D. Shop," "E.D. Storehouse," and "E.D. Repair Shop" in the map legend of Figure 2-2. The smaller structure just above them and behind Pit B is identified as the "Transformer and Switchboard Bldg."

Figure 2-43 This undated photo, looking west in the mid to late 1930s, provides a closer view of the four 12-inch mortars installed in Battery White in Powerhouse Canyon (US Army photo courtesy SSC Pacific).

Figure 2-44 shows the test firing of Gun 1 in Pit A for Safety Certification on 22 July 1936, with the 12-inch mortar aimed west towards the Pacific Ocean. Gun 2 is obscured from this perspective by the battery's back-up power house at lower left. The gallery at far right leads to the shell and powder magazines beneath the sloped reinforced-concrete roof (see again Figure 2-41), and continues southwest to the tunnel entrance seen earlier in Figure 2-42. Note the air vent left of top center in the earthwork above the powder magazine.

Figure 2-44. Test firing the 12-inch mortars in Pit A (upper level) at Battery White for Safety Certification on 22 July 1936 (US Army photo courtesy SSC Pacific). Note stairs leading to power-house roof at lower left.

Pit-mounted with a high elevation angle, the 12-inch mortars were obviously not equipped with telescopic sights like the 10-inch disappearing rifles at Battery Calef-Wilkeson (Stanton & Thayer, 2019a):

"Each mortar pit had a data booth at the rear that conveyed azimuth and elevation information to the gun crews and directed the firing. The data booth was attached to the plotting room. In front of the mortar pits was a concrete magazine that stored the shells and powder. The magazine was protected by a thick covering of earth. Shells and powder were wheeled from the magazine to the mortar loading platforms on shot carts."

Figure 2-45 shows shells being rammed from the shot carts into the mortar breeches in Pit B (lower level) during a live-fire practice in 1941.

Figure 2-45. Target practice with Battery White's 12-inch breech-loading mortars in Pit B (lower level), circa October or November of 1941 (US Army photo courtesy SSC Pacific).

The four 12-inch mortars of Battery White were obsolete at the beginning of World War II and scrapped in 1942, with the guns and carriages listed for disposal on 3 November that same year (Stanton & Thayer, 2019a). Figure 2-46 shows Battery White inside a fenced-in compound at Naval Submarine Base Point Loma in 1968, looking northeast towards Ballast Point. The Pit A entrance to the connecting tunnel leading to Pit B is seen just left of the power house at foreground center, with Pit B further down the ravine left of center. The transformer-and-switchboard room is just beyond Pit B on the left side of White Road.

Figure 2-46. Gun Pit A (right) and Gun Pit B (left) of Battery John White as they appeared in 1968, looking northeast towards Ballast Point (photo courtesy Al Grobmeier).

Battery McGrath

Battery McGrath was named for Major Hugh J. McGrath, who died from wounds received on 8 October 1899 in the Philippine Islands (Moore, 2019). According to the Report of Completed Works, updated to 15 September 1925, construction of this reinforced-concrete structure commenced in August 1899 and was completed just 8 months later in March 1900.

In a 21 June 1899 memorandum to Brigadier General Wilson, US Army Chief of Engineers, Captain James J. Meyler briefed his plan to save money by not having to haul in stone to the Battery McGrath construction site (Meyler, 1899b):

"It is intended, however, to allow the stone for this battery to be taken from Ballast Point spit. The amount is comparatively small and it is considered a good opportunity of determining whether or not the sea will replace such stone within a short time after it has been taken away. There is no doubt that, many years ago, thousands of yards of stone were annually taken from this point for ships,[22] and yet the strip seems not to have suffered any injury."

The battery was transferred to the Coast Artillery on 17 November 1900 at a cost of $18,203.71 (Figure 2-47).

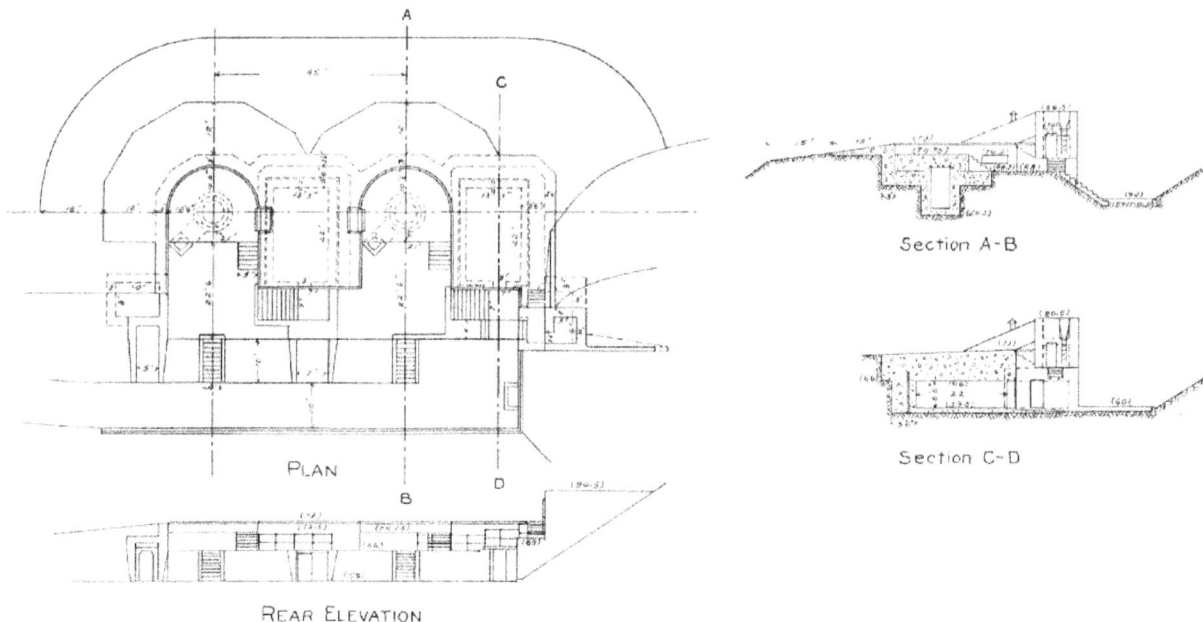

Figure 2-47. Situated just northeast of Battery Wilkeson, Battery McGrath originally had two emplacements for a pair of 5-inch rapid-fire guns on balanced-pillar mounts (NARA).

---

22 Hence the name "Ballast Point."

There is some confusion in the literature as to the type and size of Battery McGrath's guns, possibly tracing back to Meyler's memorandum to Wilson regarding the battery's construction status, which includes both the following descriptions (Meyler, 1899b):

> "...two emplacements for 5-inch rapid fire guns on balanced pillar mounts..."

> "...emplacements for the 3-inch 15-pounder rapid fire guns."

Meyler was apparently referring to two different batteries by their armament, McGrath and Fetterman, as opposed to using their names, which may have not yet been finalized. Battery McGrath, however, was actually armed with both 5-inch and then 3-inch guns during its service, as will be further discussed. The undated Report of Completed Works form shown in Figure 2-48 specifically identifies the two original 5-inch *M1897* guns and carriages of Battery McGrath by model (year) and number.

COAST DEFENSES OF San Diego, Cal.

FORT Rosecrans   BATTERY McGrath   NO. OF GUNS: 2   CALIBER: 5"

	EMPLACEMENT NUMBER 1.	EMPLACEMENT NUMBER 2.	EMPLACEMENT NUMBER 3.	EMPLACEMENT NUMBER 4.
1-a. Azimuth.	31°55'09"	174°52'30"		
b. Distance.	285.45 feet	45.0 feet		
2-a. Model of Gun.	1897 No.14	1897 No. 7		
b. " " Carriage	1896 No.29	1896 No.30		
c. Type of "	D. P.	B. P.		
3- Elevations of.				
a. Gun trunnions (in battery).	74.5	74.5		
b. Interior crest.	73	73		
c. Loading platform.	66.2	66.2		
4- Angle of depression.	5°	5°		

	FROM.	TO.	FROM.	TO.	FROM.	TO.	FROM.	TO.
5- Limiting azimuths of field of fire as determined by:	° '	° '	° '	° '	° '	° '	° '	° '
a. Construction of emplacement.	All round fire							
b. Interference by adjacent emplacements.	*							
c. Interference of other obstacles external to emplacement.	348°	180°	335°	167°				
d. Position of Ordnance stops.	All round fire							

Figure 2-48. The upper-right corner of this undated Report of Completed Works form for Battery McGrath indicates two 5-inch guns (1897 numbers 7 and 14) were installed on balanced-pillar mounts.

Battery McGrath's 5-inch guns were removed during WWI and shipped overseas to support the war effort (HCP, 2006). A 1918 Engineer's Journal entry reported that a Board of Officers convened at Fort Rosecrans recommended on 10 March of that year that the 3-inch guns that had been removed from the abandoned Battery Meed at Fort Pio Pico be transferred to Battery McGrath (Willard, undated). These relocated guns were subsequently mounted in 1919 (Figure 2-26), as previously discussed.

Figure 2-49. A member of the 19th Coast Artillery stands next to one of Battery McGraths' later 3-inch guns repurposed from Battery Meed (undated US Army photograph courtesy Al Grobmeier).

Form 6 from the Report of Completed Works, updated to 2 January 1943, indicates some additional refurbishment of Battery McGrath had been conducted in June 1938, just prior to the outbreak of WWII in Europe the following year (Figure 2-50 below):

- Five wooden doors were replaced by steel-covered doors.

- Two tile ventilator hoods were replaced by metal hoods.

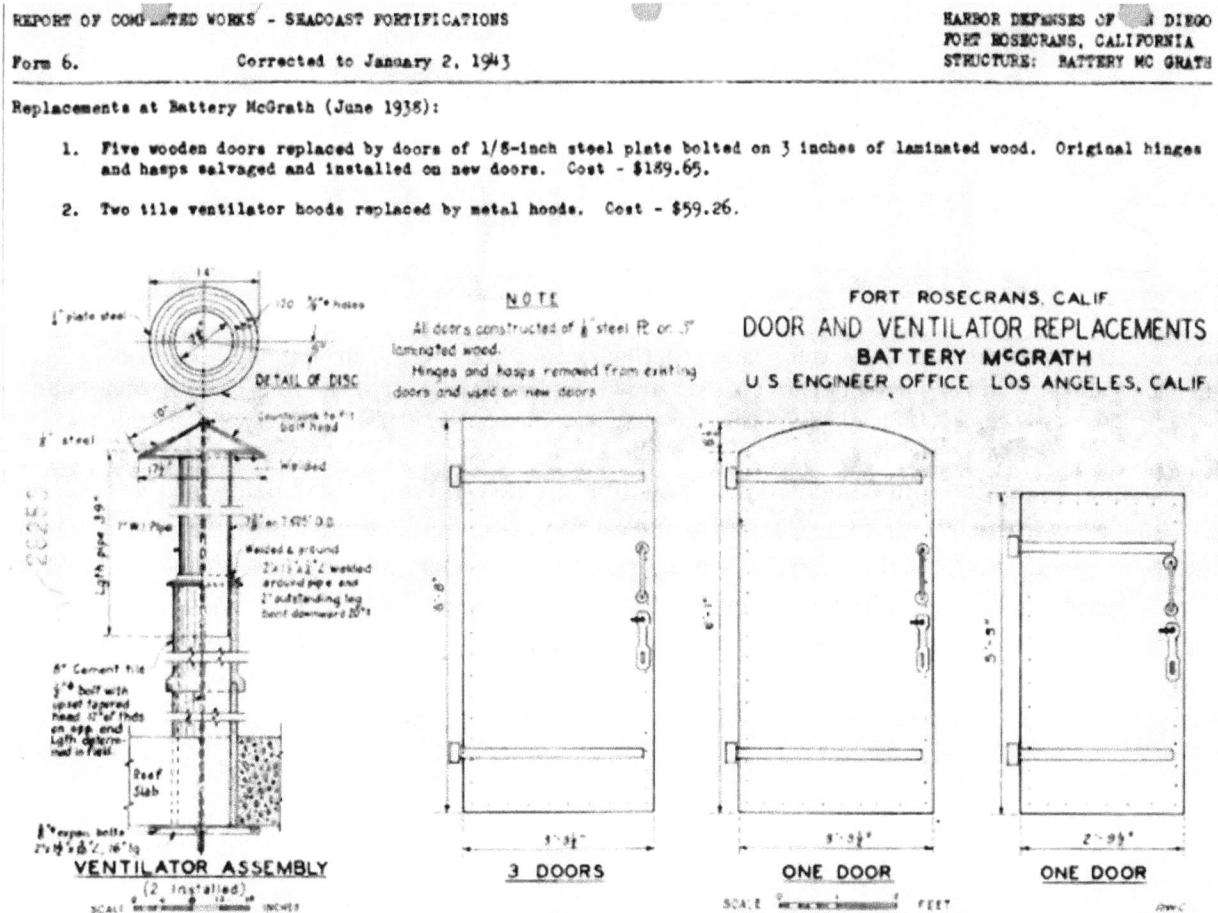

Figure 2-50. Form 6 from the Report of Completed Works, updated to 2 January 1943, shows detail regarding the replacement of five wooden doors and two tile ventilator-shaft covers at Battery McGrath in June 1938.

The undated aerial photo seen in Figure 2-51 below shows the original 3-inch Battery Fetterman, which was later demolished in 1940, just north of Battery Calef-Wilkeson at the foot of Rosecrans Street in the lower-right corner. The 12-inch-mortar Battery John White is seen in Powerhouse Canyon at upper left. Battery McGrath is on an elevated plateau to the left of Battery Calef-Wilkeson, just downslope from McClelland Road in the lower-left corner. The L-shaped mine-planter wharf is at right center, with Ballast Point out of image to the northeast at bottom right.

Figure 2-51. The two 3-inch emplacements for Battery McGrath are above and just south (left) of Gun1 of Battery Calef-Wilkeson, with an affiliated area of interest further south (up arrow). The original Battery Fetterman, demolished in 1940, is still extant in the lower-right corner, just north of Calef-Wilkeson (undated but pre-1940 photo courtesy SSC Pacific).

The zoomed-in portion of the undated aerial photo presented in Figure 2-52 below shows Battery McGrath "under treatment," with some "experimental stripping to be covered with preserved vegetation." The treatment area is above and behind the two emplacements of McGrath and east of (below) McClelland Road, which serpentines down from Catalina Boulevard in the upper-right corner. These annotations, and the fact that Battery McGrath is difficult to see, suggest this photo was taken during camouflage efforts in the early WWII timeframe. Note new excavation underway for the subterranean WWII radio-transmitter bunker at image center, just below the switchback in McClelland Road.

McGrath   cndertreatment

experimental stripping
to be covered with proper
vegetation

Figure 2-52. Battery McGrath (up arrow at bottom center) appears well camouflaged in this presumed WWII photo, compared to that seen in Figure 2-51 (undated US Army photo courtesy Coast Defense Study Group). Note excavation for the subterranean radio transmitter underway east of McClelland Road at image center.

The earlier WWI-era Fort Rosecrans radio station, constructed in 1919, was located on what is now Cabrillo National Monument property, about 100 yards north of the old Spanish lighthouse. The above-grade three-room structure, which was reconfigured as the Fort Rosecrans meteorological station during WWII, is still extant and being used as a small storage facility (Keniston, 1995). Plan and section views of the subterranean WWII-era transmitter bunker just east of McClelland Road are presented in Figure 2-53.

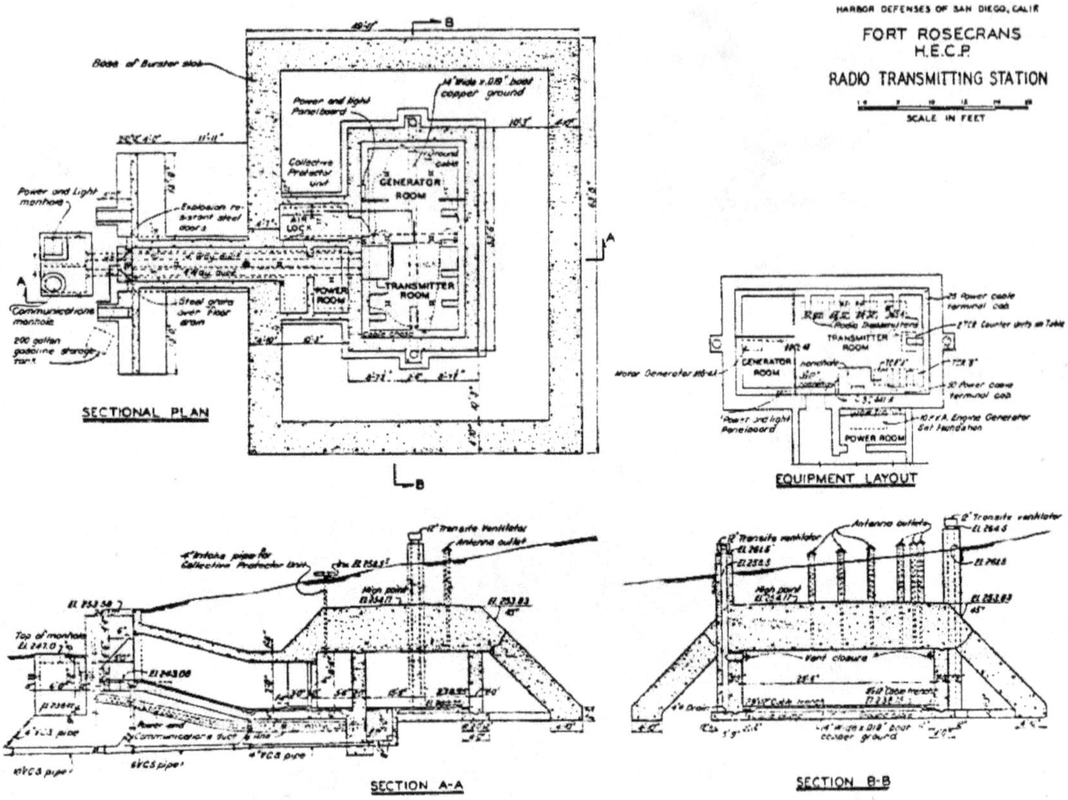

Figure 2-53. Plan and section views from the Report of Completed Works corrected to 20 July 1944 of the subterranean WWII-era bunker housing the radio transmitter station located upslope from Battery McGrath.

The associated transmitter antenna tower, apparently reconfigured sometime after the war as a watch-tower, is seen adjacent to McClelland Road in Figure 2-54 below, looking east circa May 2008. This structure was designated as "RT" (Radio Tower) in the 1 July 1945 map presented in Figure 2-55. No obvious evidence of the downslope transmitter bunker was found during this brief site visit, though it presumably is still extant. The tower itself, however, has since been dismantled.

Figure 2-54. This steel tower, seen looking east circa May 2008, is just upslope from the radio-transmitter excavation shown earlier in the WWII-era aerial view of Figure 2-52 (photo courtesy Jeff Bowen).

Figure 2-52 also showed a dirt path leading from the Gun 1 position of Battery McGrath up to the previously mentioned area of apparent activity to the south. Thompson (1991) indicates the battery commander station and a coincidence rangefinder were 150 yards south of and above the battery at an elevation of 100 feet, which correlates well with this location. The map reproduced in Figure 2-55 below, extracted from "Fire Control Stations at South Fort Rosecrans, Sites 6, 7, 8, 9" map dated 1 July 1945, shows instead an "Emerg. BC$_6$" in this area.[23] The regular battery commander station appears to have been immediately adjacent to the battery's Gun 1 emplacement, as seen later in Figure 2-56.

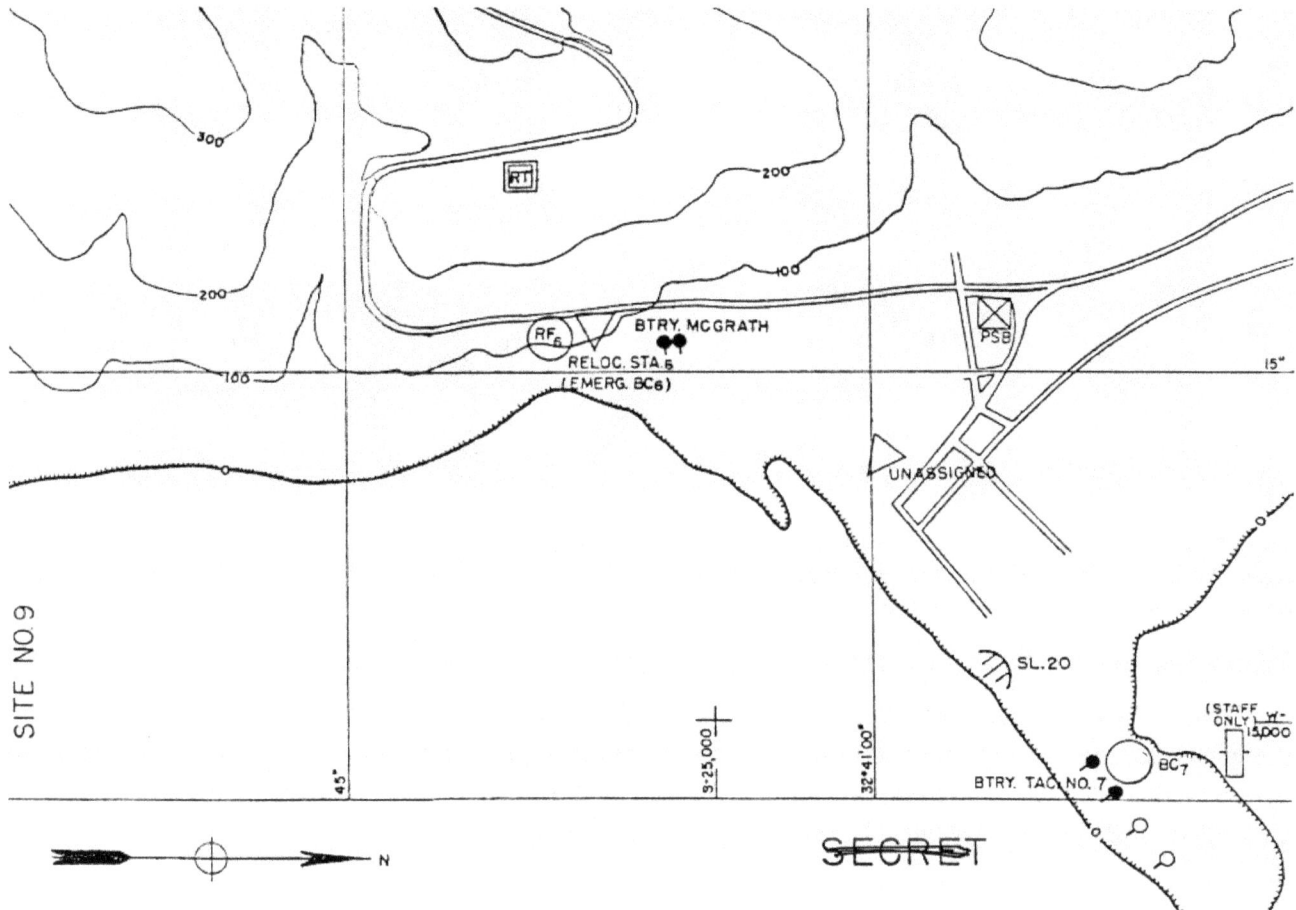

Figure 2-55. This 1 July 1945 map shows a range-finding station and an emergency battery commander station just south of Battery McGrath (Tac. No. 6); the designation "RT" (radio tower) at top center correlates with the excavation site shown earlier in Figure 2-52. Note AMTB Battery Fetterman at bottom right.

The undated aerial photo presented in Figure 2-56 below shows Battery McGrath just to the south of Battery Calef-Wilkeson, which is in the upper-left quadrant. Not shown in the battery's RCW or in Figure 2-51, the wooden structure in the bottom-right corner of the photo appears to be the battery commander station. Note lack of development along Ballast Point, with the exception of the old Coast Guard lighthouse compound, which was torn down in June of 1960 (Lougher, 2011). Accordingly, and given the battery's obvious state of disrepair, this photo was probably taken in the mid-to-late 1950s timeframe.

---

23 Emerg. BC$_6$ - Emergency battery commander station for Battery McGrath, Tac. No. 6 in WWII.

Figure 2-56. Battery McGrath is seen in a state of disrepair in this undated post-war aerial photo, looking northeast towards Ballast Point. Note the WWII-era battery commander station (lower-right corner) and Ballast Point in the upper-right background (US Navy photo LSF197 courtesy SSC Pacific).

As nicely summed up by Lewis (1979):

"The extravagant proposals put forth by the Endicott Board in 1886 were never fully realized, nor was the less ambitious project recommended by the Taft Board twenty years later, but the efforts of these two groups nevertheless led to an overlapping two-stage fortification program of a magnitude which, for the second time, gave the United States a system of harbor defenses unexcelled by those of any other nation."

When the US was drawn into WWII following the Japanese attack on Pearl Harbor some 55 years later, however, the Harbor Defenses of San Diego were woefully ill prepared.

In the 1962 aerial photo presented in Figure 2-57, the remains of Battery McGrath (up arrow bottom center) are just left of Battery Calef-Wilkeson, which overlooks San Diego Bay at bottom center. The two mortar pits of Battery John White can be seen in Powerhouse Canyon, just right of top center. Recall the original 3-inch Battery Fetterman, which is not seen in this photo, was destroyed in 1940. The subsequent AMTB Battery Fetterman that replaced it on Ballast Point is also no longer extant. Note radio-tower compound below the switchback in McClelland Road at upper left, which winds down from Catalina Boulevard on the crest of Point Loma (out of image).

Figure 2-57. The abandoned emplacements of Battery McGrath are seen at bottom center (arrow) in this 1962 aerial image (US Navy photo LSF 45 11 1962 courtesy SSC Pacific). Note Battery Calef-Wilkeson at bottom right, Battery White in Powerhouse Canyon right of top center, and the radio tower left of center.

Figure 2-58 shows a better-maintained Battery McGrath looking north towards San Diego Bay, with the four 10-inch disappearing-rifle emplacements of Battery Calef-Wilkeson in the immediate background. The WWII-era battery commander station south of the Gun 1 position is in a state of decay, and a collapsed support structure behind this same emplacement can be seen in the lower-left corner. On the left side of the first curve in the dirt road behind the north end of the battery is what appears to be a concrete ammunition bunker.

Figure 2-58. Battery McGrath as seen in April 1968, with Battery Calef-Wilkeson just to the north in background center (photo courtesy Al Grobmeier). Note the wooden battery commander station south of the Gun 1 position in the lower-right corner, and the collapsed support structure at bottom left.

The October 2006 photo presented in Figure 2-59 shows the mounting bolts for the two 3-inch rapid-fire guns that had been removed from Battery Meed at Fort Pio Pico in 1918, then remounted in Battery McGrath in 1919. The former WWII-era battery commander station to the right of the Gun 1 emplacement has been torn down, and the entire compound is now freshly painted and surrounded by a chain-link fence. Note the upper-level lateral access between the Gun 1 and Gun 2 positions.

Figure 2-59. An October 2006 photo of Battery McGrath inside its fenced compound at Naval Base Point Loma. The dilapidated wooden battery commander station seen earlier in Figure 2-58 (circa 1968) is no longer extant (photo courtesy Jeff Bowen).

3

Military Threats to San Diego

The growing conflicts in Europe and Asia seemed far away from San Diego in the late 1930's, and the prevalent isolationist attitude in the United States had left the country fully unprepared for war. Only in 1939-1940 did US armed forces start planning for possible conflict, mobilizing manpower, and building new defenses. The shock of the 7 December 1941 Japanese attack on Pearl Harbor, however, brought into sharp relief the full range of military threats the city would face to during World War II.

Shortly after the Pearl Harbor attack, British Field Marshall John Hill arrived in Washington, DC to serve as senior liaison officer. Shocked by what he encountered, Dill immediately cabled London with the following report (Breur, 2000):

> "This country is the most highly organized for peace that you can imagine. This country has not – repeat not – the slightest conception of what war means, and their armed forces are more unready for war than it is possible to imagine... The whole organization belongs to the days of George Washington."

When World War II broke out in Europe, the United States had hoped its geographic isolation would allow it to remain neutral for the duration. Significantly improved military technology, to include high-performance aircraft and long-ranged carriers to bring them to bear where needed, soon rendered such an isolationist policy rather moot. Japan was a small insular country with limited natural resources, and therefore pursued a ruthless foreign policy of aggressive colonization. Their highly trained, technically advanced, and well-equipped military met little resistance from surprised and ill-prepared victims in its brutal advance across Asia and the Pacific.

Meanwhile, according to the California State Military Museum, the US Army Air Forces had only 16 modern fighter planes available to defend the entire state in early December 1941 (MM, 2007). The mood at Fort Rosecrans was consequently rather apprehensive, as recalled by Fred J. Weist of Battery C (Overton, 1993):

> "After December 7, for a couple of months there, they didn't know where the Jap fleet was. They thought it was headed for the States. So they had us 4 hours on and 4 hours off. That first winter in '42 it was cold and wet. The wind blew, rain. You couldn't even get your clothes dry in between shifts."

In retrospect, the chances of another surprise carrier attack after the Pearl Harbor fiasco were greatly reduced. No longer blissfully asleep, the US Navy had patrolling ships and aircraft looking for enemy fleet movements, and radar was a promising new means for locating incoming aircraft. The *SCR-270* installation on Oahu, HI, had in fact detected the incoming Pearl Harbor raid at a distance of 132 miles (Malone, 2004), but it was confused by inexperienced operators with an expected flight of American *B-17s*.

Detection of the enemy fleet was only part of the problem, however. Evolving battleship designs had for some time been following a trend toward higher firing angles, which increased both the range and curvature of the projectile trajectory (Lewis, 1979):

> "By 1916, several foreign battleships could already outrange by a substantial margin any harbor-defense weapon within the country's continental limits. Moreover, the plunging type of fire these vessels could now deliver nullified the principal protective advantages inherent in the disappearing carriage on which seven of every eight heavy American seacoast guns were mounted."

While long-term construction of more capable defense batteries to address this situation under the Modernization Program was viewed as a high-priority need,[1] the more-immediate west-coast threats were Japanese submarines and incendiary balloons.

Submarine Threat

Following the outbreak of WWII in Europe, the German submarine threat on the US east coast was severe, with Allied ships brazenly torpedoed and sunk, often in sight of horrified beachgoers up and down the Atlantic seaboard. There was little that could be done at first to counter this threat, as years of apathy and neglect had taken a heavy toll (Breur, 2000). The west coast, on the other hand, had remained relatively calm. That changed overnight following the 7 December sneak attack on Pearl Harbor, which caught the United States completely off guard.

Several Japanese submarine attacks on US shipping quickly followed off the California coast (Ellis, 2006):

- On 18 December 1941, the *I-17* attacked the San Diego-bound freighter *Samoa*, which escaped undamaged.

- On 20 December, the *I-17* attacked the oil tanker *Emidio* off Cape Mendocino, CA, forcing it aground, and the *I-23* attacked the tanker *Agwiworld* off Santa Cruz, which escaped by zigzagging.

- On 22 December, the *I-19* fired three torpedoes at the oil tanker *H.M. Storey*, all of which missed.

- On 23 December, the *I-17* twice attacked the tanker *Larry Doheny* southwest of Cape Mendocino, which also escaped. The *I-21* sank the tanker *Montebello* off Cambria, CA, but later that day failed to damage the tanker *Idaho*.

- On 24 December, the *I-19* attacked the steamer *Dorothy Philips* in Monterey Bay, forcing her aground.

- On 25 December, the *I-19* attacked the lumber carrier *Absaroka* 3 miles off Point Vincente, causing extensive damage. The *I-19* then unsuccessfully attacked the *Barbara Olson* off San Pedro.

- On 28 December, the *I-25* attacked the tanker *Connecticut*, which subsequently ran aground but was later salvaged.

San Diego Bay was protected by two anti-submarine nets (Figure 3-1), but the Fort Rosecrans commanding officer, Colonel Peter Hill Ottosen,[2] was quite leery of a submarine sneak attack via La Jolla Canyon.[3] In his 13 June 1992 oral-history interview with Chief Ranger Howard Overton of the Cabrillo National Monument, John Huntoon of the Headquarters Battery, Fort Rosecrans, recalls Colonel Ottosen's concern (Overton, 1993):

> "Well, his theory was that a Japanese submarine could sneak in that deep water and surface and shell San Diego or whatever. As I remember, they had a 155 battery up there overlooking that. There were other 155s up at Torrey Pines and Camp Callan."

Due to their extremely low profile, surfaced submarines were hard to detect by early radar equipment. Furthermore, the *SCR-582* surveillance radar atop the Signal-Station tower by the old Spanish lighthouse would not become operational until August 1944, almost 3 years after the Pearl Harbor attack.

---

1   The bigger HDSD threat was a massive air attack like Pearl Harbor, as opposed to a fleet bombardment.

2   Colonel Ottosen served as commanding officer from 21 July 1940 until retiring on 28 December 1945.

3   This extensive and deep underwater canyon approaches the shoreline near the beach town of La Jolla, CA, just north of San Diego.

Figure 3-1. a) A powerful searchlight beam sweeps the harbor entrance near the northern submarine net (far left) guarding San Diego Bay. **b)** A second submarine net to the south extended across the harbor entrance at Ballast Point (US Army photos courtesy Cabrillo National Monument and SSC Pacific).

Similarly, the small canvas-covered aircraft that could be carried onboard I-class Japanese submarines presented a low radar cross section in flight. The *I-7*, for example, had launched its reconnaissance seaplane at dawn on 17 December 1941 to ascertain what damage had been done to Pearl Harbor, which went undetected despite significantly heightened security (Boyd & Yoshida, 1995). The *I-19* conducted follow-up missions by moonlight on 4 January and 23 February, while other sub-launched reconnaissance flights over Alaska also took place (Webber, 1997).

Following the Pearl Harbor attack, nine *I-class* submarines began patrolling off the US west coast, with the *I-10* taking station near San Diego (Boyd & Yoshida, 1995). On Christmas Eve, 1941, several submarines reportedly surfaced with the intent to "shell Point Loma and other west-coast installations on Christmas Day" (Showley, 1996).[4] Lockwood (2003) claims the attack was aborted when the *I*-class boats were called back to Japan. Boyd and Yoshida (1995), however, report Rear Admiral Sato in the *I-9*, which had joined the *I-10* off San Diego, had elected to head south on 21 January 1942 to intercept a rumored US Navy carrier task force in route to San Diego after transiting the Panama Canal.

A month later, the *I-17*, skippered by Commander Nishino Kozo, surfaced and shelled an oil field in Goleta, California, (CSMM, 2006):

> "Contemporary newspaper accounts describe the attack as off the Ellwood oil fields 12 miles north of Santa Barbara, and report 16 shells fired, beginning at 7:15 PM on the 23rd of February 1942. Three shells struck near the Bankline Company oil refinery, the apparent target of the shelling. Rigging and pumping equipment at a well about 1,000 yards inland were destroyed but otherwise no damage was caused."

On the evening of 21 June 1942, the submarine *I-25* shelled a coast-defense installation at Fort Stevens in Oregon (Edwards, 1942):

> "Nine high-velocity projectiles, probably fired from a five-inch rifle on the deck of a large Japanese submarine, fell screaming near the Fort Stevens Military Reservation Sunday night, signaling Japan's first attack on a primary military objective in continental United States."

This attack reportedly began at 2330 (1130 PM), but also did no real damage.

---

4   This appears to be an exaggerated and unsubstantiated claim by a San Diego newspaper reporter.

According to Webber (1997):

"Although there was supposed to be a dimout along the coast because a submarine had shelled the Estevan Point lighthouse on Vancouver Island the night before,[5] lights blazed in Seaside, a resort town on the beach two miles from Gearhart."

No doubt this brightly illuminated area served as a useful point of reference for the enemy submarine during its nighttime attack.

Webber (1997) reports the *I-25* had actually fired 17 as opposed to 9 shells, the closest of which landed "70 or 80 yards south of Battery Russell" (Figure 3-2), an Endicott Period Battery armed with two 10-inch *M1900* rifles on *M1901* disappearing carriages (Stanton & Beck, 2019). The submarine itself could not be seen,[6] and difficulty tracking its intermittent gun flashes prevented the Fort Stevens base-end stations from getting consistent course data for a reliable fire-control solution.[7] Erroneously believing the submarine to be out of range, the Commanding Officer of the Columbia River Defenses ordered the battery to not return fire (Webber, 1997).

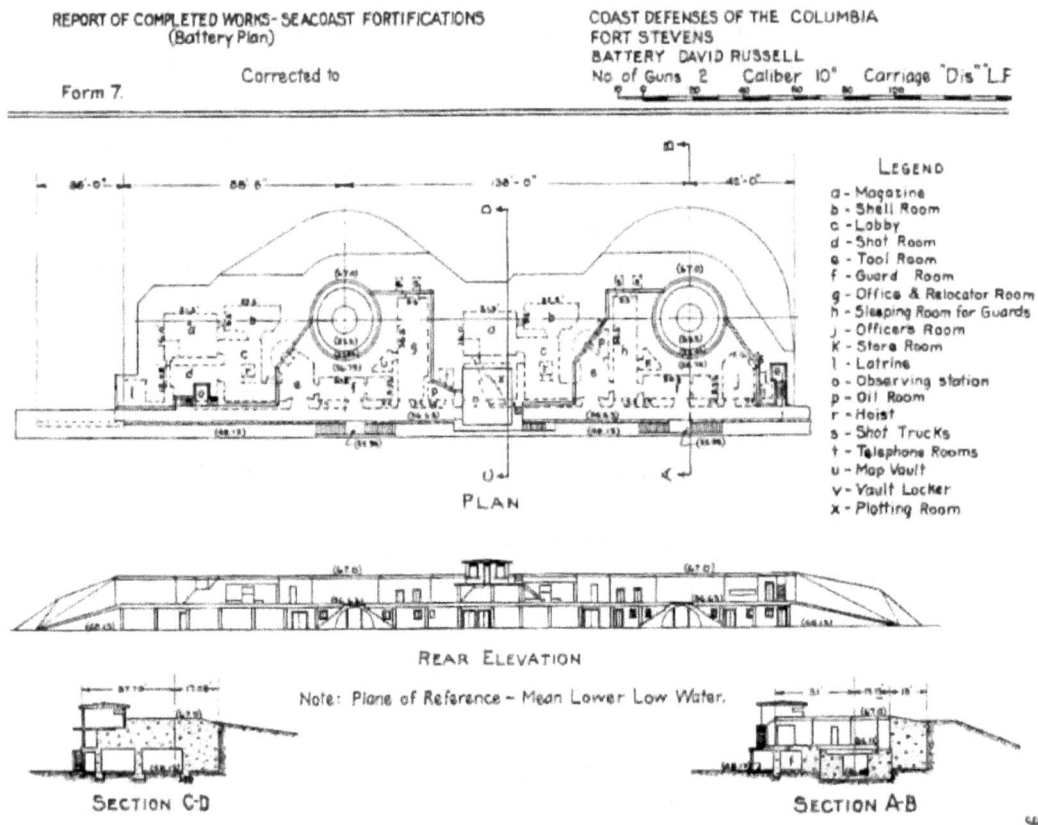

Figure 3-2. Completed in 1904, Battery David Russell at Fort Stevens, Oregon, was armed in 1907 with a pair of 10-inch *M1900* rifles on *M1901* disappearing carriages, which had a maximum range of 14,000 yards. As more recently determined by Webber (1997), the *I-25* opened fire from 10,936 yards.

One rather chilling post-war account describes a plan, code-named "Cherry Blossoms at Night," to use submarine-launched Kamikaze pilots to dump plague-infested fleas on the city of San Diego. According

---

5   Boyd and Yoshida (1995) report the *I-26* shelled a radio station on Vancouver Island, British Columbia, on 20 June 1942.

6   Permission was never granted to energize the 60-inch searchlights at Fort Stevens, for fear they would enable more accurate enemy fire.

7   A Depression Position Finder operator, for example, had to place the instrument's vertical crosshair on the target's stack and the horizontal crosshair on the waterline, which in this case was impossible to see.

to Toshimi Mizobuchi, an instructor at *Unit 731*,[8] the target date was set for 22 September 1945 (BWU, 2001). Similar "plague-bombs" had previously been air-dropped on Chinese civilians in Quzhou, with devastating results (McNaught, 2002). The host submarines for this mission were fortunately diverted to engage the US Fleet at Ulith just prior to the Japanese surrender on 14 August 1945 (BWU, 2001).

Incendiary Balloons

Following General Jimmy Doolittle's *B-25* raid on Tokyo in the spring of 1942, the Japanese began preparing a rather unorthodox retaliatory strike involving large numbers of windborne hydrogen-filled balloons.[9] Military engineers reasoned the 200-mph winter jet stream, about which the rest of the world knew very little, could theoretically push at least 10-percent of these weaponized balloons across 5,000 miles of Pacific Ocean in about three days. In addition, the predictable nature of these high-altitude winds significantly reduced the chances of an errant balloon landing on friendly soil.

Constructed of an impermeable paper made from the mulberry tree, each 33-foot-diameter "Fu-Go" balloon could carry a 1000-pound payload of control equipment, ballast, and incendiary/fragmentation bombs. The first armed launchings took place from the east coast of Honshu on 3 November 1944 (Rizzo, 2013), with a reasonably intact balloon recovered by a US Navy patrol craft on 4 November (Figure 3-3),[10] some 66 miles southwest of San Pedro, CA (Mikesh, 1973). For the next 6 months, relentless *Fu-Go* balloons crossed the North American coastline from Mexico all the way up to the Aleutians, with some traveling eastward as far as the outskirts of Detroit (Goebel, 2006).

Figure 3-3. A US Navy sailor holds the relief valve of a "Type B" rubberized balloon found out at sea (53° 3'N, 135° 52'W) on 23 April 1945, which is the same type found off San Pedro, CA, on 4 November 1944 (US Navy photo, NARA).[11]

---

8   Unit 731 was the infamous "Asian Auschwitz" that conducted Japan's extensive biological warfare program in occupied Manchuria (Harris, 1994).

9   The first ever use of such weaponized balloons appears to have been during Austria's siege of Venice during the Italian War of Independence in 1849 (Mikesh, 1973). The British attempted similar action against Germany in WWII (Everett, 2015).

10   This was a rubberized-silk "B-type" balloon developed by the Imperial Japanese Navy, eventually discontinued, which carried a smaller payload than the paper "A-type" balloon developed by the Army. Launched on 3 November 1944 (Japanese date), it was recovered 2 days later on 4 November (US date).

11   The original caption reads: "Japanese rubberized balloon recovered at sea on 23 April, 1945."

The biggest technical challenge in this complex undertaking was compensating for the thermal contraction and expansion of hydrogen due to day/night temperature variations. As the balloon lost altitude from nighttime cooling, a barometric sensor would trigger a pair of "blowout-plug" charges to release some ballast. To preserve balance, the sandbags used for this purpose were alternately dropped from opposite sides of the supporting aluminum ring as shown at upper right in Figure 3-4. For redundancy, three aneroid switches were wired in parallel to perform this function whenever the balloon descended below 30,000 feet. A spring-loaded valve at the bottom of the gas envelope would open if the balloon rose above 38,000 feet, venting hydrogen to the atmosphere (Mikesh, 1973).

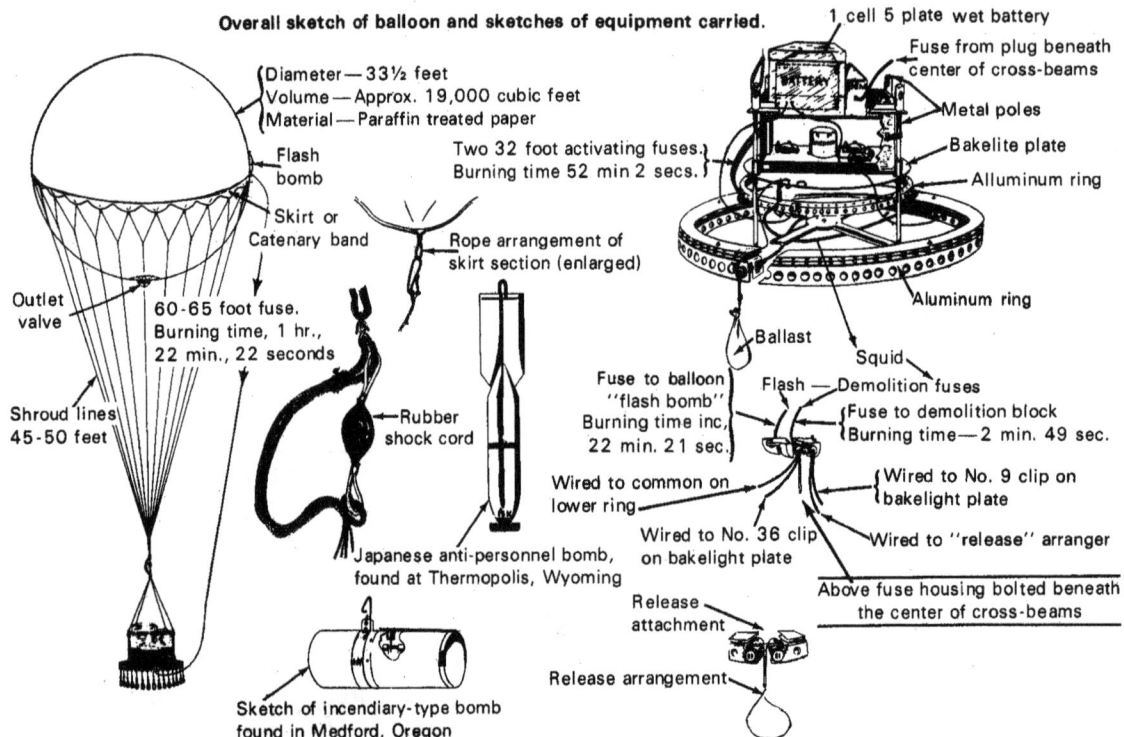

Overall sketch of balloon and sketches of equipment carried.

Diameter — 33½ feet
Volume — Approx. 19,000 cubic feet
Material — Paraffin treated paper

Flash bomb

Skirt or Catenary band

Two 32 foot activating fuses. Burning time 52 min 2 secs.

Rope arrangement of skirt section (enlarged)

Outlet valve

60-65 foot fuse. Burning time, 1 hr., 22 min., 22 seconds

Shroud lines 45-50 feet

Rubber shock cord

Japanese anti-personnel bomb, found at Thermopolis, Wyoming

Sketch of incendiary-type bomb found in Medford, Oregon

1 cell 5 plate wet battery

Fuse from plug beneath center of cross-beams

Metal poles

Bakelite plate

Alluminum ring

Aluminum ring

Ballast

Squid

Fuse to balloon "flash bomb" Burning time inc, 22 min. 21 sec.

Flash — Demolition fuses

Fuse to demolition block Burning time — 2 min. 49 sec.

Wired to common on lower ring

Wired to No. 9 clip on bakelight plate

Wired to No. 36 clip on bakelight plate

Wired to "release" arranger

Release attachment

Above fuse housing bolted beneath the center of cross-beams

Release arrangement

Figure 3-4. Originally published in *Signal* in April, 1970, this diagram illustrates the various systems that controlled the balloon's altitude to keep it in the jet stream, released the incendiary bombs over the target, then caused the entire configuration to self-destruct (US Army diagram adapted from Zahl, 1972).

This automatic altitude-control system was activated by a pair of 32-foot delay fuses ignited at time of launch, which allowed 52 minutes for the balloon to reach its initial cruising altitude before dropping any ballast. The aneroid-based control scheme engaged at that point, powered by a 2.3-volt lead-acid storage battery. Each of the blowout plugs that released the sand ballast also lit a 24-inch delay fuse, providing about 2 minutes for the balloon to ascend, at which point more ballast was dropped if still below 30,000 feet.

A total of 70 blowout plugs were arranged in pairs around the aluminum ring to accommodate 35 ballast loads, the first 32 of which were sandbags. The final stations held thermite incendiary bombs weighing either 5 or 12 kilograms, which also doubled as ballast. Lastly, a 15-kilogram *Type 92* high-explosive bomb was released from the center of the aluminum ring, followed by delayed detonation of self-destruct charges attached to the control mechanism and the balloon envelope.

An onboard continuous-wave radiosonde transmitted wind and telemetry data to seven plotting stations arrayed along the coasts of Japan (Mikesh, 1973):

"The primary purpose of the radio equipment would be to report the balloon's flying course, its altitude, and to measure the balloon's inside pressure. In addition, it would also provide data on the balloon's descending and ascending flight. This would give an added indication of when the ballast was being dropped."

On 19 January 1945, a Western Defense Command radio operator tracking a *Fu-Go* radiosonde some 2,700 miles west of San Francisco found the balloon had been moving at an average speed of 174 miles per hour over a period of 9 hours (Coen, 2014).

Postwar estimates vary as to the number of *Fu-Go* balloons manufactured and how many were subsequently launched (Webber, 1997):

> "General Kusaba,[12] while collaborating with Neal Conley in Japan in 1963, wrote to Conley that there were no records due to the combination of *B-29* raids and the purposeful destruction of the scientific and operating-cost records at the end of the war. General Kusaba's best estimate was that about 9,000 were built and about 6,000 were launched."

According to Webber (1997), there were 361 confirmed *Fu-Go* arrivals in the United States and Canada,[13] where fortunately it was the rainy season, with the threat of forest fires substantially reduced. The Japanese carefully scrutinized the American media for any mention of balloon-related incidents, but the US Office of Censorship had requested that no reports be issued. This ban on press coverage was eventually lifted after a pregnant Sunday-school teacher and five children were killed on 5 May 1945 when they tried to move the remains of a balloon found in Bly, OR (Mikesh, 1973).

John Huntoon describes the recovery of one of these balloons near Descanso, CA,[14] about 40 miles east of San Diego (Overton, 1993):

> "This one had picric acid, I think, it was the explosive on it.[15] But it came down somewhere near Descanso, and it was made of rice paper, the balloon part of it.[16] The way they controlled it was they had a barometer on it and they had bags of sand and some valve in the bottom of the bag. If it got too low, the barometer would drop these bags of sand and if it got too high the valve would open and release the gas, so that it stayed at a fairly constant altitude. Apparently, the intention was to start fires with the explosives. We picked it up and brought it back to Fort Rosecrans and it was in some secrecy because we didn't want them to know how successful the operation was."

There was growing anxiety by this point that the balloons might also be used for germ warfare, since the Japanese were aggressively pursuing biological weapons at the *Unit 731* site in Pingfang, Harbin, in occupied Manchuria (Williams & Wallace, 1989).[17] Post-war analysis later revealed this fear had been valid, as several Japanese generals had been strongly advocating precisely that very action (BWU, 2001). Lieutenant Colonel Murry Sanders, an Army bacteriologist, was dispatched to Japan immediately after the war ended to investigate *Unit 731* and track down its infamous director, Lieutenant General Shiro Ishii.[18]

Some 40 years later, in an interview with investigative journalists Peter Williams and David Wallace, Sanders put the balloon-delivery issue into chilling perspective as follows (Williams & Wallace, 1989):

> "The only explanation I had, and still have, is that Ishii wasn't ready to deliver what he was making in Pingfang; that he hadn't worked out the technology. If they had been, we were all at Ishii's mercy."

---

12   Major General Suiki Kusaba, who headed the 1st Division of the 9th Army Technical Research Laboratory, was in charge of the *Fu-Go* program.

13   Mikesh (1973) reports a total of 285 landed in the United States.

14   Mikesh (1973) claims the balloon landed in Julian, just north of Descanso, on 31 January 1945.

15   The demolition charge for destroying the flight-control mechanism after the *Type 92* high-explosive bomb and four *Type 97* incendiary bombs were dropped was a 2-pound block of picric acid.

16   The balloon envelope was constructed from an impermeable paper made from the mulberry tree, not rice.

17   More than 10,000 Chinese, Russian, and Korean prisoners of war were slaughtered as human lab rats in *Unit 731*'s experiments to develop and refine biological weapons; some 300,000 Chinese civilians were later killed by the weapons themselves (McNaught, 2002).

18   As a young Army doctor, Major Shiroh Ishii turned "Manchuria into one huge biological-warfare laboratory during the Japanese occupation" (Harris, 1994).

Acutely aware of Japan's germ-warfare transgressions against China, the US government attempted to implement radar intercept of incoming balloons and warned the public to quickly report any unusual signs of human or livestock sickness (Mikesh, 1973). Earlier tests conducted by the Naval Research Laboratory at Anacostia, VA, had determined the paper balloons could not be reliably detected by radar at any range, but Air Force radar successfully tracked an incoming *Fu-Go* flight on 13 April 1945 at a distance of 85 miles (Coen, 2014).

Meanwhile, in one of the first wartime applications of forensic geology, the *Military Geology Unit* of the *US Geological Survey* identified some recovered ballast sand as endemic to the northeastern coast of Japan (Rogers, 2006). Detailed photo reconnaissance of that area in early 1945 revealed two of the three plants producing hydrogen for the project, which were destroyed by *B-29* raids in April. Further discouraged by no indication the balloons were getting through to US soil, the Japanese government decided to abandon the project as a costly failure.

Ironically, on 10 March 1945, one of the descending *Fu-Go* balloons had short-circuited a transmission line from the Bonneville Dam hydroelectric plant on the Columbia River, blacking out parts of the nearby Hanford Engineering Works (Webber, 1997). This secret Department of Energy facility was processing plutonium for the atomic bomb later dropped on Nagasaki, and the power outage caused the first reactor SCRAM (emergency shutdown) in history. Fortunately, the elaborate safeguards for shutting down the pile worked as intended, preventing a catastrophic nuclear meltdown (Groueff, 2000). Had that not been the case, the strategic impact of the Japanese *Fu-Go* program could have turned out quite differently indeed.

4

Temporary Batteries

> "You cannot invade the mainland United States. There would be a rifle
> behind each blade of grass."
>
> *Admiral Isoroku Yamamoto*

According to the "History of Harbor Defenses of San Diego," the following armament was in place at Fort Rosecrans as of 7 December 1941 for repelling attacks from the sea (HSD, 1945):

- Batteries Calef and Wilkeson, 1900-vintage 10-inch disappearing-rifle batteries of two guns each, with limited coverage to the south.

- Battery McGrath, installed in 1919, two 3-inch guns that covered the harbor entrance channel on the east side of Fort Rosecrans.

- Batteries White and Whistler, 1920-vintage 12-inch mortar batteries of four guns each, extremely limited in both range and rate of fire.

- Battery Strong, completed summer of 1941, two unshielded 8-inch Navy guns on barbette carriages, facing westward.

- Battery Point Loma and Battery North, each with four 155-millimeter guns mounted in field emplacements to cover the southwest, west, and northwest approaches to the harbor.[1]

Given this woefully inadequate state of defensive affairs, a natural consequence of America's isolationist attitude following the First World War, Pearl Harbor came as quite a wake-up call.

Immediately after the Japanese surprise attack, a platoon of Battery A troops from Fort Rosecrans was dispatched to man a temporary coast-defense position at Camp Callan (Figure 4-1), just north of La Jolla. While there were thousands of soldiers stationed at the Coast Artillery Replacement Training Center there, the organization and support infrastructure were set up for training, and not immediately adaptable to 24-hour tactical manning of alert batteries (HSD, 1945). Accordingly, this platoon remained in place to assist in that capacity for several months, manning a training battery of 155-millimeter *Grande Puissance Filloux (GPF)* guns.[2]

---

1   These two 155-millimeter batteries were not identified by name in the cited reference.
2   *Filloux* was the name of the gun's designer, while *Grande Puissance* is French for "great power."

Figure 4-1. One of several unmounted 155-millimeter *GPF* guns located at Camp Callan Coast Artillery Replacement Training Center (CARTC) in La Jolla, CA (US Army photo courtesy Cabrillo National Monument). Length of the gun tube was 19 feet 6 inches.

During the First World War, the United States had purchased a number of these tractor-drawn field artillery pieces from France, model *M1917* (model year 1917). A slightly improved version was later manufactured in this country as the *M1918*. The proven performance and ready availability (approximately 3,000 on hand) of this superb weapon led to its subsequent post-war adoption for coastal defense (CAJ, 1929). As they had originally been designed to fire upon stationary land targets, these guns were often installed on "Panama mounts" to increase their accuracy and ensure a consistent gun location for fire-control solutions. An equally important advantage of this mount was an increase in the allowable traverse, normally carriage-limited to 60 degrees, to enable full 360-degree coverage.

Developed in the Canal Zone (hence the name) during the 1920s, the Panama mount consisted of a raised concrete pedestal that supported the gun, surrounded by a circular steel ring for the trails (Figure 4-2).[3] The conventional spade plates were removed from the *M1918* carriage trails and replaced by a special design that was matched to the rail curvature (TM 9-2300, 1949). Numerous modifications to the standard implementation (without raised pedestal, less than 360-degree traversal, etc.) were introduced. The portable "Kelly Mount," for example, was designed by Colonel P.E. Kelly, CAC (Lafrenz, 1944).

---

3   The photographs, drawings, and accompanying descriptions for the cited 1929 article were submitted to the *Coast Artillery Journal* by Major Paul H. French, CAC, and Captain Ben Bowering, CAC, who were both serving in the Canal Zone at the time (CAJ, 1929).

Figure 4-2. Photo of a typical 155-millimeter *GPF* installation on a 360-degree Panama mount (US Army photo courtesy Sitka Historical Society). The elevated center pedestal created a suitable recoil pit for firing at high elevation angles.

The *M1918 GPF* fired separately loaded ammunition consisting of a 155-millimeter projectile and a "base-and-increment" propellant charge packaged in silk bags. The most commonly used projectiles for coast defense were armor piercing (100 pounds) and high-explosive (95 pounds), although the gun was also designed to fire chemical and shrapnel rounds in other employments (TM 9-2300, 1949). The weapon was hand loaded and employed a percussion-type primer, which was fired by a lanyard. Two 155-millimeter batteries of four guns each were operational on Fort Rosecrans when war broke out in December 1941, Battery North and Battery Point Loma, as discussed in the following sections.

Battery North

On the *American Forts Network* website, Payette (2006) lists the dates as "1930s-1937" for Battery North,[4] a little-known fortification apparently built by troop labor. The 1958 aerial photo shown in Figure 4-3 shows the general location of this battery in the northwest corner of Fort Rosecrans, approximately 150 yards south of the northern fence line. The U.S. Navy Electronics Lab (NEL) is just to the right of this same fence line on the east side of Catalina Boulevard at the top of the Point Loma Ridge, with North Island Naval Air Station in the background.

---

4   This battery, informally referred to as "North" due to its location at the north end of Fort Rosecrans, was still operational in December 1941.

Figure 4-3. Battery North was located on the east (far) side of the horizontal dirt road at lower right (up arrow). The U.S. Navy Electronics Lab is right of top center, with the North Island Naval Air Station across the harbor entrance in the background (1958 US Navy photo courtesy SSC Pacific).

Arranged north-to-south, the four 155-millimeter *GPF* gun emplacements of Battery North are clearly seen in Figure 4-4, just south of the ravine running parallel to the northern fence line of Fort Rosecrans at lower left, circa December 1941. The emplacements were grouped in pairs, with a slightly wider lateral separation between Guns 2 and 3 than was used between Guns 1 and 2 or Guns 3 and 4. Note also what appears to be a fifth location to the east behind Gun 2 (up arrow), which possibly served as a maintenance emplacement.

Figure 4-4. This zoomed-in view of the northwest corner of Fort Rosecrans, taken just after Pearl Harbor, shows four emplacements for Battery North at photo center, with a fifth emplacement further east (up arrow) (US Army photo courtesy Cabrillo National Monument). Note the northern fence line in lower-left corner.

In response to Pearl Harbor, these four *GPF* guns were moved to Coronado Heights during the period 9 to 14 December 1941, where they were re-emplaced in what came to be known as Battery Imperial. Battery Gillespie later incorporated 5-inch naval guns in three of the vacated Battery North emplacements, which brings up an interesting question. Were the original North emplacements in fact modified Panama mounts with 270-degree coverage as is generally believed (Figure 4-5), or did they simply have one rail section removed in 1942 to allow easier access later for the 5-inch gun installation?

Figure 4-5 a) An apparent 270-degree traversal was afforded by the modified Panama-mount rail installations at Battery North, with the steel rail supported on 8-inch wooden ties. **b)** Close-up view of the stepped concrete anchor that secures the southwest end of the modified Panama-mount traverse rail. The wooden beam atop the anchor presumably supported the gun tube when fully depressed

The bigger mystery regarding Battery North, however, is when and by whom it was actually built. An interesting clue is provided in Part IV of Appendix II of "Harbor Defenses Included in the San Diego - San Pedro Area" (HDP, 1933):

> "Additional secondary weapons are necessary to cover other water areas to prevent the close approach of small vessels. 155-millimeter guns will be satisfactory for this purpose. One battery will be located in the vicinity of the 12-inch mortar battery, Whistler, and one in the vicinity of Coronado Heights (northwest of South San Diego). Concrete platforms to permit wide fields of fire will be provided for both these batteries though the one at Coronado Heights will not be constructed until an emergency arises."

The above wording assigns first priority to the northwest Fort Rosecrans battery near Battery Whistler (Battery North), yet the associated 1933 map depicting existing armament and armament to be installed showed only the proposed Coronado Heights location (Battery Imperial).

In addition, the "Exhibit 3-A" HDSD map from the 1936 Annex shown in Figure 4-6 makes no mention of Battery North, but shows instead a proposed 155-millimeter emplacement just north of the new lighthouse at the southern tip of the point.[5] As discussed at the beginning of this chapter, however, the post-war "History of Harbor Defenses of San Diego" stated two batteries of four 155-millimeter guns each were operational at Fort Rosecrans on 7 December 1941 (HSD, 1945). Accordingly, it can be reasonably assumed that Battery North was installed sometime after 1936 but before December 1941.

---

5   This emplacement was later installed as Battery Point Loma.

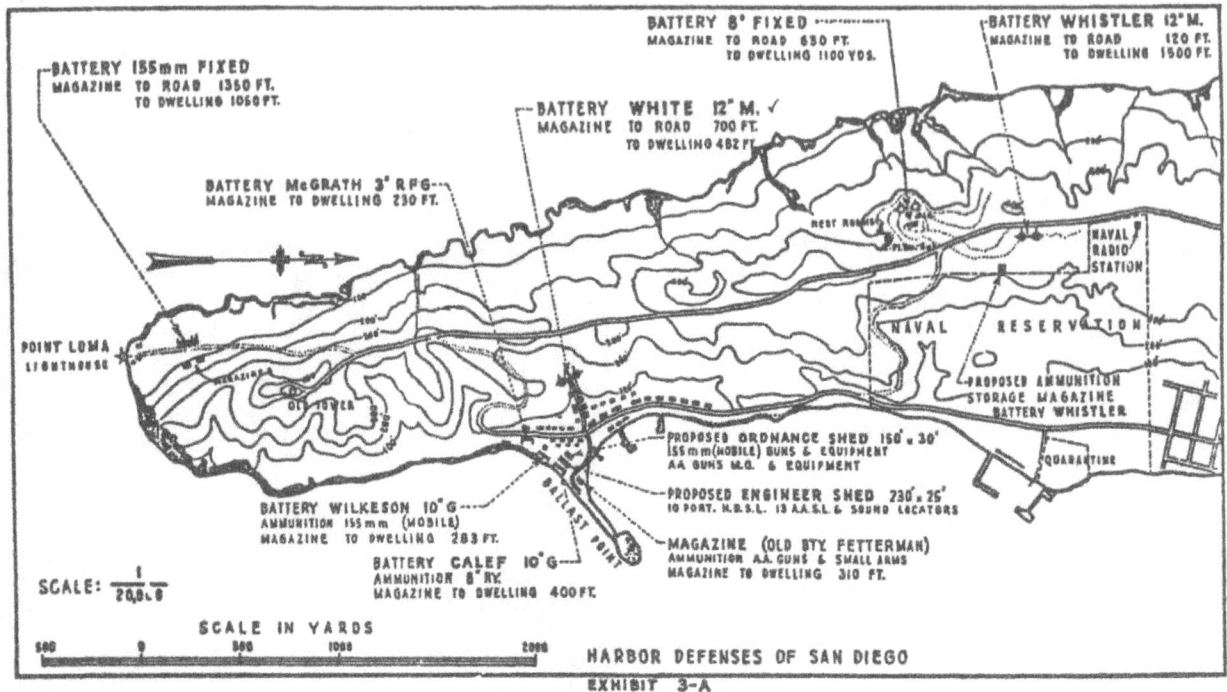

Figure 4-6. The proposed location for the "Battery 155-millimeter Fixed" (i.e., Battery Point Loma) is shown at upper left in this HDSD map, Exhibit 3-A from the 1936 Annex. Note absence of any reference to Battery North near coastline above Battery Whistler at upper right.

A second possible clue is provided in a 1937 *California Historical Landmark Series* article on Fort Rosecrans, a transcribed portion of which was provided me by Coast Defense Study Group member Al Grobmeier in July 2008 (Tays, 1937):

"At present the fortifications at Fort Rosecrans consist of a row of strong reinforced concrete emplacements for huge disappearing guns of 12- and 14-inch caliber. There are a number of smaller guns also, one group on the harbor side of the peninsula. These command the entrance to the bay just south of Ballast Point. Another group faces towards the Pacific Ocean on the west side of the peninsula and commands the approach from the north, south, and west."

It is hard to assign much credibility to this description, given its obvious ambiguity, especially in light of the errors associated with the only specifics provided. Batteries Calef and Wilkeson were equipped with 10-inch disappearing rifles, for example, and the 12-inch armaments of Batteries White and Whistler were in fact mortars. Nevertheless, the last sentence is rather intriguing, since both Battery Strong and Battery Point Loma were not brought on line until 1940. So, if Tays was correct in stating "another group faces towards the Pacific Ocean on the west side of the peninsula" in 1937, it could potentially have been Battery North. On the other hand, he may have been referring to one or both of two USMC practice batteries (5-inch and 7-inch) on the southern end of the Point, as discussed later.

Another clue is perhaps found in the construction techniques associated with the three 155-millimeter batteries included in the Harbor Defenses of San Diego. Recall the two proposed 155-millimeter emplacements mentioned in "Harbor Defenses Included in the San Diego – San Pedro Area" (HDP, 1933) were to be mounted on "concrete platforms to permit wide fields of fire." While Battery Point Loma and Battery Imperial were equipped with 360-degree Panama-mount rails set in concrete, the steel rails for the modified 270-degree Panama mounts of Battery North were reportedly attached to wooden ties.

In addition, Batteries Point Loma and Imperial were virtually identical in layout and construction techniques, and officially documented in pre-war Army planning documents. Battery North, on the other hand, is of markedly different construction and unmentioned in pre-war records discovered so far.

Flower & Roth (1982) reported the following conditions during their historic-site survey of the Battery North/Gillespie area:

> "The battery is intact although in a state of disrepair. Features include panama (sic) mounts for 155 mm guns and underground bunkers for quarters and maintenance. All underground quarters are connected by tunnels. Many contain furniture, rifle racks, etc. Three emplacements still exist. The fourth, on the north end, has been destroyed.[6] Bunker facilities include kitchen, mess, sleeping quarters, and generator emplacement. Approximately 50 yards south of the southern installation is a cement mount for 5-inch guns. Inscribed in the center and northwest corner of the mount are the words: 'Btry I - 19th C.A., 6-9-42.' Directly west of the battery on the cliffs are several wooden boxes built into the bluffs."

Figure 4-7a shows this so-called "cement mount," which is ridiculously too small for a 5-inch naval gun, and appears instead to have been a concrete foundation for some type of flanged pole, probably for an azimuth instrument. The pitch circle for the embedded mounting bolts, within which is found the inscription cited above, measures just 20 inches across (Figure 4-7b). This same inscription is repeated on the lower-right corner, somewhat more weathered due to greater exposure. A lifting ring was more recently substituted for one of the original square nuts to attach a guy wire for a nearby utility pole.

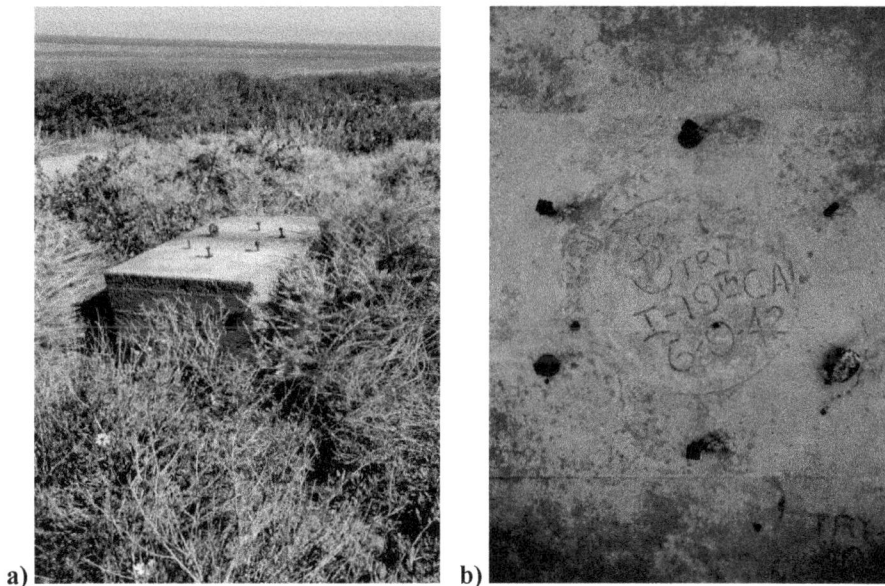

Figure 4-7. a) The inscribed azimuth-instrument foundation measures 45 inches long by 37 inches wide, with an exposed height of 32 inches. **b)** The inscription within the bolt circle reads: "Btry I – 19th CA 6-9-42" (Battery I of the 19th Coast Artillery).

The "wooden boxes built into the bluffs," small subterranean enclosures for lookouts, were still in reasonably good shape when I first saw them. While Flower and Roth (1982) reported "several" of these present, I only encountered two in this area following my arrival at the Center just 4 years later in 1986 (Figure 4-8), both of which were sufficiently intact so as to permit my entry. The drainage gully seen at upper right in the photo has been filled in north of the three former Battery North emplacements,[7] allowing the dirt road to continue beyond the northern fence line.

---

6   In actuality, it appears the Gun 4 position on the south end was destroyed, as further discussed later.

7   These three former Battery North emplacements were later repurposed for Battery Gillespie.

Figure 4-8. Wooden observation posts were dug into the cliff projections indicated by up arrows in this view of the Building F-28 complex, looking north circa 1988. Three of the former emplacements for Battery North are east (right) of the dirt road and north of the utility pole at right center (US Navy photo courtesy SSC Pacific).

The outboard hatches of these two observation bunkers had originally been equipped with glassed-in cupolas that allowed the observer a 360-degree view without being exposed to the elements. The natural soil erosion left these wooden structures more vulnerable with each passing year (Figure 4-9), and when I attempted to photograph them in 2006, there was but one lone holdout (Figure 4-10), which had fallen into the surf when I revisited the area just two years later in April 2008.

a) b)

Figure 4-9. a) Observation bunker as seen in April 1971, almost 30 years after construction, with field wire running to the telephone post in background left. **b)** The same site minus its telephone post in 1992, with the subterranean portion almost fully exposed on the south side (photos courtesy Lee Guidry).

Figure 4-10. The eastern half of the dual-compartment observation structure seen previously in Figure 4-9b was all that remained when I took this photo in 2006. Behind this observation bunker on the northern side of the finger was a 0.30-caliber machine-gun pit, as further discussed in Chapter 6.

Battery Gillespie

For reasons previously discussed in conjunction with Battery North, similar confusion exists regarding Battery Gillespie, which later occupied the same site. Payette (2006), for example, lists the dates for Gillespie as 1937 through 1943, which includes a portion of the time this site was actually Battery North. Claiming Gillespie had "modified Panama mounts," Payette describes its 2006 condition as "one emplacement destroyed, one buried, and one partially uncovered." Joyce (1995) reports Gillespie was used by the Marines for gunnery practice in the southwest corner of Rosecrans until turned over to the Army, then moved "north of the cemetery."

Other accounts indicate that while the guns used in Battery Gillespie were indeed relocated from the southern tip of Fort Rosecrans, the battery itself did not exist in name prior to its establishment at the former location of Battery North in 1942. Because Battery North was constructed during peacetime by troop labor and never officially named, it is often confused with Battery Gillespie, which occupied the same site during WWII. Citing Springer (1930), the Flower and Roth (1982) survey, for example, says Gillespie was "an anti-aircraft battery of 155-millimeter GPF guns on Panama mounts installed prior to 1930," which is incorrect on several counts.

For starters, Springer (1930) makes no specific mention of Battery North or Gillespie, describing the Fort armament only generally as follows:

> "The armament of Fort Rosecrans consists of batteries of disappearing rifles, of mortars, and of anti-aircraft guns, these last having been installed since the World War."

Furthermore, Battery Gillespie did not exist until after the 7 December Pearl Harbor raid in 1941, as will be discussed later.

Finally, with a maximum elevation of only 35 degrees, and for several other reasons, the 155-millimeter guns of Battery North were obviously never intended for an antiaircraft role. Traversal of the *M1918* carriage was limited to 60 degrees, which was sufficient for static or slow-moving targets, as previously noted. Beyond this, the entire piece had to be rotated in azimuth by moving the trails, which for land warfare typically happened only when targeting a new objective. Shifting the arc of fire in this manner was no trivial task, requiring 1 to 2 hours to accomplish (CAJ, 1929). The steel tracks of the Panama mount made it faster and easier to make course azimuth adjustments for slow moving targets such as enemy ships, but certainly could not accommodate much faster aircraft.

On 26 July, 1994, Chief Ranger Howard Overton interviewed Jack R. Skaggs, who in September 1940 was a young Marine assigned to the 2nd Defense Battalion at the Marine Corps Base in San Diego (Overton, 1995). According to Skaggs, the 2nd Defense Battalion left for Iceland in 1941, except for the artillery batteries, which were reorganized as the 2nd Artillery Group and continued to practice on Point Loma prior to the war. On 25 September 1995, Al Grobmeier sent a hand-drawn map of Fort Rosecrans to Skaggs requesting information on the nature and location of USMC gun batteries (Grobmeier, 1995).

In his oral history interview with Overton, Skaggs stated he was in "A Battery" of the Second Artillery Group (Overton, 1995). The term "battery" is used by the military in a couple of different ways, depending on context, as explained by Overton (1988):

> "A battery, if named, such as Battery Point Loma, consists of one to four guns mounted to direct fire at a point. If it is an alphabetic designation, such as Battery E, it was a group of men adequate to fire the number of guns assigned to it."

Grobmeier's map is reproduced in Figure 4-11.

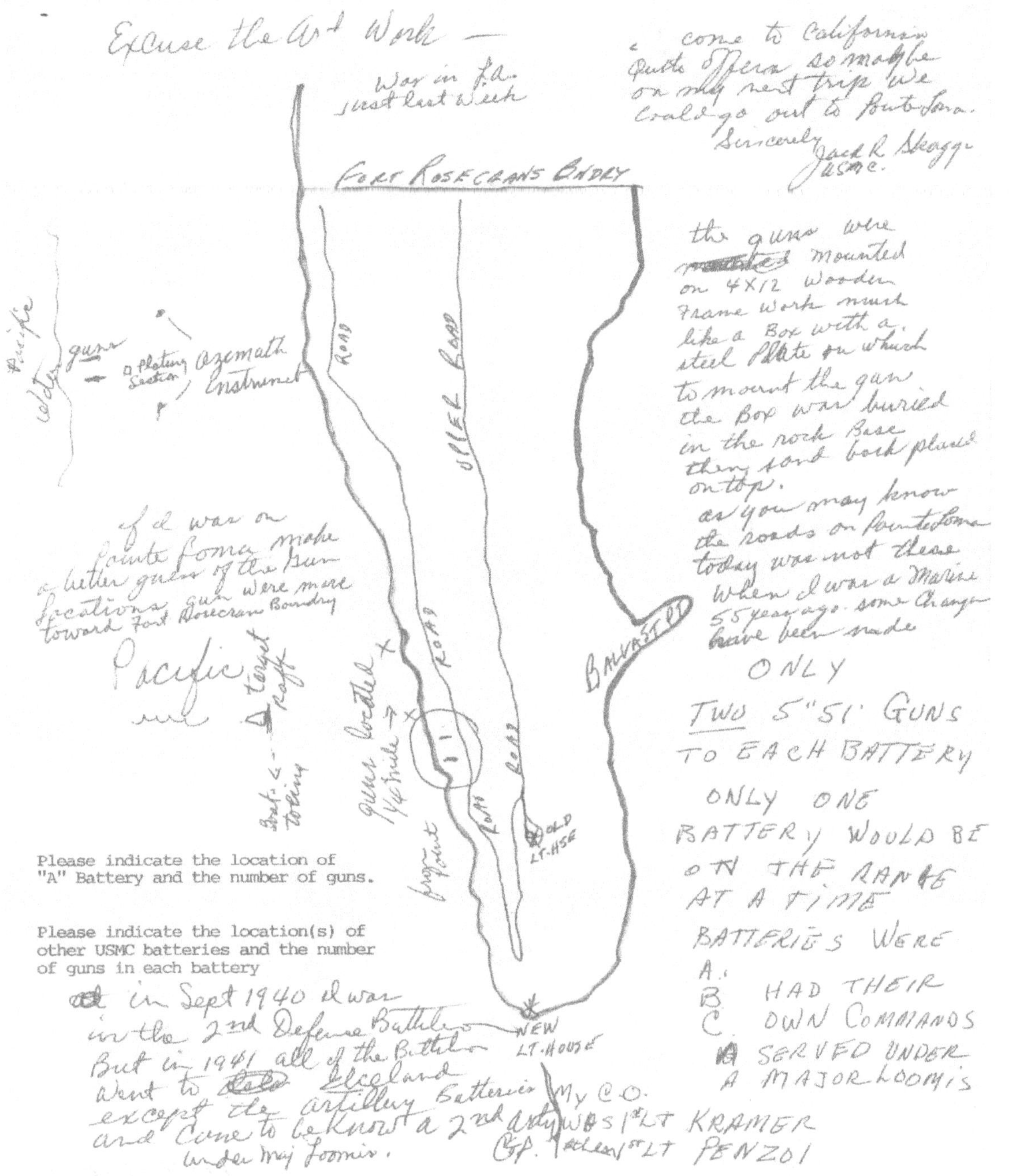

Excuse the Art Work —

War in L.A. just last week

come to California quite often so maybe on my next trip we could go out to Point Loma. Sincerely Jack R. Skaggs USMC.

FORT ROSECRANS BNDRY

the guns were ~~mounted~~ mounted on 4×12 wooden frame work much like a Box with a steel Plate on which to mount the gun the Box was buried in the rock Base they sand back placed on top.

as you may know the roads on Point Loma today was not these when I was a Marine 55 years ago. some changes have been made

ONLY TWO 5"51' GUNS TO EACH BATTERY

ONLY ONE BATTERY WOULD BE ON THE RANGE AT A TIME

BATTERIES WERE
A.
B   HAD THEIR
C.   OWN COMMANDS
D   SERVED UNDER
A MAJOR LOOMIS

Please indicate the location of "A" Battery and the number of guns.

Please indicate the location(s) of other USMC batteries and the number of guns in each battery

in Sept 1940 I was in the 2nd Defense Battln But in 1941 all of the Bttln went to Iceland except the artillery Batteries and came to be known a 2nd Arty under Maj Loomis.

My C.O. was 1st Lt KRAMER Cpt. ___ 1st Lt PENZOL

Figure 4-11. The above hand-written annotations were added in October 1995 by Jack R. Skaggs, USMC, to this map of Fort Rosecrans, which was drawn in September 1995 by Al Grobmeier.

Other sources indicate the USMC had two practice batteries near Battery Pont Loma, one equipped with two 7-inch guns that were later transferred to Battery Zeilin, and the other equipped with three 5-inch guns that were transferred to Battery Gillespie. Skaggs' recollection of only two 5-inch guns is therefore somewhat suspect. When asked if he had been assigned to Battery Zeilin or Gillespie, Skaggs responded thusly (Overton, 1995):

"Battery what? No. Now these were 5-inch 51s... They came right out of the Marine Corps Base over here which is now MCRD. They didn't have Pendleton then, that wasn't built. These guns didn't have names on them like the Army. It was strictly just USMC."

Skaggs' spotty interview with Overton is rather hard to follow, but it appears he had actually trained on the unnamed group of three 5-inch 51 naval guns located west of Gatchell Road on the southern tip of Point Loma, their presumed emplacements circled in Figure 4-12 below. Skaggs' cryptic note on the right margin of Grobmeier's map (see again Figure 4-11) describes their mounting (Skaggs, 1995):

"The guns were mounted on a 4X12 wooden framework much like a box with a steel plate on which to mount the gun. The box was buried in the rock base then sand back-placed on top. As you may know the roads on Point Loma today was (sic) not there when I was a Marine 55 years ago. Some changes have been made."

Figure 4-12. Prior to the war, the USMC used three 5-inch (circled) and two 7-inch (up arrows) naval guns for training. The two 7-inch guns were located just north (left) of the four original 155-millimeter emplacements of Battery Point Loma (undated US Army photo courtesy Cabrillo National Monument).

The Flower and Roth (1982) survey stated these three 5-inch naval guns were moved from near the new Coast Guard lighthouse in June of 1942 and "installed in 155-millimeter emplacements." These 155-millimeter emplacements were the former Panama mounts of Battery North,[8] which subsequently was renamed Battery Gillespie by the Army.[9] Since the USMC only had three 5-inch guns, one of the original Battery North emplacements was no longer needed and presumably removed or covered over with fill.

Adding to the confusion, the "Historic CA Posts" website seems to have the dates transposed (Ruhlen, 2006), with Battery Gillespie listed as constructed in 1937 and Battery North in 1942. The likely reason for all this uncertainty is the fact that Battery North and Battery Gillespie were temporary installations built with troop labor, and hence lack formal documentation.

---

8   Recall Battery North's 155-millimeter GPF guns had been relocated to Battery Imperial at the start of the war.
9   Named for Lieutenant Archibald H. Gillespie, USMC.

One surviving and very credible account, however, is provided in the "History of Harbor Defenses of San Diego" (HS, 1945):

"Prior to the war, the U.S. Marine Corps had mounted three 5" Navy guns near Battery Point Loma for training purposes. When the 155-millimeter battery was moved from the northwest corner of Fort Rosecrans to Coronado Heights, no armament was left in this area. Request was made then to move these three 5" guns to the old 155-millimeter emplacements and use them as interim Harbor Defense armament until the modernization battery in that area could be built. This request was approved and in June 1942 the move was completed and the manning of the battery taken over by Harbor Defense troops. The battery was named Gillespie and provided an addition to the defense against submarines and light surface units for well over a year."

After considering all the various sources of information, it seems clear that Battery Gillespie was established at the former site of Battery North when the three 5-inch Navy guns used by the Marine Corps were relocated in June of 1942 from the location shown previously in Figure 4-12. The center pedestals of the original Panama mounts were removed to accommodate construction of wood-beam foundations, but the outer steel rails were left in place.

The only image I have been able to locate of Gillespie's 5-inch guns is a digital scan of a photocopy of a microfiche printout of a small wartime photograph (Figure 4-13). Considering the convoluted path from the original print, the quality is understandably very poor, but the information content merits inclusion anyway. The gun appears to be pointing northwest, based on the northern fence line of Fort Rosecrans seen in the upper left corner, indicating the photographer would have been looking northeast.

Figure 4-13. Camouflage netting for one of the 5-inch emplacements at Battery Gillespie, with the hinged forward section dropped to the ground in the firing position (US Army photo from Wolfgast, 1943). Note the rope and pulley for this purpose attached to the pipe frame of the overhead net near the vertical support pole at right.

The photo presented in Figure 4-13 above provides a possible clue as to the function of the concrete foundation seen in Figure 4-14 below. The small hole in the upper right corner might suggest this foundation possibly served as an anchor for the one of camouflage support poles, but no other similar pole foundations were found to be present. The location of the hole in the upper right corner, as opposed to centered, also casts some doubt on this theory.

Figure 4-14. a) A small concrete foundation was uncovered in the brush behind the partially buried Panama-mount rail for the Gun 2 emplacement in May 2006. **b)** View of the exposed concrete foundation in May 2008, with a section of the dirt road seen in Figure 4-15 below just visible at extreme upper left corner.

Figure 4-15, a zoomed-in portion of a 1988 aerial photo, shows the Gun 2 and Gun 3 emplacements of Battery Gillespie looking southwest towards the F-28 complex. The concrete object this side of the utility pole at upper left, previously shown in Figure 4-7, was erroneously reported by Flower and Roth (1982) as the foundation for a 5-inch gun. The wooden structure at bottom right, which does not appear in earlier photos of this area, appears to be a more recently discarded shipping container.

Figure 4-15. Aerial view of the two southernmost 5-inch gun emplacements of Battery Gillespie, looking southwest toward the Building F-28 complex in 1988 (US Navy photo courtesy SSC Pacific). Note traces of the abandoned Gun 4 emplacement this side of the previously discussed concrete foundation block near the utility pole at upper left.

While walking this area during lunch one day with outside plant engineer Jeff Bowen, we ran across a discarded 5-inch shell casing in a rubble pile about 100 yards southwest of the Gun 3 emplacement (Figure 4-16a). Interestingly, this casing was made of galvanized steel as opposed to the more traditional brass, which was in short supply during the war. Randy Peacock of Facilities later told Jeff there were originally two such casings in this area, but we saw no sign of the second one. The stamped inscription on the end (Figure 4-16b) reads "SPDN-813??," with the last few digits obliterated by rust.

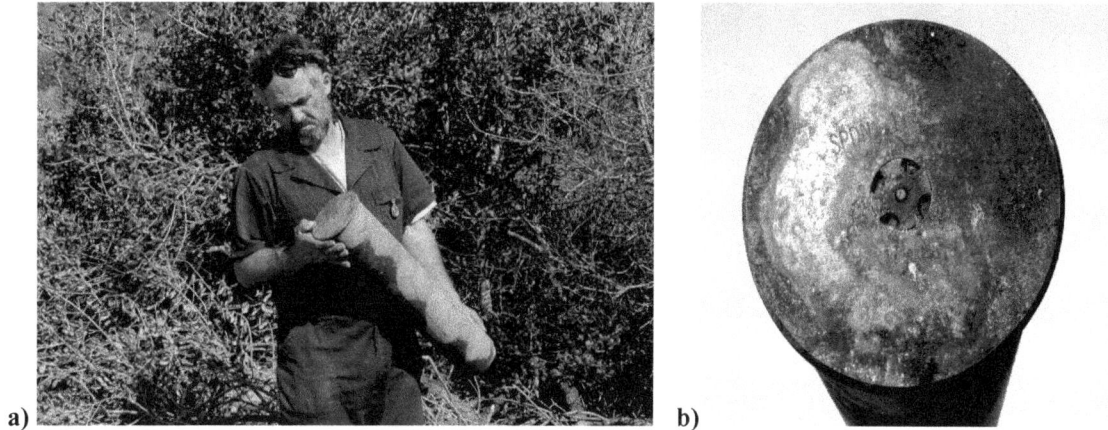

a)   b)

Figure 4-16. a) Jeff Bowen examines a rusty galvanized-steel 5-inch shell casing found in May 2006 in a rubble pile approximately 100 yards southwest of the southern (Gun 3) emplacement of Battery Gillespie. **b)** The stamped inscription on the end denotes the type of propellant charge.

According to Tony DiGiulian (www.navweaps.com):

"In the USN, smokeless powder is designated as "SP" and is usually a uniform ether-alcohol colloid of purified nitrocellulose with a quantity of diphenylamine (D suffix) or ethyl centrality (C suffix) added for stability... Moisture or heat speeds its deterioration and the combination of the two is extremely damaging to the propellant. *SPDN* is a diphenylamine-stabilized smokeless powder to which nonvolatile materials have been added to reduce its hygroscopic tendencies. The N stands for nonhygroscopic."

As previously indicated, the general consensus has been that Guns 1 through 3 of Battery Gillespie were installed in the original emplacements for Guns 2 through 4 of Battery North (Flower & Roth, 1982; Keniston, 1996). In other words, Battery North's old Gun 1 emplacement supposedly had been covered over or removed. Careful examination of Figure 4-17, however, shows Gun 1 for Battery North in 1941 to be immediately adjacent to the large ravine running east to west down the left side of the image.

A fifth and possibly a maintenance emplacement is located behind the battery in the upper-left corner of this photo, just south of this ravine at the north end of Woodward Road. Note also the strong correlation between Figures 4-8 and 4-9 with respect to the foot-path trace from the Gun 1 emplacement to the cliffside wooden observation station that was previously shown in Figure 4-9, and also seen in the lower-left corner of Figure 4-17.

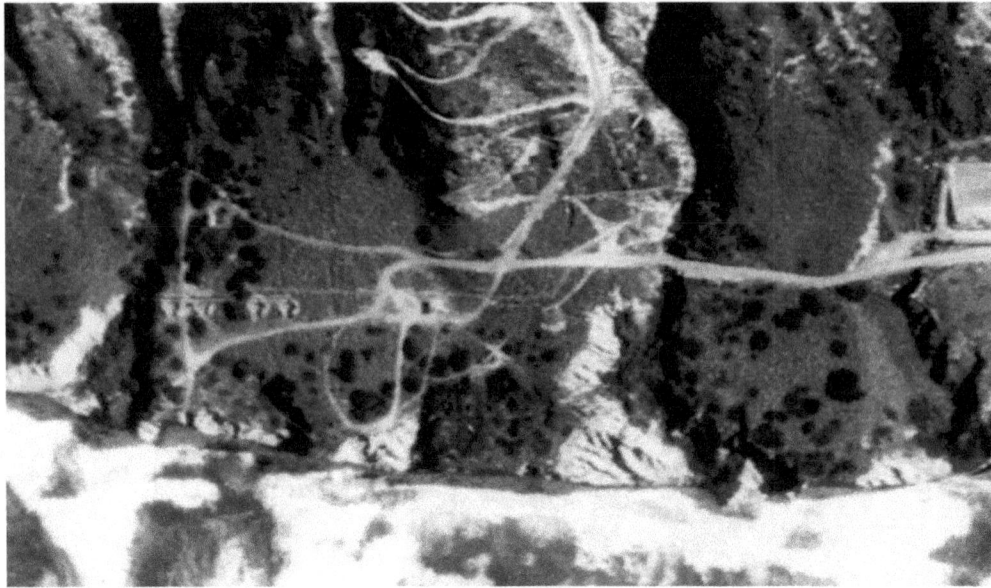

Figure 4-17. This December 1941 aerial photo of Battery North shows the Gun 1 emplacement was immediately adjacent to a large ravine to the north (US Army photo courtesy Cabrillo National Monument). Note the possible maintenance emplacement in upper-left corner at the north end of Woodward Road.

Figure 4-18 below shows the Gun 1 position for Battery Gillespie to also be situated immediately adjacent to this same ravine in 1954, the former location of Gun 1 of Battery North. Note also the small wooden structure just to the right of this ravine on the east side of Woodward Road, just left of what appears to be a recently excavated area due east of Gillespie's Guns 2 and 3. This structure was not present in the earlier December 1941 photo seen in Figure 4-17 above. The second wooden structure indicated by the up arrow south (right) of Gun 3, also not seen in Figure 4-17, will be discussed later in this section.

Figure 4-18. Looking east in April 1954, Gun 1 for Battery Gillespie is situated immediately adjacent to the east-west ravine, and the spacing between Guns 2 and 3 appears slightly greater than that between Guns 1 and 2 (see also Figure 4-17). Note small wooden structure south (right) of the ravine on the far side of Woodward Road, and the collapsed tunnel behind Gun 3 (US Navy photo courtesy SSC Pacific).

There is a wealth of information available in Figure 4-18. Woodward Road now crosses over the east-west ravine in the upper-left quadrant, presumably filled with dirt from the excavation at top center. The full extent of the Battery Gillespie complex is suggested by the area of reduced vegetation at image center, with possible inclusion of the adjoining area on the far side of Woodward Road. The dark line running east from the Gun 3 position at image center is a collapsed tunnel section that eventually turns north, connecting to one of two 30-foot underground Multi-Plate shelters directly behind Gun 2 (Figure 4-19a).

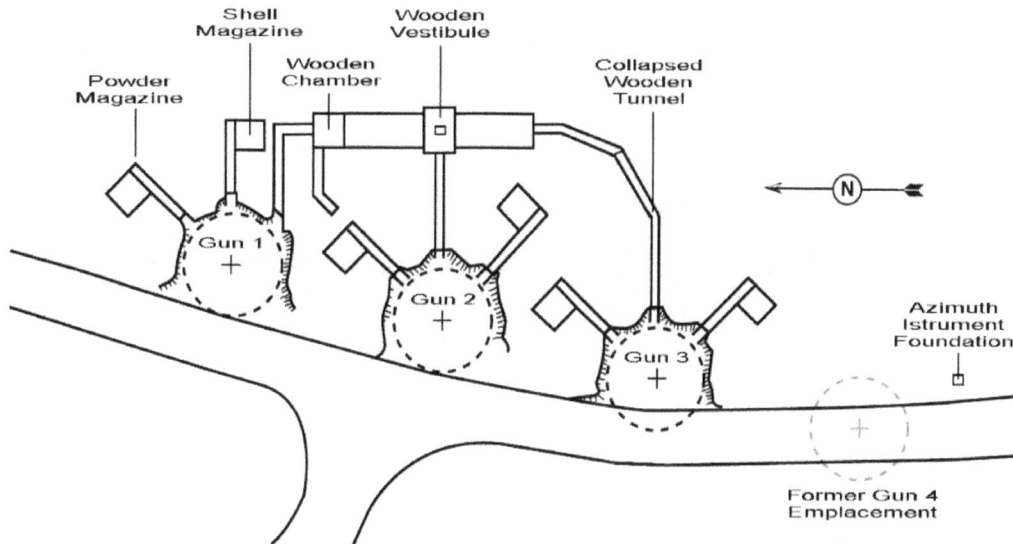

a)

b)

Figure 4-19. a) Layout of Battery Gillespie (modified with permission from 1995 field sketches prepared by Lee Guidry). Note the 65-foot buried shelter with vestibule skylight behind Gun 2. **b)** Remains of the skylight above wooden vestibule behind Gun 2, circa 1971 (photo courtesy Lee Guidry).

Figure 4-20 provides another view of this area in 1958, with evidence of continued excavation of fill dirt east of Woodward Road. Closer inspection reveals a distinct rectangular shape in the left side of the excavation pit, with a wooden building hidden close by in the brush to the north (up arrow), approximately 10 yards east of the previously mentioned wooden shack on the far side of Woodward Road. There appears to be trace evidence of a footpath from the excavation pit leading to this wooden building and continuing north through the brush to the white sandy area in the ravine at upper right.

Figure 4-20. Battery Gillespie looking east in 1958 (US Navy photo courtesy SSC Pacific). Note roof joists for a wooden building (up arrow), further-collapsed tunnels behind the Gun 3 position (see again Figure 4-18), and what appears to be the overgrown remains of the former Gun 4 position just south (right) of Gun 3.

This brush-covered wooden building was identified by Keniston (1996) as a utility structure, but no mention was made of the adjacent excavation, nor what appears to be the remains of a footpath running north to the nearby ravine. Several roof-support joists for this structure are visible in Figure 4-20 (up arrow). While the wooden building has been reduced to rubble, its 10-foot by 15-foot concrete slab is still in place, upon which still stands a 28-inch-tall worktable, 31 inches wide by 68 inches long (Figure 4-21). Directly below this well-preserved table were two hemispherical wooden objects that Lee Guidry identified in 1995 as part of a hand-crafted expanding barrel plug for a 5-inch gun.

Figure 4-21. Remains of a worktable and assorted debris at the collapsed utility structure site in October 2006, looking south (see again up arrow in Figure 4-20). Note the two hemispherical wooden components of an expanding barrel plug for a 5-inch gun below the table at center foreground (up arrows).

Figure 4-22a shows Lee Guidry trimming some of the brush back from the surface of this work table in April 2008, with the wooden barrel-plug assembly now resting on the table top below his left hand. Figure 4-22b provides a close-up view of the plug detail. A recreation of the original design, redrawn from Lee Guidry's 1995 field sketch, is presented in the plan and section views of Figure 4-22c. The moveable plug piece at the top of this drawing (both views) is driven outward by the wedge block as the threaded bolt is tightened, thereby securing the plug in the end of the barrel.

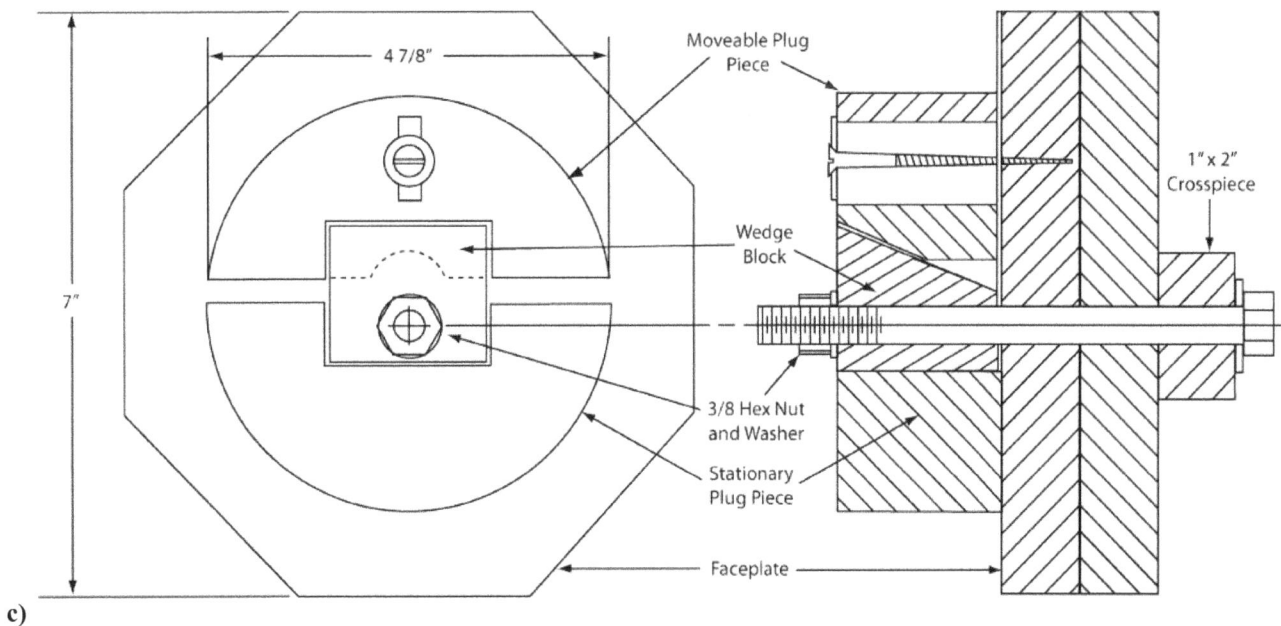

Figure 4-22. a) Lee Guidry probes for the foundation slab extents in April 2008. **b)** Remains of expanding gun-barrel plug, minus the center wedge mechanism. **c)** Detail of original barrel-plug configuration (redrawn by Lindsay Seligman from original field sketches prepared by Lee Guidry).

In May 2008, Lee Guidry and I returned to clear the 12x15-foot utility-structure slab of years of accumulated debris (Figure 4-23). One intriguing find was a large disk-shaped wooden object laying on the ground just west of the work table, which unfortunately was seriously decayed due to prolonged ground contact. The size and shape of this disk, combined with its proximity to the presumed maintenance emplacement of Battery North seen in Figure 4-17, raised the possibility it may have been part of the center Panama Mount pedestal for that site. The remains appeared to be about 6 feet in diameter, but the advanced state of decay precluded any accurate measurement.

Figure 4-23. After clearing away surface debris in May 2008, the 12x15-foot concrete slab of the collapsed utility structure is fully exposed for inspection. The decaying remains of a large disc-shaped wooden artifact were found on the ground just west (right) of the table in the upper-right corner.

J.R. Skaggs describes USMC fire-control procedures used with the three 5-inch Naval guns before they were moved from the vicinity of Battery Point Loma (Overton, 1995):

> "We set up our base-end stations, three of them. We used triangulation firing back in those days. I was on one of those stations. We used to park the trucks down at the bottom of the hill and walk back up there...
>
> They even used a deal that came off a ship. You cranked in your information. Our guns were set up like they were aboard ship. They would get all their azimuth readings to fire with off that machine."

The USMC reportedly used portable azimuth instruments at this earlier practice location, described by Skaggs as "just a wooden box with a tripod," packed in for the firing exercise and removed immediately afterwards. Figure 4-24 provides evidence of a fixed wooden observation station at Battery Gillespie, however, just left of the power pole at image center. This photo also shows dirt from the excavation area at top center apparently being used to further fill in the ravine in the upper-left corner (see again Figure 4-18 and Figure 4-20). The sandy patch indicated by the up arrow likely indicates the presence of an underground berthing structure. The stepped concrete block in the Gun 2 emplacement seen earlier in Figure 4-5b is just beyond Woodward Road at the left edge of the photo.

Figure 4-24. View looking east during site preparation for Building F-28 in April 1954, with a wooden observation structure left of the utility pole at photo center (US Navy photo courtesy SSC Pacific). The sandy patch indicated by the up arrow this side of Woodward Road was the site of an extensive collapse in 2005.

Figure 4-25 shows this area after being cleared for an antenna ground plane in November 1962. The wooden observation structure has been removed, but the concrete foundation for its azimuth device (Figure 4-7) is still extant, repurposed as a guy-wire anchor for the small power pole to its right. Note also the stepped concrete block (up arrow) in the Gun 2 emplacement, and the entrances to the shell and powder magazines behind all three emplacements at photo left center. Excavation continues on the far side of Woodward Road, presumably to provide dirt for building up the road and filling in the ravine.

Figure 4-25. Looking east in November 1962, showing the stepped concrete block (up arrow) and inscribed instrument foundation left of the same utility pole at the former site of the wooden fire-control structure (US Navy photo courtesy SSC Pacific). Note new Building F-28 at lower right and enlarged excavation area behind Woodward Road at top center.

The interior of the southernmost 30-foot bunker depicted in the battery layout of Figure 4-19a is seen in Figure 4-26 below, circa April 2008. The 10-foot-by-10-foot wooden vestibule that once adjoined the north end of this structure is no longer extant, destroyed by decay and attempts to seal off access in 1989. Note the galvanized nuts and bolts joining the Multi-Plate-Arch sections of the overhead. The interior of the northernmost 30-foot bunker on the other side of the wooden vestibule is seen in Figure 4-27.

Figure 4-26. View looking north towards the vestibule sinkhole from inside the southernmost 30-foot bunker in April 2008, showing fill dirt used to seal off the south end in 1989 (photo courtesy Lee Guidry).

Figure 4-27. View looking south in the northernmost 30-foot bunker towards the vestibule sinkhole in April 2008. The large concrete object in foreground left appears to be debris from the collapsed skylight over the vestibule (see again Figure 4-19a).

Referring again to the layout of Battery Gillespie in Figure 4-19a for reference, the entrance to the powder magazine for Gun 1 is seen in Figure 4-28, looking northeast in 1971, with the northern fence line of Fort Rosecrans in the upper-left corner. The previously mentioned fill-dirt-excavation area east of Woodward Road is at left center. Figure 4-29 shows the entrance to the shell magazine of Gun 1, looking east.

Figure 4-28. The well-preserved entrance to the powder magazine behind Gun 1 is seen looking northeast towards the excavated area beyond Woodward Road in 1971 (photo courtesy Lee Guidry). Note wooden stairs and boardwalk along northern fence line of Fort Rosecrans in the upper-left corner.

Figure 4-29. The entrance to the shell magazine behind the Gun 1 emplacement, looking east in 1971 (photo courtesy Lee Guidry). The entrance to the southernmost 30-foot bunker is just out of image to the right.

In May 2008, Lee Guidry and I enlarged a small sinkhole found at the east end of the collapsed tunnel from the Gun 1 emplacement leading to the shell magazine directly behind it (Figure 4-30). Rather than a wooden overhead like the rest of the tunnel, this portion was covered with a single section of elephant iron, the south edge of which rested upon the elephant-iron overhead of the adjoining magazine (Figure 4-31). Decayed remains of the northern end wall of the magazine are seen in the lower-right corner.

Figure 4-30. Entering the enlarged sinkhole on the east end of the tunnel to gain access to the adjoining shell magazine for Gun 1 (May 2008 photo courtesy Lee Guidry). Note decayed remains of the wood shoring supports on both sides, and the edge of the elephant iron used for the tunnel roof just above my left arm.

Figure 4-31. Interior view (looking north) of the shell magazine for Gun 1, with the collapsed tunnel in background left, leading back to the emplacement. Note wooden remains of the northern end wall at bottom right, and dangling section of elephant iron used for the east end of the tunnel overhead in background.

Another view of the interior of the Gun 1 shell magazine is seen in Figure 4-32, this time looking south. The end wall seen in this view is made of concrete, for better protection against incoming rounds, with a well-preserved wooden shell-storage bench still present. The remains of a second storage bench are strewn about the concrete floor, mixed in with spillage from the collapsed connecting tunnel previously seen in Figure 4-31 above.

Figure 4-32. Interior of the Gun 1 shell magazine looking west towards a wooden shell-storage bench in front of the concrete end wall, and material from a second bench strewn about the concrete floor.

The collapsed magazine and tunnel entrances behind the Gun 2 emplacement are seen in Figure 4-33 below, looking east in 1971. A recently salvaged tower section of the *SCR-296-A* fire-control radar for Battery Woodward has been temporarily laid between the two power poles in the upper-left corner. This radar tower had previously been located just left of the battery commander station for Battery Gillespie seen at upper right (down arrow).

Figure 4-33. Collapsed magazine and tunnel entrances behind the Gun 2 emplacement looking east in 1971 (see again Figure 4-19a). Note the recently salvaged section of the *SCR-296-A* radar tower for Battery Woodward laying horizontally between the two power poles at upper left (photo courtesy Lee Guidry).

The remains of the Gun 3 powder magazine and the tunnel entrance leading to the northernmost 30-foot corrugated section of the 80-foot shelter behind the Gun 2 position are seen in Figure 4-34. The small white pyramid-shaped object on the ridgeline at top center is one of four concrete support pillars for the *SCR-296-A* fire-control radar tower for Battery Woodward. Slightly to its right is the wooden battery commander station for Battery Gillespie, previously shown in Figure 4-33.

Figure 4-34. The powder magazine entrance of the Gun 3 emplacement is at extreme left, looking southeast in 1971, with the wooden-tunnel entrance to the 80-foot corrugated shelter behind Gun 2 at far right (photo courtesy Lee Guidry). Note battery commander station for Battery Gillespie along ridgeline just right of top center, and the concrete support pillar for the *SCR-296-A* radar tower to its left.

Figure 4-35 shows a sinkhole that has revealed a portion of the corrugated-roof over the northeast end of the shell magazine for Gun 3 (see again battery layout of Figure 4-19a). Armed with a flashlight, shovel, and a safety escort, I was easily able to enlarge this opening and gain entry in April of 2008. The interior of this 10-foot by 10-foot bunker is seen in Figure 4-36, looking northeast towards the collapsed connecting tunnel, with sunlight spilling down the entry hole. The presence of discarded concrete debris in the spillage suggests the connecting tunnel had been intentionally filled in to preclude entry.

Figure 4-35. A growing sinkhole from heavy rains during the winter of 2005 has exposed the northeast end of a Multi-Plate-Arch bunker that served as the shell magazine for Gun 3, as seen in April 2008.

Figure 4-36. Interior of the Gun 3 shell magazine, looking northeast in April 2008 towards the sinkhole shown in Figure 4-35. Note the considerable presence of concrete debris in the spillage.

A southwest view of the shell magazine interior is seen in Figure 4-37. When I first saw these wooden assemblies, I thought they were bunks, but later realized they were used for ammunition storage. The raised shelves helped keep the projectiles dry in the event of seepage during the rainy season, with the slotted wooden posts on the sides supporting a safety rail to keep things from falling over. The bunks found in the underground berthing shelters at Battery Point Loma (see next section) were made from much larger wood frames strung with fence wire, as opposed to solid wood.

Figure 4-37. View looking southwest of the remains of the two projectile storage shelves, partially covered in spillage from the collapsed communication tunnel. Note the whitewashed concrete end wall and overhead sections of elephant iron.

Battery Point Loma

Battery Point Loma is located on the west side of Cabrillo National Monument property at the southern tip of the point (Site 7). Towards the end of 1940, four 155-millimeter GPFs were operational in this area just north of the new Coast Guard lighthouse, although not mounted due to lack of funds (Figure 4-38). Eventually $6,000 was allocated to build four Panama mounts with troop labor (Thompson, 1991), which were situated immediately behind these temporary emplacements, between what are now Cabrillo and Gatchell Roads.

Figure 4-38. a) The unmounted 155-millimeter GPF emplacement at Battery Point Loma prior to the war, looking northwest towards the Pacific Ocean. **b)** Easterly view of the emplacement, with the Point Loma ridgeline in the background (late 1940 or early 1941 US Army photos courtesy Al Grobmeier).

These Panama mounts were completed by September 1941 (Figure 4-39), whereupon the unmounted *GPFs* were relocated slightly eastward to their new site. The concentric steel rail sections that supported the twin gun trails at this location were embedded in a 3-foot-wide concrete perimeter ring for improved stability. A 10-foot-diameter center pedestal with a 4-inch-high raised hub served as a wheel guide to prevent the modified trail plates from binding. The repositioned battery became operational in September of 1941.

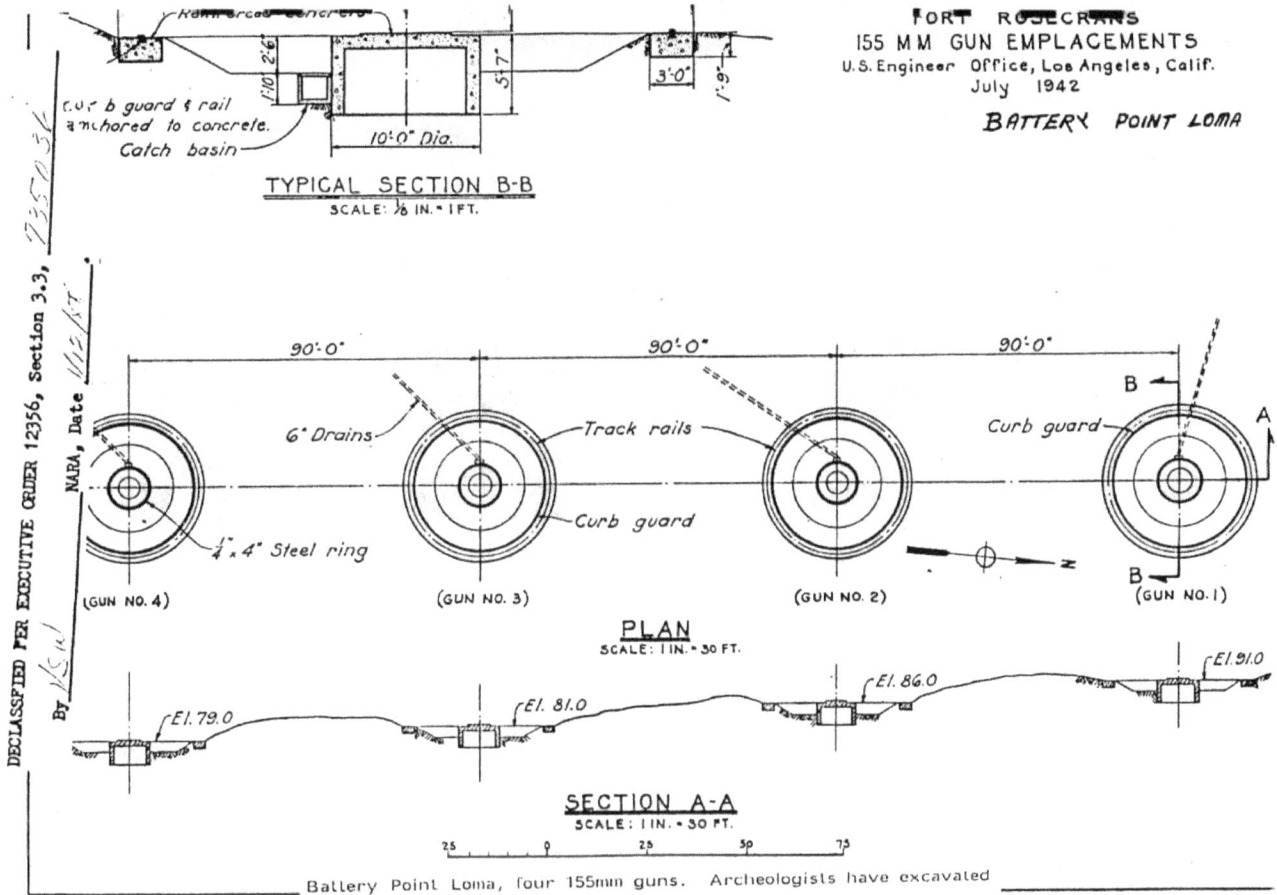

Figure 4-39. July 1942 Panama-mount construction details for the 155-millimeter *GPF* guns at Battery Point Loma, arranged north-to-south with a lateral separation of 90 feet (NARA).

These new Panama mounts and bunkers are seen in the hairpin loop between Gatchell and Cabrillo Roads (Figure 4-40) circa January 1942, directly behind and slightly north of their original locations west of Gatchell Road (up arrows). The two additional emplacements seen to the north were for a pair of 7-inch Navy guns used by the USMC, which were remounted in Battery Zeilin shortly after Pearl Harbor. Note the line of pyramidal tents at image center along both sides of Optic Road leading to Searchlight 17, presumably for temporary berthing until the underground bunker complex of Battery Point Loma was completed.

Figure 4-40. The original emplacements (up arrows) of Battery Point Loma are seen just west of Gatchell Road, looking east January 1942, with the new emplacements behind them to the east, between Cabrillo and Gatchell Roads (US Army photo courtesy Cabrillo National Monument). The operating position for Searchlight 17 is at far right, with the abandoned WWI Searchlight 4 silo above (see also chapter 8).

Jerry Harrison, Battery E, recalls an unusual target practice with the 155-millimeter GPFs at Battery Point Loma (Overton, 1993):

> "When we fired them, they used a tow truck, I mean a tow boat out there. We fired on the target. You bracket the target, one short, one far, and the third one was down the chute, you know. One time we targeted the tow boat. Ole boy cut his line and took off because he knew the others were going to come right down the tube. I wasn't on the gun that done that but I was there when it happened, you know, on another gun. It was kind of comical."

According to Hollis Gillespie, Battery B, the US Army consequently had to purchase an aging ferry boat from Seattle to provide their own subsequent tow services for coast-artillery practice (Gillespie, 1979):

> "We did hear that one day a sleepy eyed gunpointer sighted on the tug boat instead of the target and fired a 95-pound 155-millimeter projectile through the smoke stack, mast, and upper rigging of the tug boat. It is believed the tug boat captain uttered a profane exclamation as he cut the tow line, abandoning the target to drift aimlessly in the Pacific while he hurriedly returned to the safety of San Diego, claiming loudly that he would never again tow a target for those crazy artillerymen on Point Loma."

The four 155-millimeter GPF guns were removed from Battery Point Loma in 1943 (Thompson, 1991), replaced by the two 90-millimeter emplacements in anti-motor torpedo boat (AMTB) Battery Cabrillo just across Gatchell Road to the west (chapter 6).

Figure 4-41. The southernmost (Gun 4) Panama mount following excavation in 1984, showing the concrete and steel center pedestal that supported the gun carriage, surrounded by the outer steel rail for the twin gun trails (photo courtesy Cabrillo National Monument).

Referring again to Figure 4-41 above, the center pedestal is 10 feet in diameter with a 4-inch-high raised hub that measures 6 feet 4 inches across and is surrounded by a steel band. This raised hub served as a wheel guide as previously mentioned. The steel curb guard along the inner diameter of the concrete perimeter footing is notched as seen in Figure 4-42 below to facilitate moving the gun trails with a pry bar (CAJ, 1929):

> "The trails are moved by hand. When the target appears to be approaching the limit of traverse permitted by the top carriage, the gun crew is directed to man the trails. Four men on each trail are necessary to move the trails in the desired direction."

Figure 4-42. The steel curb guard along the inner diameter of the concrete footing is notched to facilitate moving the gun trails, which rested on the smooth concentric outer rail in background (circa May 2006).

The layout of Battery Point Loma is depicted in Figure 4-43 below, which shows the four 155-millimeter gun emplacements arranged north to south between Cabrillo and Gatchell Roads. Buried behind each of these emplacements were a pair of corrugated-arch ammunition magazines for projectiles and powder. A bit further to the east were two underground 38-foot corrugated-arch berthing shelters with wooden end walls. A 19-foot wooden underground berthing shelter was further north on the west side of Cabrillo Road, with a 40-foot berthing shelter north of the ravine across the road.

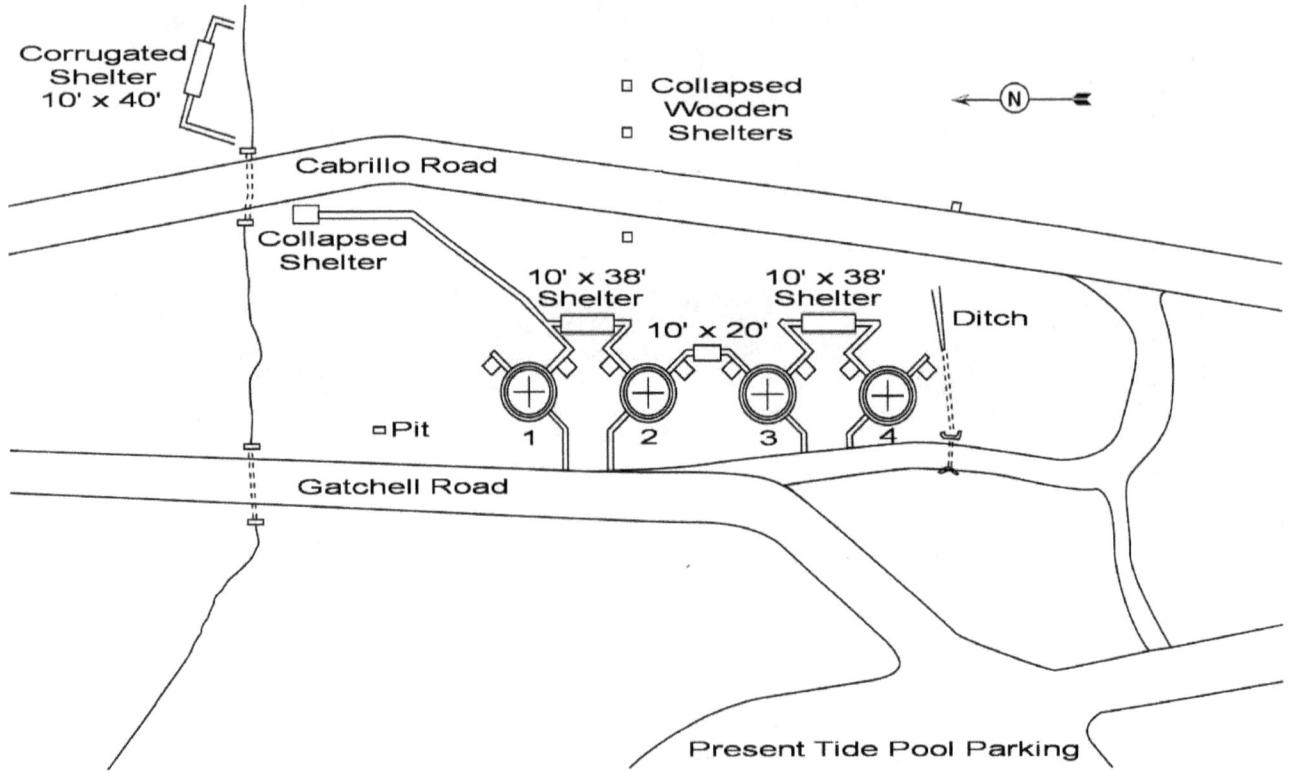

Figure 4-43. The approximate layout of Battery Point Loma (redrawn by author from original January 1992 field sketch prepared by Lee Guidry).

Figure 4-44 shows the southern ammunition magazine behind Gun 4, with the collapsed communication tunnel from the emplacement leading off to the right, and the Pacific Ocean just visible behind the brush in background left. Note the small sinkhole at the rear of the bunker in the upper-left corner. A sturdy wooden bench-like structure within each of these bunkers kept the powder or projectiles from getting wet in the event of seepage.

Figure 4-44. Exposed by 60-plus years of erosion, this subterranean ammunition bunker is approximately 10 feet long and 10 feet across, constructed of ¼-inch-thick galvanized sections of corrugated "Elephant Iron." The collapsed tunnel at upper right leads to the Gun 4 emplacement (circa May 2006).

The rear of this bunker is made of wood. The manufacturer's stamp, which appears on almost every curved section throughout the Battery Point Loma compound, can be seen on the left side of the interior wall in Figure 4-45. Since the stamp was applied to a corrugated surface, it is somewhat difficult to read. A careful piecing together of the more legible portions at several different locations revealed the words "MULTI PLATE" circularly inscribed around a triangular logo, below which is spelled out "INGOT IRON."

Figure 4-45. This close-up view of one of the magazines shows the substantial horizontal cross supports for the heavy shelving. The notched vertical pieces attached to the front edge of the shelf supported removable rails that kept the stacked projectiles from tumbling over in the event of a nearby concussion.

The southernmost 38-foot Multi-Plate berthing shelter behind Guns 3 and 4 is seen in Figure 4-46 below, fully exposed by sinkholes at both ends, circa May 2006. The wooden end walls, floorboards, support joists, and several of the bunk-support posts have been removed, probably for salvage, as wood was often taken from abandoned structures for reuse on new construction during the war. It is also possible this wood was salvaged or perhaps even stolen after the war. In either event, the bunk frames have fully collapsed on the east side and are seen precariously tilting to the right on the west side.

Figure 4-46. View facing north of the southern underground berthing shelter, fully exposed on both ends in May 2006, which once held 24 wood-frame chicken-wire bunks stacked two high along either side of a center aisle. Note the 12-inch layer of concrete added to the top for extra protection against incoming rounds.

Figure 4-47 shows the northern berthing shelter situated behind Guns 1 and 2, which has also been exposed to view by long-term erosion on both ends. This otherwise identical 38-foot concrete-covered shelter was found in slightly better shape than the southern shelter behinds Guns 3 and 4, but with bunk frames 1 through 4 missing or displaced. The sand-bagged communication tunnels leading from both ends of this shelter had collapsed some time ago and were for the most part filled in with spillage.

Figure 4-47. The northern berthing shelter, situated behind Guns 1 and 2, is also outfitted with bunk frames and covered in concrete, circa May 2006. Although otherwise in better shape than the southern shelter (see again Figure 4-46), bunks 1 through 4 have been displaced or destroyed.

During our 18 May 2006 site visit, my son and I found the tattered remains of a canvas duffel bag lying on top of the remains of bunk 1, which had been placed atop bunk 5 in the northern shelter as seen in Figure 4-48 below. The wooden floor boards were missing, exposing some underlying floor joists and causing the bunk assemblies to sag. Bunks 2 through 4 were no longer extant, possibly removed during an attempt to salvage the flooring. The bunk numbers, painted in the center of each frame and on the vertical support posts with orange paint, were still visible.

Figure 4-48. Bunk 1 was found resting on top of bunk 5 (upper left) in May 2006, while bunks 2 through 4 were missing. Note the canvas duffel-bag scraps on bunk 1 at foreground left (up arrow), the steel fencing stretched beneath the chicken wire for added support, and the exposed floor joists in background.

On 10 May 2008, Lee Guidry, Jeff Bowen, and I were inspecting restoration work at Battery Point Loma at the invitation of Ranger Charles Schultheis, Chief of Maintenance at Cabrillo National Monument. As we approached the southern berthing shelter behind guns 2 and 3, Ranger Schultheis noticed a tattered piece of canvas protruding from the soil near the north entrance. My initial interpretation of this item was biased by the earlier discovery of the canvas duffel-bag remnants at the northern shelter seen in Figure 4-48, so I assumed this was just more of the same. As Lee cleared away some dirt, however, it became obvious to me this item was something altogether different (Figure 4-49).

Figure 4-49. Remains of a canvas satchel charge showing a loop of det-cord at lower left, above which can be seen the edge of the wax-paper wrapping of the deteriorated cast explosive, circa 10 May 2008.

I had to get forceful with all present at that point, as no one seemed to believe this long-buried relic was in any way dangerous, and advised Ranger Schultheis to call the county bomb squad. They referred him instead to the Navy Explosive Ordnance (EOD) Detachment at Naval Submarine Base Point Loma, which was closer. An EOD technician removed the item for destruction, having identified it as "a WWII-era cast explosive and about 12 inches of det cord." The origin of this device is perhaps explained in an article entitled "Battery Point Loma," published in the *Fort Guijarros Quarterly* (Overton, 1988):

> "The sandbagged trenches and steel bunkers were, unfortunately, detonated by the Government and bladed over in the early 1960s in fear of visitors injuring themselves. In the mid-1980s, Cabrillo National Monument uncovered and stabilized the Number 4 position for interpretation."

Referring again to the battery layout presented in Figure 4-43, an underground 19-foot wooden berthing shelter was located further to the north at the end of a long wood-lined tunnel from the Gun 1 emplacement. Though traces remain, both this tunnel and the associated berthing shelter had long since collapsed from decay. There was yet a fourth underground berthing shelter at Battery Point Loma, this one on the east side of Cabrillo Road, with entry doors on both ends. Lee Guidry's field sketch of this 40-foot concrete-covered corrugated-arch structure is presented in Figure 4-50.

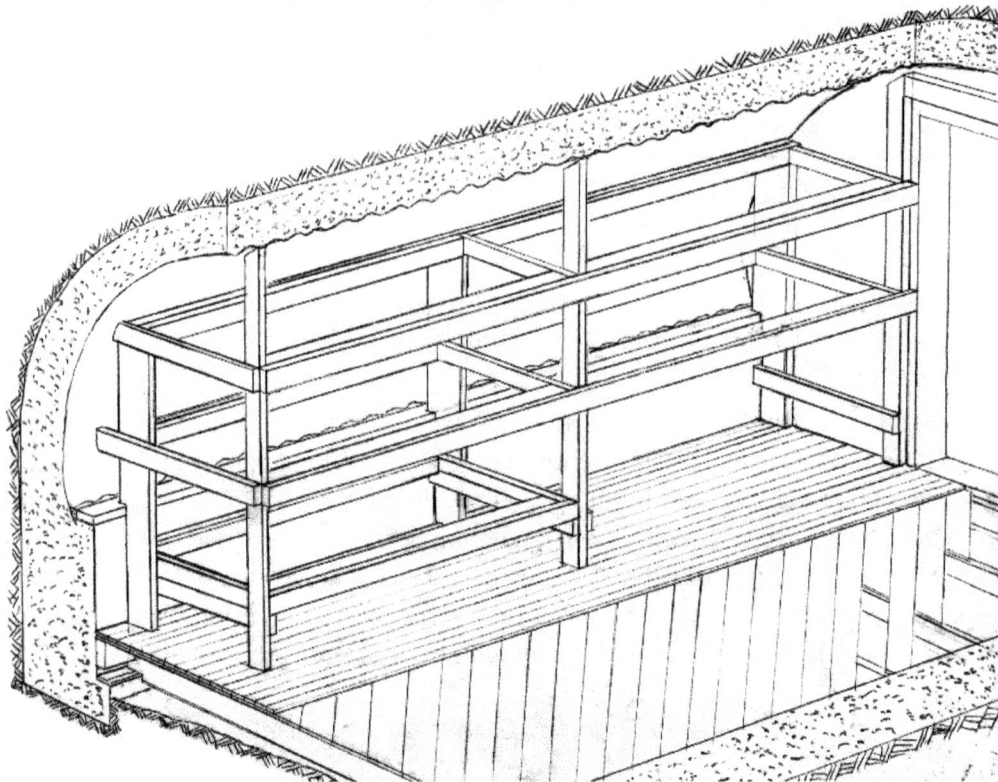

Figure 4-50. Cutaway-view (looking northeast) of the east end of the fourth berthing shelter on the east side of Cabrillo Road, as shown earlier in the site map of Figure 4-43 (February 1992 field sketch courtesy Lee Guidry). Note one of the two wooden entry doors (east end) at upper right.

This fourth underground berthing shelter east of Cabrillo Road was easily located during our site exploration on 10 May 2008 (Figure 4-51). The wood-lined tunnel leading to the east entrance to this shelter was found collapsed and obstructed by spillage and dense underbrush. The west end, however, was fully exposed at the bottom of a large sinkhole, which allowed easy access. The interior of this 40-foot Multi-Plate structure was found to be in similar condition to the two 38-foot shelters across the road, with the wood flooring removed and the collapsed bunk frames piled haphazardly about.

Figure 4-51. **a)** Ranger Charles Schultheis (foreground) inspects the obstructed east end of the underground berthing shelter, while Jeff Bowen (background) examines an armored cable splice and junction box. **b)** Interior view of the collapsed berthing racks, looking east from the sinkhole at the west entrance in May 2008.

Figure 4-52 shows the interior of this fourth underground berthing shelter looking west towards the sunlit sinkhole at the end of the collapsed connecting tunnel leading back to Cabrillo Road. The bright noonday sun has washed out the background seen through the open doorway, giving the false impression the shelter is above ground level. While the previously shown 38-foot berthing shelters at Battery Point Loma had wooden end walls due to their north-south orientation, this east-west-oriented shelter had concrete end walls for increased protection from incoming rounds.

Figure 4-52. Interior view of the eastern berthing shelter looking towards the doorway in the concrete wall on the west end of the berthing shelter. The entry sinkhole flooded with noonday sun gives the illusion this subterranean structure is at ground level.

Battery Imperial

Following the 7 December 1941 Pearl Harbor attack, a concerted effort was made to also beef up defenses along the Coronado side of the San Diego harbor entrance. As previously mentioned, the four 155-millimeter GPF guns of Battery North on Fort Rosecrans had been relocated to Battery Imperial on Coronado Heights by 14 December. Camouflaged Panama mounts for these new emplacements were completed the following month (Figure 4-53), providing additional defensive coverage to the south, which was felt to be the most vulnerable direction of attack (HSD, 1945).

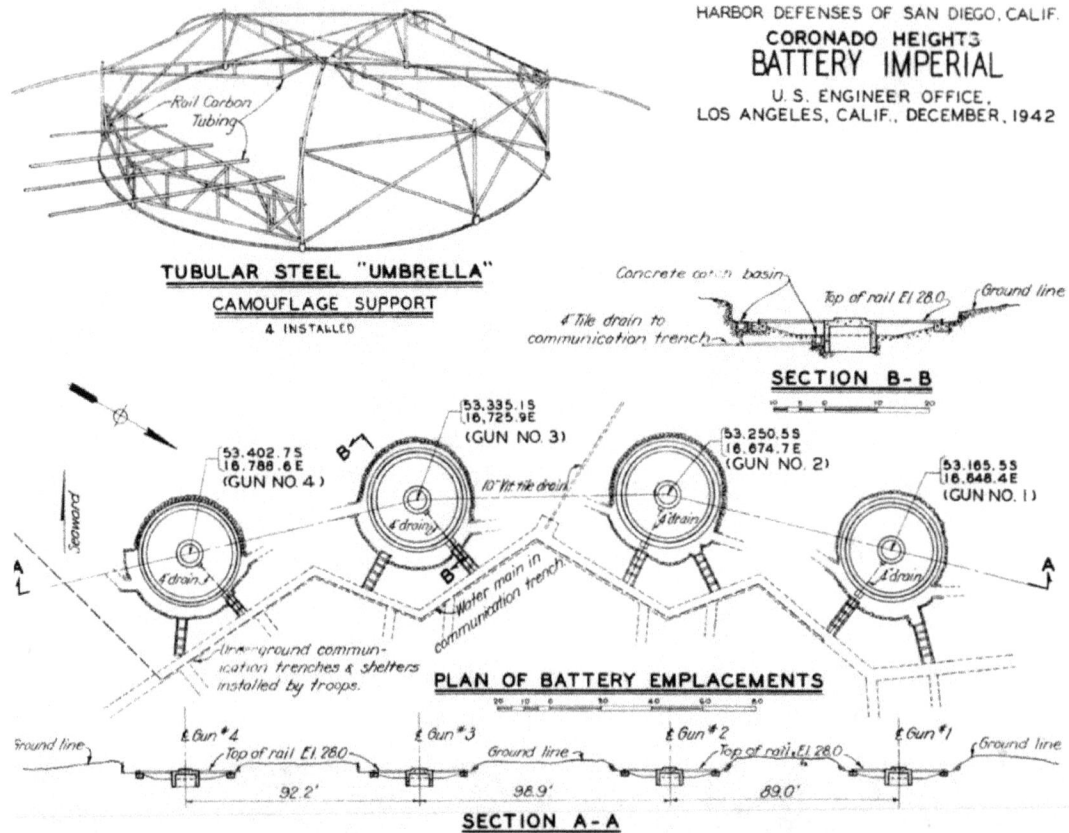

Figure 4-53. Plan and section views of the four Panama Mount emplacements from the Report of Completed Works for Battery Imperial on Coronado Heights, dated December 1942, which became operational in January of 1942.

In an interview with John Huntoon of Battery D, 19th Coast Artillery Regiment, Chief Ranger Howard Overton (1993) of Cabrillo National Monument made the following statement while discussing the unfinished 16-inch Battery 134 (Gatchell):

> "They had a battery of 155s there also. And they weren't installed where the 1936 plans said they would be installed, and I know a person who does this stuff down to the finest point and he found the 155 Panama gun mounts for Battery Imperial." [10]

By "plans," Overton is presumably referring to the battery's Report of Completed Works (Figure 4-53), though it is not clear this was available in 1936.

Figure 4-54 shows part of a 1 July 1945 HDSD map entitled "Fire Control Stations at Fort Emory, Site 12, sheet 9 of 15, Exhibit 6-8." Note handwritten annotations for a battery of guns numbered 1 through 4, right to left in accordance with standard Army practice. Just to the right of this unidentified battery is another handwritten annotation reading "BC STA, 50' TWR," the associated battery commander station

---

10 The person Overton is referring to here was very likely Coast Defense Study Group member Lee Guidry.

atop a 50-foot tower.[11] These hand-drawn symbols appear to correlate well with the location of Battery Imperial presented in the undated post-war map of Figure 4-55, albeit flipped in perspective.

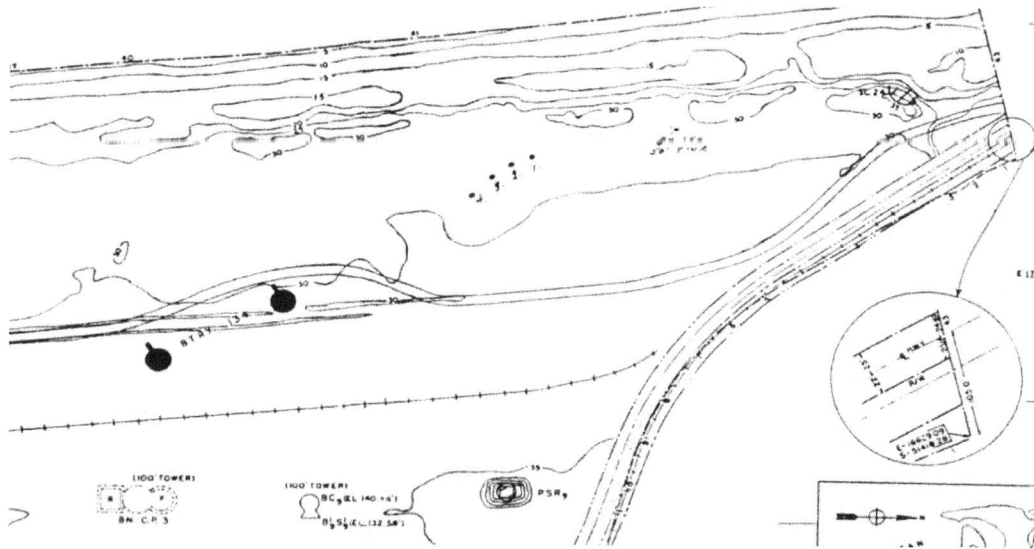

Figure 4-54. Handwritten annotations presumably indicating the location of Battery Imperial and its associated battery commander station are seen in the upper-right quadrant of the "Fire Control Stations at Fort Emory, Site 12" map of 1 July 1945, with the Pacific Ocean across the top.

In his 1992 interview with Chief Ranger Howard Overton (1993), Sergeant Major John P. Hennely of Battery C made the following curious statement about his WWII service as an observer in one of the water-tower-disguised base end stations at Fort Emory:

"We had twelve one-five-five Long Toms, GPFs. Everything was underground."

Strangely, Overton asked for no elaboration regarding the twelve versus four GPF guns.

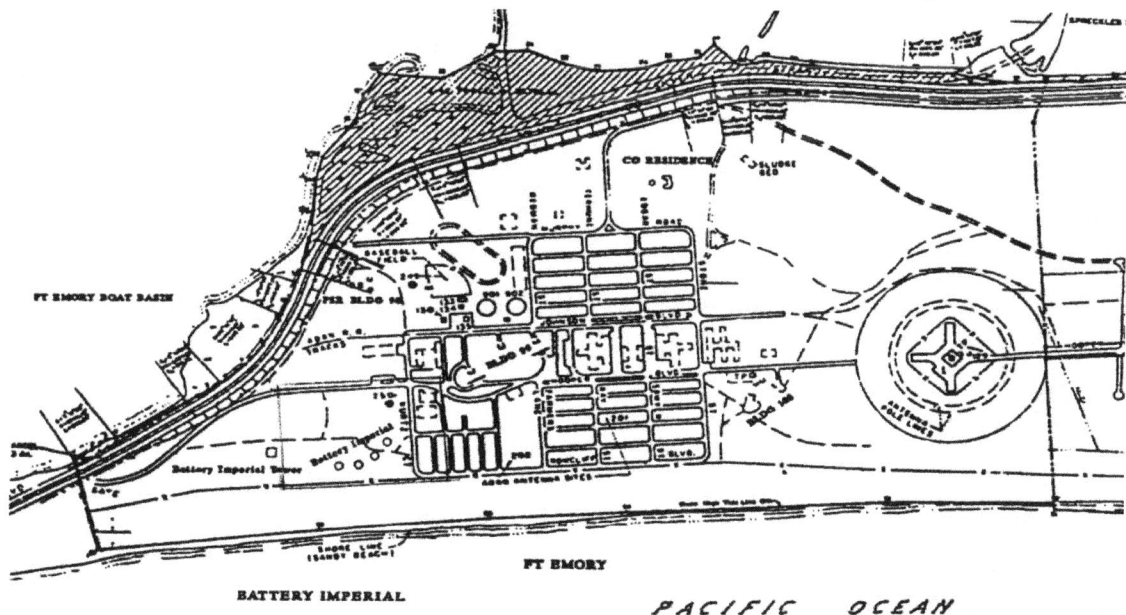

Figure 4-55. This undated post-war map of Fort Emory shows Battery Imperial at bottom left, with Battery Gatchell (Building 99) to its right and slightly above. The circular "Elephant Cage" antenna in lower-right corner was an AN/FRD-10 Wullenweber Antenna (Grobmeier, 2007), as further discussed in Chapter 5.

11 Due to the low elevation of this area, many fire control structures along the Silver Strand were mounted on such towers for extended visibility, which is restricted by the curvature of the earth.

Lee Guidry's detailed October 1992 field sketch of the remains of Battery Imperial, just west of the intersection of Kurtz Court (A Avenue) and Hooper Boulevard, is presented in Figure 4-56 below. The Panama mounts for Guns 1 and 3 were found to be partially exposed, but the emplacements for Guns 2 and 4 were buried. Note the four 21-by-56-foot concrete foundations immediately behind (east of) the battery, which possibly supported above-ground Quonset-hut berthing shelters.

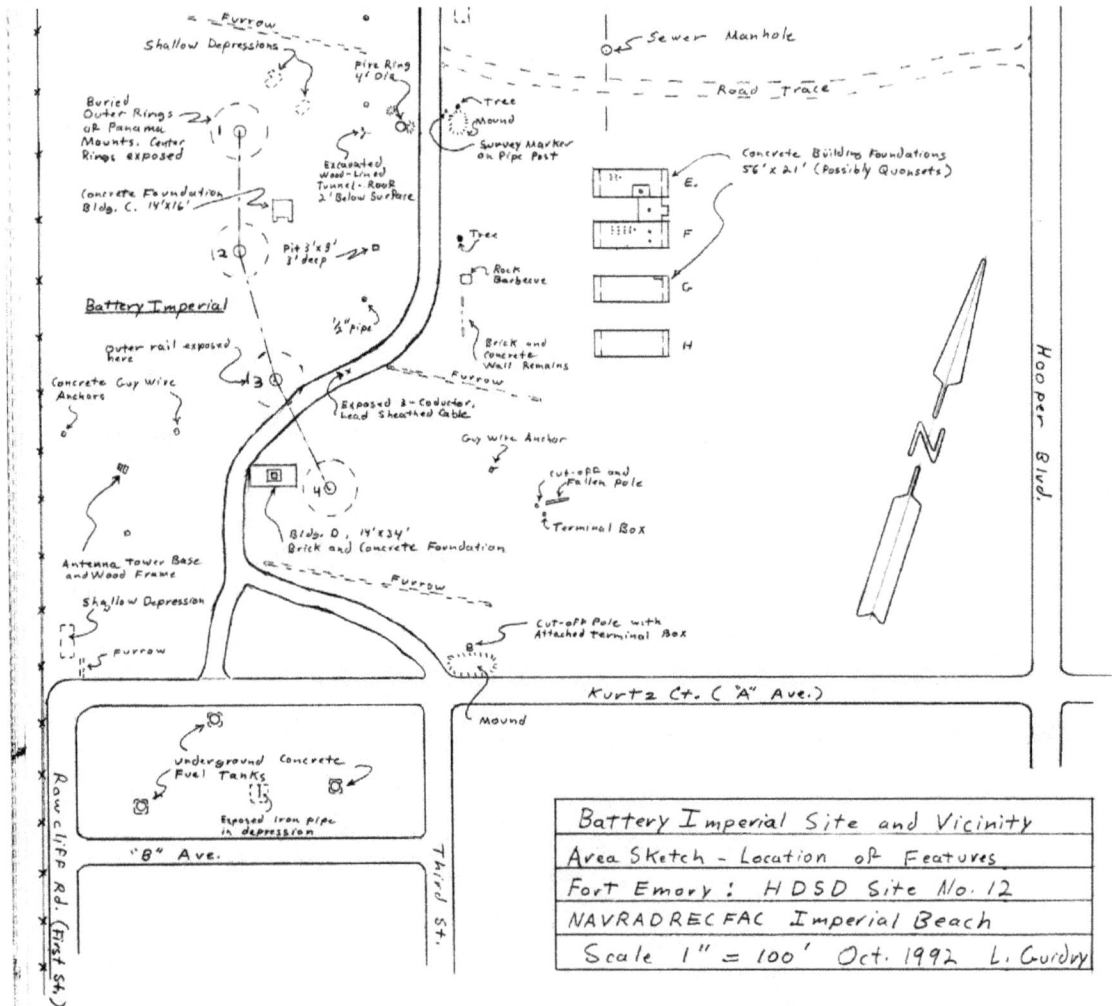

Figure 4-56. A 1992 field sketch prepared by Lee Guidry during an October 1992 site survey of Battery Imperial (upper left) at Fort Emory shows its location relative to Hooper Boulevard and Kurtz Court.

Battery Zeilin

To further augment the westerly defenses until the modernization batteries (Chapter 5) could be completed, another interim battery was constructed by troop labor at Fort Rosecrans. This 7-inch battery was named after Brigadier General Jacob Zeilin, USMC, who later became Commandant of the Marine Corps (Keniston, 1995). As with Battery Gillespie, the guns were relocated from an existing USMC practice site just north of Battery Point Loma. The following post-war historical summary was recorded (HSD, 1945):

"The Harbor Defense also obtained the loan of two 7" Navy guns from the U.S. Marine Corps. The emplacement of these was completed in July 1942 and the battery named Zeilin. This battery strengthened the density of the defense to the west, although not extending the range. It was manned for over a year."

As reported by Keniston (1995), these were 7-inch 45-caliber Mark 2 single-mount open-pedestal Navy guns (Figure 4-57), which had been repurposed for shore defenses. They had originally been employed on *Connecticut (B-18)* and *Mississippi (B-23)* class battleships in the early 1900s timeframe (NavWeaps, 2019).

Figure 4-57. The USMC's two 7-inch Navy guns repurposed for Battery Zeilin (up arrows) were originally installed just north (left) of the initial unmounted locations of the four 155-millimeter GPF guns at Battery Point Loma, as seen in this 24 January 1942 aerial photo (US Army photo courtesy SSC Pacific).

Lee Guidry's 1996 field sketch presented in Figure 4-58 below provides a section view of Battery Zeilin's southern Gun 2 emplacement, looking north. Compare the massive concrete gun block to the inscribed azimuth-instrument foundation shown earlier in Figure 4-7, which the Flower and Roth (1982) *Cultural Resource Inventory* mistook for "a cement mount for 5-inch guns," as previously discussed. For perspective, the below 7-inch gun base is 34 feet in diameter, whereas the misidentified azimuth-instrument foundation measured just 45 inches long by 37 inches wide, with a depth of 32 inches.

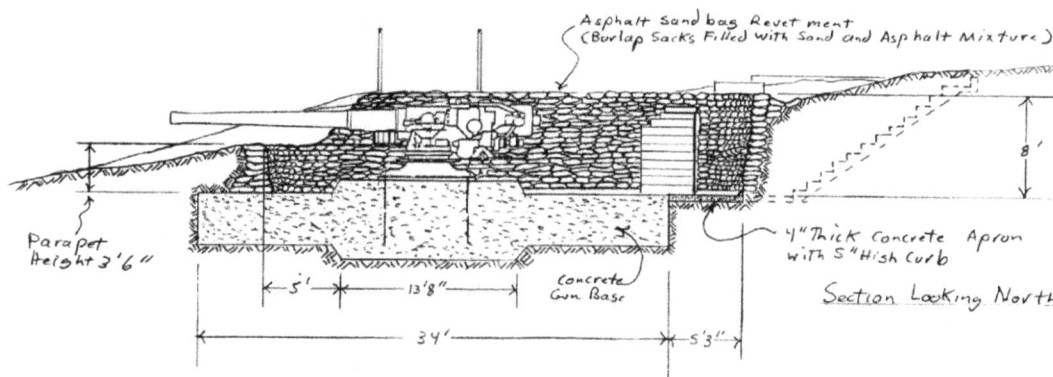

Figure 4-58. Schematic drawing of a 7-inch Mark 2 Navy gun in a sandbagged revetment for Gun 2 (1996 field sketch by Lee Guidry). Note the massive concrete gun block, and the narrow stairway at far right.

Figure 4-59 (undated) shows the asphalt sand-bagged revetments for Guns 1 and 2 of Battery Zeilin, which have both been mounted, with a large drainage gully just to the north (left). Construction on Battery Woodward has not yet begun west of (below) Woodward Road at lower right, which dates the photo somewhere between July 1942 and March 1943. Note what appears to be three defensive machine-gun emplacements about halfway between the battery and the coastline at bottom center, the pistol range on the east side of Woodward Road at top center, and the auxiliary range in the bottom-right corner.

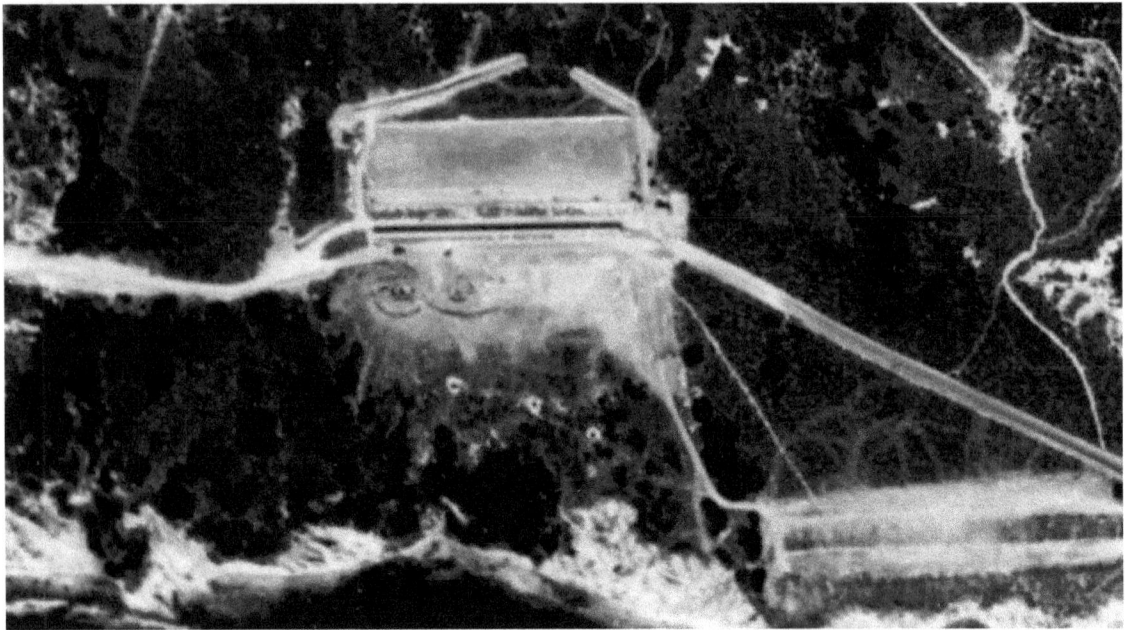

Figure 4-59. The 7-inch gun emplacements of Battery Zeilin are west of (below) Woodward Road in this undated aerial photo, directly across from the north end of the pistol range (US Army photo courtesy SSC Pacific). Note diagonal line of three defensive machine-gun positions along the cliff edge at bottom center.

The 1967 aerial photo in Figure 4-60 below shows the northern (Gun 1) emplacement for Battery Zeilin (photo center) to be still extant just west of Woodward Road at photo center. The southern (Gun 2) emplacement has been buried, with its approximate location indicated by the black up arrow at top center. The battery's Plotting and Spotting Room (PSR) is directly below Gun 1, its airshaft/skylights on either end indicated by the two circles. The roof of the latrine for Battery Zeilin can be seen in the dark shadows on the south (far) side of the drainage gully in the lower-right corner of the photo.

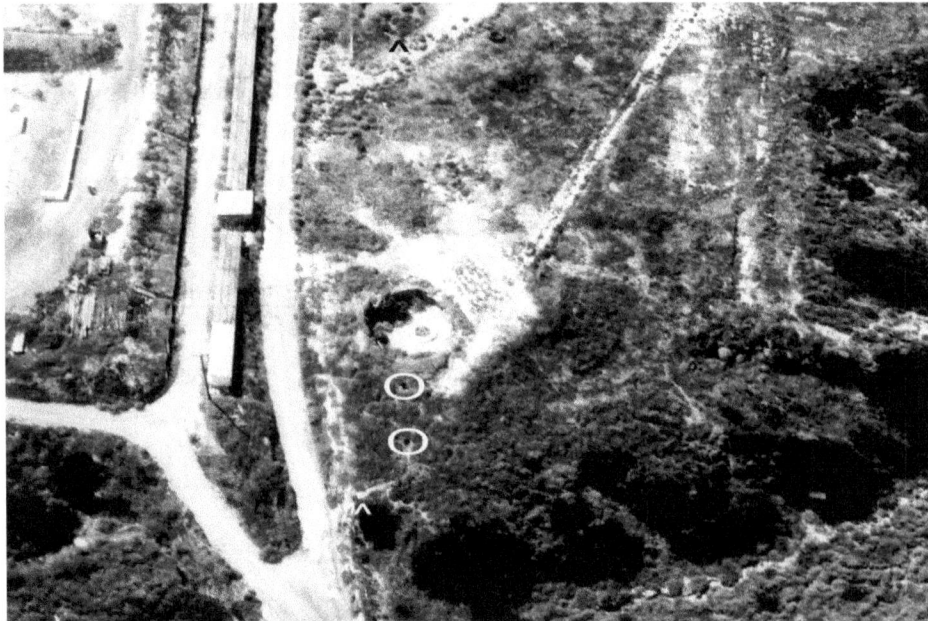

Figure 4-60. The airshafts/skylights (circled) on either end of the underground PSR are just below the Gun 1 emplacement, with the PSR access tunnel entrance (white up arrow) in the gully just west of Woodward Road, circa 1967. The buried Gun 2 position is indicated by the black up arrow at top center. Note rectangular roof of latrine structure in shadows at lower right (US Navy photo courtesy SSC Pacific).

In 1989, we built a volleyball court near the extreme lower-right corner of the above photo, and bermed off the adjacent gully to prevent erosion. When serving the ball from the west side of the court, one would be standing about 25 feet from the remains of the Battery Zeilin latrine, which by that point was completely hidden by brush. After noticing the unidentified roof seen in Figure 4-60 above, I began searching for this structure in 2008, which was located in about 2 minutes (Figure 4-61a). Excavation of the rusted urinal tray, which had originally been about 3 feet up on the west wall, confirmed it as the latrine.

Figure 4-61. a) View looking east of the collapsed latrine in April 2008. **b)** Lee Guidry is preparing a field sketch of the structure, with a section of the excavated urinal tray (dark rectangle) in the southwest corner.

As seen in Figure 4-62 below, the northern Gun 1 emplacement of Battery Zeilin was still visible in 1969. Building F-33 of the present-day Unmanned Systems Group is at lower right, with the remains of the pistol range at image left. Note the serpentine access road to the more recent pumping station for the TRANSDEC pool on the projection of land in the upper-right corner, which was the former location of the auxiliary range seen earlier in Figure 4-59. The purpose of the post-war steel tower slightly above and to the left of photo center (up arrow) will be discussed in Chapter 5.

Figure 4-62. Looking south in 1969, the concrete gun block in the northern Gun 1 emplacement is just west of Woodward Road, left of photo center (US Navy photo courtesy SSC Pacific). The twin 6-inch gun emplacements for Battery Woodward are seen at top center, with the WWII pistol range at image left.

Rear access to the underground bunker complex of Battery Zeilin was via a set of concrete stairs just west of Woodward Road, which runs north-south on the east side of the battery, as seen in Figure 4-62 above. The unusually narrow width of these stairs (Figure 4-63), and the fact that they are not centered between the two emplacements, are just two of many design oddities of this troop-built structure, as will be further discussed. All things considered, this stairway was probably intended to be more of an emergency escape route than an entryway.

Figure 4-63. The narrow concrete stairway to the underground bunker complex of Battery Zeilin is seen looking west from Woodward Road in 2006.

The stairway shown above is depicted just right of top center in the schematic plan view of Battery Zeilin presented in Figure 4-64 below. The stairs lead down to a north-south communication tunnel that connects to both emplacements, but with a curious design quirk that requires passage through the shell magazine for continued access to Gun 1. Note also the redundant tunnel connections between Gun 1 and the tunnel leading to the corrugated PSR shelter on the north side of the emplacement, possibly intended to provide alternate escape routes from this 30-foot underground bunker.

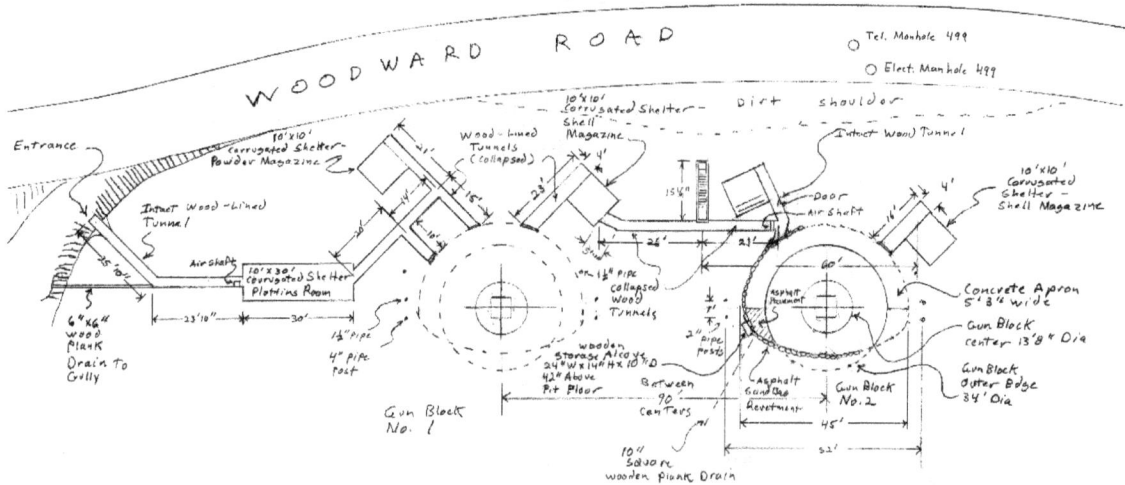

Figure 4-64. Planview layout of Battery Zeilin, with north to the left (extracted from 1996 field sketch by Lee Guidry). Note PSR entrance tunnel in the drainage gully west of Woodward Road at far left.

The narrow stair access from Woodward Road seen in Figure 4-63 leading to the tunnel layout of Figure 4-64 was clearly not used for ammunition resupply, as the tunnel did not connect to the powder magazine for Gun 1 or the shell magazine for Gun 2. Replenishment instead took place from the west (front) side of the battery. The entrance to the powder magazine for Gun 1 is at background right in Figure 4-65, with that for the shell magazine out of image further to the right. Note the three white support poles for camouflage netting at background left.

Figure 4-65. The northern Gun 1 emplacement as it appeared in 1972, with tunnel entrances to the PSR at background center and to the Gun 1 powder magazine at background right (photo courtesy Al Grobmeier). This emplacement remained accessible until covered over sometime in the mid-1980s.

The 1995 Keniston Survey reported both Battery Zeilin gun emplacements had been covered over when the entire area was apparently bulldozed. During this survey, Lee Guidry located and determined that the concrete gun blocks for both emplacements were buried some 4 feet below surface grade. A partial excavation of the gun block at the southern Gun 2 emplacement was conducted, exposing several rusted 1&7/8-inch bolts (Keniston, 1995), as seen in Figure 4-66. This exploratory dig was filled in afterwards and both emplacements have remained buried ever since.

Figure 4-66. Exposed mounting-plate bolts on the excavated gun block of the southern Gun 2 emplacement (July 1995 photo courtesy Al Grobmeier).

Lee Guidry's field sketch presented in Figure 4-67 below shows an expanded view of the Gun 2 emplacement layout, with the narrow stairs west of Woodward Road, previously seen in Figure 4-64, depicted at upper right. These stairs lead down to the wood-lined north-south communication tunnel that runs to both emplacements, intersecting the short service tunnel between Gun 2 and its powder magazine to the east. Note the air shaft to the surface directly beneath this point, and the 7-inch Navy gun on the massive concrete gun block.

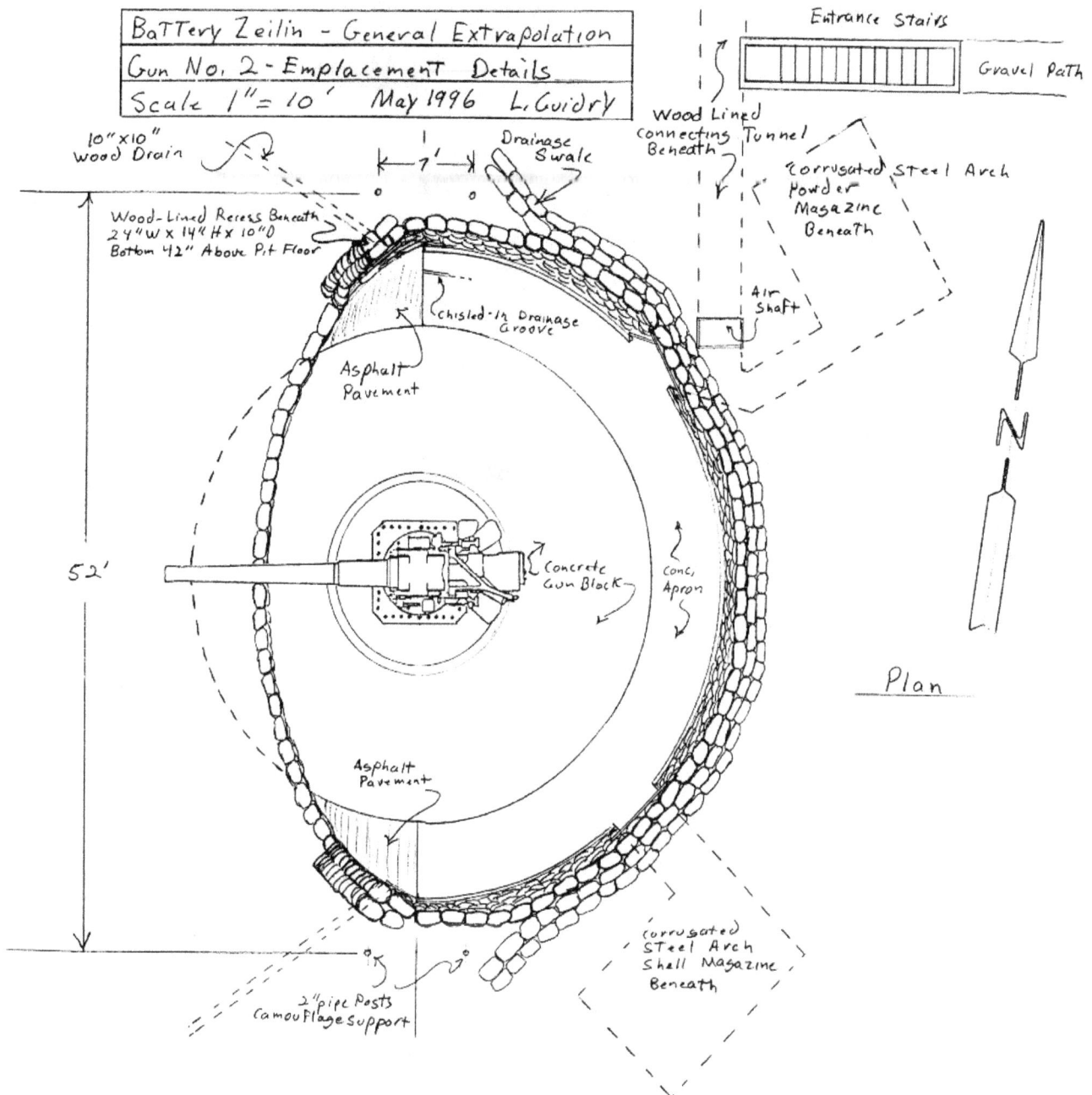

The sketch contains the following handwritten labels:

Battery Zeilin - General Extrapolation
Gun No. 2 - Emplacement Details
Scale 1" = 10'   May 1996   L. Guidry

10" x 10" Wood Drain

Drainage Swale

Entrance Stairs

Gravel Path

Wood Lined Connecting Tunnel Beneath

Corrugated Steel Arch Powder Magazine Beneath

Wood-Lined Recess Beneath 24"W x 14"H x 10"D Bottom 42" Above Pit Floor

Chiseled-In Drainage Groove

Air Shaft

Asphalt Pavement

52'

Concrete Gun Block

Conc. Apron

Plan

N

Asphalt Pavement

Corrugated Steel Arch Shell Magazine Beneath

2" Pipe Posts Camouflage Support

Figure 4-67. Plan view of the Number 2 (southern) gun emplacement (1996 field sketch by Lee Guidry).

A surviving 7-inch Mark 2 Navy gun, which during WWII had been mounted at Battery Harbor on the Sand Island Military Reservation, Harbor Defenses of Honolulu, is on display at Battery Randolph, Fort DeRussy, Oahu, HI (Figure 4-68). A comparison of the bolt pattern of the square base plate in this photo with that shown in Figure 4-67 above suggests this is the same 7-inch Navy gun as those once mounted at Battery Zeilin.

Figure 4-68. Detail of the pedestal baseplate and mounting bolts on a 7-inch Mark-2 Navy gun at Battery Randolph, Fort DeRussy, Oahu, HI, circa April 1993 (photo courtesy Lee Guidry).

The interior of the powder magazine for Gun 2 is presented in Figure 4-69a below, circa 1992, with an improvised wooden storage pallet seen in the west corner. Figure 4-69b shows the connecting tunnel that leads to the narrow stairway off Woodward Road, looking southwest from inside the Gun 2 powder magazine. The hinged door with the small window (up arrow) on the left side of this tunnel leads to the Gun 2 emplacement (see again Figure 4-67). Note sunlight entering through the collapsed tunnel roof at top center, which provided easy access to the magazine.

a)   b)

Figure 4-69. a) Improvised pallet in the southern (Gun 2) powder magazine as seen looking west in 1992. **b)** View of the connecting tunnel to the narrow stairway, looking southwest from the Gun 2 powder magazine. Note small window (up arrow) in the door leading to the Gun 2 emplacement (photos courtesy Lee Guidry).

Referring as needed to Figure 4-67 for reference, Figure 4-70a shows a portion of the short tunnel that leads from the Gun 2 powder magazine towards the doorway leading to the Gun 2 emplacement, looking southwest. The spillage on the right side of this image is from the collapsed airshaft and tunnel section that ran north to the entrance stairs and beyond to the Gun 1 emplacement. Figure 4-70b provides a further look beyond this door (previously seen at photo left in Figure 4-70a) towards the former opening to the Gun 2 emplacement, which has been blocked off from the outside with fill dirt.

a)   b)

Figure 4-70. a) Tunnel section leading from powder magazine (portion of arch visible at lower right) towards doorway to Gun 2 emplacement, looking southwest in 1992 (see Figure 4-67). **b)** View of tunnel section beyond the doorway (note window opening in door at left) leading to Gun 2 (photos courtesy Lee Guidry).

The interior of the fully exposed Gun 2 shell magazine at the southern end of Battery Zeilin is presented in Figure 4-71, looking southwest in May 2007. The wooden rear wall of this structure has completely collapsed, with the resulting spillage nearly filling the back half of the 10-foot corrugated-roof shelter. Similarly, the front wall and associated short tunnel leading to the Gun 2 emplacement have also collapsed. As there appeared to be several underground beehives in this area, I did not conduct a comprehensive search on this follow-up visit in May 2007.

Figure 4-71. Interior view looking southwest of the southern Gun 2 shell magazine, showing spillage from the collapsed rear wall, circa May 2007. The powder and shell magazines for the northern Gun 1 emplacement remain completely buried from the bulldozing and were therefore inaccessible.

Each of Battery Zeilin's four corrugated-metal shelters for powder and shells measured 10 feet by 10 feet, while the plotting and spotting room (PSR) bunker was 10 feet by 30 feet. The recently revealed south entrance to this latter structure, following heavy rains in the winter of 2005, is shown in Figure 4-72 below. The exposed lip of the elephant-iron roof of the PSR was approximately 12 inches below ground level, with a small gap above the fill-dirt spillage inside. I enlarged this opening a bit with a shovel to facilitate access.

Figure 4-72. The Multi-Plate-Arch roof section over the south end of the PSR shelter for Battery Zeilin as found in May 2007, exposed by a sinkhole from heavy rains. A similar sinkhole at the north end of the bunker was less accessible due to dense brush and a collapsed air shaft.

The PSR interior had been photographed by Lee Guidry in 1995 (Figure 4-73), but I didn't meet him until sometime in mid-2008, so I had no idea what to expect upon entry in January of that same year. The left side of the wooden ladder in the southern airshaft/skylight structure is barely visible through the doorway at background center in Guidry's photo, behind which is the connecting tunnel to the northern Gun 1 emplacement. Note the unusual door-frame construction to accommodate the overhead-arch curvature, since the doorway for some reason was not centered. Various field-telephone cables are looped over the top of the door, with chunks of asphalt from the bulldozed Gun 1 emplacement seen in the spillage.

Figure 4-73. Interior of the PSR as seen facing south in 1995. Note the camera-flash illumination of an extensive spider-web network along the Multi-Plate-arch overhead at right, and the unusual slanted door frame leading to the airshaft/skylight structure in background (photo courtesy Lee Guidry).

As there were quite a few beehives and yellow-jacket nests in and around the Battery Zeilin complex, not to mention rattlesnakes and spiders, it seemed advisable to use a robot to remotely inspect the PSR interior before entry. This option also provided good training for our *Unmanned Systems Combat Support Team*, which periodically deployed to Iraq and Afghanistan. Figure 4-74 shows FC1 Mendell Baker, FC2 James Galvan, and YN1 Christina Reed of the Center's Unmanned Systems Reserve Unit setting up a video-equipped Inukton robot for bunker reconnaissance. Once we had remotely viewed the interior, I slithered down through the sinkhole for a first-hand look.

Figure 4-74. a) FC1 Baker (foreground left), FC2 Galvan (background), and YN1 Reed (right) of our Unmanned Systems Reserve Unit prepare a tethered Inukton robot for bunker insertion in January 2008. **b)** The teleoperated robot approaches the sinkhole leading to the exposed south entrance of the PSR.

After I swept the bunker overhead clear of cobwebs, YN1 Christina Reed, who is deathly afraid of spiders (I'm no fan myself), bravely joined me via the southern sinkhole (Figure 4-75a). The extension cord seen to her left provided power to a pair of halogen work lights from a portable generator above. Note the wooden debris from the doorframe seen earlier in Figure 4-73, which likely collapsed when the fill dirt used to seal the airshaft/skylight structure progressively settled during years of rain. Figure 4-75b shows a much relieved YN1 Reed safely back on the surface.

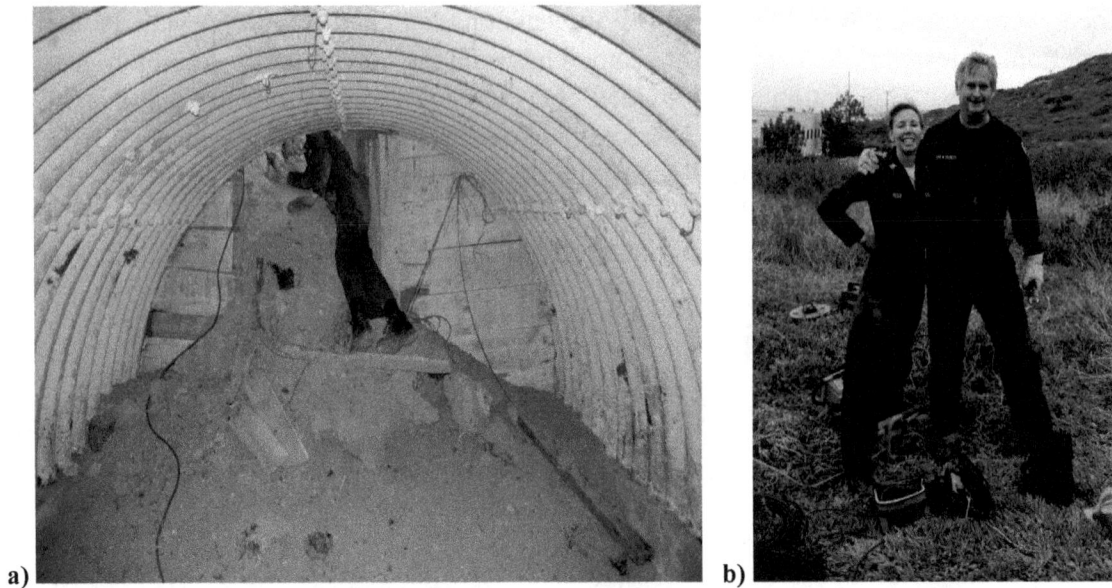

Figure 4-75. a) YN1 Christina Reed of our Unmanned Systems Reserve Unit squeezes through the narrow opening at the south end of the PSR, with the collapsed doorframe debris and attached field-telephone cables beneath her feet. **b)** Congratulating YN1 Reed on temporarily overcoming her fear of spiders (January 2008).

Figure 4-76a below provides a closer look at the south end of the PSR bunker, with the overhead sun illuminating the enlarged sinkhole in the upper-left corner that enabled our entry. Apart from the collapsed field-telephone lines, the end bulkhead on the right is little changed from that seen in Guidry's 1995 photo presented earlier in Figure 4-73. Note the painted inscription "B.C. Order" (Battery Commander Order) just left of the field wire. Another inscription, "COR. RANGE" (Corrected Range), was found on the eastern wall as seen in Figure 4-76b.

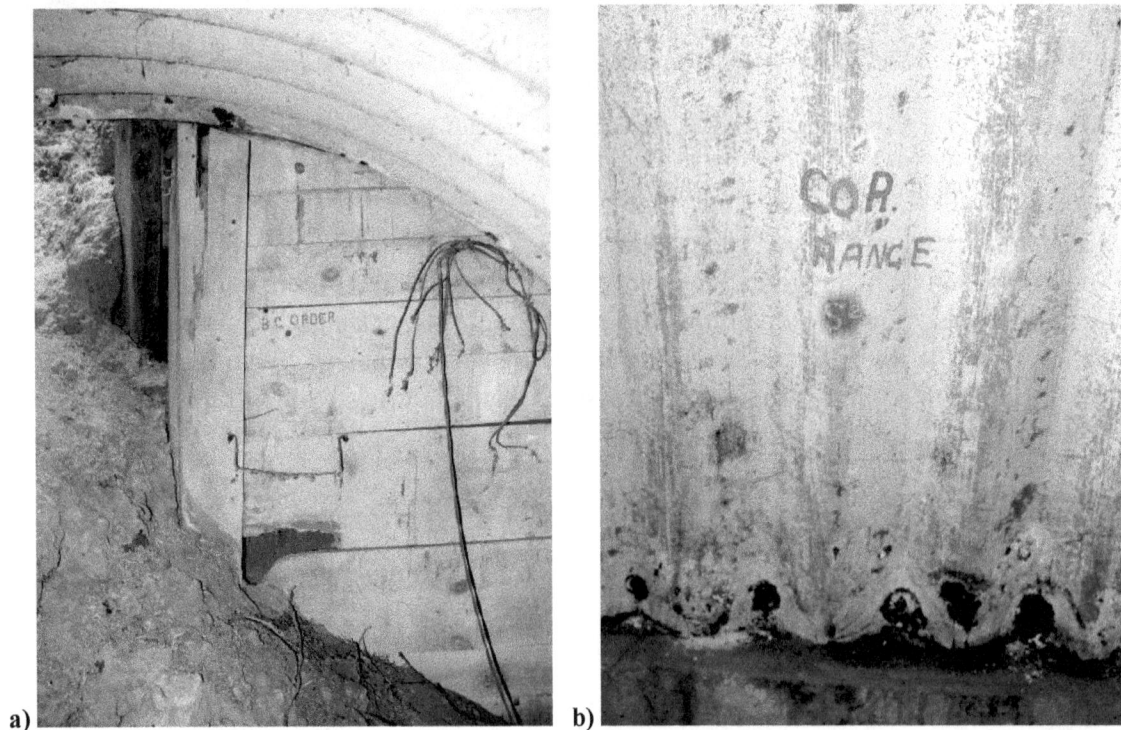

Figure 4-76. a) The field-telephone lines previously seen over the doorway in Figure 4-73 are partially buried in the spillage below. **b)** The painted inscription "COR. RANGE" (corrected range) was found at the approximate horizontal midpoint of the eastern wall, roughly 18 inches above the concrete curb.

Figure 4-77 shows the well-preserved PSR interior looking north towards the remains of the collapsed combination skylight/airshaft/escape structure leading up to the surface in January 2008. A similar configuration had once been located at the southern end of the PSR, but was apparently destroyed at some point after the war when the bunker was sealed off with fill dirt, which spilled into the structure at both ends. This latter feature was apparently not extant during Guidry's site survey of Battery Zeilin in 1995, as there is no mention of it on either of his meticulous field sketches presented earlier in Figure 4-64 and Figure 4-67.

Figure 4-77. Interior view of the PSR looking north towards the partially collapsed skylight/airshaft, showing the concrete floor and foundation curbs supporting the arched overlapping Quonset-like sections of "Elephant Iron" that formed the roof. There was no sign of any electrical lighting or outlets in this bunker.

A detailed plan view of the PSR layout and connecting tunnels is seen in Figure 4-78 below. The location of the collapsed skylight/airshaft structure is documented on the north end, but there is no indication of the similar configuration that had once been at the south end. Note Guidry's meticulous annotation of the inscription locations along the western side of the corrugated shelter, one of which was shown previously in Figure 4-76b. The gully entrance off Woodward Road to the northern PSR access tunnel is seen at upper left.

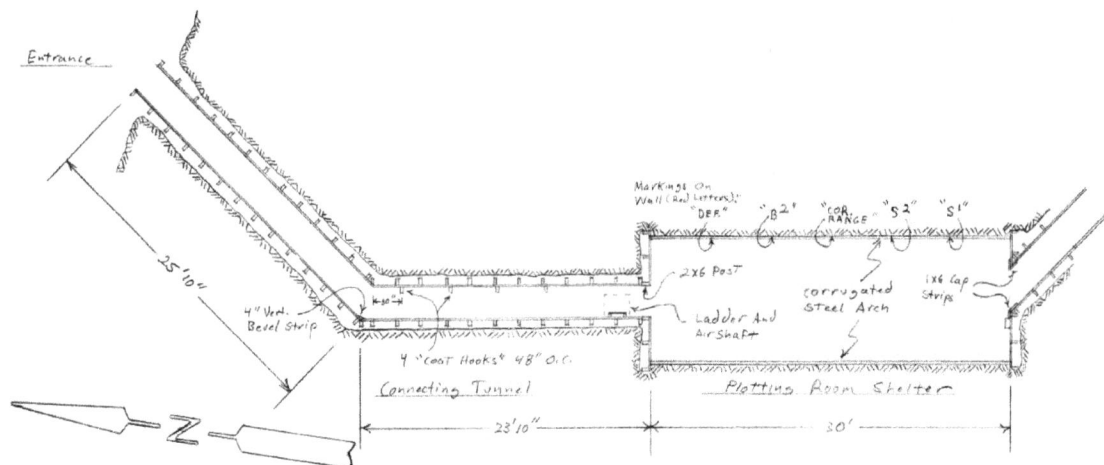

Figure 4-78. Plan view of the PSR and associated connecting tunnels. The tunnel on the right (south) end leads back to the Gun 1 emplacement, while the tunnel on the left leads to the gully entrance just west of Woodward Road, as seen earlier in Figure 4-64 (1995 field sketch by Lee Guidry).

A close-up view of the wooden skylight/airshaft structure in trhe access tunnel at the north end of the PSR is presented in Figure 4-79a, the walls and roof of which were made of wood, with a floor of concrete as seen in Figure 4-79b. Note the upper rung of the escape ladder at ground level on the left side of the image, the rest of which has been completely buried in fill dirt sometime after the war. The first section of the adjoining PSR access tunnel runs north from this skylight for about 23 feet (Figure 4-80a), as previously shown in the plan view of Figure 4-78.

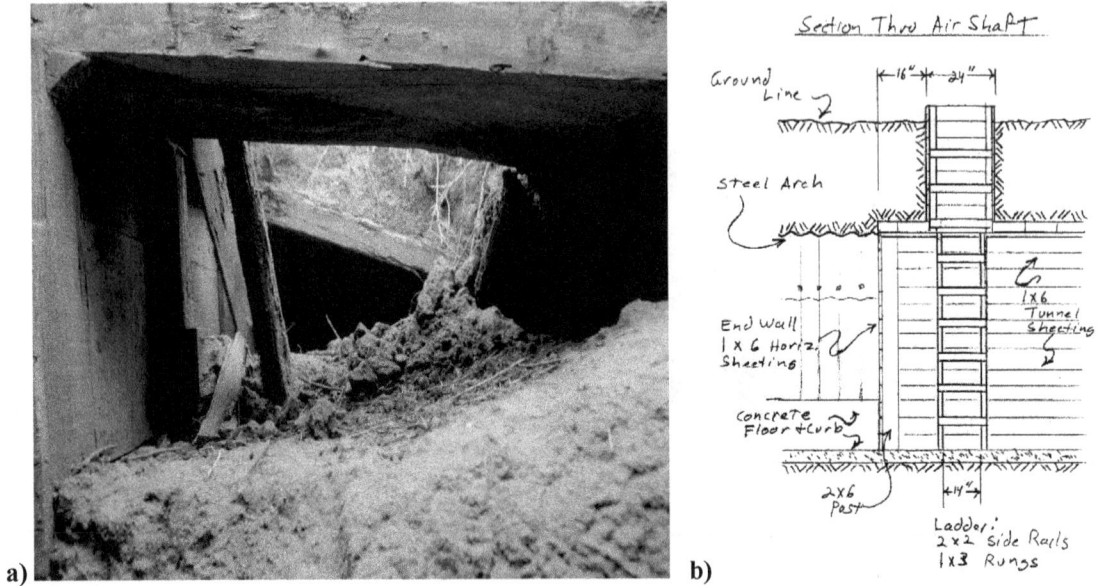

Figure 4-79. a) View of the wooden skylight remains as seen from the north end of the PSR in January 2008, beyond which is part of the collapsed tunnel that provided an underground access from the nearby gully. **b)** Section view of the wooden skylight/ladder/airshaft at the north end (field sketch by Lee Guidry).

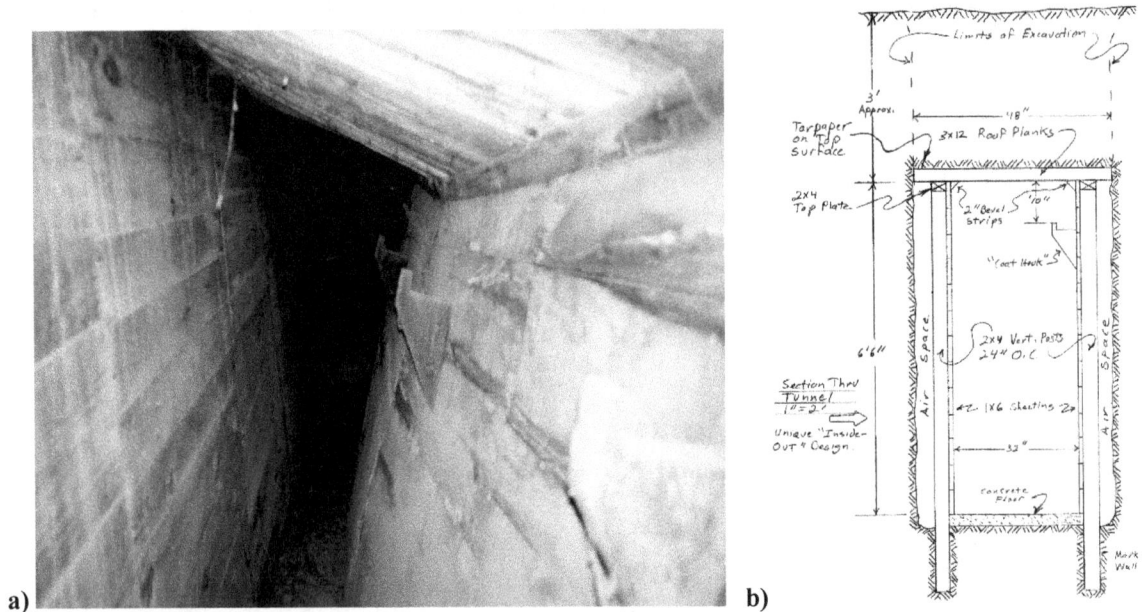

Figure 4-80. a) View of the PSR access-tunnel interior looking north from beneath the skylight in 1992 (photo courtesy Lee Guidry). Note the pair of wooden coat hooks on the displaced east wall just right of photo center. **b)** Section view of the tunnel construction (extracted from field sketch by Lee Guidry).

As seen in Figure 4-80b above, there was originally an air space between the earthen tunnel sides and the wood-framed tunnel structure within, which was never expected to last more than a few years.[12] Eventually the surrounding soil settled into this void, and the east wall was ultimately pushed inward due to downslope pressure. Construction details of the approximately 75-degree turn leading to the final leg of the access tunnel are seen in Figure 4-81 below. The overhead 3x12 roof planking is radially reinforced below by a number of wooden 4x4 beams attached to the 2x4 top plates of the tunnel walls.

Figure 4-81. Construction details of the access-tunnel overhead, where the north leg turns approximately 75 degrees easterly in the direction of Woodward Road (1995 photo courtesy Lee Guidry). Figure 4-82 below shows this location from above in 2007.

Figure 4-82. Top view of the north-to-east turn in the now collapsed PSR access tunnel, circa May 2007.

---

12 Completed in July 1942, Battery Zeilin was superseded by Battery Woodward in August 1944.

Figure 4-83a shows the final leg of the access tunnel leading to the drainage gully. As seen earlier in Figure 4-78, this section originally exited on the south side of the gully west of Woodward Road, but the road has since been widened to where the tunnel now ends right at the road shoulder (Figure 4-83). As the elevation of the roadbed is somewhat higher than that of the nearby gully, however, only a small upper portion of the tunnel is exposed. Shrouded in heavy brush, its location remains unnoticed, despite this stretch of road's popularity with hikers and joggers during lunch hour.

a)   b)

Figure 4-83. **a)** View of partially collapsed tunnel segment looking northeast towards the entrance on Woodward Road in 1992 (photo courtesy Lee Guidry). **b)** Remains of the tunnel entrance on the west side of Woodward Road as it appeared in December 2007.

5 Modernization Batteries

> "There are more things in heaven and earth than are dreamed of in our present-day science. And we shall only find out what they are if we go out and look for them."
>
> *Freeman Dyson*

The new 6-inch and 16-inch modernization batteries at Fort Rosecrans and Fort Emory began replacing the temporary batteries discussed in Chapter 4 in late 1943. The following discussion is ordered in accordance with battery location, north to south, starting at Fort Rosecrans on the Point Loma Peninsula.

Battery Construction Number 237 (Battery Woodward)

Construction began in March 1943 on Battery Construction Number 237 (Figure 5-1), which replaced the temporary Batteries Gillespie and Zeilin when proof fired in November of that same year (HSD, 1945). The battery was to be named for Colonel Charles Woodward, who commanded Fort Rosecrans in the early 1900s. The name was never officially bestowed by a Army General Order, but the Navy continued to refer to this battery as "Battery Woodward" once they took over the reservation. This 200-series battery was officially transferred to the Coast Artillery on 31 August 1944. The *M1903A2* 6-inch guns manufactured by Watervliet Arsenal were mounted on shielded *M4* barbette carriages built at Watertown Arsenal (Thompson, 1991), as shown in Figure 5-2.

Figure 5-1. The empty gun pits can be seen in this March 1944 aerial view of Battery Woodward prior to installation of its 6-inch guns. Note dark shadows from the morning sun rising in the east (US Army photo courtesy SSC Pacific).

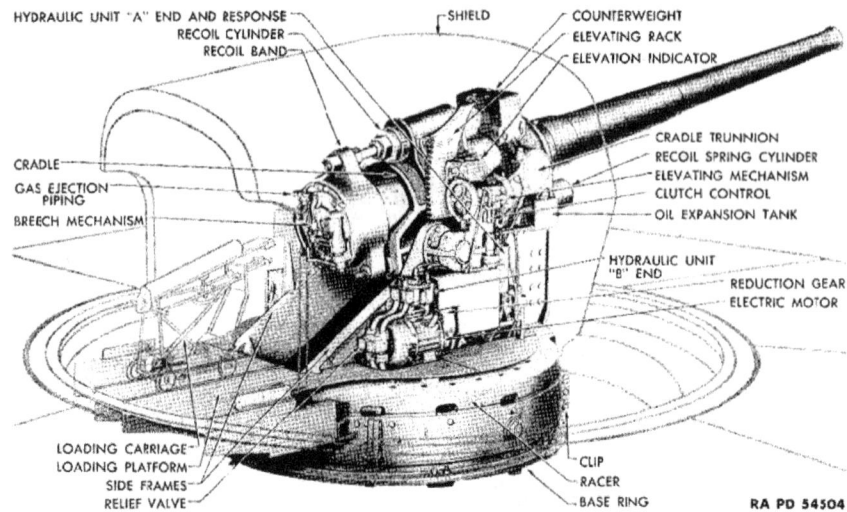

Figure 5-2. *M1903A2* 6-inch gun mounted on a shielded barbette carriage *M4* (US Army diagram adapted from *TM 9-428*, 1943). The gun was manually trained in azimuth and elevation as directed by the *M7* data-transmission system.

The 6-inch *M1903A2* gun employed separately loaded ammunition, either a 90-pound high-explosive or a 105-pound armor-piercing projectile, respectively propelled by a single 32- or 37-pound powder bag. Each gun tube was 25 feet long, weighed about 20,700 pounds, and had a maximum range of around 27,500 yards (15.6 miles). The gun was normally fired electrically, with a back-up magneto circuit in the event of power loss, but could also be fired with a lanyard. Two of these types of guns and carriages are currently emplaced at Fort Columbia, Washington (Figure 5-3).

Figure 5-3. Typical 200-Series 6-inch gun on a shielded barbette carriage at Battery Construction No. 247, Fort Columbia, WA (photo courtesy Fort Columbia Historical State Park).

An automatic gas-ejection system was provided to clear the barrel of any smoldering debris after firing, thereby eliminating the need to swab the powder chamber between rounds. Compressed air was plumbed up through the center of the base ring and racer, through the trunnion on the left side of the gun cradle, then along the gun cradle to the breech. The air valve was activated by the tripper on the breechblock carrier whenever the breech was opened, forcing 150-psi air through three drilled holes in the breech recess. The valve was then manually closed by an operator as the projectile was being rammed.

According to the Report of Completed Works (RCW), the plotting and spotting room of Battery Woodward was air conditioned, unlike the other Harbor Defense San Diego 200-series Batteries Humphreys and Grant. The layout of Battery Woodward seen below in Figure 5-4 is nearly identical to that of the 6-inch Battery Humphreys (Construction No. 238) across the front corridor, but varies slightly in terms of the physical dimensions of the power room and the plotting and spotting room. There is also a separate muffler gallery adjacent to the Power Room that is not incorporated into the Battery Humphreys design.

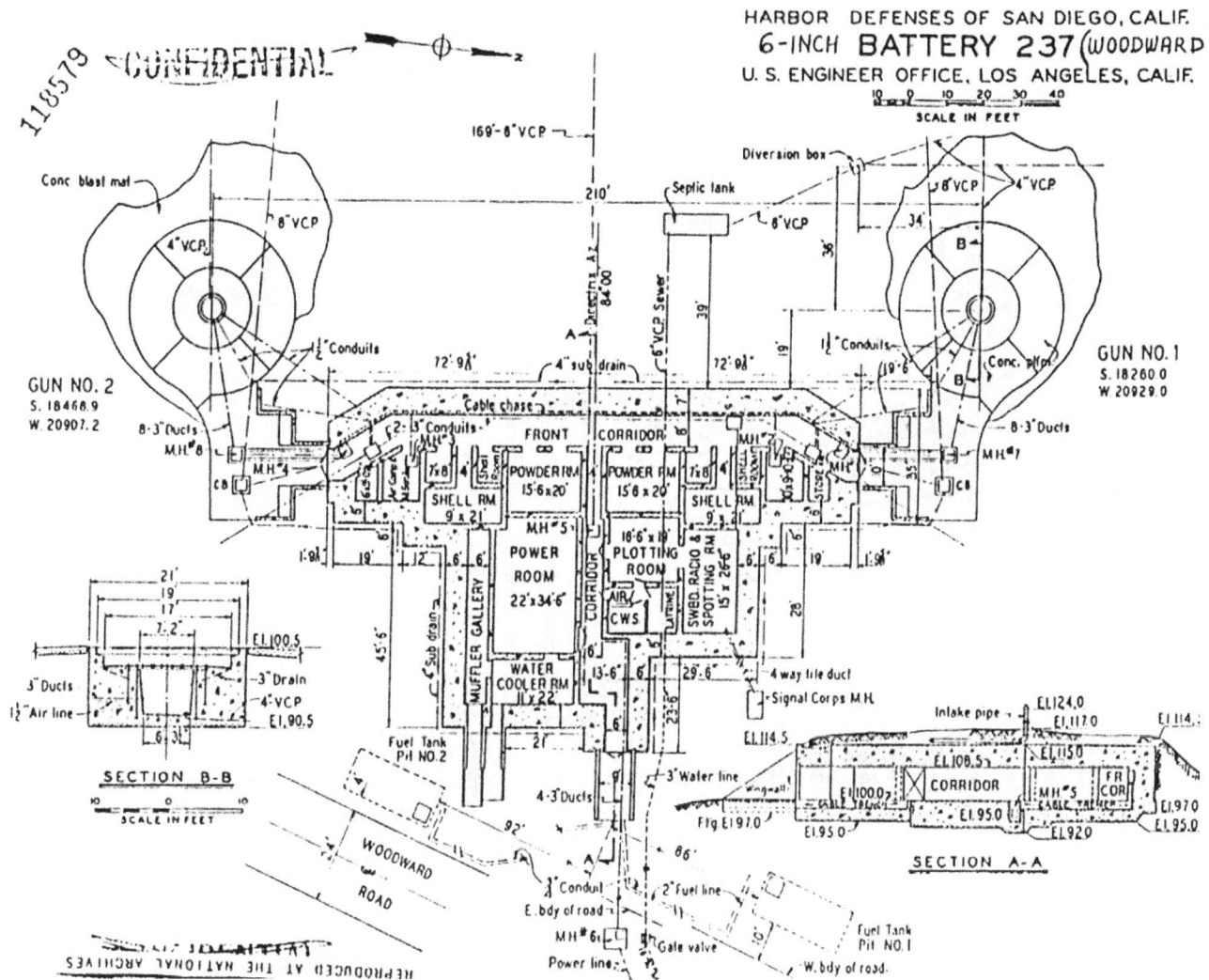

Figure 5-4. Construction details of 6-inch Battery Woodward, completed in November, 1943 (NARA). Note enlarged "Switchboard, Radio, and Spotting Room" relative to the "Spotting Room" shown later for 6-inch Battery Humphreys (Construction No. 238) in Figure 5-72.

The Gun 1 emplacement of Battery Woodward can be seen in the upper-left corner of the undated wartime aerial photo presented in Figure 5-5. Emplacements 1 and 2 for the temporary Battery Zeilin, its two 7-inch guns no longer mounted, are just across Woodward Road opposite the north end of the Pistol Range at top center. The 12-inch-mortar Battery Whistler is partially visible in the extreme lower-right corner, with the Topside base chapel seen at bottom center. The standby Harbor Defense Command Post is on the far side of Catalina Boulevard across from Battery Whistler (bottom-right corner).

Figure 5-5. This undated photo shows Gun 1 installed at Battery Woodward (up arrow) at upper left, looking west, with Gun 2 out of image (US Army photo courtesy Cabrillo National Monument). Note pistol range this side of Woodward Road at top center, across from the two Battery Zeilin emplacements (down arrows).

The reduced west-coast threat following the resounding Japanese defeat at Midway in June 1942 is reflected in the recollections of Fort Rosecrans veteran Verner B. Sorensen (Battery E), who served at Battery Woodward, shown in fully armed in Figure 5-6 below (Overton, 1995):

"Towards the end of the time we were there, we were putting in the 6-inch guns. That must have been in '43. I was assisting some others put the guns in, which consisted mainly in cleaning the parts up and putting the gun together. About the time we were ready to settle in there they decided to ship the entire battery out as a unit and go into field artillery... We left in April of '44."

Figure 5-6. Aerial view of Battery Woodward looking east in March 1944, showing both guns turned in towards each other at bottom right, just west of (below) Woodward Road, with the two abandoned emplacements for Battery Zeilin below the pistol range at bottom left (US Army photo courtesy SSC Pacific).

Figure 5-7 shows the northern Gun 1 emplacement of Battery Woodward as seen in May 2006. The shielded *M4* barbette carriage and 6-inch *M1903* gun have long since been removed, with the mount filled in and paved over with asphalt. The ready-projectile and powder-storage shelves are on either side of the north entrance to the transverse corridor at background left, with the Pacific Ocean out of image to the right. The purpose of the more recent improvised black wooden structure atop the ready-service shelf on the right side of the north entrance is unknown.

Figure 5-7. The northern Gun 1 emplacement looking southeast in May 2006, with the north bunker entrance to the transverse corridor at background left.

The remains of the southern (Gun 2) emplacement for Battery Woodward are seen in Figure 5-8 below, circa May 2006. The wooden stairs leading to the various antennae towers atop the bunker were constructed by a Cold War tenant and are no longer in use. Note the San Pedro tower along the ridgeline just left of center, and the more recently added floodlight pole in the upper-right corner.

Figure 5-8. The southern Gun 2 emplacement is seen circa May 2006, with the San Pedro tower along the ridgeline in background. The stairs leading to the antennae on top of the bunker are a post-war addition.

A close-up view of the southern entrance to the transverse corridor of Battery Woodward is seen in Figure 5-9a, circa May 2006. Note the post-war wooden storage locker added to the ready-service ammunition shelf at far left, which is further discussed later in this section. Our Unmanned Systems Group made extensive use of this facility for several years in the testing of indoor robotic solutions for various military applications. Figure 5-9b shows an autonomous robot equipped with a 360-degree laser scanner entering the battery to explore and map the interior with no human assistance.

a)   b)

Figure 5-9. a) Close up of the southern entrance to Battery Woodward in May 2006, flanked by ready projectile and powder shelves. **b)** An iRobot *PackBot*, upgraded with our autonomy hardware/software package, enters the bunker for autonomous exploration and mapping tests in 2011.

We also used Battery Woodward for testing the indoor security robots developed in collaboration with Cybermotion, Inc.,[1] for the US Army *Mobile Detection Assessment Response System (MDARS)* program (Laird & Everett, 1999). Unlike the fully autonomous robot shown in Figure 5-b above, which created and maintained its own world model, the semi-autonomous commercial robots seen in Figure 5-10 required an *a priori* human-generated map. The robots then used this map of potential route segments to execute 24-hour security patrols within the bunker, returning as needed to their respective battery chargers. Following this feasibility testing, production versions were installed for further evaluation in a large Army warehouse in Anniston, AL.

---

1   Cybermotion, a small business founded by John Holland in Roanoke, VA, initially marketed material-handling robots before switching their focus to semi-autonomous security robots.

Figure 5-10. Semi-autonomous security robots developed for the *Mobile Detection Assessment Response System (MDARS)* program automatically recharge in Battery Woodward's former shell room for Gun 2, circa 2012. These robots patrolled pre-determined routes and could detect fire, smoke, or human presence.

The most notable feature of Battery Woodward is the pristine condition of the power room, still outfitted with its original equipment, which was apparently retained to provide back-up AC power for various post-war tenants. Three Worthington diesel engines mated with three Westinghouse AC generators are seen in Figure 5-11 below. The following information is listed on each of the generator nameplates, dated June 1943: "Westinghouse Electric and Manufacturing Company, 126 KVA, 460 Volts, 157 Amps, 80% Power Factor, 3-Phase, 60 Cycles." The electrical switchboard can just be seen along the western bulkhead in the upper-right corner.

Figure 5-11. As seen in December 2007, the power room still contains the original three Worthington diesel engines (background) coupled to Westinghouse AC generators (foreground). Note the three exhaust lines that penetrate the back bulkhead into the muffler alley, and the chain-hoist maintenance rails above each unit.

The various electrical panels of the switchboard are shown in Figure 5-12a. On the left (south) end are the three generator control panels, labeled "1," "2," and "3." To the right of these are the two mount-feeder panels that supplied the north and south gun mounts (see also Figure 5-13). The final panel at far right provided electrical distribution to all other circuits throughout the battery, such as lighting, receptacles, and miscellaneous support equipment (e.g., pumps, fans, air compressors, communications, and plotting gear, etc.) The inscription at bottom left in Figure 5-12b indicates a cannibalized breaker had been repurposed for use in the power room of Battery Humphreys in 1977.

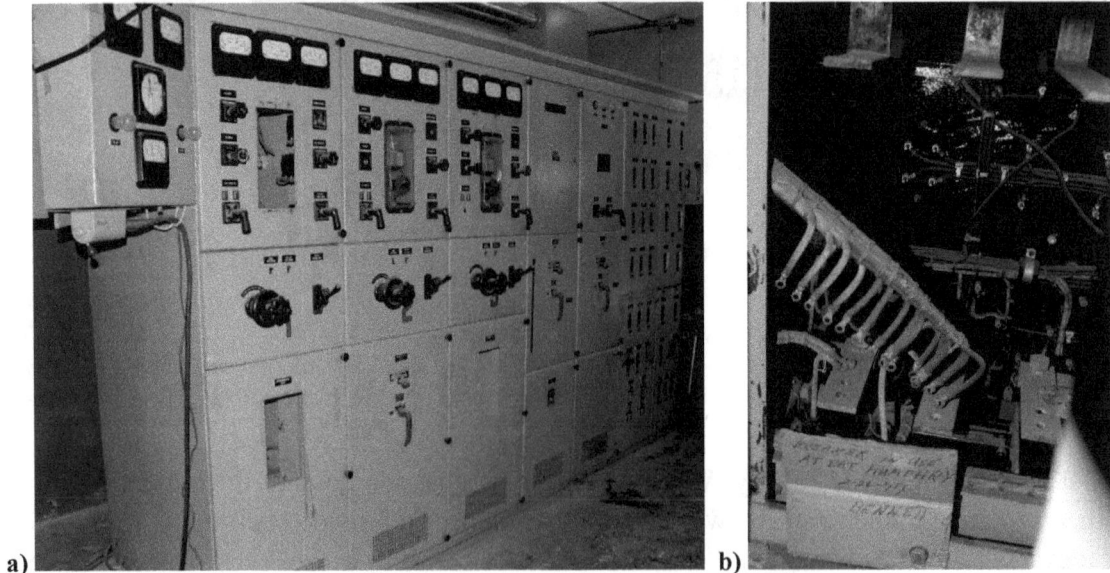

Figure 5-12. a) Front view of the switchboard along the west wall, with the three generator-control panels on the left, and the two "mount feeder" panels to their right. b) Rear view of switchboard with a note indicating a cannibalized breaker had been repurposed for Battery Humphreys on 26 February 1977.

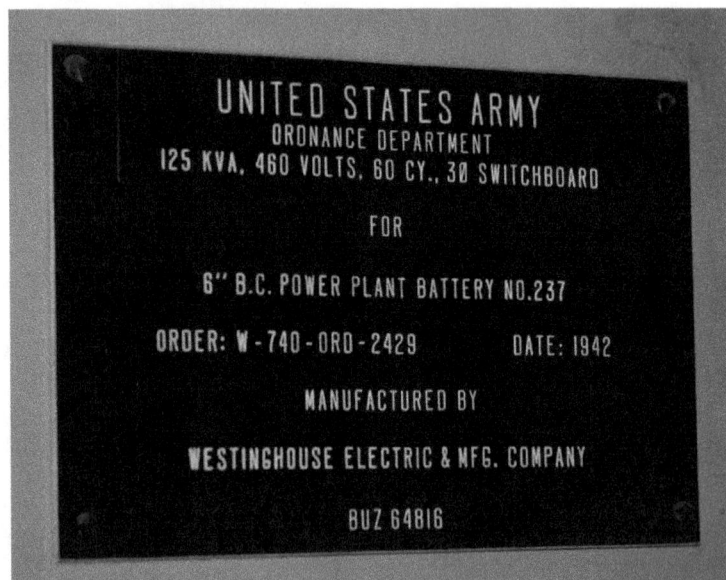

UNITED STATES ARMY
ORDNANCE DEPARTMENT
125 KVA, 460 VOLTS, 60 CY., 3Ø SWITCHBOARD

FOR

6" B.C. POWER PLANT BATTERY NO.237

ORDER: W-740-ORD-2429　　　　DATE: 1942

MANUFACTURED BY

WESTINGHOUSE ELECTRIC & MFG. COMPANY

BUZ 64816

Figure 5-13. This engraved nameplate for a 125-KVA, 460-volt, 60-cycle 3-phase switchboard was affixed to the "Mount Feeder No. 1" electrical panel for the "6-inch battery-construction power plant for Battery No. 237" (Woodward).

The Worthington diesel engines were started by compressed air stored in the three dedicated accumulators seen in Figure 5-14a, which were pressurized by the electric Worthington air compressor seen in background right of Figure 5-14b. The gasoline-powered back-up compressor can be seen in the foreground of the same photo.

Figure 5-14. a) This bank of three accumulators, circa 2007, provided compressed air for starting the diesel engines. **b)** The accumulator bank was pressurized by an electric Worthington air compressor (background right) or the hand-cranked gasoline-powered backup compressor in the foreground.

Referring again to Figure 5-11, the diesel exhaust from each of the three generator sets was piped through the south wall of the power room into the muffler gallery (see again Figure 5-4). These three exhaust pipes ran through the overhead as seen in Figure 5-15a and exited the bunker just outside the battery's rear entrance on Woodward Road (Figure 5-16). The two galvanized conduits more recently installed beneath them provide a more recent connection to 2.4-kilovolt three-phase commercial power. Figure 5-15b shows various components of the CWS system that were removed by post-war tenants.

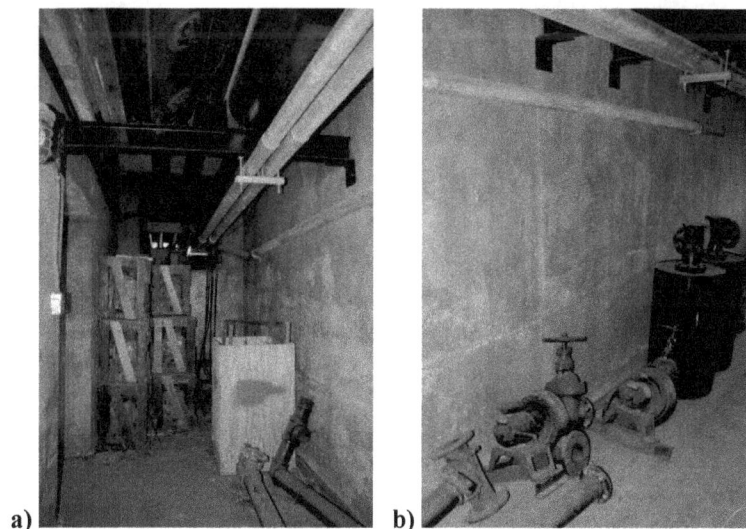

Figure 5-15. a) View inside the muffler gallery looking east towards Woodward Road, with crated spares in background. Note the three diesel exhaust lines supported by crossbeams in the overhead. **b)** Various components of the CWS system have been removed to accommodate post-war tenants.

Figure 5-16. The three generator exhausts exit the muffler gallery (background left) and connect to three vertical mufflers hidden by the underbrush, circa 2006. The opening on the right is the air intake to the water cooler (radiator) room (see again Figure 5-4). The rear entrance to the battery is out of image to the right.

Figure 5-17 shows the rear entrance to Battery Woodward on the right (west) side of Woodward Road, just beyond which is the muffler gallery exit, looking south towards the Pacific Ocean. Directly across the street from the battery's rear entrance is a long single-story dug-in wooden bunker that served as a combined berthing and messing facility. A one-room kitchen was included on the south end of this structure, equipped with water, sewer, and electricity.

Figure 5-17. Looking south towards the Pacific Ocean along Woodward Road in 2006, with the rear entrance to Battery Woodward off to the right, and its dug-in berthing and messing facility (now Building F-16) directly across the street on the left.

The floorplan of this berthing and messing facility, now designated Building F-16 at the Center, is shown in Figure 5-18 below, with the kitchen area at far left and the three street entrances along the east side of Woodward Road seen across the top of the drawing. The berthing/dining area is the long rectangular room to the right of the kitchen, which like the kitchen had a concrete floor with wooden walls and overhead. The entire structure was dug into the hillside and covered with several feet of dirt, with a number of protruding wooden air vents above.

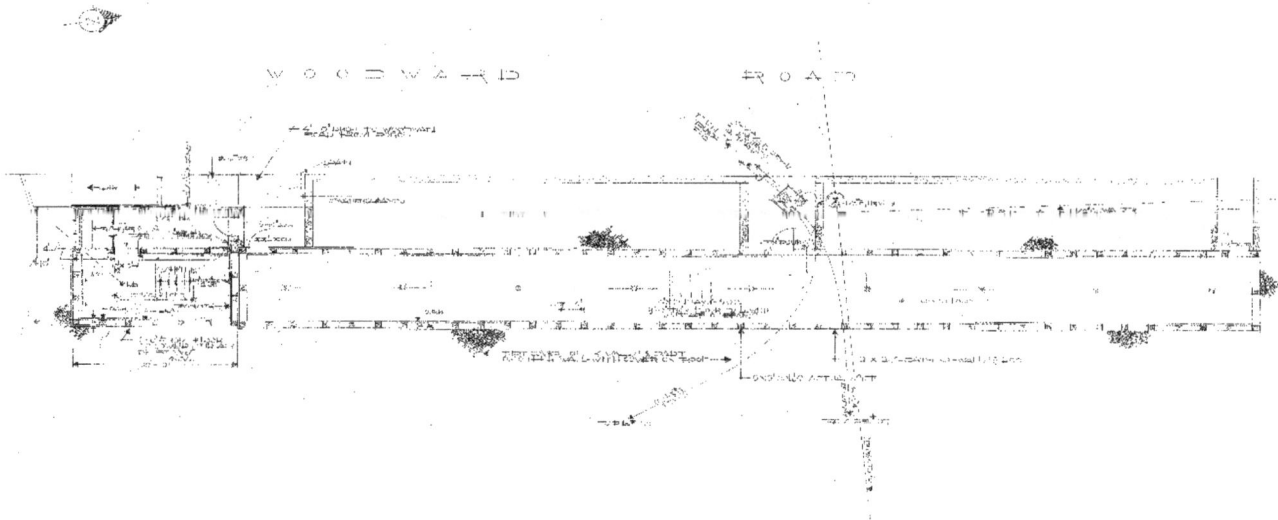

Figure 5-18. Faded floorplan of Building F-16, showing the kitchen area at the south end (far left), with the three street entrances along Woodward Road across the top of the undated drawing.

The wooden façade of the kitchen at the south end of the dug-in berthing/messing facility on Woodward Road behind Battery Woodward is seen Figure 5-19a. The three windows that provided ventilation and lighting have been boarded over, but multiple openings in the decaying walls encouraged several beehives (Figure 5-19b), which posed a hazard to nearby pedestrians and were removed. The concrete stairs just left of the shuttered windows lead to the southwestern entrance (leftmost in Figure 5-18 above), with the other two entrances indicated by the up arrows at image left.

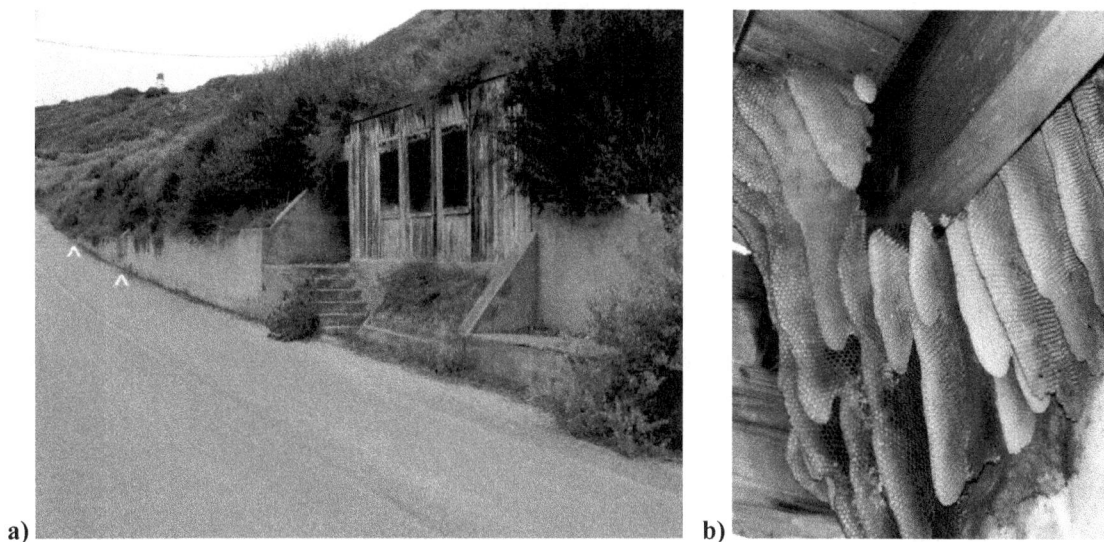

Figure 5-19. a) View looking northeast towards the berthing and messing facility on the east side of Woodward Road, circa 2006. The southern (kitchen) entrance is at image center, with the middle and northern entryways indicated by up arrows. **b)** One of several beehives in the kitchen area, soon removed.

The kitchen interior is seen in Figure 5-20, looking south from the entrance, with the food-preparation area at far left, the serving counter at foreground left, and a food-storage closet at background right. The wooden shipping crates on the counter and in the closet are addressed to the "Naval Radio and Sound Lab," suggesting this former Army support structure for Battery Woodward had also been used by the Navy laboratory, at least for storage. A 17 January 1951 memorandum from the District Public Works Officer to the Commandant, Eleventh Naval District, provides a post-war indication of continued Navy interest in both Battery Woodward and Battery Strong (Johnson, 1951):

"The Navy Electronics Laboratory contemplates using the Woodward and Strong locations in the area which is under a "use" permit from the United States Army."

Figure 5-20. View of the kitchen area, looking south circa August 2006. A shelved closet with door ajar is seen at the south corner (background right), with a built-in serving counter to the left. The wooden shipping crates on the counter, floor, and in the closet are addressed to the "Naval Radio and Sound Lab."

Entry to the kitchen proper was via the opening seen just beyond the serving counter, which is at foreground left in the above photo. A large sink was attached to the far wall (Figure 5-21), with an overhead cold-water service line dropping down to feed two spigots. Note the raised concrete perimeter footing beneath the large vertical wooden posts supporting the overhead. Two of four light fixtures can be seen attached to the ceiling joists. There was no obvious evidence of any type of cooking equipment or electrical connections for such, suggesting the food was probably prepared topside and trucked in for serving.

Figure 5-21. The remains of a large sink installation extend west from the left corner along the far wall, plumbed for two cold-water spigots. A rusty cast-iron sewer connection for the drain is seen rising from the concrete floor below. The corner of the serving counter is at foreground right.

My initial interest in this elongated support structure was not the kitchen but the adjacent dining/berthing section to the north, which we wanted to use as a subterranean test facility for our man-portable robotic systems. We had discovered this tunnel-like section before we ever gained access to the kitchen, as the south entry door had at some point been removed from its hinges (Figure 5-22), which are still extant on the doorframe at background right. The concrete floor in this section was covered with a layer of dirt, and there were a number of interesting artifacts to serve as realistic obstructions for the robots.

Figure 5-22. The entry door leaning against the far wall of the berthing/dining section was loaned to Cabrillo National Monument to serve as a pattern in their 2007 restoration of the powerhouse for WWII Searchlights 18 and 19 (see Chapter 8). Note dirt spillage covering the concrete floor surface.

Figure 5-23a shows the interior of the berthing/dining section as seen from just inside the south entrance, looking north, showing both side walls to be significantly canted by downslope pressure. Randy Peacock of our Facilities Department told me that a 0.30-caliber water-cooled machine gun, complete with a live ammo belt, had reportedly been found inside this area in the 1970s timeframe. Figure 5-23b suggests there may have been a small barber shop associated with this support structure, although these chairs were possibly relocated from elsewhere and simply stored in this location.

Figure 5-23. a) View of the berthing/dining section, looking north in August 2006, which we used for training man-portable-robot operators in tunnel-like conditions. **b)** Four old barber chairs were found stored inside the dug-in structure.

There is considerable evidence of continued activity within and around Battery Woodward after the war, but the details remain rather elusive, and in some cases border on the mysterious. One of the first such puzzles I noticed following my transfer to the Center in 1986 was a cylindrical concrete tank situated outside the rear entrance near the muffler-gallery discharge. I had wondered about the nature of this item long before I ever saw the inside of the battery, in that I drove by it several times a day going to and from my office just a few hundred yards north on Woodward Road. My curiosity at the time was further piqued by the structural damage seen in Figure 5-24.

Figure 5-24. Damaged remains of a cylindrical concrete tank on the west side of Woodward Road, just outside the rear entrance to Battery Woodward, circa 1996 (US Navy photo courtesy Cabrillo National Monument). Note the apparent drain outlet right of center above the hexagonal concrete base.

In the above photo, the tank appears to have been constructed from a prefabricated section of concrete pipe set vertically into a hexagonal concrete foundation. Keniston (1996) curiously describes this strange artifact in the caption of an undated photo of the battery's easterly face as:

> "Circular concrete form in left foreground is a no longer extant exterior capacitance well, used in the electrical system as an emergency power supply."

Suffice it to say this speculative explanation has no technical substance whatsoever.

During one of our many book-progress discussions, Coast Defense Study Group member Al Grobmeier provided me with a 1972 picture of this same area, which included a portion of this concrete tank on the far-left side of the image (Figure 5-25). What looks like a 1-inch fill line can be seen running from above and behind the tank, upslope and over the east (rear) entrance to the battery, then down to the main water-supply manifold at lower right. There also appears to be an almost makeshift steel superstructure for supporting something suspended inside the tank, at the very top of which is a vertically mounted electric motor.

Figure 5-25. Partial view of the concrete tank (lower left corner) behind Battery Woodward, with a steel superstructure for supporting an electric motor driving something inside the tank, with the battery's east (rear) entrance at right (1972 photo courtesy Al Grobmeier). Note the tank's fill-water line running up the slope.

Al also recalled seeing a similar configuration outside of Battery 241 (unnamed) at Fort MacArthur in San Pedro, CA. Battery 241 was a "folded" 200-series design with the muffler gallery adjacent to one of the gun emplacements, in this case Gun 2. After searching his extensive collection, he graciously loaned me several photos of this tank setup, two of which are shown in Figure 5-26, circa 1977 and 1982, respectively. Note the large vertical electrical conduit with power cables leading down to the top of the wood-stave tank, and the apparent fill-water pipe running from the muffler gallery. The tank has since been removed, but the electrical conduit is still extant.

a)

b)

Figure 5-26. a) A taller wooden-tank configuration is seen adjacent to the Gun 2 emplacement at Battery Construction No. 241 at Fort MacArthur, circa 1977. **b)** Heavy-gauge cables from the electrical conduit attach to three equilaterally spaced electrodes supported inside the tank by a metal ring, circa 1982 (photos courtesy Al Grobmeier).

In August 2007, our Facilities Department loaned me a copy of a 1957 blueprint of Battery Woodward that showed the electrical connections to a "Water Rheostat Tank" situated outside the east entrance. On that portion of this print reproduced in Figure 5-27 below, the electrical wiring leading to this tank is specifically described in the upper-left corner as:

- 1 – 3/C #14 LIMIT SW. CONTROL – 115V – 1Ø
- 1 – 3/C #12 440V – 3Ø TO MOTOR ON TANK
- 3 – 1/C #2/0 TO GENR. BUSS

Translated to less cryptic terminology, the device was serviced by:

- One three-conductor 14-gauge cable for a 115-volt single-phase limit-switch.
- One three-conductor 12-gauge power cable to the 440-volt three-phase motor.
- Three single-conductor double-aught-gauge power cables to the generator buss.

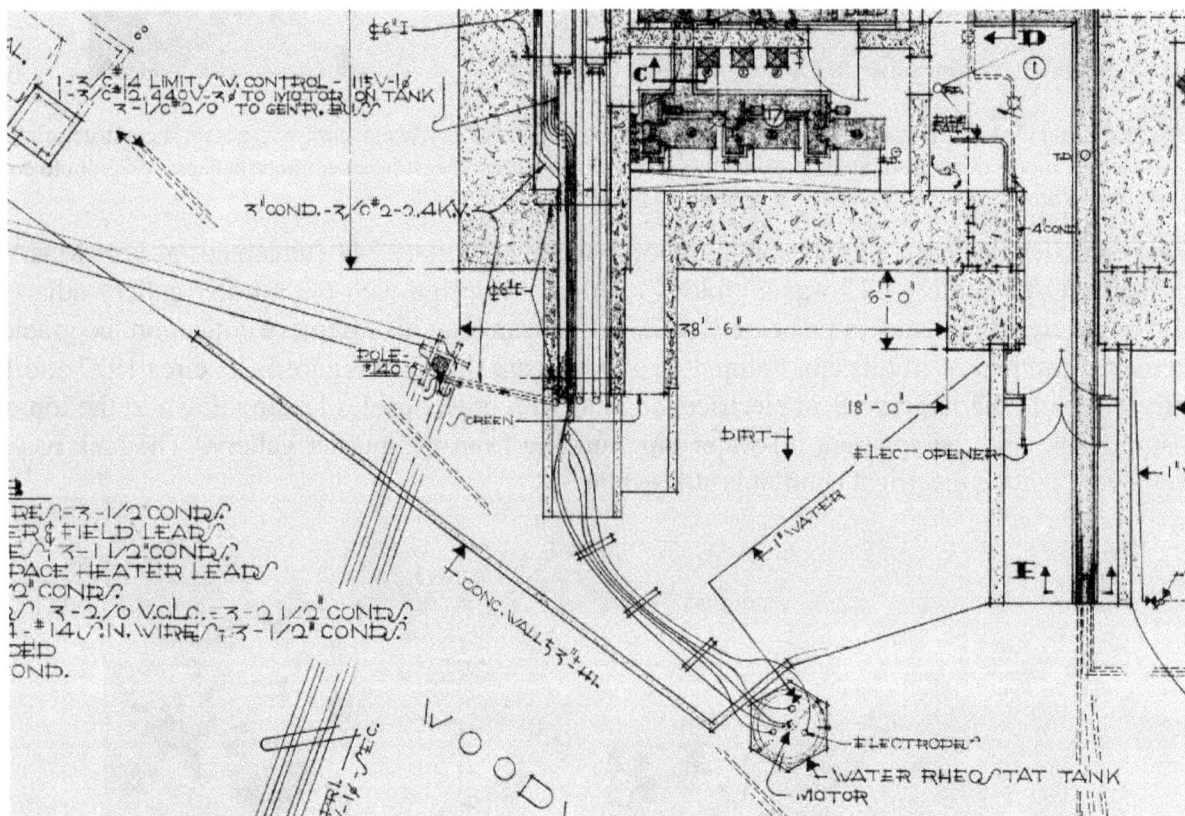

Figure 5-27. Lower-left portion of "Sheet 1 of Alteration A" to Navy Electronics Lab "Structure F-12 Woodward Floor Plan" blueprint dated 2 October 1957, showing the "Water Rheostat Tank" (bottom right), with three-phase electrical connections through the muffler gallery to generator buss (courtesy SSC Pacific).

From an engineering perspective, these wiring specifications provide considerable insight into the functional nature of the device itself. The direction of rotation of a three-phase electric motor can be reversed simply by switching any two of its three power leads. In addition, a three-conductor limit-switch control would be sufficient to automatically stop the motor rotation and effect such reversal at preset clockwise and counter-clockwise travel limits. At this point in my investigation, I began to suspect the rheostat tank was set up to serve as a remote-controlled water-cooled load for testing the generators.

Two observations regarding the precise nomenclature employed on the blueprint are worth further consideration, however. First, the tank is actually labeled as a "Water Rheostat Tank," as opposed to a "Water-Cooled Rheostat Tank." Secondly, the three heavy-gauge cables from the generator buss are shown connecting to three "electrodes" arranged in an equilateral-triangle configuration, looking down into the tank. Since the term "electrode" can be used to describe a means for passing current into or out of a fluid, such terminology suggests the water in the tank served not only for cooling, but was in fact the resistive element itself.

If such were indeed the case, what then was the purpose of the three-phase motor, which does not appear in the Fort MacArthur setup of Figure 5-26? For additional clues to this mystery, I decided to hunt down any remaining evidence of the electrical cables that once connected the tank equipment to the electrical switchboard. The muffler gallery access to the battery had for years been obscured by an overgrowth of dense brush, but once this was cleared away, the original cables were found to be still present as seen in Figure 5-28.

Figure 5-28. The control and power cables for the "Water Rheostat Tank" were uncovered under considerable brush and debris outside the muffler gallery for Battery Woodward in December 2007.

These five cables enter the bunker through a hardware-cloth screen behind the generator exhausts, then run along the floor on the south side of the muffler gallery as shown in Figure 5-29a. At the west end of the gallery, they loop up through the ventilation fan opening into the generator room, then disappear behind the electrical switchboard (see again Figure 5-14b). All five cables enter the fifth panel of the switchboard, where the three double-aught power leads are connected to what had previously served as the "Mount Feeder No. 1" circuit breaker. The inscription "Water Rheostat Load Switch" has been written in grease pencil on the front panel just beneath this breaker (Figure 5-29b), indicating its new function was to connect the water-rheostat load to the generator buss.

The motor and limit-switch cables are both connected to a pair of electromechanical relays inside the switchboard enclosure just below the "Load Switch." These relays appear to be wired together to form a set/reset flip-flop that reverses the motor's direction of rotation, triggered by the limit-switch feedback. No connections were seen between the relays and any board-mounted control input. Such a flip-flop configuration would cause the motor to continuously rotate back and forth (like a washing machine agitator) between the two extremes defined by the clockwise and counterclockwise limit-switch positions.

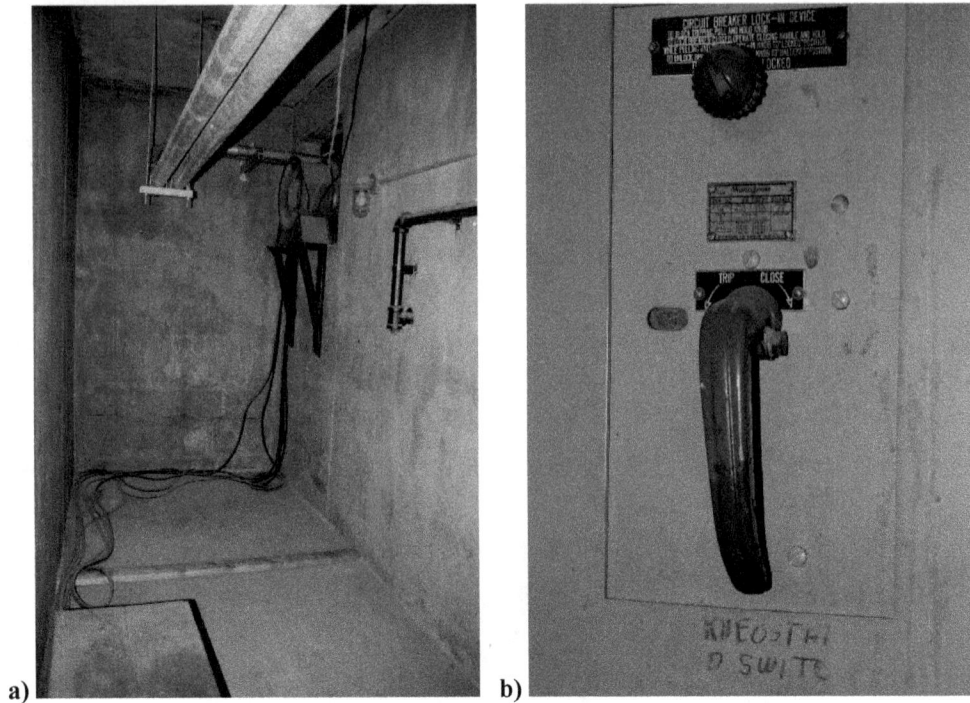

Figure 5-29. a) The rheostat cables run along the floor of the muffler gallery and into the generator room via the ventilation fan opening at upper right. **b)** The original "Mount Feeder No. 1" circuit breaker has been reconfigured as a "Water Rheostat Load Switch" for connecting the rheostat to the generator buss.

The motor and its relay controller are apparently powered up as soon as the water rheostat is connected to the generator buss as a load. One possible explanation for such an arrangement would be to cyclically vary the generator load as a function of time, perhaps via a three-pole rotating armature within the tank that altered the gap geometry between the three suspended electrodes and perhaps a fourth static element.[2] As no trace evidence of a fourth static electrode was found, however, a more likely purpose of the cyclic rotation was simply to increase heat transfer for cooling purposes.

All things considered, the apparent function of the water rheostat tank was simply to serve as a three-phase test load for the generators. As recounted by Lyndon (1916):

> "Rheostats for absorbing the full-load energy of a generator may be of two kinds, one being the well known 'water rheostat' in which the liquid forms the resistance and energy-absorbing medium. The other is the submerged iron-wire rheostat, in which the resistance is made up of iron-wire coils."

The fact that a similar configuration existed behind Battery 241 at Fort MacArthur suggests this was not part of an independent NEL research project, however, as I had initially surmised. Both setups show a rather crude steel superstructure temporarily affixed to an improvised water tank, with heavy-gauge electrical cables connected to some apparatus that appears to be suspended in water. But why had these nearly identical post-war modifications been implemented at two different batteries 10 years after the war?

CDSG member Mark Berhow (2008), who served as a volunteer at the Fort MacArthur Museum from 1985 to 1994, suggests the following possible connection to Battery 241:

> "The Navy had a radar post there from 1950 (or so) to the 1960s some time. They used Battery 241, as it had its power equipment for back-up power. That is why the generators are still in place in 241. As the Navy was also in charge of Battery Woodward, my guess is that it had something to do with the Navy radar equipment in use there as well."

---

2   In the absence of a static (fourth) electrode, however, the cyclic back-and-forth motion of the three suspended electrodes did not alter their relative spacing, hence the generator load, as their collective geometry remained fixed.

Battery Woodward had in fact served as the postwar Harbor Entry Command Post at the U.S. Navy Electronics Lab (Fisher, 1951). We will examine this water-rheostat topic again in the next section on Battery Ashburn, which was similarly equipped.

Another mystery that eluded me for the first 2 years of my research was Building F-19 (Fieldhouse 19), a small one-room bunker just north of Battery Woodward and across from the Pistol Range on the west side of Woodward Road. Surface marks on the concrete indicate horizontal board (versus plywood) form construction, which remained common into the '50s. Various conduit runs and a large covered cable trough suggest the structure was functionally associated with a steel tower some 40 yards due west.

Careful study of my growing collection of period photographs showed no evidence of either the tower or the bunker during the war, but both appear in post-war images beginning in 1948, as for example Figure 5-30, circa 1969. The tower has long since been removed, but its distinctive concrete support slab remained in place until a few years ago. Continued development of long-range high-resolution radar was a key focus at the Navy Electronics Lab after the war, with the entire Pacific shore area west of Woodward Road designated as the "Electromagnetic Radiation Equipment RDT&E Range." [3]

Figure 5-30. Building F-19 is just west of the bend in Woodward Road, with the mystery tower some 30 yards due west, circa 1969. The northern (Gun 1) emplacement for Battery Woodward is at upper right (down arrow), with both of Battery Zeilin's gun emplacements (up arrows) at upper left, and the WWII pistol range on the far side of Woodward Road (US Navy photo courtesy SSC Pacific).

Accordingly, I first assumed this tower and Building F-19 (see Figure 5-31) were part of some post-war radar project in the early-1950s timeframe. Obviously, any Navy employees that might have been involved have long since retired, making it difficult to pursue details from a project-history standpoint. In addition, while our Facilities Department keeps extensive records on the Center's infrastructure, they were unable to locate any structural blueprints or historical documentation at all for this particular building, which struck me as strange.

---

3 RDT&E – Research, Development, Test, and Engineering.

Figure 5-31. Period photos and form marks suggest the original concrete structure (Building F-19, far right) adjoining this recently erected maintenance shed was built shortly after WWII. Our Unmanned Systems Branch used this area to support the *Tactical Amphibious Ground System (TAGS)* robot (left of center).

Building F-19 is a steel-reinforced concrete structure measuring 15 feet across and 12 feet deep, with two 10-inch I-beams supporting a concrete roof. The most intriguing aspect of this otherwise nondescript bunker is the presence of six 2-inch-thick glass windows on the western face (Figure 5-32). Stains on the concrete wall beneath these windows suggest the original ground level was near the bottom of the rectangular opening at center, which appears to be the case in Figure 5-30 on the preceding page. The combination of bullet-proof viewing ports and semi-submerged 12-inch-thick concrete walls strongly suggests Building F-19 was specifically designed for safe observation of some inherently dangerous operation, perhaps somehow connected to the mysterious steel tower.

Figure 5-32. The western side of the Building F-19 bunker is equipped with six bullet-proof viewing ports facing west towards the tower location and the Pacific Ocean.

In December 2007, acting on a tip from Dr. Doug Gage, a former member of our Unmanned Systems Group, I purchased a copy of *Project Orion: The True Story of the Atomic Spaceship*, by George Dyson (2002). Project Orion was General Atomic's rather radical concept for a 4,000-ton nuclear-powered rocket in the late '50s to early '60s timeframe. Doug felt I might find the book insightful, especially one chapter entitled "Point Loma," given my suddenly acquired interest in local history. As it turned out, this casual recommendation led to my first hint that General Atomic (later General Atomics) had used part of Convair's Point Loma Atlas-missile facility (Figure 5-33) to conduct tests of their Orion prototype.

Figure 5-33. Convair's southern Point Loma Test Facility for the Atlas missile was located on the west side of Gatchell Road, just north of the San Diego Sewage Treatment Facility (undated US Navy photo courtesy SSC Pacific). This is the current site of our remote Building 651 and Robotic Test Area South.

I was aware of Convair's extensive Atlas-missile complex that had been located down by the sewer plant on Gatchell Road, so this was where I initially assumed General Atomic's *Orion* tests also took place. Upon closer inspection of several photos in Dyson's book, however, I recognized a couple of terrain and structural features that clearly suggested otherwise. Figure 5-34, for example, shows General Atomic personnel inspecting their 1-meter launch-vehicle prototype in front of "a high-explosive storage bunker at the Point Loma test site" (Dyson, 2002). Note the small triangular concrete projection atop the wall just right of the wooden locker at upper left.

Figure 5-34. Project engineer Jerry Astl (foreground) and technician Jim Morris inspect the "pusher plate" of a 1-meter proto-type of the Orion vehicle in 1959 (adapted from Jerry Astl film footage courtesy George Dyson). Note the triangular concrete projection right of the wooden storage structure at upper left and the electrical conduits at lower-left corner.

The wooden lap-siding addition adjacent to the concrete retaining wall on the left side of the above image immediately caught my eye. Purely by chance, I had the very day before closed the tattered doors of that same storage locker in preparation for a robotics documentary we were filming at Battery Woodward. This wooden locker was situated atop the ready-service ammunition shelf for Gun 2, which meant Astl and Morris were inspecting their prototype just outside Battery Woodward's southern entrance. I quickly pulled up a recent photo of this area on my computer for comparison, seen in Figure 5-35 below, and sure enough, it was a perfect match!

Figure 5-35. The southern entrance to Battery Woodward, circa 2006, showing the same electrical conduits from the cable trough entering the post-war lap-siding storage locker on the Gun 2 ready-service ammunition shelf. The triangular concrete projection just right of the top of this locker was also seen in Figure 5-34.

I took Dyson's book down to the F-19 area the next day and excitedly compared his launch-sequence photographs with the remains of the concrete foundation that once supported the steel tower. It soon became unmistakably clear that I was standing in the very same spot where these historic flights had taken place almost 50 years earlier! Had Doug Gage not pointed me to this book, I'm quite sure I would have never made this amazing connection to *Project Orion's* use of Building F-19 and its associated tower.

Further investigation revealed both structures had been originally been built by Convair for static engine tests of their *Project MX-744 "HIROC"* missile in the late 1940s.[4] In April 1946, the Army Air Force had awarded a $1.4M contract to Consolidated Vultee Aircraft (i.e., Convair) Corporation of San Diego to perform a study of two proposed designs for long-range (5,000-mile) missiles (Chapman, 1960). The effort was led by Karel "Charlie" Bossart of Convair's Vultee Field Division near Downey, CA (ASME, 1985). An additional $493K was added to the contract in June of that same year.

Convair's subsonic cruise-missile option was eventually dropped in favor of its supersonic ballistic-missile approach, which showed enough promise for further authorization to develop 10 research prototypes. Widespread defense cutbacks under Secretary of Defense Louis Johnson and the Truman administration, however, caused the Air Force to reluctantly terminate the contract in July 1947 (Chapman, 1960; Boyne, 1998). With sufficient funds remaining to complete only three *MX-744* missiles, Convair relocated Bossart and his Downey group to San Diego and quickly set up a static test facility in the northwest corner of Fort Rosecrans. They brought with them an oil derrick to serve as a captive-missile support tower (Figure 5-36), and constructed a concrete "blockhouse" (now Building F-19) from which engineers could safely observe the tests (Chapman, 1960).

Figure 5-36. a) The Convair *MX-744* missile is being raised into position under the test tower, looking southwest in 1947 (photo courtesy NASA (AP, 2017)). **b)** Later view of missile in an almost vertical orientation (photo courtesy ASME National History and Heritage Committee). Note the counterweighted missile-access gantry attached to the second horizontal cross member in both images.

Convair's long-range missile design had built upon the numerous technical achievements of Germany's infamous *V-2* missile in WWII (Table 5-1), just as German engineers had capitalized on the earlier successes of the American rocket pioneer, Dr. Robert Goddard (Swenson et al., 1989). One of the biggest design challenges facing Convair was how to reliably steer a supersonic missile over a significantly extended flight path. With a maximum range of only 200 miles, the *V-2* had employed moveable car-

---

4   The "HIROC" (High-altitude Rocket) project name chosen by the Air Force was not extensively used.

bon-vane control surfaces in the nozzle exhaust, which suffered appreciable ablative wear during a relatively short burn time.

Missile	Purpose	Length	Diameter	Weight (empty)	Thrust
V-2	Weapon	46 feet	66 inches	8,900 pounds	56,000 pounds
MX-744	Research	31.5 feet	30 inches	1,200 pounds	8,000 pounds

Table 5-1. Comparative specifications of the *MX-744* and its German *V-2* predecessor (Chapman, 1960; Zaloga & Calow, 2003; Dungan, 2005). Note the similarity of the *MX-744* to the *V-2* in Figure 5-37.

Bossart proposed swiveling the engines as an alternative steering approach,[5] a previously untried concept that would require extensive ground evaluation of the entire *MX-774* control system. This live-fire testing was the fundamental purpose of the mysterious tower shown previously in Figure 5-36 (Chapman, 1960):

> "The MX-774 was suspended by its waist at the center of the tower, resting on a gimbal ring so that the missile was free to tilt in any direction in response to engine movement. The four-finned base of the missile was about seventeen feet off the ground, placed directly above a water pipe (better known as 'the enema tube'), which was to spray the engine area in case of fire."

In addition, the exposed concrete slab beneath the tower was protected by the application of cooling water from a fire hose (Walker & Powell, 2005):

> "Cooling was provided by a 4-inch water line just hooked up to city water. The water was aimed at the concrete pad and turned on when the motor was ignited, splashing all over the pad and up the tower."

Compared to the earlier version seen in Figure 5-36, the tower shown in Figure 5-37 below has been structurally augmented with sections of steel plate, presumably to shield the four support legs from the intense heat generated by the missile exhaust.

Figure 5-37. Liquid oxygen is pumped to a captive *MX-744* missile prior to its static firing in 1948 (photo courtesy Convair San Diego). Note technician standing on the missile-access gantry (in the down position) behind and to the right of the tail fins.

---

5   The gimbaled engine-exhaust nozzle was developed for Convair by Reaction Motors, Inc. (Lethbridge, 1998).

As Chapman (1960) further described events leading up to the first scheduled hot firing, a rather interesting detail caught my eye:

"Outside, a Navy guard sealed off the area by parking a jeep across the entry road. Observers retreated to their 'outpost' – an old gun pit about four hundred feet away."

Presumably this "gun pit" would have been the southern (Gun 2) emplacement of Battery Zeilin, which was about 320 feet to the northeast (see again Figure 5-30).

In the latter half of 1948, the three *MX-744* prototypes underwent flight tests from a modified *V-2* flame-deflector launch pad at White Sands Proving Ground, NM (Table 5-2). While the three launches at White Sands were certainly not without problems (the first missile crashed in a fiery explosion just 200 yards from the blockhouse), they nonetheless provided convincing proof that Convair's innovations were on the right track. In retrospect, the *MX-744* development was a revolutionary undertaking that in just two short years introduced three major contributions to the fledgling field of rocketry (Chapman, 1960):

- Integral fuel tanks for significant weight savings.
- Gimbaled engines for improved control of pitch, roll, and yaw axes.
- A detachable nosecone to eliminate the need for missile reentry.

Vehicle ID	Static Test Complete	Launch Date	Altitude Attained
No. 1	26 May 1948	13 July 1948	6,200 feet
No. 2		27 September 1948	29 miles
No. 3		2 December 1948	30 miles

Table 5-2. Summary of captive tests at Point Loma and test flights at the Army's White Sands Proving Ground for the three *MX-744* prototypes (Chapman, 1960).

During the austere funding climate that followed, Convair stayed the course with internal funds, maintaining a skeleton crew consisting of Bossart and two supporting engineers (Walker & Powell, 2005). With the September 1949 announcement that the Soviet Union had successfully tested an atomic bomb, thus ending the US monopoly on nuclear weapons, the proverbial spending pendulum started to swing back the other way. It gained further momentum with the communist invasion of South Korea in June 1950, and a new Air Force contract was awarded to Convair in January 1951 for "Project *MX-1593*," which soon came to be known as "Project Atlas" (Chapman, 1960). Charlie Bossart and Convair began to ramp up for what was to become an unprecedented undertaking of monumental proportions.

To put things somewhat into perspective, the relatively small *MX-774* missile was built to serve as a research prototype, never intended to ascend more than 100 miles (Chapman, 1960). *Atlas*, on the other hand, was a very large intercontinental ballistic missile (ICBM), specifically designed to deliver an 8,000-pound nuclear warhead to within 500 yards of a target some 5,000 miles away (Walker & Powell, 2005). In addition to its cold-war role as a nuclear deterrent, *Atlas* was also adapted to carry our early *Mercury* astronauts into space, beginning with John Glenn in 1962. The overall program, which at its peak employed some 100,000 workers, would ultimately cost US taxpayers somewhere between 6 and 8 billion dollars (Walker & Powell, 2005).

Convair, acquired by General Dynamics in 1953, had obviously outgrown their small test facility near Battery Woodward, and in June of that same year began looking for other options. The *Atlas* rocket was way too big for live-fire engine testing on the Point Loma peninsula, so two new test stands were constructed north of the city in Sycamore Canyon on the remote eastern fringes of Camp Elliott.[6] Some 5

---

6   The first *Atlas* missile was delivered to Sycamore Canyon for systems testing on 29 August 1956 (Lethbridge, 1998).

acres of land on the Pacific shores of Fort Rosecrans (see again Figure 5-33), approximately a mile south of Building F-19, were also leased from the Army for structural and cold-flow fuel tests (Walker & Powell, 2005). The Army transferred Fort Rosecrans to the Navy in 1957, the principal tenant at that time being the U.S. Navy Electronics Laboratory.

Convair's new Point Loma Test Facility, situated on Navy property just north of today's San Diego sewer treatment facility, underwent continuous incremental expansion over the next several years as the program increased in scope. On 12 May 1959, for example, installation of a specialized test tower for the *Centaur* second stage (Figure 5-38) was authorized by the Air Force Plant Representative for Convair Astronautics, Colonel H.E. Moose (1959):

> "After a review of all the facts presented by references a. and b. above with respect to the location of the Centaur Test Stand, and in view of the need for an immediate decision for go-ahead of A&E design, you are directed to proceed to locate the Centaur Test Stand to the south of the existing Point Loma test installation and within the permitted area."[7]

Figure 5-38. View of the *Centaur* Test Stand, looking south, showing liquid nitrogen being dumped over the cliff following cold-flow testing (photo courtesy Convair San Diego).

About this same time, General Atomic was looking for an area more suitable than the company's new headquarters in La Jolla to test prototypes of their radical Orion propulsion concept using conventional explosives. In August of 1958, the company hired Lieutenant Commander (Retired) Carroll Walsh, who had worked at the U.S. Navy Electronics Lab for 8 years. Walsh was aware of the abandoned Convair tower and Building F-19 blockhouse by Battery Woodward, and "persuaded his former colleagues" to grant General Atomic the use of these facilities for Project Orion (Dyson, 2002).

In a "License for Use of Real Property by Other Federal Agencies" dated 18 May 1959, the District Public Works Office, Eleventh Naval District, authorized the Air Force a 12-month extension for continued use of this site, described as follows (Brown, 1959):

> "One MX-744 Static Test Tower and Blockhouse (Bldg F-19) located on the west shore of the Naval Electronics Laboratory."

The nature of the new work for which this extension was required was rather subtly worded in the following fashion (Brown, 1959):

---

7 References "a." and "b." were Convair letters 110-76B dated 3 April 1959 and 110-1577 dated 6 May 1959, respectively.

"Tests involve the use of tower and blockhouse to determine the effects of small explosions on certain materials. The explosions generally will be of the order of ten grams to a kilogram of slow-burning propellant. The tests will be conducted on an intermittent basis."

Led by former Los Alamos standout Ted Taylor, the General Atomic team didn't waste any time. A 1959 aerial photo taken that same month shows full-height scaffolding erected on the south side of the Convair tower, with a large construction crane situated on the north. The mid-level gimbal platform and missile-access gantries seen earlier in Figure 5-36b were removed to allow reasonably unobstructed vertical displacement of the tethered prototype during single-shot testing using conventional explosives (Figure 5-39).

The objective of this test phase was to validate the concept of propelling a spacecraft by ejecting a series of explosive charges in its wake, literally blasting the craft along a controlled trajectory. Single-shot tests were performed to measure the ablative wear on the "pusher plate," which was shock-mounted to cushion the explosive impact and prevent damage to the vehicle.

Figure 5-39. View looking northwest of an engineer atop a stepladder inspecting the tethered model prior to a single-shot explosive test in May of 1959 (Jerry Astl film footage courtesy George Dyson). Note scaffolding for tower modifications in place at left, and steel shields on the tower supports.

The next challenge involved developing a reliable means to eject a series of spherical charges for sustained flight and ensuring they consistently detonated at a fixed distance from the pusher plate. Each 2.3-pound ball of C-4 was attached to a carefully measured length of Primacord that uncoiled from a cylindrical propulsion canister as the charge was expelled by compressed nitrogen (Dyson, 2002). As the descending charge exited the ejection tube and reached the desired offset, the Primacord drew taught, actuating a switch on the ejection stack that triggered a small detonator, which in turn fired the Primacord to explode the C-4. The individual propulsion canisters were carefully mated together to form a vertical charge-ejection stack as shown in Figure 5-40.

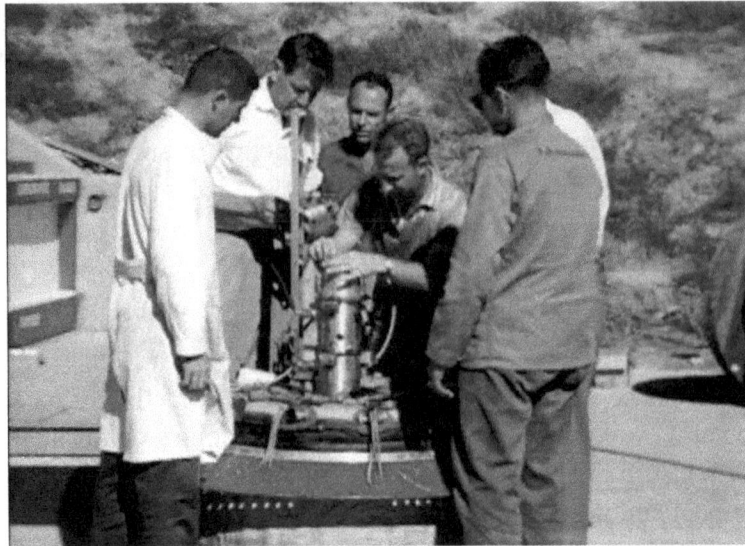

Figure 5-40. General Atomic personnel load propulsion canisters of C-4 onto the charge-ejection stack of a 1-meter prototype outside the south entrance to Battery Woodward (Jerry Astl photo courtesy George Dyson). The ready-service ammunition shelves on the east side of the south entrance to Battery Woodward appear to have been repurposed as an explosives locker.

In addition to firing the Primacord to explode its charge, each canister detonator also triggered a nitrogen release that ejected the next charge in the stack, resulting in six successive detonations at quarter-second intervals.[8] The final detonation would trigger a shotgun-shell squib that released a 14-foot recovery parachute. Figure 5-41 below shows the Orion prototype being prepared for launch beneath the former *MX-744* Static Test Tower.

Figure 5-41. Richard Goddard (left), Walt England (background) and W.B. McKinney (right) preparing the Orion prototype for flight (Dyson, 2002), with the Gun 1 emplacement for Battery Woodward in upper-left corner. A portion of the fiberglass "hull" is seen at lower right (Jerry Astl photo courtesy George Dyson).

---

8   The design and implementation of the innovative C-4 propulsion system for this proof-of-concept prototype was largely the work of General Atomic employee and explosives expert Jaromir (Jerry) Astl, a Czechoslovakian aeronautical engineer who immigrated to the US in 1949 (Wiki, 2019).

Jump-starting the 270-pound Orion prototype from zero initial velocity required a gentle boost from a one-pound toroidal charge of gunpowder in the bottom of a tub-like launch fixture as shown in the lower frame of Figure 5-42. Once clear of the launch fixture, the onboard charge-ejection stack would sequentially release the stacked C-4 canisters to progressively accelerate the untethered prototype in its vertical ascent. The upper frame of Figure 5-42 shows the proof-of-concept capsule has successfully cleared the top of the steel tower.

Figure 5-42. Free-flight test of 1-meter prototype from a "launch tub" near the Point Loma test tower in October 1959, looking northwest (Jerry Astl photos courtesy George Dyson).

Interestingly, the Convair *MX-744* and General Atomic *Orion* flight-test area was the same place we later tested our own vertical-takeoff UAVs. The foundation remains for the tower are just a couple hundred feet from our tethered-flight setup for the Allied Aerospace *iStar* ducted-fan UAV, developed under the DARPA Organic Air Vehicle (OAV) program. The 29-inch version of the *iStar* can be seen sitting on the concrete pad behind my pickup truck in the background of Figure 5-43a, and later during autonomous free flight in Figure 5-43b. The "Hot Rod" version of the 1-meter Orion model (i.e., without its ballistic shroud) that flew in this same area now hangs in the Smithsonian Air and Space Museum.[9]

Figure 5-43. a) The previously discussed "enema tube" cooling outlet is seen in the center of the test tower foundation, circa 2005, with the Allied Aerospace *iStar* UAV tethered-flight test area in background center, and the number 1 gun emplacement for Battery Woodward to its left. **b)** The ducted-fan *iStar* UAV in free-flight, looking east towards the new maintenance shed adjacent to Building F-19, with the Point Loma ridge on the far side of Woodward Road.

Battery Strong

Battery Strong is a bit of an enigma in that it was certainly not a temporary battery, but not a modernization battery either. Situated at an elevation of 350 feet, this two-gun emplacement was approximately 180 yards west of what is now Gate 5 at the Center as shown in Figure 5-44. Construction began in 1937, but the Army Ordnance Office soon estimated a 12-month lead time to accommodate a needed design change for the battery's 8-inch guns, with a best-case delivery date of May 1938 for the barbettes (Thompson, 1991). To facilitate ammunition transport, the guns were served by 2-foot narrow-gauge rail tracks, which are still extant.

---

9   Digitized film footage of this historic prototype can be found at http://www.nuclearspace.com/hotrod.htm.

Figure 5-44. The distinctive horseshoe approach to the two 8-inch gun emplacements of Battery Strong is seen in the center of this undated WWII aerial image (US Army photo courtesy SSC Pacific). Gate 5 is at the intersection of Woodward Road (lower left quadrant) and Catalina Boulevard, just below the parade ground right of the battery.

Battery Strong was named for Major General Frederick S. Strong, West Point class of 1880, who organized the 40th Division, California National Guard, at Camp Kearny, CA, which later deployed under his command to France in 1918 (Thompson, 1991; Stanton and Thayer, 2019). Construction details are depicted in the Form 7 plan view of Figure 5-45, taken from the Report of Completed Works, corrected to 21 July 1942. The battery was armed with two 8-inch Navy *MkVIM3A2* guns (45 calibers) on unshielded *M1* barbette carriages (Berhow, 2004), which were test fired in 1941 (Overton, 1993).

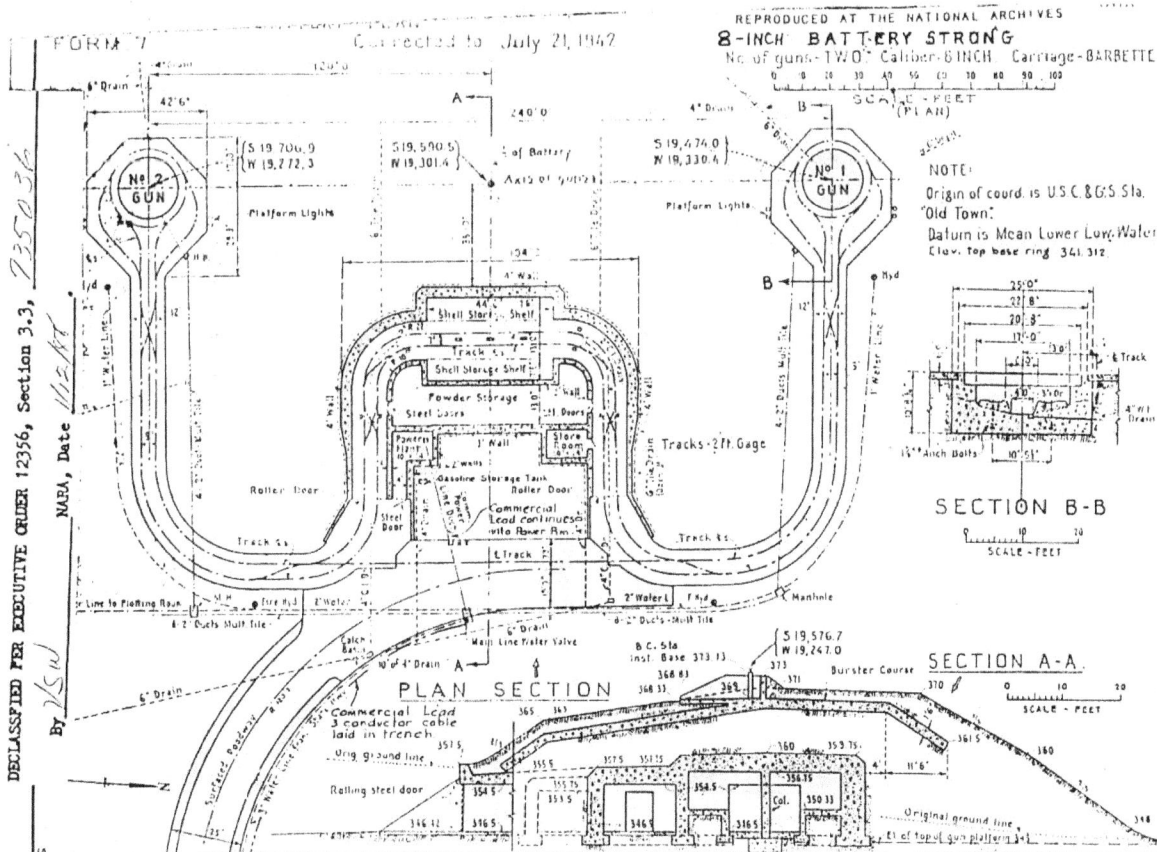

Figure 5-45. Construction details of Battery Strong, begun in 1937, corrected to 21 July 1942 (NARA).

Dwight Smith, who initially served in Battery C on the 12-inch mortars at Battery Whistler, recalls his experiences at Battery Strong in 1941 (Overton, 1993):

"After a few days they put us on machine guns, digging holes for machine guns. As high as your head. After that they transferred a bunch of us to D Battery on those 8-inch (Battery Strong). So they could man those 24 hours, around the clock. That was the most powerful gun in San Diego. When we first got there, they didn't have the guns on there yet. They set them sometime in '41, and test fired them."

The 8-inch Navy guns had been mounted and were officially approved for use in April 1941, but their protective shields were never installed. A chain-linked fence was erected around the compound (Figure 5-46), which was accessed via the parade ground to the east (upper-right corner), where a row of parked trucks can be seen along its west side. Additional security features included pole-mounted floodlights along the fence line and stucco-covered sandbag walls on either side of the access lane to the Gun 2 emplacement at foreground center (see also Figure 5-48 and Figure 5-52). A similar access-control point would soon be constructed for Gun 1, as seen later in Figure 5-51.

Figure 5-46. The two 8-inch Navy guns, minus their protective shields, are pointing westward on either side of the center bunker, which is fully covered over with dirt (undated US Army photo courtesy SSC Pacific). Note the Fort Rosecrans Parade Ground behind the battery in the upper-right corner.

The 1967 photograph presented in Figure 5-47 below shows the curved access routes to emplacements 1 and 2 of Battery Strong at top center, directly east (left) of ongoing construction for the Naval Electronics Laboratory Center's expanding Model Range in the upper-right corner. Catalina Boulevard / Cabrillo Memorial Drive heads south (upward) toward the Cabrillo National Monument in the lower-left corner, with Woodward Road branching off to the southwest in the upper-left corner. Note the south end of the fresh-water reservoir at bottom center, with the access road to Battery Whistler just visible in the bottom-left corner.

Figure 5-47. This 1967 aerial view of the Model Range shows Battery Strong just right of top center, with Woodward Road in the top left corner and Catalina Boulevard / Cabrillo Memorial Drive at lower left. Note Model Range excavation in upper-right corner just west of the battery.

Dwight Smith (Battery D) recalls the daily routine at Battery Strong (Overton, 1993):

"[If] we had an aircraft carrier left there, those guns were pointing right at that aircraft carrier until it went over the horizon... If anything happened, the gun was in position to go. If there was anything out there, you wouldn't have to move it much. That one got old, every morning and every night, in fact all day long."

Figure 5-48. The southern Gun 2 emplacement in 1968 (lower right), after removal of the barbette carriage but before it was filled in (US Navy photo courtesy SSC Pacific). The emergency battery commander station atop the bunker is in foreground center. Note the Building 584 microwave tower in the background.

Figure 5-49 shows the south entrance to Battery Strong, just to the right of which is a steel doorway for personnel access; the north entrance at far right has no such doorway. At some point during the Cold War, the earth-covered bunker was provided with additional ventilation and used as a fallout shelter. The US Navy repurposed this facility in 1972 for its current use as an Extremely High Frequency (EHF) laboratory. Some of the various antennae configurations associated with this effort are seen in background right.

Figure 5-49. View of the south entrance to Battery Strong's earth-covered ammunition bunker (left), looking northwest (1972 photo courtesy Al Grobmeier). Note the Cold War fallout-shelter and the Navy designation as Building 397 signs above the door. The north entrance to the bunker is seen at far right.

Figure 5-50 shows this bunker as it appeared in May of 2006, with the now-enclosed Building 584 microwave tower seen in the background at upper left. Note the upgraded ventilation systems and split-plant air-conditioning compressors above each bunker entrance, and the heavy coax cables running up the bank to the new wooden structure above. The north end of Building 401, which originally served as the battery's latrine, is just visible at far left.

Figure 5-50. View of Building 397, circa May 2006, which is still in use by the Navy as a millimeter-wave laboratory. The latrine (Building 401) can just be seen behind the stairs at far left.

Figure 5-51 below shows the remains of the Gun 1 emplacement, circa 2006. The narrow-gage ammunition-cart rail tracks are seen encircling the emplacement, which is overgrown with brush. The curved Model Range tower in the background allows a simulated satellite antenna to be positioned at various elevation angles for communications testing with a ship model below. Note the crumbling stucco gate walls in foreground. The narrow-gauge rail tracks from the ammunition storage bunker have been buried beneath a dirt access road that now leads to the top of the bunker.

a)   b)

Figure 5-51. a) The remains of the northern Gun 1 emplacement and rail tracks as seen in 2006, beyond which is Model Range Antenna Tower 423, with the stucco gate walls seen in foreground. **b)** Close up view showing the remains of the telephone communications station seen at far right (up arrow) in Figure 5-51a.

Figure 5-52 shows the southern Gun 2 emplacement as seen from on top of the earth-covered bunker, with the base of the Building 584 microwave tower in the upper-left corner. Note the stucco-covered-sandbag gate walls either side of the emplacement access lane at lower left. The long rectangular concrete locker just left of foreground center provided storage for the rammer and sponge poles.[10] The wooden stairs at far right provide access to the antenna ground-plane area of the Model Range further west. The faded sign adjacent to the stairs reads: "USN Electronic Lab Center,[11] Experimental Development Antenna Test Range."

---

10   The crews that used these implements were known as the rammer detail and the sponge detail (Berhow, 2020). A similar locker for Gun 2 is seen in the lower-left corner of Figure5-51a.
11   The Naval Command, Control, and Communications Laboratory Center changed its name to Naval Electronics Laboratory Center in 1967 (see also Appendix).

Figure 5-52. Remains of southern Gun 2 mount in May 2006. Note the crumbling stucco access-lane gate walls at lower left, and the enclosed lower section of the Building 584 microwave tower in background left.

The interior of the bunker just inside the north entrance is seen in Figure 5-53a, looking west in May 2006. The sheet-metal doorway in the transverse corridor at background left leads to an anechoic chamber. Figure 5-53b shows the view continuing around the corner to the left, now looking south. The ammunition-transport rail tracks were removed and the concrete floor patched, with partitions installed to support the Navy tenants of the millimeter-wave laboratory.

a)   b)

Figure 5-53. Interior of the Battery Strong bunker, circa 2006: **a)** looking west just inside the north entrance; **b)** continuing around the corner and facing south.

Battery Ashburn

Named for Major General Thomas Quinn Ashburn, US Army, Battery Ashburn in Site 6 was a standard *100-Series* design (Construction Number 126), outfitted with two 16-inch *Mark II Mod 1* Navy guns mounted on *M1919M4* barbette carriages with frontal shields. With a lateral separation of 500 feet, the two casemated gun positions were connected by a transverse corridor as shown in Figure 5-54. Battery Ashburn was one of 27 such installations authorized by the War Department in 1940, with funding made available in November 1941, just a month before WWII would come to American shores.

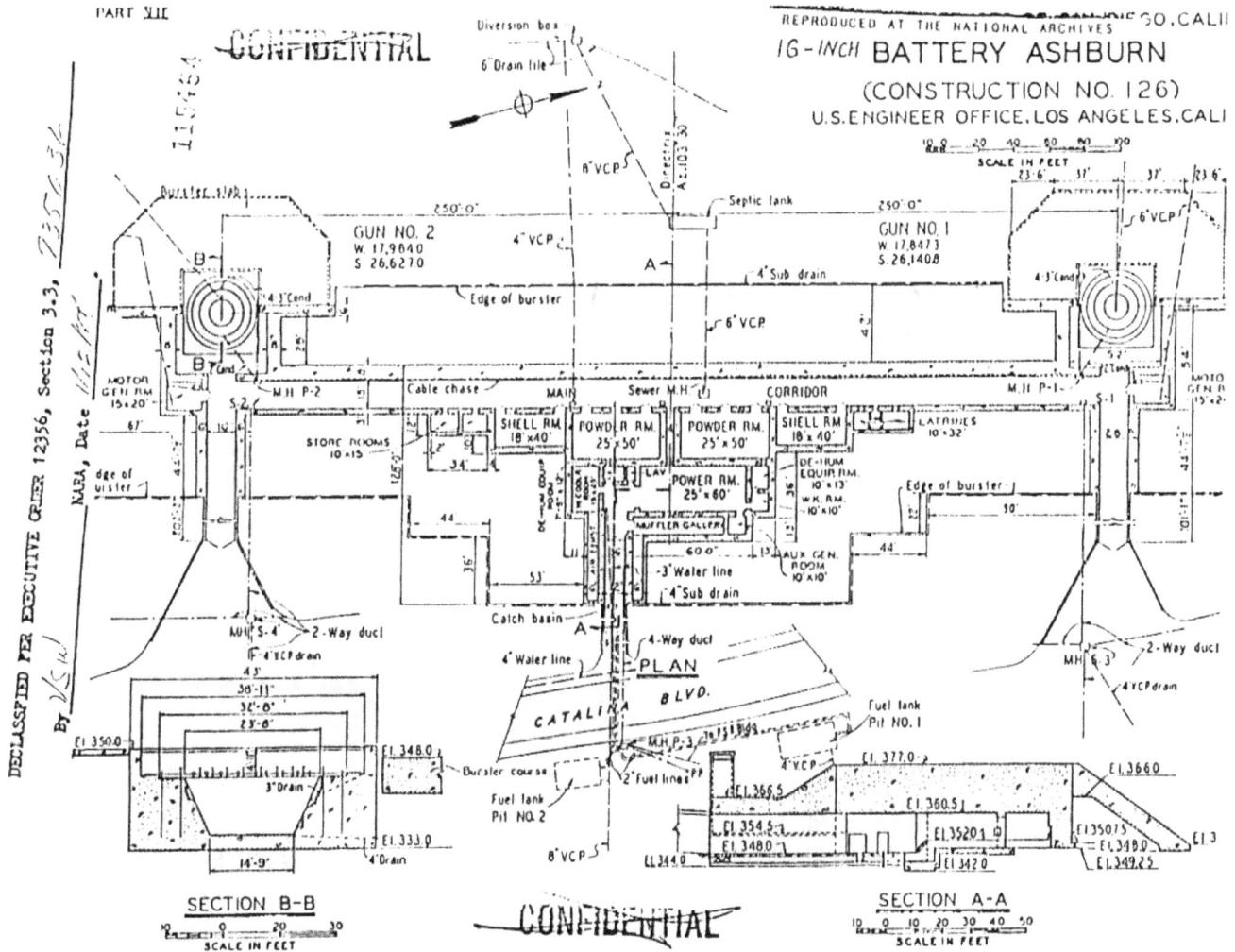

Figure 5-54. The lateral gun separation on the *100-Series* bunkers was 500 feet, as opposed to 210 feet for the 6-inch *200-Series* (NARA). The southern extension of Catalina Boulevard (bottom center) through Navy property was more recently renamed Cabrillo Memorial Drive.

Battery Ashburn was located west of Catalina Boulevard / Cabrillo Memorial Drive, just above the current San Diego sewer plant down below on Gatchell Road. According to Part I of the Report of Completed Works for this battery (Figure 5-55), construction did not commence until 12 June 1942 and took almost 2 years, essentially completed by 16 March 1944 when the guns were mounted. The facility was officially transferred to the Coast Artillery on 19 October 1944, although the gun shields were not installed until February of 1945. The total cost of construction in 1944 dollars was a staggering $1,323,912.

600. 7/4 (San Diego) 115464

REPORT OF COMPLETED WORKS – SEACOAST FORTIFICATIONS
(Batteries)

115464

HARBOR DEFENSES OF SAN DIEGO
FORT ROSECRANS
BATTERY ASHBURN (126)   No. of guns 2
Caliber   16-INCH   Carriage Barbette

S. M-1

Part I   Corrected to 24 November 1944

GENERAL:
    Battery commenced: 12 June 1942
    Battery completed: 16 March 1944 (see Remarks)*
    Date of transfer: 19 October 1944
    Cost to that date: $1,323,912.38
    Materials of construction: Reinforced concrete
    Battery new or modernized: New

    Trunnion elevation in battery: 355.67
    Datum plane: Mean Lower Low Water

UTILITIES:
WATER SUPPLY
    Source of: City of San Diego Water Department
    Alternate source: None
    Size of main: 4-inch and 6-inch
SEWER:
    Connected to sewer: No
    Type of disposal: Septic tank
    Type of latrine: Syphon-jet

UTILITIES: (Contd.)
ELECTRIC POWER
    Sources of: Main generators (in power room)
    Procured & installed by: Ordnance
    Characteristics: Voltage 460 AC   3 Phase   60 cycle
    Number of units and capacity: 3 at 375 KVA; 80% P.F.
    Max. K.W. required for utilities: 18.00
    Max. K.W. required for nonbattle conditions: 20.00
    Commercial power provided: Yes Capacity 12000/480/240
**Auxiliary power unit provided: No
    Type of lighting fixtures: Direct
    Dehumidifying unit: None installed
    Rooms wet or dry: Dry
    How ventilated: Not ventilated
    How heated: Not heated
DATA TRANSMISSION:
    Type: M-5

REMARKS:
    * Gun shields not installed – scheduled delivery date:
       15 February 1945
    ** Commercial power is auxiliary source.

ARMAMENT

Emplacement Number	Cal.	Length	Guns Model	Serial No	Manufacturer	Mounted	Type	Carriage Model	Serial No.	Manufacturer	Serial No Traversing Motor
1	16"	816"	I	71	Bethlehem Steel	March '44	Barbette	M-4	31	Wellman Engr Co	FX9896
2	16"	816"	Mark II Model I	97	Watervliet (N. Y.), Arsenal	"	"	M-4	39	Watertown Ars.	HX9731

[Declared Surplus]   Disposition Form from Off. h... ... 10 May 45, f. R. CSGSP/1 21000
... ... No Map Installation... (... Surplus) includes Fire Control

Figure 5-55. The Report of Complete Works, Part I, for Battery Ashburn (126), corrected to 24 November 1944, shows construction began on 12 June 1942 and the battery was transferred on 19 October 1944.

Battery Ashburn was one of only two San Diego batteries to be casemated (guns enclosed within a reinforced concrete structure for protection from bombardment) as seen in Figure 5-56, the other being the 16-inch Battery Gatchell (134) at Fort Emory. The tradeoff was the reduced range of motion, in that the 360-degree traversal of the *M1919* carriage was constrained to 145 degrees by the casemate walls; the maximum elevation of 67-degrees was also reduced to 47 degrees by the casemate overhead. For the assigned coast-defense roll of engaging enemy warships at extended over-the-horizon ranges, however, these tradeoffs were deemed acceptable in light of the perceived defensive advantages.

Figure 5-56. Undated wartime aerial photo of the southern (Gun 2) casemate with the 16-inch gun installed (US Army photo courtesy Cabrillo National Monument). Note the three concrete approach aprons (up arrows) for the rear entrances to the battery along Catalina Boulevard / Cabrillo Memorial Drive.

Trucking in a 69-foot tube that weighed over 150 tons was no trivial matter, however. Charles Schiller, who served at Fort Rosecrans in Battery C, First Battalion, 19th Coast Artillery, recalls some of the equipment and challenges involved (Overton, 1993):

> "When they brought that gun tube in they were coming up the hills down in Point Loma and they had one of these big west-coast diesels, tandem axle, and that is where they had the front of the barrel on it, and the back end was three axles, solid rubber tires. They made a track in that blacktop about 3 inches deep."

According to Schiller, the truck's transmission eventually failed, and it had to be towed the remaining distance by two International Harvester *TD-18* tractors brought in from the Marine Base.

The guns were finally mounted and test fired in July of 1944 (Figure 5-57). According to Army veteran Oliver Unruh, who served as a gun mechanic at Battery Ashburn, each gun fired only one round (Overton, 1993):

> "It was fired along the coast. It went over Mission Beach if I'm not mistaken, a little town there. And spotter came back and they never found the splash. Everybody says the spotters were looking at the people who came out of their houses and apartments. For the commotion that was going on, it might have been an attack of some kind."

Figure 5-57. Imagery apparently extracted from 16-millimeter wartime footage showing the muzzle blast from what appears to be Gun 1 during test firing in July 1944 (US Army photo courtesy Cabrillo National Monument).

Unruh further explained that initially only one 16-inch gun could be operated at a time (Overton, 1993):

"First the governors they had on them pumps, when both guns go in, the motors would slug down just like that. They couldn't take it. They had electrical governors; they wouldn't open up fast enough. The elevating mechanism was an 80-horse motor and they were instant starting. You hit that button and they were there. And if both of them would do it, boom, boom, there was no way they could handle it."

This design oversight was eventually resolved by upgrading to a new governor.

The 16-inch 50-caliber Navy guns of Battery Ashburn had originally been intended for installation in the *USS South Dakota*-class battleships and the *USS Lexington*-class battle cruisers. When construction of these ships was halted in compliance with the Washington Naval Treaty of 1922, some 20 of the already completed guns were transferred to the Army for coast-defense use. The first six of these were installed at Fort Kobbe, Panama, and Fort Weaver in Oahu, HI. Serial numbers 71 and 97 were later assigned to Guns 1 and 2, respectively, at Battery Ashburn. Serial number 138 is on display at the former site of US Army Ordnance Museum in Aberdeen, MD (Figure 5-58).[12]

Figure 5-58. Beth Gwaltney (left), Test Director at US Army Aberdeen Test Center, and former Test Director Cindy Van Seeters (right), stand next to a 16-inch Navy gun at the US Army Ordnance Museum, Aberdeen, MD, circa July, 2007 (US Army photo courtesy Beth Gwaltney).

12   The Army Ordnance Museum finalized its move from Aberdeen, MD, to Fort Lee, VA, in 2011, but this massive 16-inch gun and barbette carriage understandably remained at Aberdeen.

The term "caliber" refers to the ratio of barrel length to bore diameter, hence the total length of a 16-inch 50-caliber barrel is 50 times 16 inches, or 800 inches (66.7 feet). A wire-wound construction technique was employed, wherein a square (0.1-inch on a side) steel wire was wrapped around an inner steel tube, into which was inserted the rifled barrel liner. The outer hoops and tubes were then shrink-fitted over the wire windings (Tilden, 2004). Maximum reinforcement was obviously required at the breech, which had an outer diameter of 5 feet, tapering off towards the muzzle to yield the stepped diameter profile seen in Figure 5-58 above. The end result was an enormous amount of weight, which had to routinely be raised and lowered in elevation with considerable precision and repeatability.

To facilitate such, the location of the horizontal pivot axis of the barrel was carefully chosen to produce two delicate states of minor imbalance: 1) unloaded, the gun was muzzle-heavy, which allowed the human operator to more easily raise the breech to the proper position for loading; 2) once the 2,100-pound projectile and bags of propellant were inserted (Figure 5-59), the gun became breech-heavy, thus providing a gravity assist when manually raising the barrel to the proper elevation for firing (Tilden, 2004). Together the gun and carriage weighed over 660,000 pounds, supported by 44 conical roller bearings that allowed a single soldier to manually traverse the weapon in azimuth using a crank handle.

Figure 5-59. The 16-inch projectile weighed 2100 pounds, not counting the powder bags, and had a maximum range of 44,680 yards (22 nautical miles). Note relative size of the woman standing next to the gun barrel in the post-war photo on display in the Military History Museum, Cabrillo National Monument.

Large DC electric motors driven by amplidynes, however, were the primary means of actuation for both train and elevation on the *M1919M4* barbette carriage. As further explained by gun mechanic Oliver Unruh (Overton, 1993):

"There was not enough juice there to operate them on commercial power. So we had the diesel engines to take them over."

Unruh is referring to the six-cylinder diesel engines that drove the three 65-kilovolt AC generators in the power room, which in turn powered the amplidynes controlling the train and elevation motors of the gun carriage.[13]

13 The amplidyne was an electro-mechanical power amplifier used at the time for large DC motor control, as further discussed in Chapter 8.

To escape the severe concussions created when firing 16-inch projectiles, the underground Plotting and Switchboard Room (PSR) for Battery Ashburn was situated across the street and some distance to the north (Figure 5-60). Note the dirt access path from the PSR runs parallel to Cabrillo Memorial Drive, loops east along Ashburn Road, then turns north to Battery White in the upper-left corner. This PSR was serviced by five base-end stations spread out along the Pacific coastline from Site 1 in Solano Beach southward to Site 13 at the Mexican border (Chapter 7), with the battery-commander station in the northern part of Fort Rosecrans.

Figure 5-60. The Plotting and Switchboard Room (PSR) for Battery Ashburn is on the east side of Cabrillo Memorial Drive at photo left, circa 1942. The rectangular structure (up arrow) under construction behind the battery is fuel tank number 2, previously shown in Figure 5-54 (US Army photo courtesy SSC Pacific).

The area to be occupied by the battery was first leveled to roadway elevation prior to construction, after which the completed structure was covered with the leftover dirt. The 1942 photo above shows the partially completed Gun 1 position, while work on the Gun 2 position (out of image to the right) had not yet begun. An earth-covered 8-foot sloping concrete burster slab extended the full length of the strike face between the guns, intended to detonate incoming rounds before full penetration.

The massive amount of steel-reinforced concrete used in the construction of Battery Ashburn means it will not likely ever go away, as it would cost far more to demolish the structure than it took to build it.[14] The main roof above the transverse corridor, magazines, and power room is a staggering 16.5 feet thick, built to withstand the 18-inch guns of the *Yamato*-class battleships. The exposed frontal shields were protected from above by a semicircular conical canopy of reinforced concrete (Figure 5-61),[15] which measured 6 feet thick (vertically) at the perimeter and 12 feet at the apex.

Rendered obsolete by new technology developed during the war, Battery Ashburn was declared surplus on 10 May 1948,[16] whereupon the big guns were publicly offered for salvage. Research notes compiled by Al Grobmeier (2008) regarding a San Diego newspaper article dated 28 July 1948 provided the following details:

14 A few years after I first wrote this, the Navy elected to completely remove the identical 16-inch Battery Gatchell at the headquarters of the Navy Special Warfare Command, as discussed later (see Figure 5-92).

15 The RCW, Part I, corrected to 24 November 1944 indicated the gun shields were scheduled for delivery on 15 February 1945.

16 Per a handwritten footnote on the battery's RCW, Part I, 24 November 1944.

"Two 16-inch coastal defense guns bought by William and Jack Rubin, 733 Columbia St., for low bid of less than $6,000; will be resold as scrap."

Figure 5-61. The southern Gun 2 emplacement, showing the massive steel-reinforced concrete casemate that protected the gun. The two men standing on the almost 67-foot barrel supposedly were the scrap dealers that purchased the steel for salvage in 1948 (undated US Army photo courtesy Cabrillo National Monument).

The aerial photo presented in Figure 5-62 below shows the burn marks and residue from the northern Gun 1 salvage operation just inside the north entrance on the left side of Cabrillo Memorial Drive. Two surviving *M1919 Mark II* gun tubes are preserved at the Naval Surface Warfare Center in Dahlgren, VA, with a third at the Washington Navy Yard, all without carriages. As previously mentioned, the only fully intact specimen is the *Mark III* gun on an *M1919* carriage in Aberdeen, MD.

Figure 5-62. Looking north along Catalina Boulevard / Cabrillo Memorial Drive, the debris in the north entrance is from the gun salvage operation, circa 1948 (US Army photo courtesy Cabrillo National Monument). The battery's underground PSR is just east of the one-lane dirt road in the upper-right corner.

The other one-lane dirt road in the upper-left corner of Figure 5-62 provided access to the front of Battery Ashburn during the gun-salvage operation. The undated (late 1940s or early 1950s) photo presented in Figure 5-63 below shows evidence of considerable debris following this salvage work at the battery's southern Gun 2 position. The chain-link fence and gate were presumably erected for safety reasons to keep out unauthorized personnel during this time frame. Note spillage from above on the conical concrete canopy above the emplacement.

Figure 5-63. Post-war exterior view of the southern Gun 2 position, looking southeast, prior to renovation by the Center (then U.S. Navy Electronics Laboratory) for post-war use (undated US Navy photo courtesy SSC Pacific).

The photo presented in Figure 5-64 shows Battery Ashburn's southern (Gun 2) pit at some point after the metal-salvage operation in 1948. A portion of the overhead rail system for ammunition handling can be seen in the upper background. Note what appears to be remnants of wooden forms and shoring from the original concrete pour on both the bulkhead and overhead at far left. The large diameter of this gun pit provides perspective on the size of these massive 16-inch former-battleship guns. Note remnants of the safety chains and stanchions around the pit perimeter.

Figure 5-64. Interior view of the southern Gun 2 position following salvage of the *M1919M4* barbette carriage and 16-inch *M1919 Mark II* gun (see also Figure 5-86). The overhead rail system was used for ammunition handling around the periphery of the pit (undated US Navy photo courtesy SSC Pacific).

The 1962 aerial photo presented in Figure 5-65 below shows the ongoing construction of the new San Diego sewer plant below Battery Ashburn on Gatchell Road, just south of Center property. The battery's northern Gun 1 (left) and southern Gun 2 (right) emplacements are marked by the two up arrows at the top of this image, behind and above the sewer plant. Catalina Boulevard / Cabrillo Memorial Drive is out of image to the top on the east side of the battery.

Figure 5-65. The two 16-inch-gun emplacements for Battery Ashburn (up arrows at top center) are located upslope and east of the new San Diego sewer plant on Gatchell Road, seen here under construction in 1962 (US Navy photo courtesy SSC Pacific).

The concrete water-rheostat tank that served as a post-war test load for the generators in the power room of Battery Ashburn (recall prior rheostat-tank discussion for Battery Woodward) is seen below in Figure 5-66, looking northeast in April 1971. This tank is considerably larger than that behind Woodward (see again Figure 5-24 and Figure 5-25), probably because the generators in Ashburn were more powerful than those in Battery Woodward. Note the power poles along Catalina Boulevard / Cabrillo Memorial Drive at upper right, with San Diego Bay in the background.

Figure 5-66. The power-room air intake and exhaust vents (bottom right) are seen atop Battery Ashburn, looking north in April 1971, with the top portion of the concrete water-rheostat tank just left of center (photo courtesy Lee Guidry).

Figure 5-67 shows a new entrance to the battery constructed by the Naval Electronics Laboratory Center (NELC) at the former northern emplacement (Gun 1) of Battery Ashburn, looking northeast circa 1972. A single glass personnel door can just be seen at the very front of the semicircular addition in the lower-left corner. Note the covered air intakes around the top of the glassed-in enclosure, with the vertical exhaust ducts above the bulkhead-mounted squirrel-cage fans on either side. What appears to be two air-conditioner compressors are seen on the ground at bottom center, with a window-type unit in the wall above.

Figure 5-67. A new entryway to the earth-covered Battery Ashburn bunker has been constructed at the former northern Gun 1 emplacement (1972 photo courtesy Al Grobmeier).

The repurposed north end of Battery Ashburn was refurbished by the Naval Ocean Systems Center (NOSC) in late 1970 to host the new Microelectronics Lab. Research with thin-film silicon on sapphire (TFSOS) sub-micron-device technology began here in 1979 (Peregrine, 2020), headed by Dr. Isaac Lagnado, senior staff scientist of the Marine Sciences and Technology Department. This important work included micro-circuit processing, device modeling, materials characterization, and electrical testing (Offord, et al., 1990). A portion of the lab interior is shown in Figure 5-68, circa April 2008.

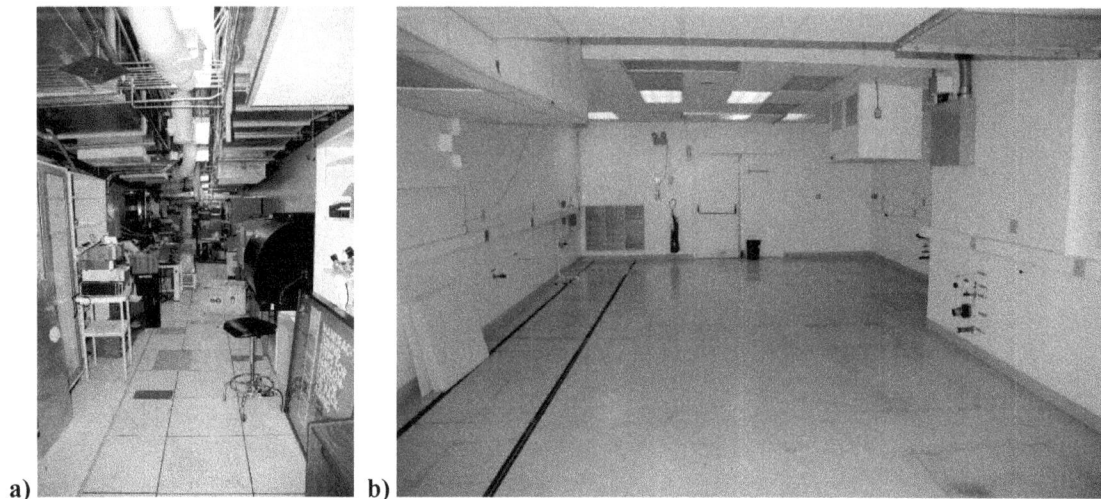

a)   b)

Figure 5-68. a) Interior view showing part of the Center's Microelectronics Lab in the north end of Battery Ashburn. **b)** This view of what once was the power room shows the original equipment removed and the floor raised to the level of the rest of the structure (April 2008 photos).

The 1989 aerial photo presented in Figure 5-69 shows the southern entrance at the former Gun 2 emplacement of Battery Ashburn, looking northeast towards North Island Naval Air Station, with the city of San Diego in the background. Note evidence of continuing development by the Naval Ocean Systems Center outside the former Gun 2 emplacement, with the new access road in foreground right. Building 555, the modern structure seen atop the bunker at image center, serves as the maintenance office for Cabrillo National Monument. Just to the left of this building are two white post-war ventilation stacks (up arrow) adjacent to the original air intake and exhaust vents for the power room.

Figure 5-69. View looking northeast of the southern Gun 2 emplacement of Battery Ashburn, with a portion of Ballast Point beyond the Point Loma ridge at upper left, and North Island and the city of San Diego in the background, circa 1989 (US Navy photo courtesy SSC Pacific). Note white post-war ventilation stacks atop the bunker (up arrow).

Figure 5-70 provides a closer view of the top of Battery Ashburn, looking southeast in 2008, with the old Spanish lighthouse at Cabrillo National Monument in the upper-right corner. The WWII-era air intake and exhaust vents for the power room are seen on either side of the newer ventilation systems for the Microelectronics Lab at image center. The large concrete rheostat tank (see again Figure 5-66) that served during the Cold War as a test load for the original diesel generators in the power room is in the bottom-right corner. Another air-conditioning compressor for the lab is seen at foreground center.

Figure 5-70. Close-up view of the post-war exhaust stacks seen earlier in Figure 5-69, looking southeast in 2008. Note the water-rheostat tank in the bottom-right corner, midway between and just west of the concrete WWII power-room air intake and exhaust vents, and the old Spanish lighthouse at extreme upper right.

Figure 5-71a below shows the inside of the water-rheostat tank, looking down from above at the remains of the wooden electrode-support disk. Three radial slots in this disk allowed adjustment of the separation between electrodes to increase or decrease the effective load. One of the cylindrical electrodes is seen laying on top of the disk, still attached to its double-aught-gauge cable, which leads back to the power-room switchboard. A second such electrode is submerged beneath the rainwater inside the tank, with the third lying on the ground next to the tank (Figure 5-71b).

a)   b)

Figure 5-71. a) Interior of concrete tank, circa 2008, showing the remains of a plywood support disk and a severely corroded electrode. **b)** Another electrode was found lying on the ground at the base of the tank.

Battery Humphreys

Named for Captain Charles Humphreys, the first commanding officer of Fort Rosecrans, Battery Humphreys was another 200-Series design. A total of 70 such installations were constructed, of which 48 were in the continental United States, 18 in the territories, and 4 overseas. Three of these were assigned to the HDSD project, two at Fort Rosecrans (Battery Woodward, No. 237, and Battery Humphreys, No. 238) and one at Fort Emory on the Silver Strand (Battery Grant, No. 239). Located above and to the east of Battery Point Loma, Humphreys was the first of the modernization batteries in San Diego to be completed, mounting two 6-inch 50-caliber rapid-fire guns (Figure 5-72). It provided substantially improved defensive coverage to the south, which was felt to be the most vulnerable direction of attack.

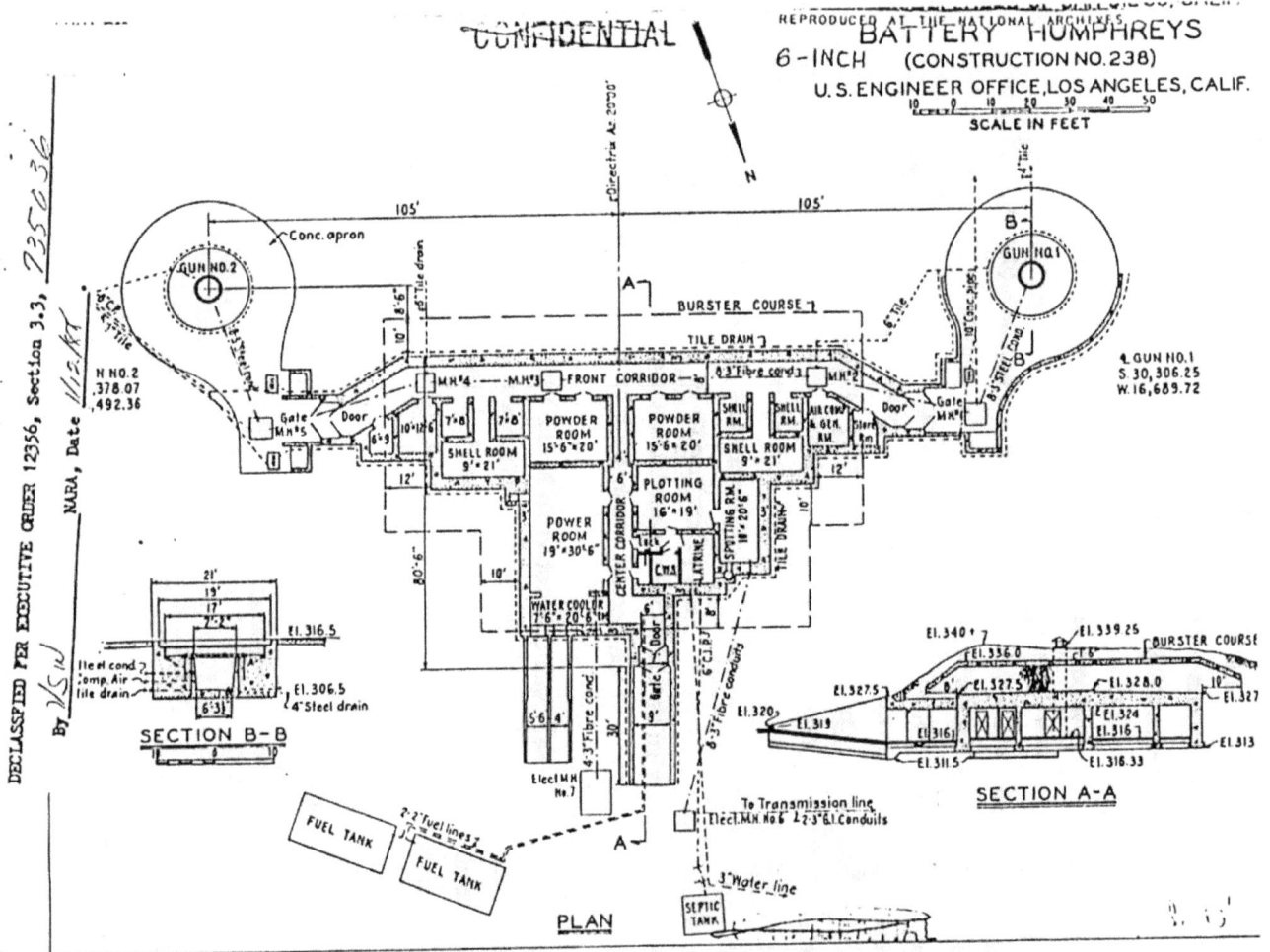

Figure 5-72. Design details of Battery Humphreys, which lacked a muffler gallery adjacent to the power room, as was seen earlier for Battery Woodward (NARA). Compare fuel tank locations behind the battery at lower left to the ongoing construction shown in Figure 5-73.

A 1942 aerial photograph of the battery during construction is shown in Figure 5-73. The 4-foot-thick steel-reinforced concrete roof, which has been completed on the western side of the structure, ultimately was covered by 6 to 7 feet of earthen fill. Above this was poured an 18-inch-thick burster slab that overhung the bunker perimeter by 10 feet on all sides, over which was then piled a minimum of 3 feet of dirt. Twin fuel tanks for the diesel generators are seen above and to the right, just across the dirt access road. Note the line of parked cars along the northern side of Humphreys Road, and the base-end station for Battery Strong under construction across the road and just south of the fence line.

Figure 5-73. Battery Humphreys is seen under construction at the southern tip of Point Loma on 5 May 1942. The wooden roof of the underground silo for abandoned WWI-era Searchlight 4 is partially visible in the extreme lower-right corner (US Army photo, LRO 110-04-67, courtesy Cabrillo National Monument).

The access hatch and roof for Searchlight 4 (up arrow) are again seen just south of the construction site in the lower-left quadrant of Figure 5-74, with a pathway and cable trace at upper left from the power station for Searchlight 17 (out of image). Since the 6-inch *M1903A2* guns for Battery Humphreys were not yet available when construction was completed in June of 1942, two 155-millimeter *GPFs* were borrowed from Camp Callan and temporarily emplaced beside the gun emplacements (HSD, 1945). Manned in July of that year by troops of Battery F, Battery Humphreys at this point replaced Battery McGrath as the "examination battery," covering the outer channel approaches to San Diego Bay.

Figure 5-74. Battery Humphreys has been covered with dirt, but the temporary 155-millimeter GPF guns have not yet been emplaced, circa June 1942 (US Army photo courtesy Cabrillo National Monument). Note the shelter for Searchlight 4 (up arrow), and the base-end station for Battery Strong in the upper right corner.

The *M1903A2* guns and their barbette carriages were finally mounted and proof fired in July of 1943, whereupon the 155-millimeter guns were removed, after serving as temporary stand-ins for just over a year (HSD, 1945).[17] Manned now by Battery A, Battery Humphreys replaced Battery Point Loma in the tactical plan for harbor defense (Gaines, 1993; Overton, 1995). Figure 5-75 shows the battery with its 6-inch guns mounted in 1943. Note the exposed rectangular barracks just across Humphreys Road to the northeast, directly behind Gun 2. The abandoned silo for Searchlight 4 to the south (down arrow) has by this point been converted into a support structure for the adjacent Battery Cliff.

Figure 5-75. Battery Humphreys with both guns mounted (1943 US Army photo courtesy Cabrillo National Monument). Note barracks behind Gun 2 on the far side of Humphreys Road, WWI base-end stations for Batteries White and Whistler at upper right (up arrows), and the repurposed Searchlight 4 silo (down arrow). Battery Cliff is the small rectangle just outboard (south) of the silo roof.

The 6-inch 50-caliber gun tube was approximately 300 inches long (for a total gun length of 310 inches), and achieved a maximum range of 27,500 yards with a muzzle velocity of 2,800 feet per second. The "estimated accuracy life" was 1,000 rounds, after which the gun tube had to be removed and sent to a suitably equipped arsenal for relining. A 41-ton metal shield provided protection from small-arms fire as well as bomb and shell splinters (Figure 5-76), its curved surfaces designed to better deflect incoming rounds. Total system weight (gun, carriage, and shield) was approximately 83 tons.

---

17 From 15 July 1942 until 30 September 1942, while equipped with 155-millimeter *GPFs*, Battery Humphreys had been manned by Battery F, 19th Coast Artillery Regiment (Gaines, 1993).

Figure 5-76. Shielded 6-inch *M1903* gun on an *M4* barbette carriage of the type installed at each end of Battery Humphreys, with its loading carriage on the radial rail tracks at lower right (US Army photo adapted from *TM 9-428*, 1943).

The *M1903A2* gun was manually loaded using a four-wheel ammunition carriage that moved radially along short tracks on the loading platform, which extended from the rear of the gun shield as previously shown in Figure 5-76 above. The inclination and height of the loading carriage ensured proper alignment with the breach when the gun was elevated to the prescribed loading position of 177.8 mils (10 degrees up elevation). The upper rotating turntable of the gun platform was supported by 40 conical rollers riding along the lower base ring (Figure 5-77), which was bolted to the foundation (TM 9-428, 1943).

Figure 5-77. The 40 conical rollers arranged radially along the lower base ring supported the racer (not shown), which was the upper rotating turntable of the gun platform (US Army photo courtesy Sitka Historical Society).

The projectile was hand-carried from the ammunition cart to the rear of the gun shield and placed on the concave surface of the loading-carriage tray, with the rammer head positioned immediately behind the projectile base. The loading carriage was then rolled up to the breech, at which point the projectile was rammed forward and securely seated in the forcing cone. The prepared powder charge was then placed on the loading carriage, its red igniter pad to the rear, and similarly inserted, after which the carriage was withdrawn and the breech closed.

The only serious Harbor Defenses of San Diego accident of WWII occurred on 29 January 1944, when a defective fuse on a high-explosive projectile caused a premature detonation at the Gun 2 position (Figure 5-78) of Battery Humphreys (Grobmeier, 2007a). One member of the gun crew was killed instantly and four died the next day (Overton, 1993), while nine more were wounded (Gaines, 1993). The gun tube and cradle were damaged beyond repair and had to be replaced, which took several months (HSD, 1945).

Figure 5-78. Post-war modifications to the outer doorway are evident at the east entryway to the transverse corridor of Battery Humphreys, as seen in 2006 from the adjacent Gun 2 emplacement area, which has been covered over with an asphalt parking lot.

Shortly after his June 1992 oral-history interview with Richard Markham, Battery A, 19th Coast Artillery, Howard Overton appended the following summary regarding this incident (Overton, 1993):

"Mr. Markham told me after our talk that he had been an observer at Battery Humphreys when the round accidentally exploded in the breech. The breechblock was not closed, it was still open. They rammed a round in and they hit it twice. The second time they hit it the round exploded. The rammer came back along with the ammunition tray and blew that man to pieces. The powder man was just coming out with two bags of powder in his hands and he got burned pretty badly and the other people were injured with the shrapnel. Markham was not hurt."

Charles Schiller, assigned to Battery C at Fort Emory in Imperial Beach, also shared his memory of the incident with Overton, as had been described to him by eyewitnesses at the time (Overton, 1993):

"Another guy that had the rammer in his hand, it hit him so hard against the concrete wall on the north side of it, it just left his imprint there. The guy that had the powder in his hand burned up right there with that powder. When you push that projectile in with the rammer and it don't seat, you come back a little and you push them again, when you hit it the second time, she went... Ordnance pulled that gun out and put a different one in. I would have liked to have (seen it), I was gun mechanic at that time over at Emory."

Julius Gary, Battery B, provided the following recollection (Overton, 1993):

"We had a guy down there named Alton. Remember Alton? He was my gun pointer down there and I loaned him to the 6-inch gun and he was on that 6-inch gun when it blowed (sic) up, but he was inside the turret and it didn't do anything but beat the thunder out of his legs, under the leggings."

First Lieutenant Hazen White, Battery A, had recently been temporarily reassigned to Fort Rosecrans when Fort Emory was downgraded to "semi-deactivated status." White had participated in only one gunnery exercise at his former duty station (where he served on Battery Grant), and that involving only sand-filled practice rounds. Upon hearing that Colonel Ottosen had ordered a live shoot at Humphreys with high-explosive projectiles, the young artillery officer had positioned himself on a nearby bluff to observe first hand.

First Lieutenant White recalled the following details (Berhow, 1994):

"It was first reported that when the projectile had been rammed home the first time, the faulty nose fuse became prematurely armed due to the jolt of the ramming. Subsequently it was extremely sensitive to any additional jarring or bumping. After the rammers, as they had in practice, rammed the base of the projectile the second time, the pin in the fuse set off the cup of fulminate of mercury, which in turn set off the explosive powder in the projectile. The projectile exploded in the tube, resulting in both a muzzle blast and a breach blast."

Figure 5-79. View of the former Gun 1 emplacement as seen from the battery's west entrance, circa May 2006, with the Pacific Ocean obscured by the thick marine layer in background.

The preceding Figure 5-79 shows the western Gun 1 emplacement of Battery Humphreys, looking southwest from the west entrance in May of 2006, with the Pacific Ocean obscured by the heavy marine layer in the background. The undated aerial photo presented in Figure 5-80 below, obviously taken some time afterwards, shows new construction underway at this former Gun 1 location. The eastern Gun 2 emplacement was beneath the two side-by-side trailers just right of image center. The post-war structure atop the battery more recently served as a Harbor Entry Command Post for the Naval Inshore Undersea Warfare Unit manned by the US Naval Reserve (Zink & Grobmeier, 1991).

Figure 5-80. The Gun 2 position beneath the two trailers just right of center is now covered over with an asphalt parking lot in this undated but fairly recent view of Battery Humphreys. New construction is underway at the former Gun 1 emplacement at center left (US Navy photo courtesy SSC Pacific).

As seen in Figure 5-81, the interior of Battery Humphreys is in remarkable shape, having been upgraded and well maintained by various tenants over the years. The cable chase in the transverse corridor has been covered over and the floor leveled and tiled. All of the original diesel and electrical equipment has been removed from the power room, with a false floor installed to bring it level with the rest of the structure. Like Battery Woodward, Battery Humphreys has been repurposed over the years for use by several different groups at the Center. Our Unmanned Systems Branch conducted autonomous mobile-robot testing here before gaining access to Battery Woodward.

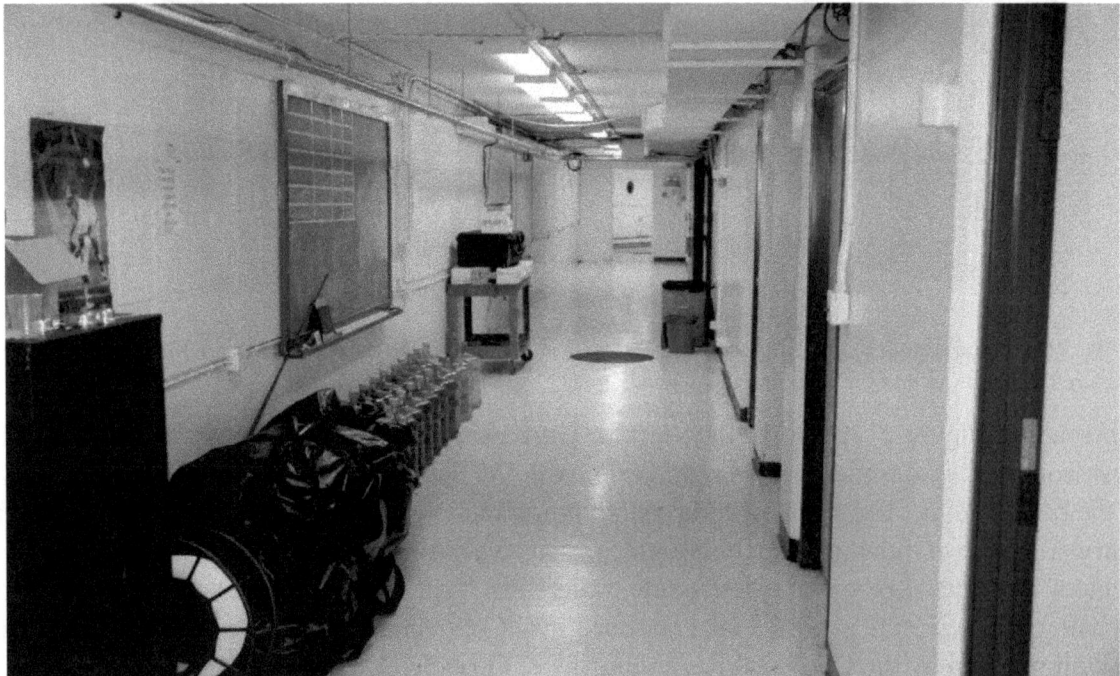

Figure 5-81. The interior transverse corridor of Battery Humphreys, looking west towards the Gun 1 entrance, circa May 2006.

Battery Construction Number 134 (Battery Gatchell)

Battery Construction Number 134 (Gatchell), a *100-Series* design located in Site 12 at Fort Emory on the Coronado Heights Military Reservation, was built between March 1943 and February 1944. Like the Plotting and Switchboard Room (PSR) for Battery Ashburn (Construction No. 126), which was dug-in and 875 feet away from the battery to minimize concussions when firing, the Plotting and Switchboard Room for Gatchell was similarly positioned. Due to the low terrain elevation and close proximity to San Diego Bay, however, the PSR bunker for Gatchell was constructed above ground and then covered with earth, as was the battery itself.

The completed battery was to be named for Brigadier General George Washington Gatchell, US Army, who served during the late 19th and early 20th centuries, but the name was never offically bestowed on the uncompleted battery. With the diminished Japanese threat following their resounding defeat at Midway in June 1942, however, the 16-inch guns were never delivered, as documented in the 1946 Annex A (Annex, 1946):

> "Construction terminated and installation of armament deferred on Battery Const. No. 134 in compliance with letter, WD Hq ASF, file SPRMS 660.2, dated 23 November 1943, subject: 'Curtailment of Construction on 16-inch Batteries, Construction Nos. 128, 134, and 129'."

As a consequence, the battery was never officially named "Gatchell," and construction of its proposed tower-mounted battery commander station BC_9 to the east was deferred as well (Figure 5-82).

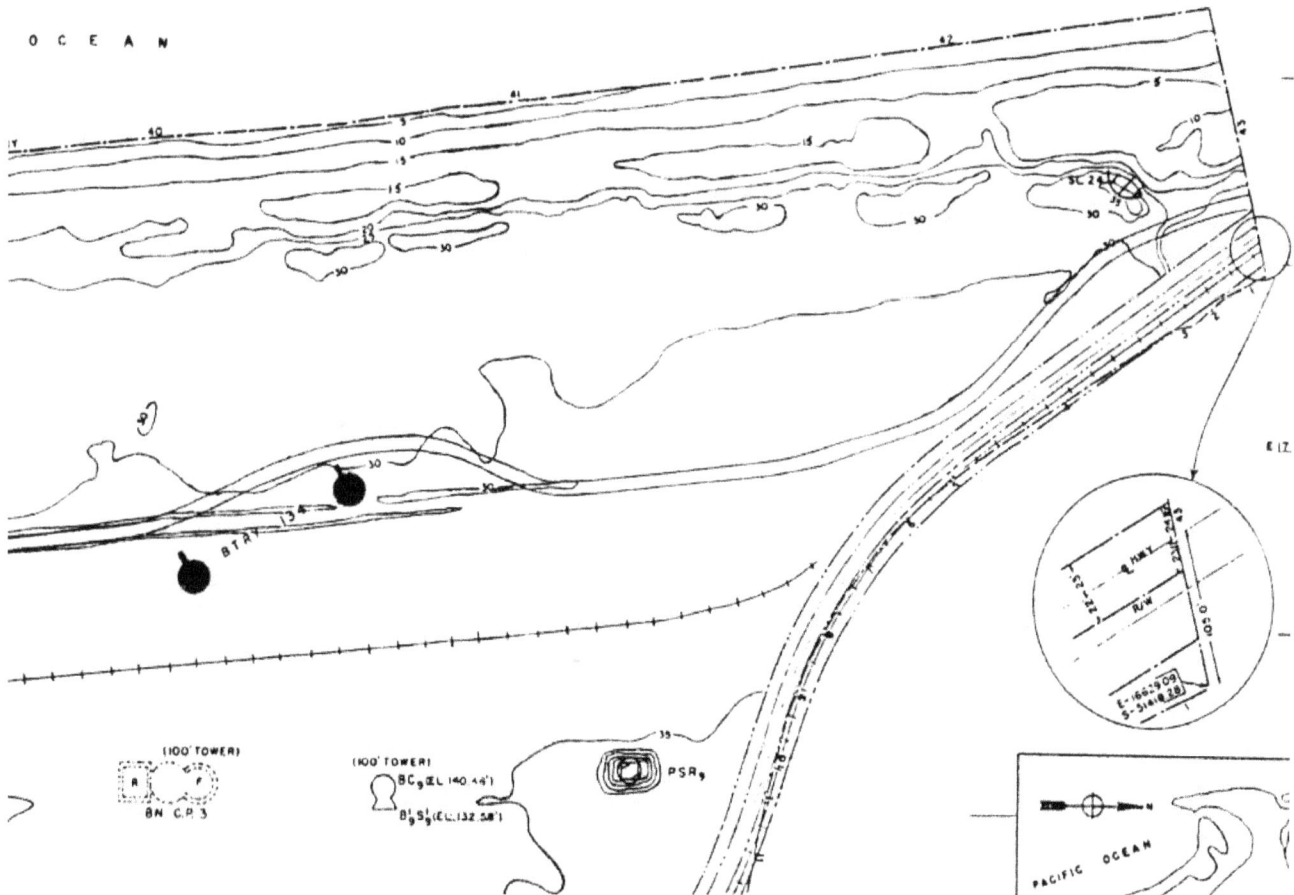

Figure 5-82. Battery 134 (Gatchell) was constructed at Fort Emory, with the proposed battery commander station BC_9 behind it to the east, midway between its Plotting and Switchboard Room (PSR-9, bottom center) and Battalion Command Post 3. The Pacific Ocean is across the top, with San Diego Bay at bottom right.

The Report of Completed Works blueprint presented in Figure 5-83 below shows the Battery Gatchell layout to be identical to that of Battery Ashburn at Fort Rosecrans. The 16-inch *Mark II Mod 1* Navy guns were to be mounted on *M1919M4* barbette carriages with a lateral separation of 500 feet and connected by a transverse corridor. As with Battery Ashburn, direct-access tunnels were provided behind each of the emplacement locations to facilitate installation and removal of the 69-foot gun tubes.

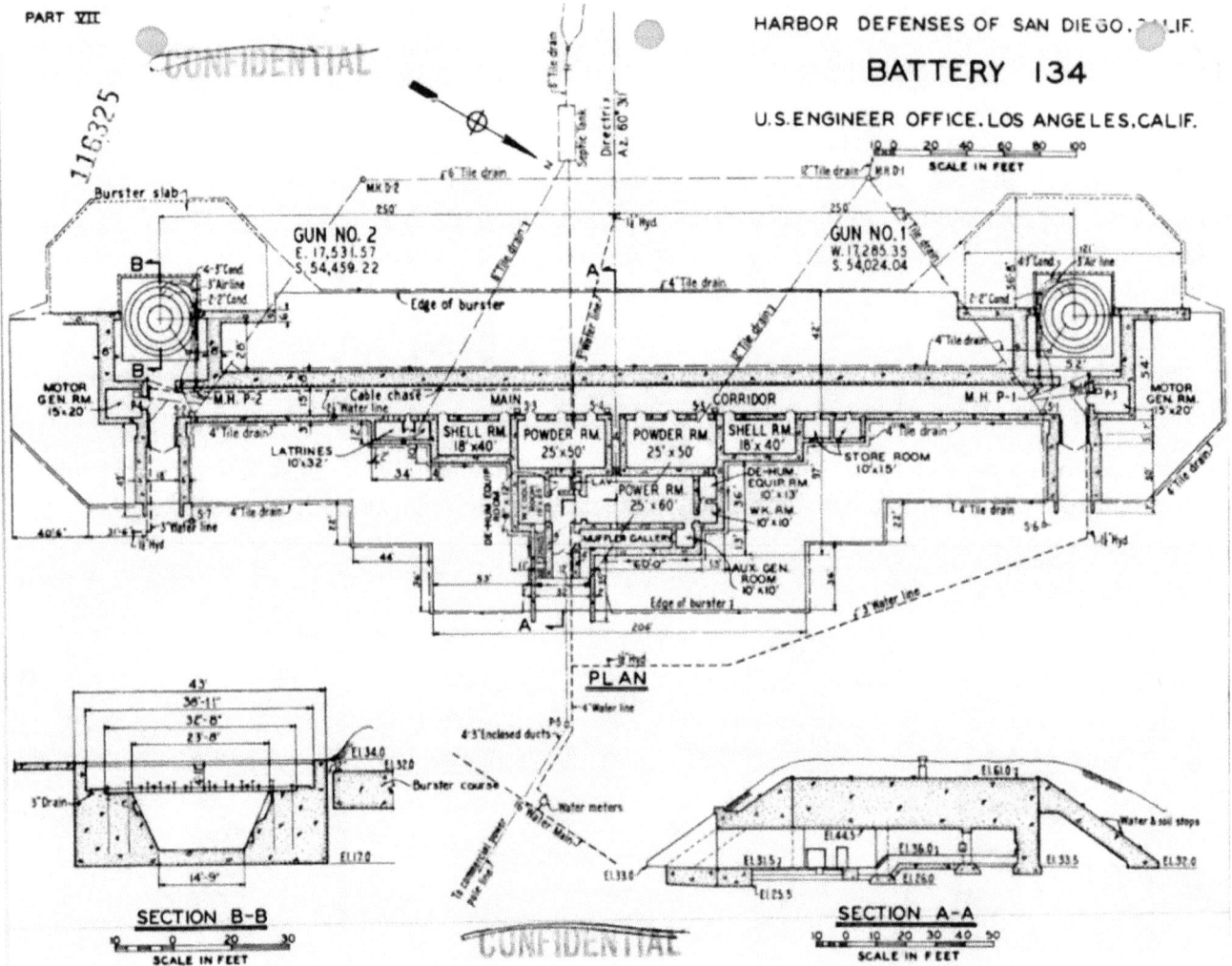

Figure 5-83. The RCW blueprint for Battery 134 (Gatchell) shows the same layout as Battery Ashburn (128) at Fort Rosecrans. Note the direct-access tunnels behind each casemate to facilitate installation/removal of the massive 69-foot gun tubes.

According to the Report of Completed Works, the unarmed Battery Gatchell (134) was officially transferred to the Coast Artillery on 11 November 1944, with a total construction cost of $1,044,907.29 in 1944 dollars. Part V of this report (see Figure 5-84), which is for the battery's standalone Plotting and Switchboard Room, contains an interesting footnote regarding that latter structure's source of electrical power:

"Note: Original Primary Source of Power: Main Battery (construction curtailed).
        Present   "   "   "   "   Commercial Power.
        "   Secondary "   "   "   Engine-Generator Set, above."

REPORT OF COMPLETED WORKS – SEACOAST FORTIFICATION   HARBOR DEFENSES OF SAN DIEGO
    (Electric Plants)   FORT EMORY

    POWER (AUXILIARY) FOR PSR BATTERY (CONST.) NO. 134

Part V   Corrected to 26 May 1944   PROCURED & INSTALLED BY ENGINEERS

95460

	Serial No.	Type	Size	Mfr.	Date of Transfer	Cost	Condition	RPM	Type of Drive	Radiator for Engine No	Fuel	Switchboard Material	No. of Panels
Engine No. 1	186521SP	D-4600	53 HP	Cater-pillar	25 Apr 44	*	Good	1200	Direct		Diesel Oil		
No. 2	--			pillar									
No. 3	--			Diesel									
No. 4	--												
Generator													
No. 1	30HK69	Alt.	37.5 KVA	Louis Allis Co.	"	*	"	"	"		--		
No. 2	--												
No. 3	--												
No. 4	--												
Radiator (or Water Cooler)													
No. 1	None	(Contained in unit)											
No. 2	--												
No. 3	--												
No. 4	--												
Motor Generator													
No. 1	--												
No. 2	--												
No. 3	--												
Switchboard	65070L	3 phase	--	Trumbull Elec.Co.	"	*	"	--	--	--			2
Storage Battery	None	--	Heavy Duty	Exide	"	*	"	--	--	--			
Transformer													
No. 1	#8019	Dry	15 KVA	Kuhlman Elec.Co.	"	*	"	--	--	--			
No. 2	"	"	"	"	"	*	"	--	--	--			
No. 3	"	"	"	"	"	*	"	--	--	--			
Auxiliaries													
No. 1	--												
No. 2	--												
No. 3	--												

Note: Original Primary Source of power: Main Battery (Construction curtailed).
    Present   "   "   "   "   Commercial power.
    "   Secondary "   "   "   Engine-Generator Set, above.

    * Cost included in Construction Contract.

Figure 5-84. This 26 May 1944 Report of Completed Works (Part V) for Battery Gatchell's nearby Plotting and Switchboard Room (PSR) contains an interesting footnote (bottom of form) regarding its secondary (backup) source of electrical power.

The above footnote in the figure outlines a history of three different sources of electrical power for the PSR (Figure 5-85):

1. The PSR was to have originally been supplied by the "Main Battery," meaning the four diesel generators in the power room of Battery Gatchell, but construction of this battery was later curtailed.
2. The PSR was eventually connected to commercial power.
3. The PSR was to have received backup power from "Engine-Generator Set, above," which is rather vague wording that seems to imply a single unit.

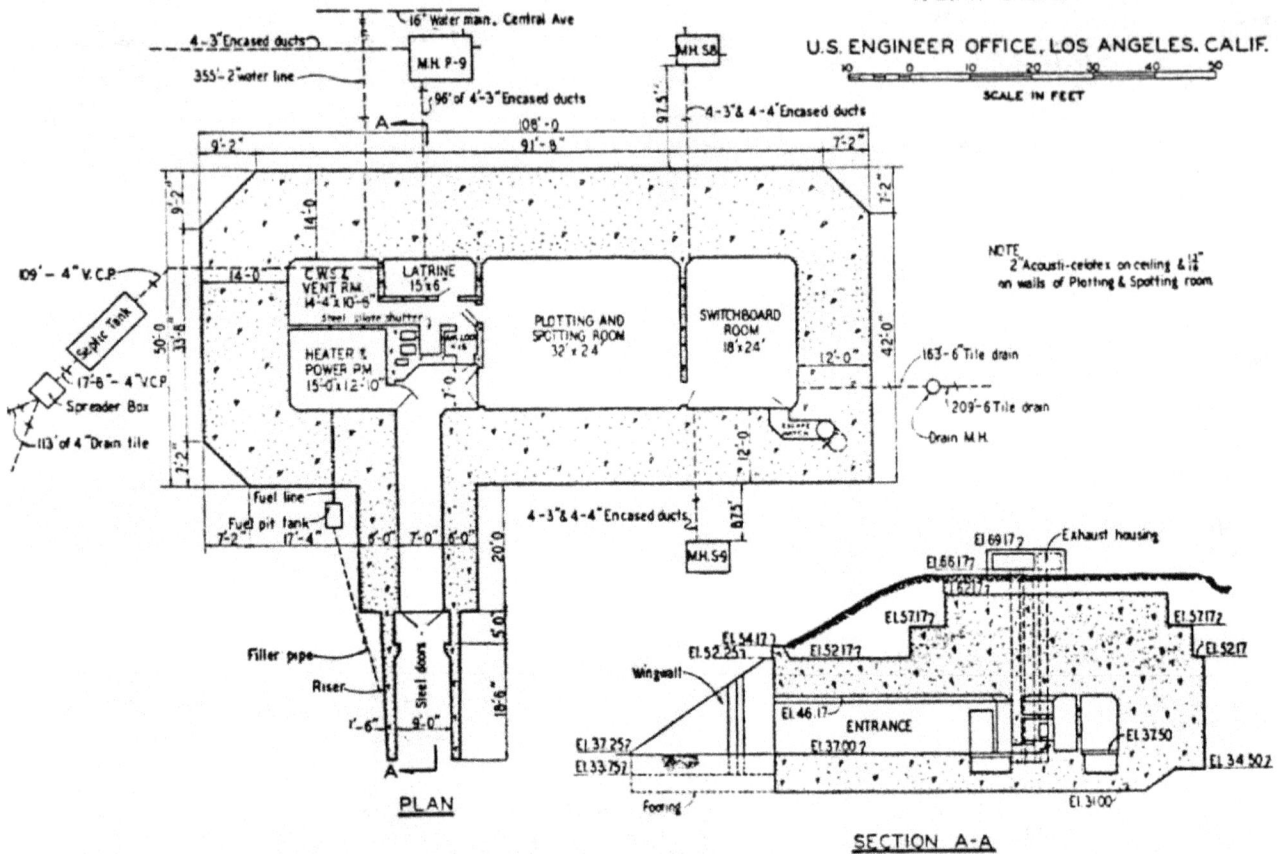

Figure 5-85. The Heater and Power Room (bottom left of internal floorplan) for Battery Gatchell's Plotting and Switchboard Room (PSR) measured 15 feet by 12 feet 10 inches. This underground structure was transferred to the 19th Coast Artillery Regiment on 25 April 1944.

The confusion here arises from the fact that there were to be four 53-horsepower 35-kilowatt diesel generators in the power room of Battery Gatchell, not one, as seemingly implied by the poorly worded Note 3 on the RCW (Part V) form. The question thus becomes, does the term "set" in line 3 of the footnote refer to a singular engine-generator "set," or to a "set" of four engine generators? The issue is rather moot, however, as the powerplant equipment in Battery Gatchell was never installed,[18] per War Department direction in February 1944 (Ruhge, 2016).

Battery Gatchell was to have been armed with the same 16-inch Navy guns as had been emplaced in Battery Ashburn at Fort Rosecrans, mounted on *M1919M4* barbette carriages with frontal shields. Figure 5-86 shows a rear view of the last surviving US example of such, minus the shield, mounted on "U.S. Barbette Carriage *M1919* Serial Number 1." This gun is actually an *M1919 Mark III*, slightly modified to incorporate minor differences from the *Mark II* barrel hoops and rings. As previously noted, this important historical artifact now resides at the former location of the US Army Ordnance Museum in Aberdeen, MD.

---

18 For confirmation, see HSD, 1945.

Figure 5-86. Rear view showing the *M1919* proof carriage (Serial 1) for the 16-inch coast-defense gun at Aberdeen, MD, circa July, 2007 (US Army photo courtesy Beth Gwaltney).

In 1920, the US Navy had established a direction-finding (DF) station on the Coronado Strand Military Reservation (NSGA, 2017), and later acquired another 145 acres in 1941 for development of a new radio receiving station at Imperial Beach (NCSSD, 2020). This new station, which would serve as the receiving terminal for all Navy ship-to-shore CW radio traffic,[19] was apparently staffed by the Fleet Radio Unit Imperial Beach, for which the 1945 wardroom roster is reproduced in Figure 5-87 below.

RMS Cl#	Name	Rank	Retired	RMS Cl#	Name	Rank	Retired
22	ABERNATHY, Marlo G	CRE	CWO2 '53		GABBARD, Eugene G	Ltjg	CWO2 '53
14	JANSAK, Joseph Jr	Ltjg	Ltjg '50				
					HAINES, Robert A	Ltjg	
	BALSLEY, Martha C (SC)	Ltjg			HARDISON, Wallace L	SClk	
	BENJAMIN, Ivan S	Ens	LCdr '56		HARRISON, Henry S	LCdr	Cdr '55
	BRONNER, Charles H	Lt					
	DAVIS,	Ltjg			NICHOLSON, Louis L Jr	Ltjg	LCdr '58
	DEARDORFF, Beverly A	Ens					
					SCHROEDER, Edward M	Ens	LCdr '58
18	EDENS, William J	Ltjg	LCdr '53		SUNDERLAND, Dorothy A	Ltjg	
					WILDER, Clifford O	Ens	Lt '56

Figure 5-87. Listing of US Naval officers assigned to Fleet Radio Unit Imperial Beach, 8 February 1945 (adapted from NCSSD, 2020).

The Navy's original radio receiving station on Point Loma began relocating its equipment to the new Imperial Beach station in 1947 (NCSSD, 2020):

"To enable the growing Navy Electronic Laboratory to expand, the disestablishment of the Point Loma station began in 1947 as receivers were transferred to Imperial Beach. The radio station at Imperial Beach, commissioned in 1920 as a direction-finder station to provide navigational aids to ships at sea, became known as Naval Radio Station (R), Imperial Beach."

---

19 CW – Continuous wave Morse Code.

Naval Radio Station (R) later relocated to the abandoned Battery Gatchell facility, but retained the Radio San Diego call sign "NPL."

As recalled by Commander (Ret.) Alvin H. Grobmeier (1990), who served as Assistant Officer in Charge at this facility from 1958-1960:

> "The 500-foot main corridor tunnel of 134 was inclined slightly upward from the emplacements at each end for about 100 feet, with the center 300 feet being level and slightly elevated... It was in this 300 feet that the Navy placed the operating positions and radio receivers for one of its west-coast radio stations. The south shell room and powder room became the officer-in-charge's office and administrative office, respectively."

To accommodate these new tenants, Battery Gatchell's gun emplacements were filled in and the casemate openings boarded over as seen in Figure 5-88 below, with emergency power provided by the diesel-generator sets in nearby Battery Grant (Grobmeier, 1990).

Figure 5-88. Front view of the northeastern casemate for Gun 1 at Battery Gatchell (134) as it appeared in 1958, the new home to the Naval Communication Station (NAVCOMSTA), San Diego (20 November 1958 US Navy photo courtesy SSC Pacific).

Figure 5-89 provides a post-war view of the former Gun 1 position at the northeast end of Battery Gatchell, with the Gun 2 position seen in background right, looking more or less southeast in June of 1972. A better perspective of the Gun 2 position is presented in Figure 5-90a. Note the various antennae atop the bunker associated with its new role as the Naval Radio Receiving Facility. Figure 5-90b shows a close-up view of the Gun 2 position, with U-shaped rebar attachment points for camouflage netting embedded in the projecting overhead.

Figure 5-89. The northwest Gun 1 emplacement, with the southeast Gun 2 emplacement in background right (3 June 1972 photo courtesy Al Grobmeier).

Figure 5-90. a) The southeast Gun 2 emplacement at image right, with the northwest Gun 1 emplacement at far left. **b)** Close up of the Gun 2 emplacement (3 June 1972 photos courtesy Al Grobmeier).

The elevation of this narrow strip of land was only 37 feet above mean lower low water (MLLW). Accordingly, Battery Gatchell had been constructed above ground and then covered over with dirt as clearly seen in Figure 5-91, which shows the repurposed bunker some 20 years later in 1992, looking northeast. Various telecommunication antennae are seen mounted on the northwest (Gun 1) end of the bunker at far left. For reference, note the relative size of the vehicles and men at photo center.

Figure 5-91. Front view Battery Gatchell looking east in 1992, showing the northwest Gun 1 emplacement at far left, with a portion of the southeast Gun 2 emplacement visible at far right (photo courtesy Al Grobmeier).

The Navy's Imperial Beach receiving site was later relocated to a two-story concrete building, with the old rhombic antennae replaced by satellite dishes and a Circularly Disposed Antenna Array (Grobmeier, 1990). The *AN/FRD-10* Wullenweber Circularly Disposed Antenna Array provided a significant improvement in high-frequency direction finding. This computer-based technology was used to track radio intercepts of Soviet naval activity during the Cold War (NSGA, 2017).

The first version of the *AN/FRD-10* Wullenweber system was introduced in 1961, having been prototyped by the Naval Research Lab and produced by ITT. The final installation took place 4 years later at the Naval Security Group Department of the Naval Communications Station San Diego in Imperial Beach (NSG, 2020), as seen in the undated photo presented in Figure 5-92. The Naval Security Group Activity San Diego was later commissioned at this location in 1998, with the array decommissioned on 30 September of the following year (NSGA, 2017).

Figure 5-92. The *AN/FRD-10* Wullenweber Circularly Disposed Antenna Array at Imperial Beach was used to track Soviet naval activity, particularly submarines, in the Pacific Ocean (undated, NARA). The various vehicles parked around the building give perspective to the massive size of this antenna configuration.

Figure 5-93 shows a more recent aerial view of a now fenced-off Battery Gatchell, which clearly reveals its identical layout to Battery Ashburn on the Point Loma peninsula. The northern Gun 1 position is seen at top center, with the Gun 2 position left of bottom center, and the rear entrance between them at middle right. The multitude of parked cars and the more recent structures erected in front of the southern emplacement and rear entrance indicate the former 16-inch gun battery was still in use by the Navy.

Figure 5-93. Looking northwest in 2014, the more recent structures erected in front of the Gun 2 position (bottom center) and near the rear entrance (center right) suggest Battery Gatchell was still in use by the Navy (photo courtesy Terry McGovern, Coast Defense Study Group).

Battery Gatchell was completely demolished in 2016 (Figure 5-94), necessitated by the lack of available real estate for a new billion-dollar Coastal Campus for the Navy Special Warfare Command (Axelson, 2016). Space is a premium on this narrow base, which is geographically bounded by San Diego Bay to the east and the Pacific Ocean to the west. According to demolition contractor Jason Golumbfskie-Jones, the base is also home to a number of protected species (Axelson, 2016), which further constrains expansion:

> "We have looked at how and where people can build or demolish during the different breeding seasons. On the south end of the campus we have 100 acres of vernal pools where we won't be developing, where there are fairy shrimp, snowy plover, and least tern."

Figure 5-94. The Gun 1 casemate on the northwest end of Battery Gatchell is seen during the Navy's demolition effort in 2016. Note the exposed portion of the massive (16-foot-thick) concrete roof at top center (photo courtesy Mark Berhow, Coast Defense Study Group).

Battery Grant

Battery Grant (239) in Site 12 at Fort Emory was named for Colonel Homer B. Grant, Coast Artillery Corps, who died in 1939. Like Batteries Woodward and Humphreys, this standard 200-series design was equipped with a pair of 6-inch *M1905* guns on shielded long-range barbette *M1* carriages. The layout of the reinforced-concrete bunker seen in Figure 5-95 was nearly identical to that of Battery Woodward (237) in Site 5 at Fort Rosecrans. Construction began on 5 June 1942 and the guns were proof fired in November the following year. The battery was completed 8 December 1943, with a total cost to that date of $218,851.95.

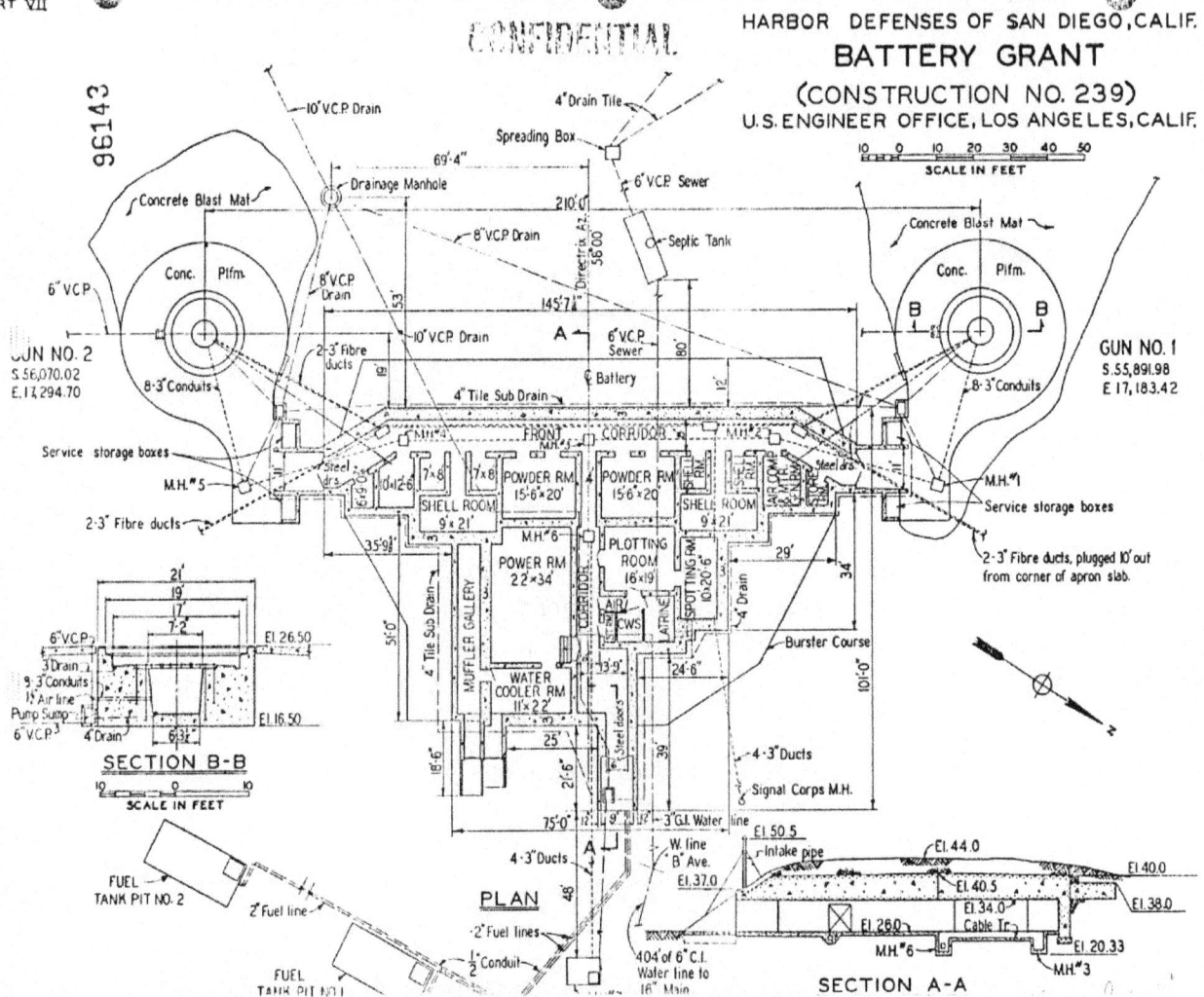

Figure 5-95. Battery Grant (239), a 200-series design at Fort Emory, had a nearly identical layout to Battery Woodward (237) over on Fort Rosecrans (adapted from Part VII of the Report of Completed Works).

The Report of Completed Works (Part I) presented in Figure 5-96 indicates the Plotting and Spotting Rooms were vented by an 18-inch-diameter pipe, as opposed to air conditioned like the PSR in Battery Woodward. Normal venting could also be accomplished by bypassing the CWS cannisters. Following transfer to the 19th Coast Artillery Regiment on 25 April 1944, this modernization battery superseded the temporary 155-millimeter Battery Imperial on the northwest side of Battery Gatchell, as was previously discussed in Chapter 4.

R_PORT OF COMPLETED WORKS – SEACOAST FORTIFICATIONS   ~~CONFIDENTIAL~~   HARBOR DEFENSES OF SAN DI
(Batteries)   FORT ~~ROSECRANS~~ (FORT EM(

BATTERY GRANT   No.
Caliber   6-INCH   Ca:

Part I   96143   Corrected to 3 June 1944   239

(handwritten: 600,914 (San Diego) 96143)

GENERAL:
   Battery commenced: 5 June 1942
   Battery completed: 8 December 1943
   Date of transfer: 25 April 1944
   Cost to date of transfer: $218,851.95
   Materials of construction: Reinforced concrete
   Battery new or modernized: New

   Trunnion elevation in battery: 31.0
   Datum plane: Mean Lower Low Water

UTILITIES:
   WATER SUPPLY
   Source of: California Water & Telephone Company
   Alternate source: None
   Size of main: 16-inch (6" line to battery-reduced to 3")
   SEWER:
   Connected to sewer: No
   Type of disposal: Septic tank
   Type of latrine: Frost proof, syphon-jet

UTILITIES: (Contd.)
   ELECTRIC POWER
   Source of: Main Generators (in power room)
   Procured & installed by: Ordnance
   Characteristics: Voltage 460 AC   3 Phase
   Number of units and capacity: 3 - 125 KVA;
   Max. K.W. required for utilities: 26.85
   Max. K.W. required for nonbattle condition:
   Commercial power provided: Yes   Capacity:
   * Auxiliary power unit provided: No
   Type of lighting fixtures: Direct (Wheele:
   Dehumidifying unit: None
   Rooms wet or dry: Dry
   How ventilated: 18" dia. vent pipe (For Pl(
             Spotting Rooms)(Normal ven1
             obtained by by-passing C.W.
   How heated: Electric (Forced-convection Ty;
   DATA TRANSMISSION:
   Type: M-7
   REMARKS:
   * Commercial power provided as auxiliary sou:
    Origin of U.S.E.D. local coordinates is U.:
    station "Old Town."

ARMAMENT

Emplacement"   '   '   '   Guns '   '   '   '   '   Carriages

Figure 5-96. Part I of the Report of Completed Works for Battery Grant (239), corrected to 3 June 1944, indicates construction started on 5 June 1942 and was completed 8 December 1943. Unlike Battery Woodward (237), the Plotting and Spotting Room for Battery Grant was not air conditioned.

The above RCW indicates the trunnion elevation of the two 6-inch guns was only 31 feet above mean lower low water, which required Battery Grant to be constructed above ground and covered with dirt (Figure 5-97), as had also been the case with nearby Battery Gatchell. This was for the most part standard procedure anyway for the 100- and 200-series batteries, as their gun emplacements, which had to be at ground level, were at the same elevation as the transverse corridor connecting them to the ammunition-storage rooms. The ready-service ammunition shelves either side of the northwest Gun 1 emplacement can be seen at photo left.

Figure 5-97. The front of the above-ground Battery Grant is seen looking northeast on 3 June 1972, with the northwestern Gun 1 emplacement at far left and part of the Gun 2 emplacement just visible at far right (photo courtesy Al Grobmeier).

The map presented in Figure 5-98 shows the location of Battery Grant (Tac. No. 10) at middle left, just south of the unfinished 16-inch Battery Gatchell (Battery Construction No. 134, Tac. No. 9) on the Coronado Beach Military Reservation in Site 12. The battery commander station BC_{10} and base-end and spotting station $B^1_{10} S^1_{10}$ for Battery Grant are behind the battery at bottom left. Note the four 155-millimeter emplacements of Battery Imperial in the upper-right corner, northwest of Battery Gatchell. The fire-control stations and Plotting and Switchboard Room for Battery Gatchell (Tac. No. 9) are at bottom right.

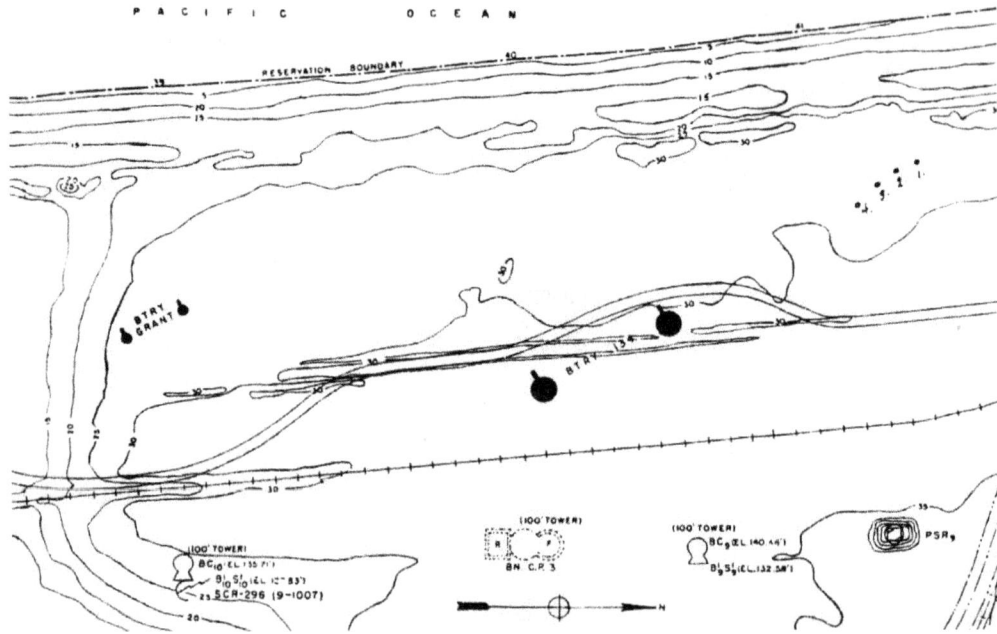

Figure 5-98. Battery Grant (left center) was located just south of Battery Gatchell (134) at image center on the Coronado Beach Military Reservation in Site 12.

Battery Grant was decommissioned in 1946, and later repurposed by the Navy. Very few post-war-era photos of this bunker have been located, presumably due to early Cold-War security measures. Figure 5-99, taken in 1972 by Coast Defense Study Group member Al Grobmeier, shows the rear entrance to the battery, looking south. What may have been the emergency battery commander station for Battery Grant is denoted by the up arrow in background right.

Figure 5-99. The rear entrance to Battery Grant is seen looking south, with the muffler gallery and air intake to the water cooler (radiator) room to its left (3 June 1972 photo courtesy Al Grobmeier). Note unidentified structure at upper right (up arrow), and the T-shaped antenna atop the bunker (down arrow).

The unknown structure marked by the up arrow in Figure 5-99 does not appear along the ridgeline in Figure 5-100 below, even though both photos were taken on the same day. This structure was probably situated slightly downslope on the bunker's rear side, and thus occluded by the peak in the front-view perspective. This theory is to some extent supported by the fact that the T-shaped antenna atop the bunker in Figure 5-99 above appears noticeably shorter in Figure 5-100 below. The consequent occluded view to the west afforded this structure from that downslope location, however, suggests it was not the emergency battery commander station.

Figure 5-100. The northwest Gun 1 emplacement of Battery Grant is seen looking southeast on 3 June 1972, with the T-shaped antenna in background center appearing much shorter than in Figure 5-99 (photo courtesy Al Grobmeier). Note ready-service ammunition shelves either side of the northwest entrance, repurposed for storage.

Due to the poor resolution and exposure of these 1972 photos, the T-shaped antenna is almost impossible to see in the printed versions of Figure 5-99 and Figure 5-100. Zoomed-in views of this antenna, extracted from these previous two images, are presented with digitally-enhanced contrast in Figure 5-101.[20] The distance between the photographer and the battery was about the same for each shot, given the similar height of the two concrete entrance surrounds. A good portion of the antenna pole seen in Figure 5-101b is clearly occluded by the earth-covered bunker, which suggests the unknown structure is similarly occluded from this same perspective in Figure 5-100 above.

Figure 5-101. a) This zoomed-in view of Figure 5-99 as seen from the rear of the battery shows a fairly tall T-shaped antenna at top center. **b)** This zoomed-in view of Figure 5-100 as seen from the front of the battery shows a much shorter version of the same antenna.

20   Images digitally enhanced by Wendy Kitchens.

6

AMTB and Machine Gun Batteries

"Goddamn it, you'll never get the Purple Heart hiding in a fox-
hole! Follow me!"

Captain Henry P. Jim Crowe

The newly authorized 6-inch and 16-inch modernization batteries (Chapter 5) were long-term construction projects, and with the threat of war looming large, America's seacoasts were woefully ill defended.
As previously discussed in Chapter 4, the Panama mount had allowed the 155-millimeter GPF guns to
quickly provide some interim measure of coast defense. A second expedient along these lines was to adapt
the existing 90-millimeter antiaircraft gun to this surface-engagement role (Hogg, 1998):

"A purely coast design was needed, and this resulted in the 90-millimeter *Gun M2* on *Mounting M3*. The
gun was in an armored turret on a pedestal mount, but was still capable of 80 degrees elevation, so that it
could still function in either the AA or the coast role. It was introduced in June 1943."

Three 90-millimeter anti-motor-torpedo-boat (AMTB) batteries had been authorized
for the HDSD project in the fall of 1942 (Figure 6-1): 1) Battery Cabrillo on Point Loma;
2) Battery Fetterman on Ballast Point; and 3) Battery Cortez at the Coronado Beach Military Reservation
(Thompson, 1991). Intended to address the growing fear of fast-moving surface craft coming in below the
coverage of the longer-range guns, all three were to be equipped with two fixed 90-millimeter (*M3*) and
two mobile 90-millimeter (*M1A1*) guns (Malone, 2004). The secondary mission for these batteries, each
of which had their own dedicated searchlight, was antiaircraft (AA) defense.

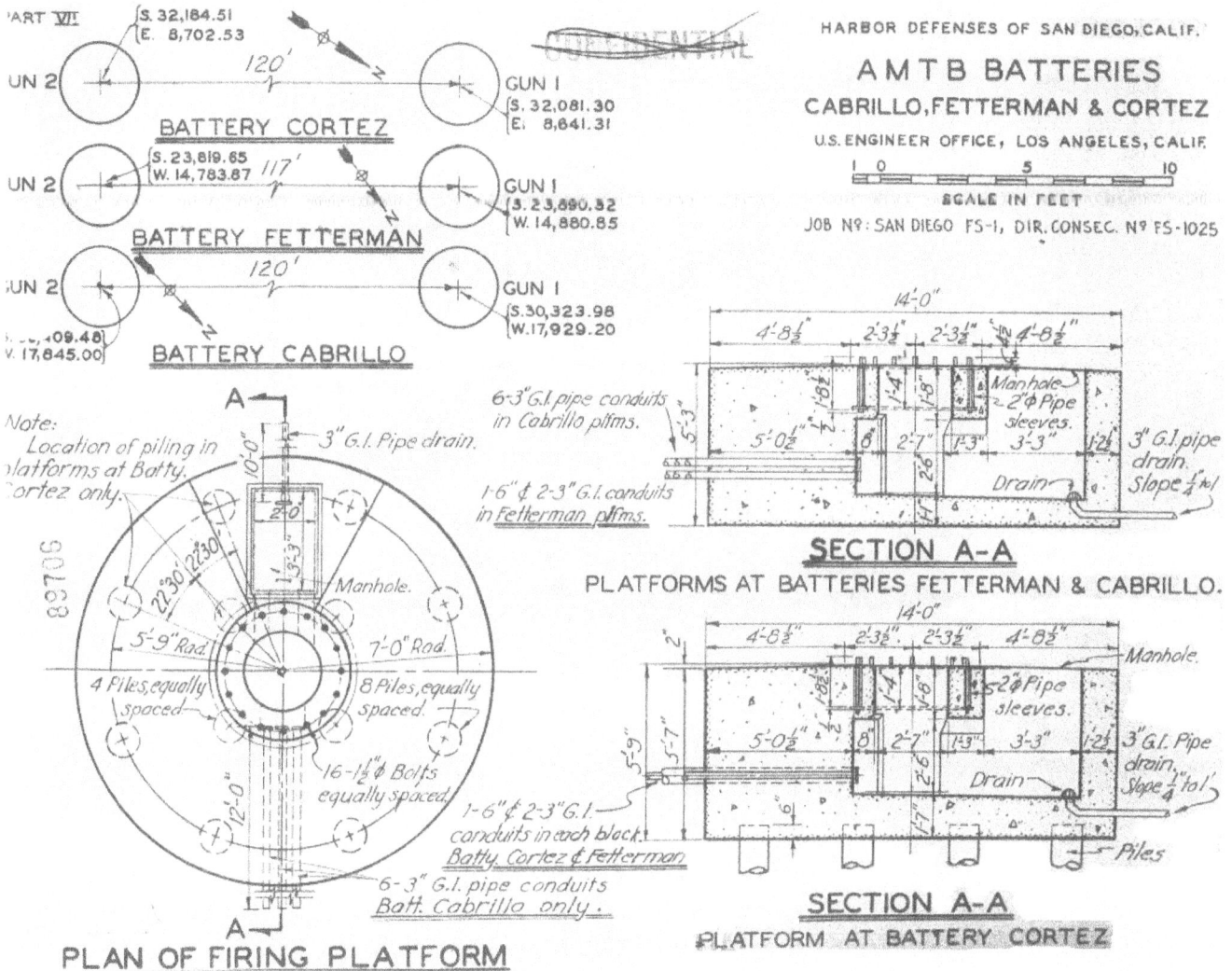

Figure 6-1. Installation plans for the 90-millimeter AMTB Batteries Cabrillo, Fetterman, and Cortez, Harbor Defenses of San Diego (NARA). These batteries were completed in August 1943, with the six fixed 90-millimeter guns *M-3* mounted the following month (Thompson, 1991).

Radar control of antiaircraft gun batteries was fairly common by this point in the war, having descended from the 1941 introduction of the *SCR-268* (Figure 6-2). This early antiaircraft searchlight radar did not have sufficient elevation and azimuth accuracy for gun laying, hence had to be augmented by optical trackers. Elevation and azimuth data were accordingly provided by either *M5* or *M7 Predictors*, which later came to be known as *Directors*. The *SCR-268* did, however, provide valuable range information, and could also be used to automatically aim a searchlight for nighttime operation.

As reported by Malone (2004):

"The principal failings of the *SCR-268* were the poor performance against low flying (500 feet or below) aircraft and the excessive interference of ground or water conditions that created clutter on the scope."

The *SCR-268* remained in service as an antiaircraft fire-control and short-range early-warning radar until 1944, by which point some 1200 were in use for automatic searchlight control (Malone, 2004).

Figure 6-2. Due to its longer wavelength, the *SCR-268* searchlight-control radar was highly susceptible to signal backscatter from the ground or water surface (undated US Army photo).

What type of radar, if any, was used in support of surface engagements by the AMTB Batteries at Fort Rosecrans is not clear. Some very general information on WWII radar-controlled antiaircraft guns comes from a Wikipedia entry entitled "90 mm Gun *M1/M2/M3*" (Wikipedia, 2019):

> "In antiaircraft use the guns were normally operated in groups of four, controlled by the *M7* or *M9 Director* or Kerrison *Predictor*.[1] Radar direction was common, starting with the *SCR-268* in 1941, which was not accurate enough to directly lay the guns, but provided accurate ranging throughout the engagement. For nighttime use, a searchlight was slaved to the radar with a beam width set so that the target would be somewhere in the beam when it was turned on, at which point the engagement continued as in the day. In 1944, the system was upgraded with the addition of the *SCR-584* microwave radar, which was accurate to about 0.06 degrees (1 mil) and also provided automatic tracking."

Malone (1990) reports the mobile *SCR-547* radar was employed in the AMTB fire-control role early in the war. This continuous-wave 10-centimeter set, which weighed 6 tons, was mounted on a trailer as shown in Figure 6-3 (FTP-217, 1943). The two diametrically-opposed parabolic antennae, which rotated on command with the central housing, could also be tilted up or down. As further explained by Malone (1990):

> "Two operators were stationed on the mount, one of which operated the A scope presentation of the oscilloscope, while the other kept the antenna pointed at the target by observing it through a sighting telescope. A third operator maintained telephonic communication with the battery command post."

The significant deficiency here was the inability to acquire target azimuth and elevation from the radar, which necessitated a human in the loop using a telescopic sight.

---

1   The *Kerrison Predictor* was produced by Singer as the *M5 Antiaircraft Predictor* (Wikipedia, 2018).

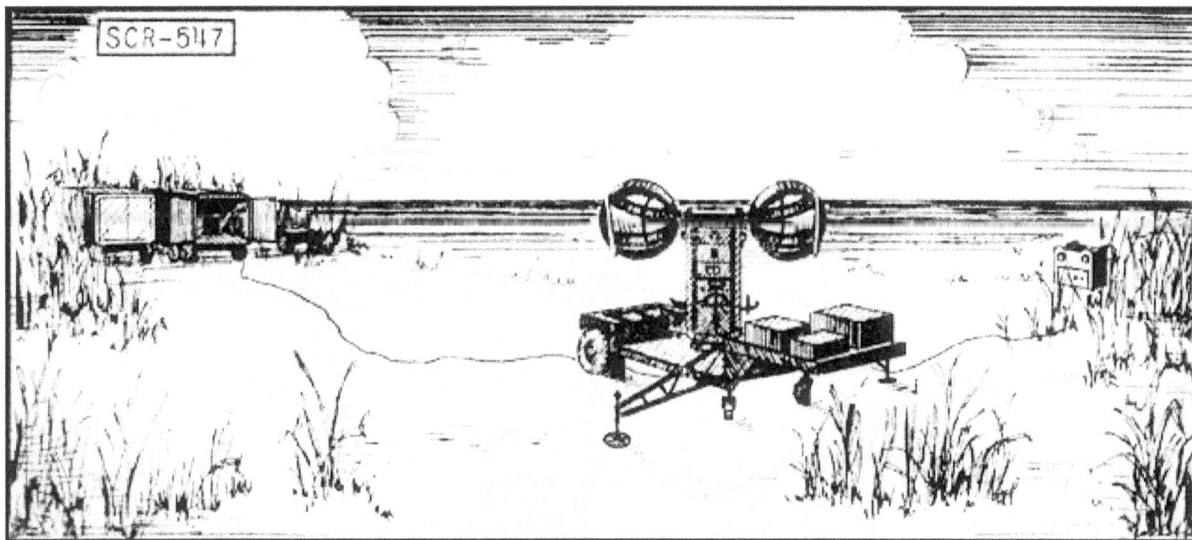

Figure 6-3. The *SCR-547* radar, which measured only range to target, relied upon an optical height finder for azimuth and elevation, but was additionally equipped with *IFF RC-148* [2] (adapted from FTP-217, 1943).

Field Manual 44-4, *Employment of Antiaircraft Artillery Guns*, lists three possible radars for the antiaircraft role circa 1945, not including the *SCR-547*, which had probably been declared obsolete by that point, for reasons stated above (FM 44-4, 1945):

> "The radar authorized for issue to each gun battalion headquarters battery and to each gun battery is either the SCR-584, the SCR-784,[3] or the SCR-545. They are mobile, short and medium range radar units, designed primarily for use with an antiaircraft artillery battery to supply present position data to the M7, M9, or M10 directors, and for medium range searching to provide advance warning of the approach of enemy aircraft... These sets are designed to automatically follow moving targets within range and to simultaneously supply present position data to the gun director."

By the summer of 1943, the radar of choice for 90-millimeter antiaircraft emplacements in general was the *SCR-584* (Figure 6-4). With an angular accuracy of about 0.06 degrees, the *SCR-584* could provide range, azimuth and elevation to the *M3 Gun Data Computer* and *M9 Director*, which then calculated the appropriate train and elevation to correctly lead the target. This continuously updated fire-control solution was electronically passed to the *M-13 Remote Control System* on the 90-millimeter guns, which in turn automatically fired as fast as they could be loaded (Moeller, 2006). The typical rate of fire for a single gun was 25 rounds per minute (FM 4-91, 1943).

---

2   IFF – Identification Friend or Foe. This directional transmitter-receiver set pinged a transponder carried by friendly aircraft to positively identify them as non-hostile targets.
3   The *SCR-784* was a lightweight version of the *SCR-584* (MMR, 2007).

Figure 6-4. A WWII-vintage 10-centimeter *SCR-584* radar being used for missile tracking at White Sands Proving Ground in 1947. Note parabolic antenna in the raised position on the trailer roof, and the two operators seated on the *Director M-9* at left (US Army photo courtesy C-E LCMC.)

While the *SCR-584* began rolling off the production line in 1943, it doesn't seem probable that harbor-defense surface engagement would have been the highest priority for deployment of these first sets. On the other hand, it would have been cheaper and perhaps beneficial to install some of these radars in the continental US versus overseas, where it was easier to train new operators and gather operational feedback. While there is photographic evidence that at least one *SCR-584* set was at Fort Rosecrans shortly after the war, wartime photos I've found so far reveal no other presence.[4]

As further reported by Malone (1990):

"As an interim set, the *SCR-584*, which had been produced for directing the 90-mm anti-aircraft guns, was modified to enable it to be used for surface fire control in the mobile 155 mm batteries. Additionally, this set was standard equipment for the 90-mm coastal artillery batteries.

A key factor enabling this role was the shorter wavelength and tightly focused beam from the parabolic antenna (Figure 6-5), allowing the *SCR-584* to minimize the surface clutter caused by wave action (Malone, 2004).

---

4   The fact that radar was highly classified during the war likely minimized the number of operational photographs taken.

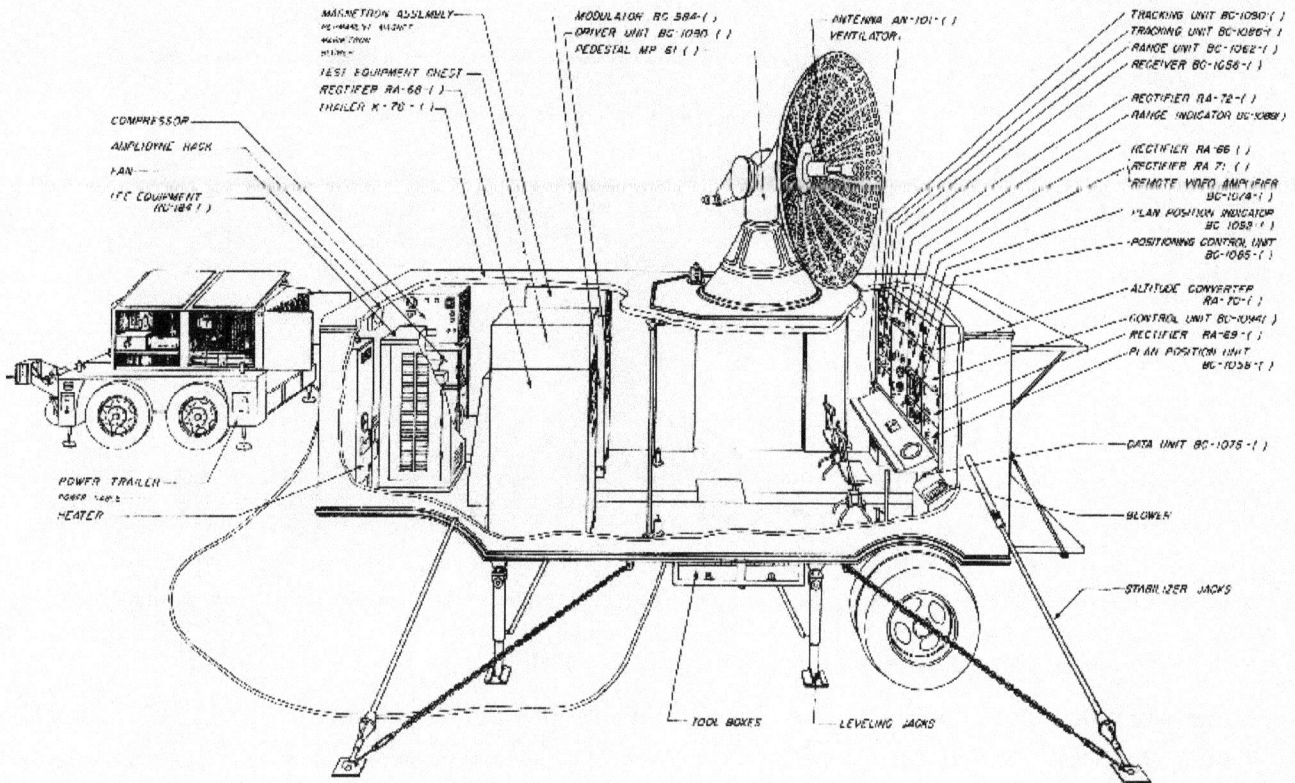

Figure 6-5. Interior view of the *SCR-584* trailer, showing the electronic equipment in front (left end) and the operators' station at rear (undated US Army diagram, NARA).

It is possible more than one type of radar was used at a given battery over the course of the war as better equipment became available. Whatever the source, radar data describing target position was electronically passed to the director, which computed the gun elevation and azimuth data for the predicted fire-control solution. The three 90-millimeter AMTB installations at Fort Rosecrans were equipped with the *Director M-9*, which had a two-seat tracker head that allowed a pair of operators to also optically follow the target in elevation and azimuth.[5]

The training and elevation commands from the director were in turn passed to the hydraulically actuated 90-millimeter guns in the form of conventional 60-Hz two-speed selsyn signals. (Before the advent of radar, target range was obtained by three men using a stereoscopic *Height Finder M-1* (Malone, 1990), which will be further discussed later.) A pair of *Indicator-Regulators*, one for azimuth and one for elevation, provided the necessary interface that allowed the artillerymen to switch between manual and automatic control of the gun (Figure 6-6).

---

5   While there are several general recollections from Fort Rosecrans veterans cited in this text, it is not clear what type radar was employed on which guns.

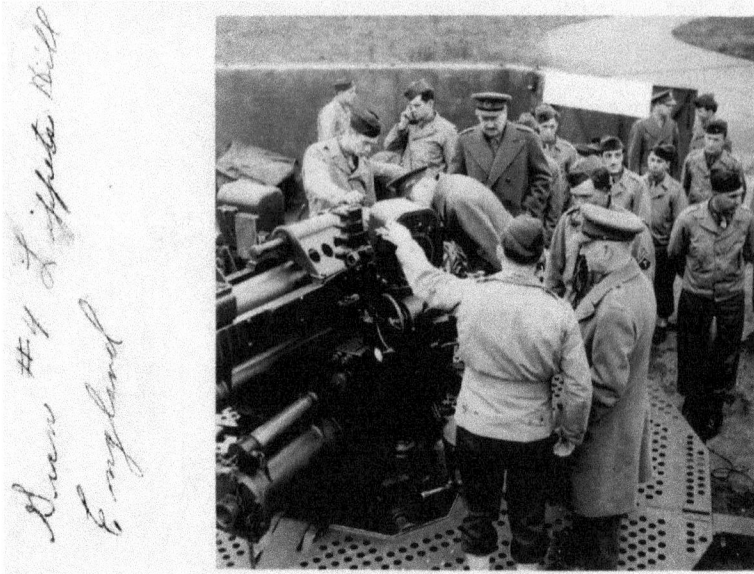

Figure 6-6. A member of Battery B, 115th AAA, (foreground) rests his hand on the *Azimuth Indicator-Regulator* for Gun No. 4 at a 90-millimeter site in Lippets Hill, England in 1944 (US Army photo courtesy Chip Chapin, 2006). The *Elevation Indicator-Regulator* can be seen on the other side of the gun carriage at far left.

In manual mode, the Gun Pointer (No. 2 man) traversed the gun in azimuth by turning his handwheel to match the concentric pointers for commanded versus actual position on the *Azimuth Indicator-Regulator*. The Elevation Setter (No. 1 man) would meanwhile perform a similar function on the right side of the carriage for gun elevation. In automatic mode, the incoming selsyn signals from the director were directly converted into closed-loop servo commands by the *Remote Control System M-13*, with no need for human assistance. A pictorial layout adapted from a Coast Artillery Training Bulletin (FM 4-91, 1943) of an *Indicator-Regulator* is shown in Figure 6-7.

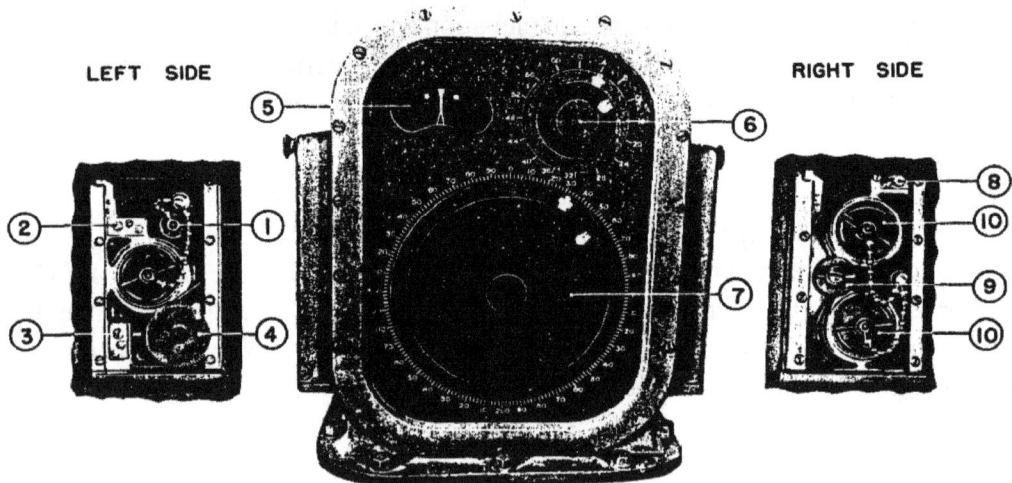

1. Lag meter adjusting shaft cover
2. Synchrotransformer adjusting shaft
3. Fine repeater adjusting shaft
4. Orientation adjusting ratchet
5. Lag meter
6. Coarse repeater
7. Fine repeater
8. Course repeater adjusting shaft
9. Fine repeater cut-out switch
10. Lamp cover

Figure 6-7 Component layout of the *Indicator-Regulator* assembly employed on the 90-millimeter gun (adapted from FM 4-91, 1943). The rightmost panel also housed the fine repeater cut-out switch (callout #9), which was used to select either manual or automatic mode of operation.

The *Indicator-Regulator* provided a more sophisticated rendition of the *Distant Electric Control (DEC)* functionality for searchlight control, which will be further discussed in chapter 8. For example, the lag meter (callout #5) is what was known as a zero indicator on searchlight equipment. The course (#6) and fine (#7) repeaters have concentric pointers representing the current position (as sensed by the local selsyn control transformers), as well as the commanded position (as requested by the director).

A recent photo of a damaged *Indicator-Regulator* unit is shown in Figure 6-8 below. Traces of the fine-repeater scale are still visible around the bottom perimeter of the display, and the distinctive remains of a zero indicator (i.e., lag meter) are clearly evident at upper left. Interestingly, the course-to-fine gear ratio for the selsyns was not the same as that employed in the searchlight designs (Davenport, 1991), which seems somewhat odd. The handwheel for manual elevation control is at far right.

Figure 6-8. A damaged *Elevation Indicator-Regulator* on a privately owned 90-millimeter gun, circa 2005 (photo courtesy Bob Meza, Santa Clarita, CA). Note handwheel for manual control at far right.

"Annex E" of the *1946 Annex Supplement to the Harbor Defense Project, Harbor Defenses of San Diego*, prepared 1 July 1945, includes the following statements:

1) "These Harbor Defenses have no antiaircraft defense other than that furnished by the local defense weapons assigned to seacoast armament."

2) "No fixed antiaircraft gun batteries exist in these Harbor Defenses, nor are any authorized. Mobile antiaircraft artillery will be provided when the situation so demands."

3) "AMTB batteries have as their secondary mission that of firing against aircraft. The 90mm guns included under the AMTB projects include six fixed and six mobile (3 batteries)." [6]

4) "There is no searchlight or radar equipment authorized for antiaircraft defense."

Number 4 above leaves open the possibility of radar for surface engagements.

---

6   These would be AMTB Batteries Fetterman, Cabrillo, and Cortez.

AMTB Battery Pio Pico

On a FortWiki page entitled "Fort Pio Pico," John Stanton (2019b) indicates a short-lived AMTB battery may have been located east of the harbor entrance to San Diego Bay on North Island:

> "There is some indication that an AMTB battery was built on the site of the old Fort Pio Pico in 1942-1943, and then transferred to the old Battery Fetterman on Fort Rosecrans in 1943. This could not be substantiated in the available records."

Ruhge (2017b) reports the original Fort Pio Pico location on North Island was demolished in 1941 when the ship channel was widened,[7] however, which suggests the AMTB site was installed a bit further inland than the WWI-era Battery James Meed (Figure 6-9).

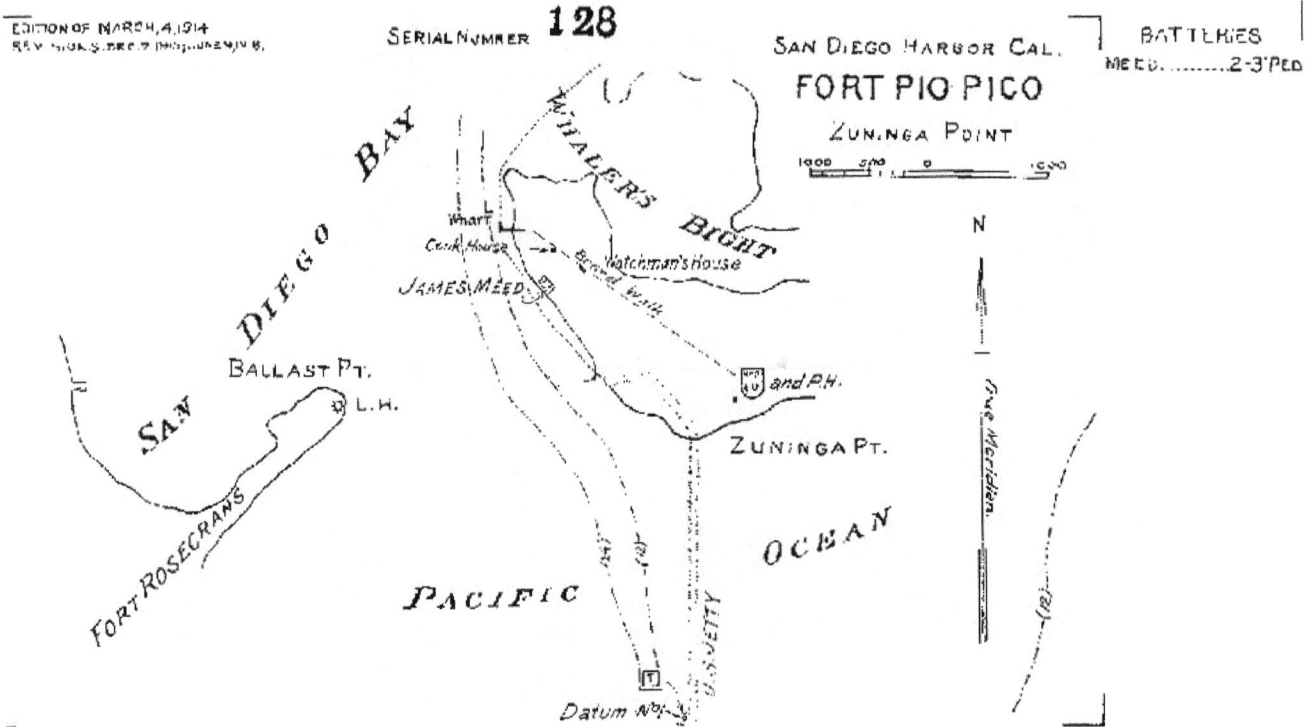

Figure 6-9. AMTB Battery Pio Pico was located inland near the site of former Battery James Meed at Fort Pio Pico (Figure 6-10), which is shown just left of top center on this 4 March 1914 map. Note alternate spelling of Zuniga Point.

According to Gaines (1993), AMTB Battery Pio Pico was manned by Battery H, 19th Coast Artillery Regiment, until 20 April 1943, then by Battery B from May through July 1943, after which date he reported "no other mention of this battery was found." A few pages later in the discussion he states "on July 6th, Battery B moves the armament and equipment of Battery Pico to Battery Fetterman and then fired the 37-milllimeter guns of Battery Pico." This wording would seemingly imply AMTB Battery Pio Pico remained equipped with at least two 37-millimeter guns that were temporarily left in place during the transition of presumably its 90-millimeter guns to Fetterman.

---

7   This North Island property had been abandoned by the Army in the 1920s and transferred to the Navy in 1935, along with the rest of the old Rockwell Air Field (Roberts, 2016), which is now part of NAS North Island.

Figure 6-10. The original Gun 1 emplacement at Battery James Meed, Fort Pio Pico on North Island, circa 1911 (US Army photo courtesy A.H. Grobmeier).

Further insight is provided by Byron D. Shutts of Battery B in his 10 June 1995 interview with Howard Overton (1995):

> "Well let's see, that would have been in, about '42 when we set up over there... But on North Island, over there we had half our sleeping quarters dug in underground. Half underground in that sand. Sand fleas was everywhere. The pits we had to work, [where] we had the 90 millimeters set up at, we had an oil base of black-top set around them. It was easier to have a blanket and sleep there in that pit than go in those huts and let those sand fleas eat you up."

In summary, it would appear AMTB Battery Pio Pico had been built by troop labor as a temporary expedient for guarding the harbor entrance while AMTB Battery Fetterman was under construction on Ballast Point. The 90-millimeter guns at Pio Pico were later relocated to Battery Fetterman, as further discussed in the following section.

AMTB Battery Fetterman

AMTB Battery Fetterman (Battery Tac. No. 7) was situated at the former site of the Endicott-era battery of the same name on the southwest end of Ballast Point. The original battery (see again Chapter 2), which had its two 3-inch *M1898MI* guns removed in 1920 (Berhow, 2004), was constructed at latitude 32.683889, longitude -117.236944, which would place it today at the very foot of Rosecrans Street. This location is the current site of the Navy Exchange on Naval Base Point Loma, just northeast of Battery Calef-Wilkeson. During WWI, Ballast Point was not as wide as today and started about where Fetterman was located (Figure 6-11), but has since been progressively filled in to provide more land for expansion.

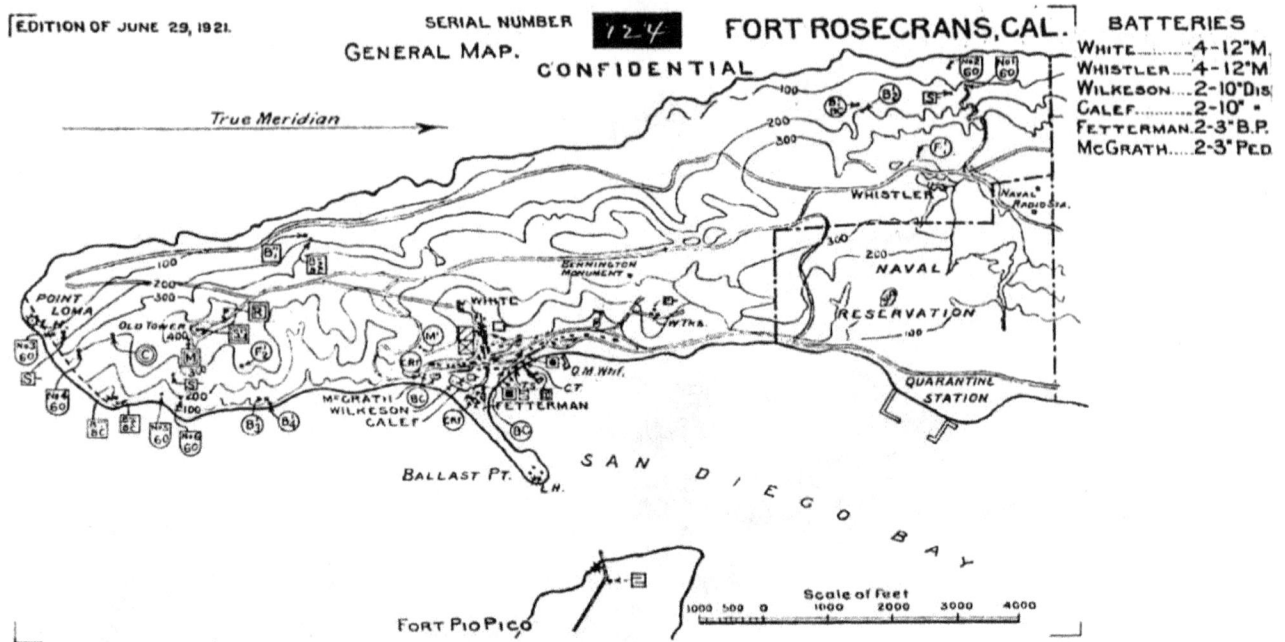

Figure 6-11. AMTB Battery Fetterman was built upon the site of the former Endicott-era Battery Fetterman at the upper (southwest) end of Ballast Point on Point Loma, which was then just northeast of Battery Calef-Wilkeson as indicated on this 1921 General Map of Fort Rosecrans, CA.

One example of such continued expansion was reported by Erwin Thompson (1991):

"In October 1940, the construction quartermaster received orders to erect temporary buildings for an increase in enlisted strength of 2,022 men. To take care of the increase in land use, a dredging project in San Diego harbor in 1940 resulted in adding about twenty acres of fill to the reservation at Ballast Point to provide a level space for drilling."

The 1936 map shown in Figure 6-12 below more clearly indicates the location of the ammunition magazine for the abandoned Battery Fetterman,[8] seen just north of the curve in the road leading to Ballast Point. The battery was still extant at the time, but not shown on the map, presumably because it was no longer in use. Authority to dispose of this obsolete battery was granted in August 1939 by Headquarters Ninth Corps Area, Presidio of San Francisco, to the Adjutant General, US Army, Washington, DC. The old battery was subsequently demolished in 1940 (Ruhge, 2017c).

---

8   Note the callout just right of bottom center on this map: "Magazine (Old Btry. Fetterman) Ammunition AA Guns & Small Arms."

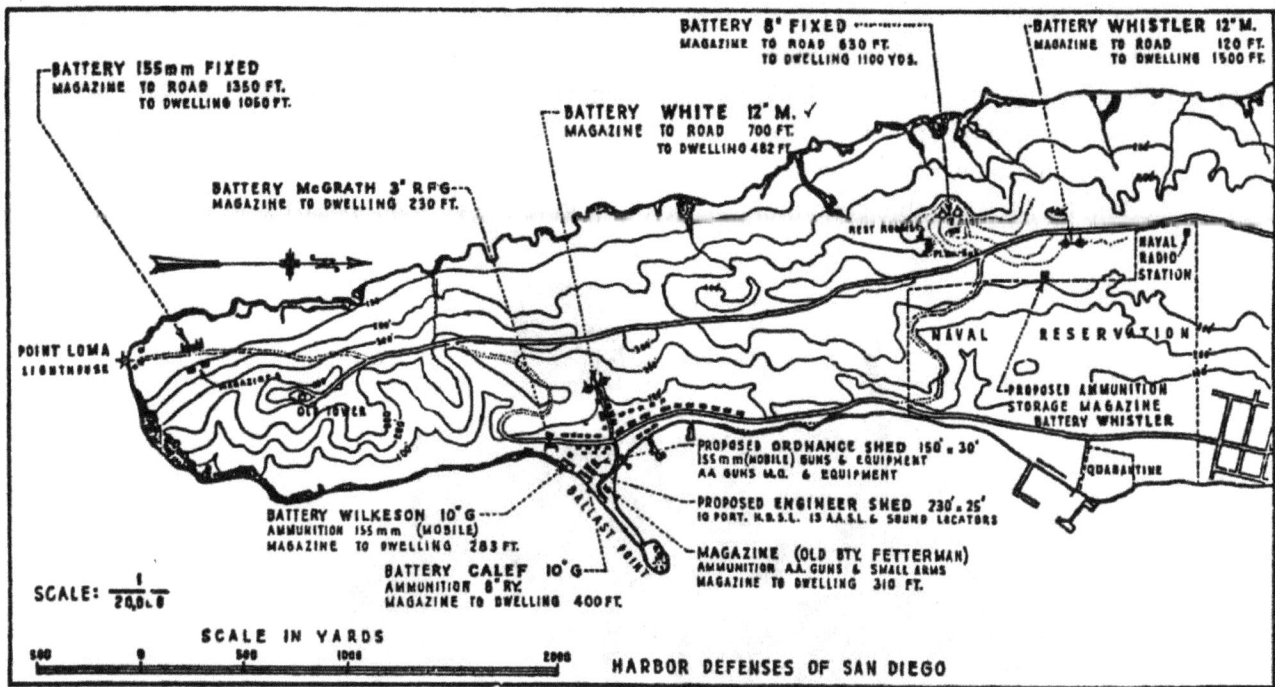

Figure 6-12. This 1936 Annex map of Point Loma shows the magazine for Battery Fetterman, which would be approved for destruction four years later, north of the road curving down to Ballast Point at bottom center.

In the 1937 aerial photo of Figure 6-13, the four 10-inch disappearing-rifle emplacements for Battery Calef-Wilkeson are in the upper-left corner. Just across the road below this massive fortification is the original Battery Fetterman, with its Gun 1 position partially obscured. The small structure directly behind this battery to the west is presumably the magazine indicated on the map in Figure 6-12 above. Note also the Coast Guard lighthouse facility at the far end of Ballast Point to the northeast (bottom left).

Figure 6-13. In this 27 November 1937 aerial image, the abandoned WWI-era Battery Fetterman can be seen just below Batteries Calef and Wilkeson (upper left corner), on the near side of the curve in the road to Ballast Point. The Gun 1 emplacement for Battery Fetterman seems to be shrouded under a tree canopy.

During WWII, AMTB Battery Fetterman was reportedly equipped with four rapid-fire 90-millimeter guns facing the San Diego Harbor entrance. Two of these dual-purpose guns were mounted 120 feet apart on *M3* fixed mounts with concrete pads, while the other two were to be installed on *M1A1* mobile gun mounts with wooden pads (FortWiki, 2019). The primary mission was defense against high-speed surface craft entering the harbor, while the secondary mission was antiaircraft. This site received commercial power backed up by *M-7* portable generators as seen in Figure 6-14 (Stanton, 2019a).

Figure 4—Generating Unit M7 and Generator Trailer M7—All Doors Open

Figure 6-14. The Portable Generating Unit *M7* on skids is seen loaded inside Generator Trailer *M7* (adapted from US Army *TM 9-618*, 1943).

Mounted in September 1943, the fixed armament consisted of the 90-millimeter gun *M-1* and top carriage *M1A1* on a shielded turret-type gun mount *M-3*, an example of which is shown in Figure 6-15, described at the time as follows (FM 4-91, 1943):

> "The mount allows 360° traverse and is capable of elevating from -8° to +80°; however, it is not contemplated firing above +15°. The maximum effective range of the gun is approximately 8,000 yards. The gun data computer, which eventually is to be supplied as part of the standard fire control equipment for this gun, provides for operation up to 12,000 yards. This provision permits tracking considerably in advance of the time when fire can be brought to bear on the target."

Recall these fixed-mount 90-millimeter guns had been relocated from AMTB Battery Pio Pico across the harbor on North Island (Stanton, 2019). This installation was confirmed by Byron D. Shutts of Battery B in his 10 June 1995 oral-history interview with Chief Ranger Howard Overton of Cabrillo National Monument. Shutts, who served as a gun commander at AMTB Fetterman, recalled the number of emplacements as follows (Overton, 1995):

> "I think we started with four [guns]. We ended there was three on Ballast Point setting there. Of course, we didn't have too much area on Ballast Point."

Figure 6-15. This photo of a 90-millimeter gun installation was taken by John Stanton on 22 July 2010 at AMTB Battery Parrot, Fort Monroe, VA. Note details of the concrete pad construction. The turret-type *M-3* mount featured a three-sided shield made of boiler plate for protection against shrapnel.

The earlier technical description from *FM 4-91* emphasized the primary AMTB mission of this weapon (high-speed surface craft). When employed in an antiaircraft role, the 90-millimeter gun could fire a 24-pound projectile to an altitude of 30,000 feet (Moeller, 2006). In his 13 June 1992 oral-history interview with Chief Ranger Howard Overton (1993), Robert E. Smith, 19th Coast Artillery Regiment, recalled:

"Around 1 March 1943, a battery of 90-millimeter high- and low-angle-fire radar-controlled guns were (sic) at Battery Fetterman. It was located about half way from the bottom-side barracks area and the Ballast Point lighthouse. It was the newest Army gun made and these were the first made. When the construction was completed, they assigned troops from Battery B to man it. When they became acquainted with the guns, they started training on them."

In a similar interview that same day, Sergeant Julius C. Gary of Battery B provided the following details regarding his experience at AMTB Fetterman (Overton, 1993):

"...we had an engineer, when we put in the 90-mms, four guns, on Ballast Point. We built a 20-foot tower. There is where we had the height finder, and the tracker, and the radar underneath,[9] and that is where the ammunition dumps were... I was Chief of Section there and of course they had proven the 90-mm over and over in the antiaircraft [role], but they never proved it on the flat trajectory. We proved it on a flat trajectory... You get your deflection and your range and all that from the tower."

---

9   Byron D. Shutts indicated the unidentified radar at Battery Fetterman did not become operational until sometime after he left in July 1943 (Overton, 1995).

Smith identified this tower location as "right there next to the lighthouse at Ballast Point," [10] and provides the following interesting elaboration on the execution of flat-trajectory sighting (Overton, 1993):

"But anyhow, what he did was open the breech block on that 90-millimeter, looked through it, and zeroed in on that target and proved its effectiveness."

Smith goes on to provide some interesting commentary regarding radar-control of this gun during a practice surface engagement (Overton, 1993):

"On the first test they brought the guns down to water as if they were firing at a water target. When they did then the radar went wild. This battery was to be the defense of the channel. Now they had none and several "Jap" submarines had been sighted. I saw three while I was at Cabrillo Lighthouse. Things became serious all at once."

The term "serious" had taken on new meaning on 23 February 1942, just a few months after the Pearl Harbor raid, when the Japanese submarine *I-17* opened fire on aviation fuel tanks at Ellwood Field near Santa Barbara, CA. According to Bert Webber (1992):

Lt. Nobukiyo Nambu, who was Torpedo Officer, recalled that the submarine reached the United States, near San Diego, about February 20. The *I-17* stood off San Diego harbor looking for any large merchantman or warship by periscope. Finding none, the big I-class submarine headed north."

The strategy behind the subsequent attack on Ellwood Field was to then lay in wait at periscope depth for responding US warships (Webber, 1992). None ever came.

When Jerry Harrison, who helped man Gun 1 at Battery Point Loma during the war, was asked by Ranger Overton if he'd ever fired at a presumed enemy vessel, he responded as follows (Overton, 1993):

"No. When that submarine came up on shore at Santa Barbara is the closest thing to it because we had to arm our guns. They figured to be an attack coming in there too. And after they got word that it shelled Santa Barbara, we were ready. Machine guns and all were ready. They had a machine-gun battery down below the one-five-fives." [11]

According to a 1943 report describing HDSD camouflage status at Fort Rosecrans, prepared by Captain Otto W. Wolgast (1943), the two additional 37-millimeter guns at AMTB Battery Fetterman were hidden under collapsible houses on Ballast Point (Figure 6-16). Each of these wooden structures was painted white with a red roof, so as to look like the rest of the buildings near the Coast Guard lighthouse on the northeast end of the Point (Wolgast, 1943). The operating position for Searchlight 20 was approximately 200 feet southwest of this battery, while the fire-control station and another searchlight were further north on Point Loma (Flower and Roth, 1982).

---

10 Smith is referring to the old Coast Guard lighthouse on Ballast Point, not to be confused with the old Spanish lighthouse at Cabrillo National Monument at the southern tip of Point Loma.

11 Harrison is presumably referring to the four 155-millimeter GPF guns of Battery Point Loma.

Figure 6-16. Looking southwest, the two 37-millimeter guns at AMTB Battery Fetterman on Ballast Point were hidden under collapsible wooden houses. Gun 2 is in the foreground, with an identical house shown erect for Gun 1 in background, behind which were the 90-millimeter guns (adapted from Wolgast, 1943).

Figure 6-17 shows the remains of the two fixed-mount 90-millimeter emplacements still extant in 1968. The Flower and Roth Cultural Resource Inventory recorded the following field-check results in 1982:

> "Battery intact and in pristine condition. Consists of two horseshoe-shaped, open-air gun emplacements separated by an underground bunker. The gun emplacements still contain the gun mounting rings. The bunker is in use by the Submarine Base. The fire-control station and searchlight could not be located."

Figure 6-17. The remains of the concrete pads for the two 90-millimeter fixed gun mounts at AMTB Battery Fetterman on Ballast Point, with the underground bunker entrance at upper left, circa 13 April 1968 (photo courtesy A.H. Grobmeier). Still extant in 1982, this structure no longer exists.

AMTB Battery Cabrillo

AMTB Battery Cabrillo (Battery Tac. No. 4) was named for the Portuguese navigator Juan Rodriguez Cabrillo, who discovered San Diego Harbor in 1542. Situated in Site No. 8 on the southern tip of Fort Rosecrans, construction of this battery commenced on 29 June 1943 and was completed on 26 August. This AMTB battery was to be equipped with four dual-purpose rapid-fire 90-millimeter guns facing the Pacific Ocean. Two of these guns were mounted on *M3* fixed mounts with concrete pads (Figure 6-18), while the other two were to be installed later on *M1A1* mobile gun mounts with wooden pads.

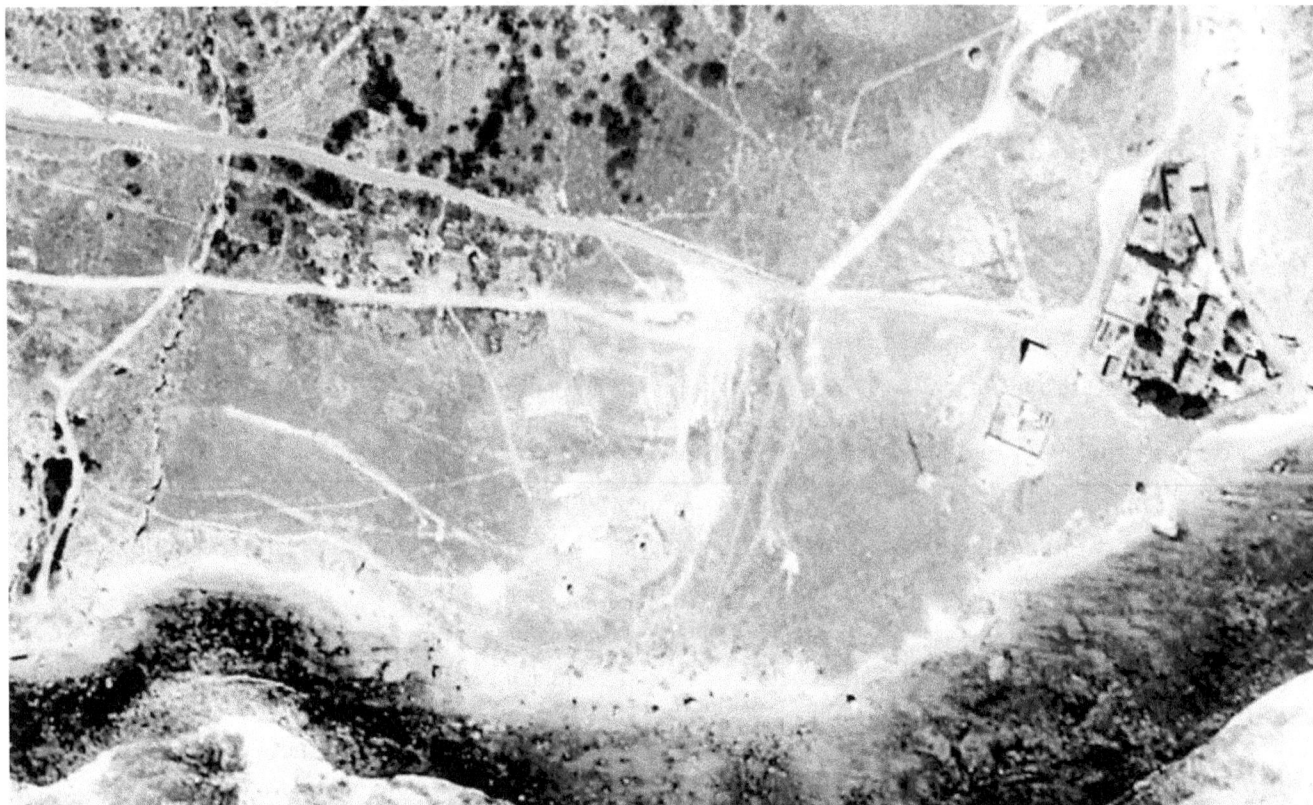

Figure 6-18. The afternoon shadows at bottom center in this 1 March 1944 aerial photo reveal the locations of the two fixed-mount 90-millimeter guns of AMTB Battery Cabrillo, which was completed in August 1943. The new Coast Guard Lighthouse facility is at far right. What appears to be additional defensive positions are seen below the battery along the cliff edge.

Each of these 90-millimeter guns was served by a crew of fifteen men, with nine in the gun squad and six in the ammunition squad (Stanton, 2019d). Commercial power was run by field wire from the unassigned manhole that had originally serviced Battery Point Loma, which in turn was fed from Manhole 105 (Figure 6-19), with backup power provided by *M-7* portable generators. The hand-written annotations on this map were added well after the war ended, when the Naval Undersea Center (NUC) Marine Mammal facility relocated to Point Loma in 1967.

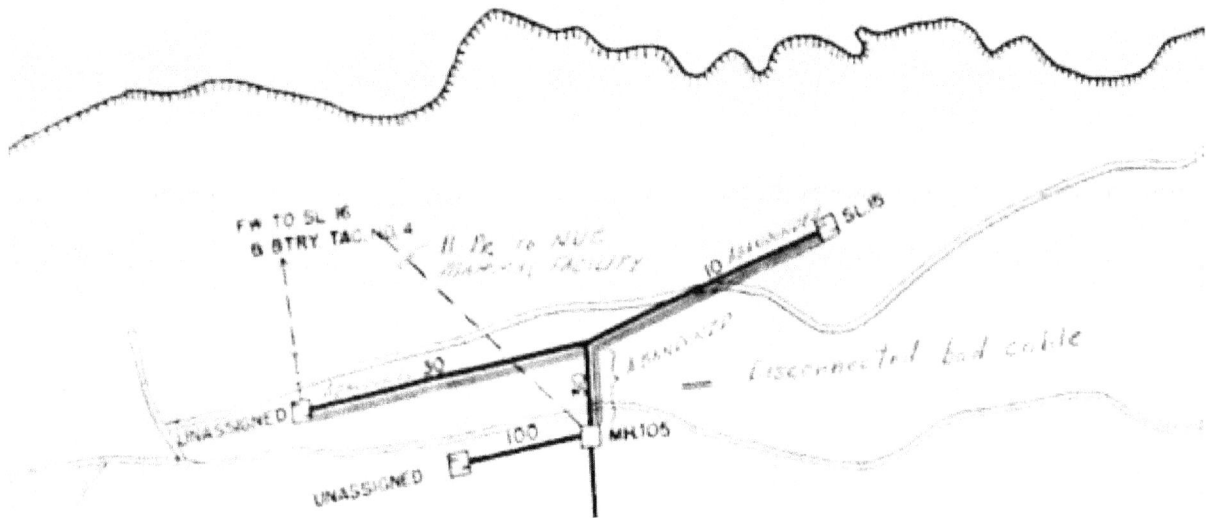

Figure 6-19. This portion of HDSD Main Cable Routings blueprint dated 1 July 1945 shows electrical power from Manhole 105, originally routed underground to Battery Point Loma and Searchlight 15 (right of center), had later been run by field wire to AMTB Battery Cabrillo (Tactical No. 4) and Searchlight 16 (upper left).

Referring now to Figure 6-20 below, the two fixed-mount 90-millimeter guns can be seen mounted on their concrete pads under partial camouflage netting in December 1945. Note the footpath along the horizontal fence line (see earlier Figure 6-18) of the Coast Guard lighthouse facility at bottom left. Some construction material appears to have been assembled behind the battery, with the beginnings of an unusual looking structure that is seen more fully assembled in the next two photos.

Figure 6-20. This 10 December 1945 aerial photo shows the two 90-millimeter guns on their fixed concrete pads of AMTB Battery Cabrillo, still under partial camouflage netting at upper left (US Army photo courtesy Cabrillo National Monument). Note construction materials arrayed on the ground behind the battery.

The Keniston (1995) historic-site survey reports the two mobile guns were never deployed:

"The mobile guns were to have remained in storage 'until an emergency occurred.' It is fairly clear from the 1945 'Annex A' that all four guns were never actually deployed at the site. However, should they have been used, their locations would have flanked (on axis with but outside) the permanent gun emplacements."

Figure 6-21 below, on the other hand, shows what may be trace evidence of mobile gun positions flanking the fixed-mount 90-millimeter guns. Construction of these two emplacements appears to have recently begun in the August 1943 photo shown earlier in Figure 6-18. It is more likely, however, that these flanking positions were for the two 37-millimeter *M1A2* guns on carriages *M3E1* that were assigned to each of the 90-millimeter AMTB batteries, versus 90-millimeter *M1A1* mobile gun mounts.

The 1946 HDSD "Annex G" subsection "Consolidated Summary of Major Items of Equipment" lists six "Gun, 90mm, M1, w/mount M3" and six "Gun, 90mm, M1, w/mount M1A1" on hand as of 1 July 1945 (Annex, 1946). A tabulation of AMTB batteries in "Annex A" of that same document, however, specifically states the six 90-millimeter guns on *M1A1* carriages were not emplaced as of that date. The lingering question is whether these mobile 90-millimeter guns had ever been deployed earlier in the war when tensions were high, then returned to storage at some later point.

Figure 6-21. The up arrows denote apparent evidence of two additional gun emplacements at Battery Cabrillo flanking the fixed 90-millimeter gun emplacements (undated US Army photo courtesy SSC Pacific). See also Figure 6-18 and Figure 6-22.

Ernie C. Johnson of Battery C related his experience at AMTB Battery Cabrillo to Ranger Howard B. Overton (1993):

"I ran the first M9 Director we got in there. She had over 100 radio tubes in it and you had to zero set every tube when you wanted that thing up."

"...our tractor head was way up on top, way up on top of the mountain. And we strung cables to offset that thing I think some 400 yards or something like that."

Each of the initial three AMTB batteries had a dedicated searchlight that was locally controlled. Battery Cabrillo was served by Searchlight 16, positioned south of the battery so as to cover the Pacific Ocean to both the west and south (Figure 6-22). This portable light would probably have been situated near the cliff edge so as to minimize reflected glare, somewhere between the battery and the Coast Guard station at lower-right in the post-war photo below. Note also the unusual post-war cube-shaped structure east of (behind) the battery, previously shown under construction in Figure 6-20 and Figure 6-21.

Figure 6-22. Battery Cabrillo is seen in the lower-left corner of this 15 July 1946 photo, with the Coast Guard Station at bottom right. Portable Searchlight 16 was located south of the battery near the cliff edge. The purpose of the post-war cube-like structure behind the battery is unknown (US Army photo courtesy Cabrillo National Monument).

Additional interesting artifacts are found in Figure 6-23, taken a few seconds after the photo shown in Figure 6-22 above, also looking east. For starters, this zoomed-in view provides a clearer image of the post-war cube-like structure behind the battery at lower right. Note also the footpath leading north along the top of the cliffs, starting from the Gun 1 position and connecting with the dirt road at lower right. A short distance to the northwest, midway into the kink in this road, the footpath branches off to the left and ends at a small partially buried structure that appears to have two rectangular openings on its western face.

Figure 6-23. Partially hidden by trees and a defensive berm, the small structure just left of the kink in the east-west road at lower left appears to have two rectangular openings on its western face. Note the two smaller structures to its right along the cliff edge, possibly observation posts, and the unusual cube-shaped structure behind the AMTB battery (15 July 1946 US Army photo courtesy Cabrillo National Monument).

AMTB Battery Cortez

AMTB Battery Cortez (Battery Tac. No. 8) was constructed in 1943 on the Coronado Beach Military Reservation, on the Pacific side of what is now Silver Strand Boulevard. The battery was located between Searchlights 21 and 22 in Site 11, but was assigned Searchlight 23 to the south in Site 12. According to Annex H (1946), Site 11 occupied 29.75 acres of undeveloped waterfront property, purchased from the state of California for $15,000. Built by troop labor, the cost of construction was approximately $20,000 (Thompson, 1991).[12]

Like AMTB Batteries Fetterman and Cabrillo, this battery featured two 90-millimeter guns installed on *M3* fixed mounts with concrete pads, with provision for two additional 90-millimeter guns on *M1A1* mobile gun mounts on wooden pads. The intended locations for both these mobile guns appear to be mid-way between the fixed mounts and Searchlight 22 to the southeast in Figure 6-24 below. Construction was completed in August 1943, with the fixed guns mounted the following month (Thompson, 1991), manned by Battery H, 19th Coast Artillery Regiment (Gaines, 1993). As with AMTB Battery Cabrillo, there is some debate as to whether the mobile guns were ever emplaced, with the general consensus being they remained in storage.

12　The "Fort Emory" Wikipedia website (http://fortwiki.com/Fort_Emory) says $19,600.

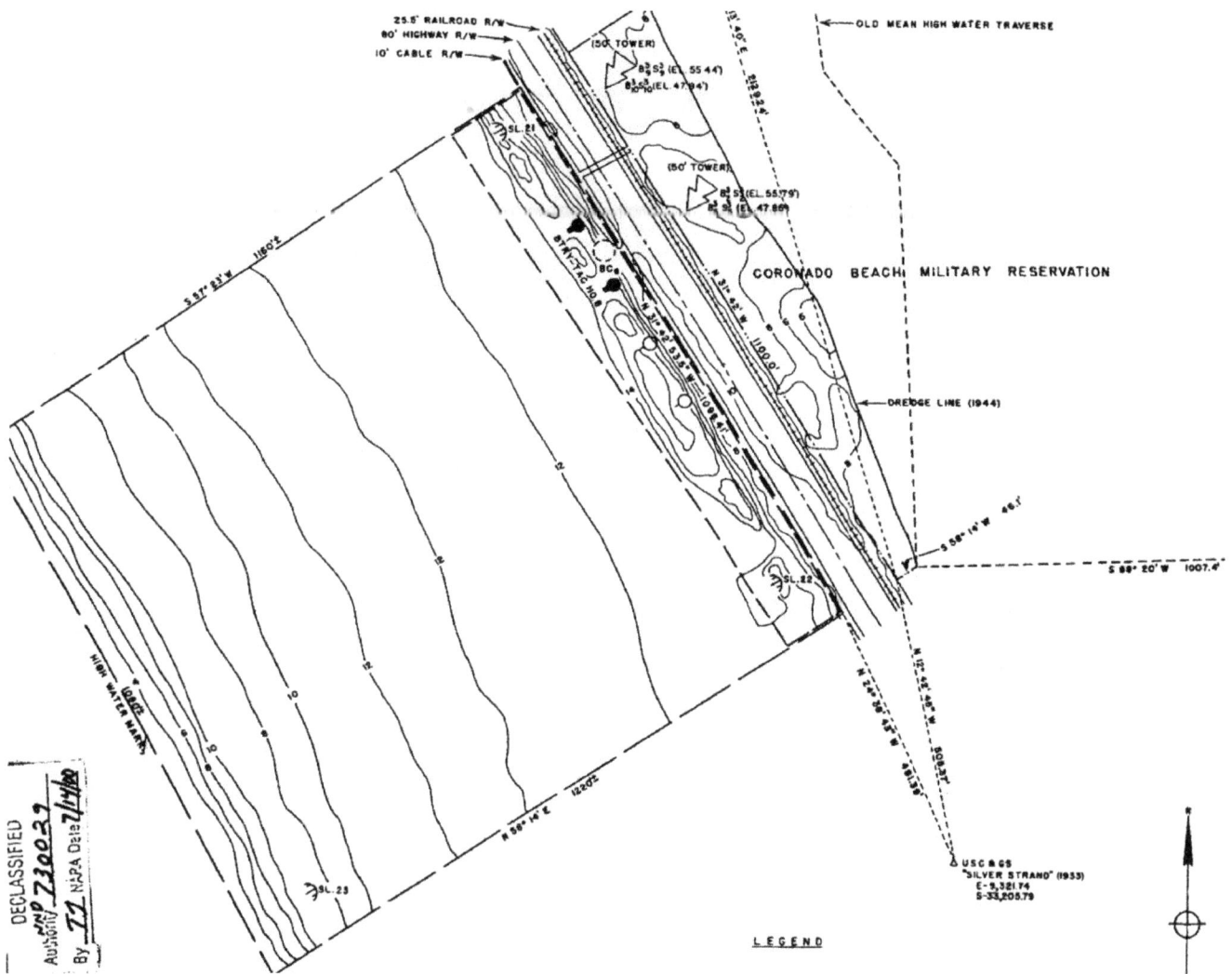

Figure 6-24. AMTB Battery Cortez, Tactical No. 8 (top center), was located on the Coronado Beach Military Reservation in Site 11 on the Silver Strand, midway between North Island (Site 10) and Fort Emory (Site 12). The Pacific Ocean is to the lower left, with San Diego Bay at upper right.

As the Keniston (1995) and Flower and Roth (1982) site surveys were restricted to "Navy and Coast Guard Lands, Point Loma," less information on and no photos of this battery have been found to date.

AMTB Battery Bluff

Battery Bluff, one of four improvised AMTB installations constructed with troop labor, was built and then manned by Battery B of the 19th Coast Artillery Regiment in July 1943 (Gaines, 1993). Like the other AMTB batteries, the tasking of Battery Bluff was to help defend against submarines or fast-moving surface craft that could potentially operate unchallenged inside the minimum effective range of the larger gun batteries (Overton, 1994). This troop-built installation was situated at Billy Goat Point on the east side of the Point Loma peninsula, just down the slope from Searchlight 19.

50 CAL. MACHINE GUNS

37 MM GUN, M1A2

RA PD 5758

Figure 2 — 37-mm AA Gun Carriage M3E1

Figure 6-25. The 37-millimeter gun *M1A2* mounted on the *AA Gun Carriage M3E1*, adapted from US Army *TM-9 235* (1944).

The dismal lack of any reliable defense against fast moving enemy vessels entering the harbor even a year after the attack on Pearl Harbor was viewed with considerable concern. Rising to the challenge, Sergeant R.E. Smith, Battery B, went to the Fort Rosecrans ordnance shop and laid claim to a couple of surplus carriage-mounted antiaircraft guns (Figure 6-25 above), each consisting of one 37-millimeter *M1A2* cannon paired with two 0.50-caliber Browning water-cooled *M2* machine guns up above (Figure 6-26 below).

Figure 6-26. Troops of Battery B are seen manning the northern Gun 2 emplacement of AMTB Battery Bluff (adapted from Overton, 1993, courtesy Cabrillo National Monument). Note water-cooled 0.50-caliber *M2* machine guns left of center, mounted atop the 37-millimeter *M1A2* cannon.

As described by Overton (1993):

"The *M1A2* was a high-velocity antiaircraft weapon of the long-recoil type. It was capable of surface and antiaircraft fire, and could be depressed minus 5 degrees and [elevated] a maximum of plus 90 degrees. It could traverse 360 degrees. The traversing, leveling, and elevating mechanisms were handwheel operated."

Sergeant Smith supervised the lowering of the mobile gun carriages down the steep grade below Sylvester Road, now the Bayside Trail at Cabrillo National Monument, using a steel cable attached to a bulldozer driven by Sergeant Julius C. Gary. (Note "trace of oiled road" right of Searchlight 19 in Figure 6-27.) He then removed the gun assemblies from their carriages and oversaw the permanent emplacement of both on improvised pedestal mounts of his own design, which were bolted securely to 2-foot-thick concrete bases. Due to the improvised nature of this battery and the steep grade which hindered access, it seems likely no radar was located at this site.

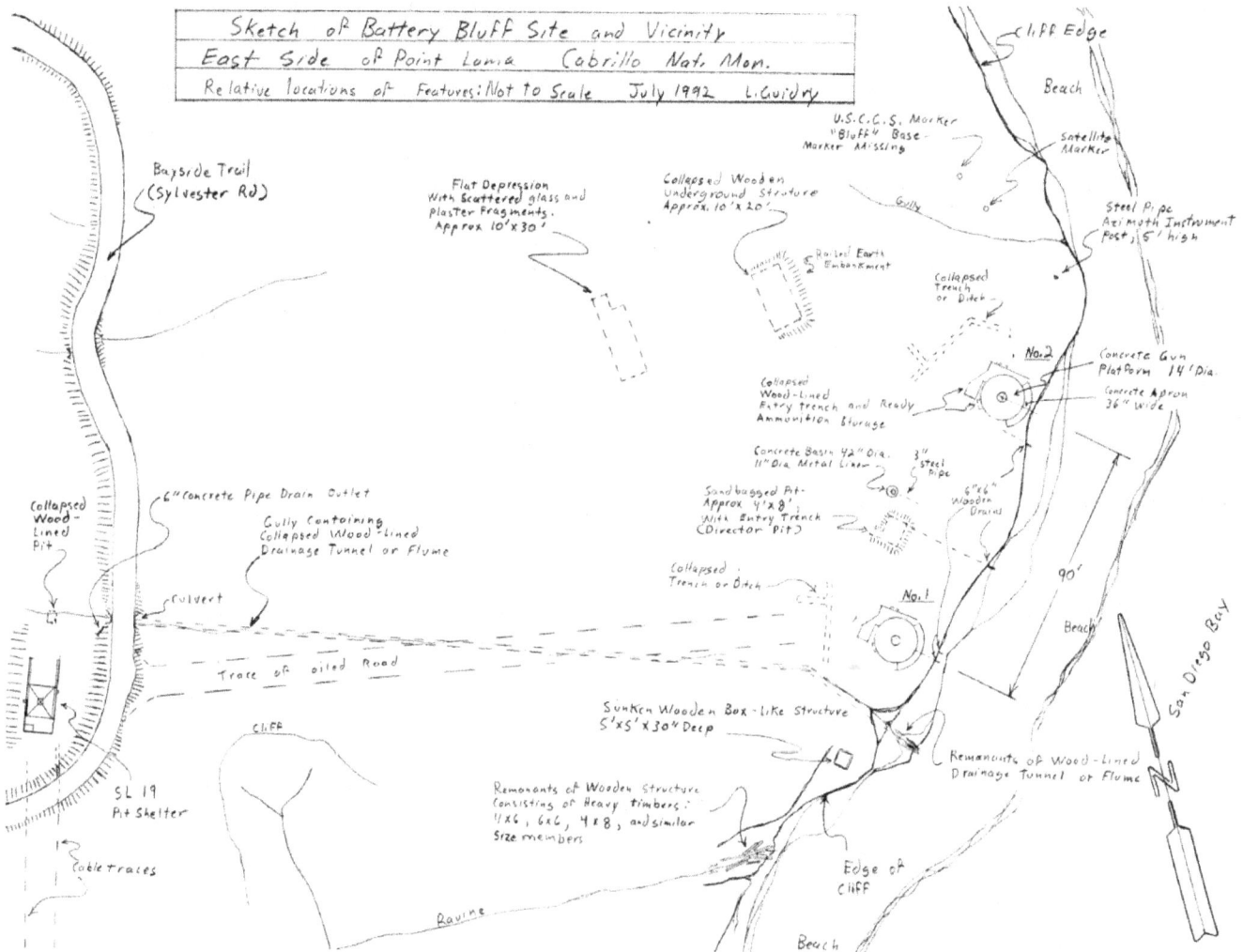

Figure 6-27. This 1992 field sketch by Lee Guidry shows Battery Bluff situated on the cliff edge (right) below the Searchlight 19 silo shelter (lower left) on the Bayside Trail (Sylvester Road). Note concrete "basin" centered behind the two pedestal-mount gun platforms, details of which are shown in Figure 6-28.

Figure 6-28 shows plan and elevation views of what Guidry described in his above field sketch as a 42-inch "concrete basin," located between the gun mounts and adjacent to the north corner of what he labeled as the "Director Pit." Given its central location between the guns and proximity to this director pit, this "basin" may have served as a foundation for some type of optical tracking instrument that provided azimuth and elevation information to the director.[13] A partial view of this "basin" is seen later in the lower-left corner of Figure 6-31a.

---

13 The mechanical *M-4* and *M-5* directors used at the beginning of the war were normally fed by optically tracking the target in azimuth and elevation, with slant range coming from an optical rangefinder (Davenport, 1991).

8 – 3/8" Bolts Evenly Spaced

Irregularly-
Shaped Hole

Top Ring:
17½" O.D.
14½" I.D.

Bottom Ring:
17½" O.D.
13½" I.D.

Both Are
¼" steel Plate

Concrete
Slab

17½"

17½"

2"x2" Angle
Sub Frame

Bolts
16"
O.C

¾"

¼"

½"

1"

¼"

2"

6"

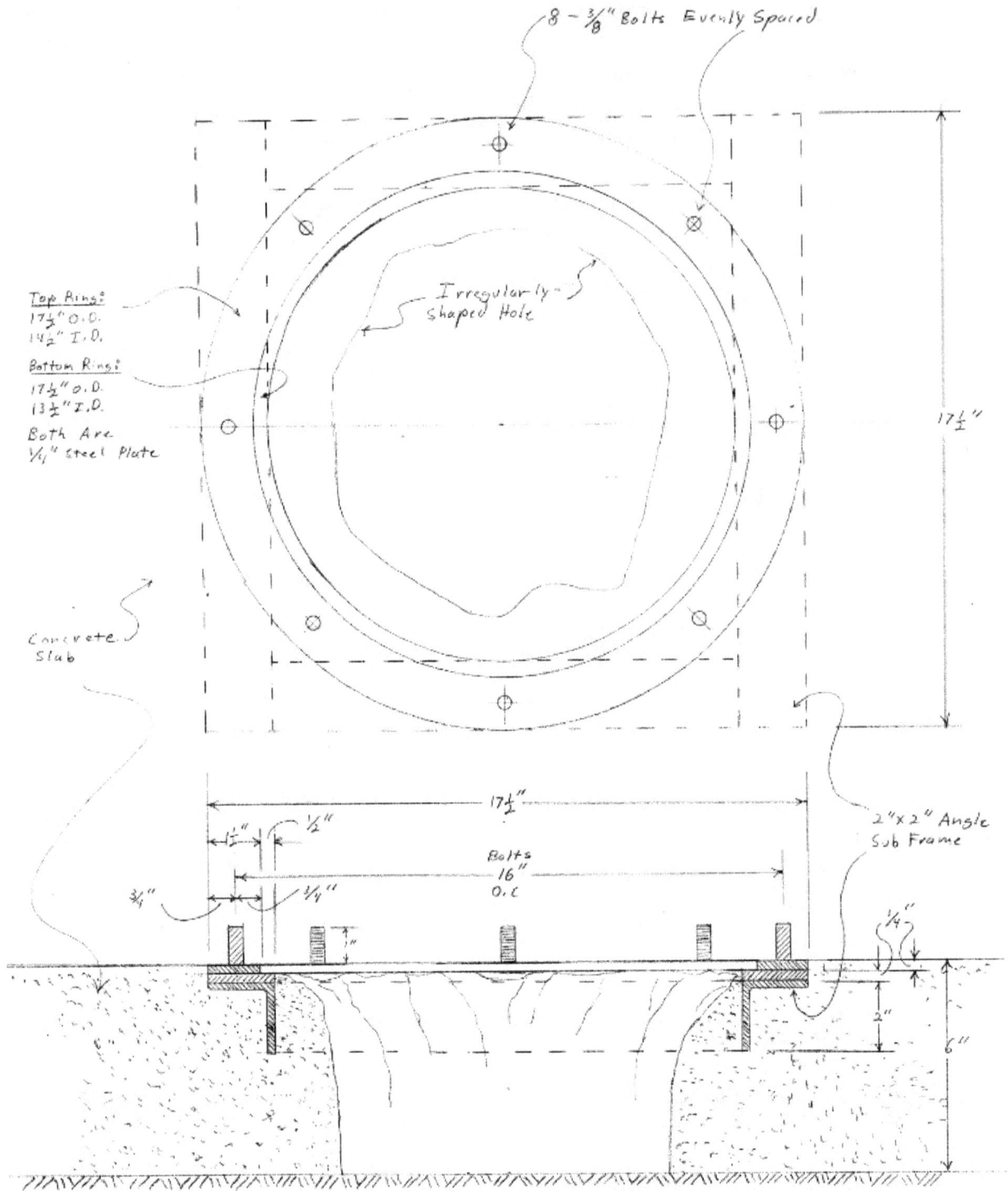

Figure 6-28. Plan (top) and section (bottom) views of the 17.5-inch mounting ring found at the center of the concrete "basin" at the Battery Bluff site (1992 drawing courtesy Lee Guidry.)

A "Historic Structures Report for Harbor Defense Structures" prepared for the US Department of the Interior reports the presence of another observation-instrument support (Carey, 2000), but provides no photo:

"Each mount held a 37-mm *M3E1* gun. An azimuth instrument was mounted on the associated four-foot steel pipe filled with concrete to the north, when the battery was active."

Lee Guidry documented this pipe on the cliff edge just north of Gun 2 in his earlier field sketch of the battery (Figure 6-27), albeit marked 5 feet tall versus 4.[14] This pipe can also be seen in the below photo of this emplacement taken at the same time by Al Grobmeier (Figure 6-29).

Figure 6-29. The concrete-filled steel pipe Guidry described in Figure 6-27 as being the support for an azimuth instrument is seen just north of the Gun 2 emplacement in July 1992 (photo courtesy Al Grobmeier).

Since the rugged terrain at this lower elevation precluded access by construction equipment, all site preparation had to be done the hard way, with pick and shovel (Overton, 1994). The excavated clay soil was used to sandbag the rear perimeter of the emplacement as shown in Figure 6-30. The troops also built a Battery Control Post, an excavated ammunition storage area, and sleeping quarters with wooden cots, all covered with camouflage netting made from chicken wire and burlap.

As recalled by Smith while reviewing wartime photos of this area during his interview with Howard Overton (1993):

> "We went to the Consolidated Aircraft plant and got large wood crates and built cots, two bunks high. Person needed a place to sleep... The next building is the Battery C.P. It had our phones in it. I had two bunks put in there and I had two guards outside. The other structure was our ammo dump, 12 feet by 12 feet by 12 feet."

The remains of some of these troop-built structures can be seen in the background of Figure 6-33, circa 1975.

---

14 The difference in reported height may be due to 8 years of soil erosion.

Figure 6-30. WWII troops manning the southern Gun 1 emplacement of AMTB Battery Bluff (undated US Army photo courtesy Cabrillo National Monument). Note camouflaged structures in background.

A sand-bagged emplacement was prepared for the gun director, which was operated by two men as shown in Figure 6-31a. A WWII-era antiaircraft *Director M-5* is seen in Figure 6-31b. The director shown in Figure 6-31a, however, is different, with its azimuth and elevation handwheels at the top versus bottom of the cubical case, and no visible indication of sighting telescopes. The comshawed 37-millimeter *AA Gun Carriage M3E1* was surplus, and as previously stated by Overton (1993), its traversing, leveling, and elevating mechanisms were handwheel operated. The apparatus seen in the director pit below was presumably of similar vintage, acquired by Smith in the same manner.

Figure 6-31. a) The sand-bagged director pit at Battery Bluff, overlooking San Diego Bay in background (US Army photo courtesy Cabrillo National Monument). Note unusual object at photo lower left, the foundation for which was shown earlier in Figure 6-28. **b)** The WWII antiaircraft *Director M-5*.

Based on Overton's earlier statement regarding handwheel operation of these 37-millimeter guns, it would appear they employed manual match-needle control similar to that used on the 60-inch carbon-arc searchlights discussed later in Chapter 8. The director was presumably connected to an optical tracking instrument mounted on a steel pole adjacent to the director pit. The *Height Finder M-1*, for example, allowed two operators to visually locate and follow a target aircraft in flight, while a third operator determined its range.

This stereoscopic-ranging device had a baseline separation of 13.5 feet between the two lenses seen on either end of the telescope assembly shown in Figure 6-32. As further explained in *TM 9-1623* (1943):

> "The azimuth tracker operates a handwheel to swing the instrument as a whole around the tripod on which it is mounted. The elevation tracker rotates the instrument on its horizontal axis, to keep his telescope sighted on the aircraft. The angular settings so obtained are indicated on two dials, and are used to insure (sic) that the height finder and antiaircraft director are sighted on the same target."

When packed for transportation, this equipment weighed 2,160 pounds, including the shipping crates.

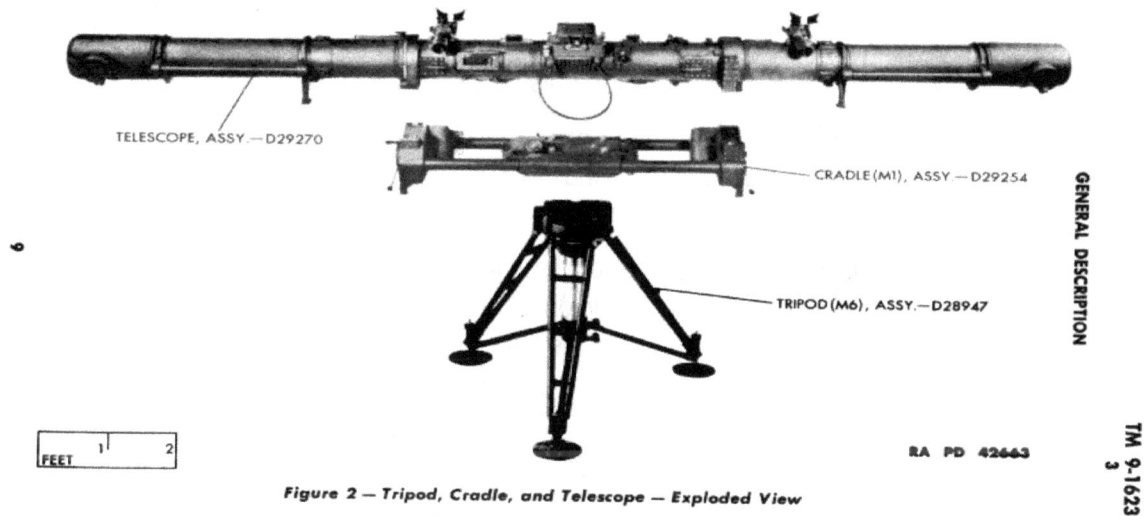

TELESCOPE, ASSY.—D29270

CRADLE (M1), ASSY—D29254

TRIPOD (M6), ASSY.—D28947

GENERAL DESCRIPTION

FEET

RA PD 42663

TM 9-1623

Figure 2 — Tripod, Cradle, and Telescope — Exploded View

Figure 6-32. Front view of the *Height Finder M-1*, a stereoscopic ranging instrument operated by three men, with a base length of 13.5 feet between lenses (adapted from *TM 9-1623*, 1943). The azimuth sighting scope is on top to the left; the elevation scope is on top to the right, with the stereoscopic eyepiece at top center.

Referring back to Figure 6-31a, note the two men on the director are standing in an excavated pit some distance from the cliff edge, from which vantage point it would be difficult to see surface craft in the water below. In addition, there is no evidence of any sighting telescopes on what can be seen of the director in this photo. War Department Technical Manual *TM 9-235* (1944) indicates the directors normally used with the 37-millimeter *Carriage M3A1* were the *M-5* and *M5A1*, and provides some interesting information relevant to this discussion:

> "The director produces firing azimuth (angle of train) and quadrant elevation data for transmission to the gun. The antiaircraft director is described in TM 9-659."

> "The *Remote Control System M1* points the gun in azimuth and elevation according to the controlling data furnished from the director. The system includes electrical and hydraulic power equipment (oil gears) mounted on the carriages and connected to the traversing and elevating mechanisms."

There is some confusion here, as the director seen in Figure 6-31a is not the *M5* or *M5A1*, and Overton (1993) earlier claimed the gun's "traversing, leveling, and elevating mechanisms were handwheel operated." Note, however, that the first line in the preceeding paragraph indicates that the *Remote Control System M1* could be added to remotely actuate the traversing and elevating mechanisms.

AMTB Battery Bluff became operational in August of 1943, going to alert status anytime a ship entered or left the harbor, or whenever the submarine nets were lowered (Overton, 1994). Sergeant Smith, one of two Section Leaders for the Battery, recalls getting a firing order late one night during a surprise towed-target drill (Overton, 1993):

> "...our men slept with their clothes on, and shoes on, everything. We had them at the guns within 2 minutes, ready to fire... And about that time, we saw the speedboat coming in, and then they give us a target, assigned us a target. Well, we opened fire at 2000 yards, what we was firing at. I doubt they went 500 feet till our target was knocked down. Our shooting was out of this world. We more than doubled the world record. The old score was 214, set on speed, number of bullets, number of hits, run-ins, and all these things. We come up with 447."

As further conveyed by Smith in his interview with Overton (1993):

> "Another plus for Battery Bluff was its effectiveness against aircraft. We were hooked up with the San Diego Air Defense Command. We had some alerts with them, but did no firing. All communication was by lights, no voice. We had a box 8 by 10 inches in our CP with lights on it. Each light meant a message."

According to Annex E of the 1946 Annex, this connection with the San Diego Air Defense Command was via the Fort Rosecrans Harbor Defense Command Post (HDCP):

> "To support the defense, it has been necessary to set up communications which keep all batteries informed of the alert status of the Los Angeles Air Defense Region through an operational agency assigned to this area. This agency has their own AAAIS [15] and also receives information supplied by the local San Diego Information Center. This information is transmitted to the HDCP by means of a light signal system. The operations officer at the HDCP transmits appropriate orders to all gun positions depending upon the situation."

Figure 6-33 shows two images taken by the Cabrillo National Monument of the Battery Bluff gun emplacements near the cliff edge below the Bayside Trail, circa 1975. Note the retaining wall to hold back the steep hillside in the left image, and various storage accommodations for ammunition and equipment. Both gun platforms appear to be in reasonably good shape at this time, though partially covered with sand and brush.

---

15 AAAIS stands for Antiaircraft Artillery Intelligence Service.

Figure 6-33. Two 1975 photos of Battery Bluff, with presumably the northern Gun 2 emplacement in the left image and the southern Gun 1 emplacement in the right (photos courtesy Cabrillo National Monument).

As described 25 years later in the *Historic Structures Report* (Carey, 2000):

"Due to erosion, these two emplacements are now perched precariously at the edge of the cliff edge. Both are partially filled with sand and vegetation. A fissure present in the soil 12 inches in front of the south gun emplacement jeopardizes the structure. Cracks in the concrete mount are visible signs of ground instability. Plant growth is present in one of the surface cracks, accelerating deterioration."

Figure 6-34 below shows a more recent photo taken by Cabrillo National Monument of the southern Gun 1 emplacement at AMTB Battery Bluff, circa 2004.

Figure 6-34. View of the southern Gun 1 emplacement, looking east towards San Diego Harbor and North Island, circa 2004 (photo courtesy Cabrillo National Monument). The Carey (2000) survey reported the concrete had settled at the front side, leaving this emplacement slightly tilted with some evidence of cracking.

The northern emplacement for Gun 2 is seen in Figure 6-35, circa 2004. As was earlier reported by Carey (2000):

> "The edge of the cliff is only two or three feet from the front of the north emplacement, and it is only a matter of time before the cliff erodes back to the face of the emplacement."

This dire prediction would unfortunately soon come to pass.

Figure 6-35. The northern emplacement (Gun 2) of Battery Bluff, circa 2004 (courtesy Cabrillo National Monument). Note the embedded gun-foundation bolts in the weeds at bottom center, and the alarming proximity of the cliff edge in this photo after 60 years of erosion.

During the unusually wet San Diego winter of 2005, the cliff erosion reached a tipping point and the norther position crumbled away, sending its rather substantial concrete mount for Gun 2 tumbling into the sea below as seen in Figure 6-36. Figure 6-37 shows the remains of this same emplacement a little over a year later, as photographed by the Cabrillo National Monument staff in August 2006. For perspective, compare this view with the 1975 photograph previously shown in Figure 6-33.

Figure 6-36. Inverted remains of the northern concrete gun block as seen in August 2006, which finally succumbed to the elements during the winter of 2005 and tumbled to the shallows below (photo courtesy Cabrillo National Monument).

Figure 6-37. The remains of the Gun 2 emplacement are seen in August 2006 following the cliff collapse in 2005 (photo courtesy Cabrillo National Monument).

AMTB Battery Cliff

AMTB Battery Cliff, another improvised 37-millimeter installation, was excavated and documented by Coast Defense Study Group members Lee Guidry and Al Grobmeier in conjunction with the Keniston (1995) site survey. Ruhge (2017a) rather vaguely indicates Battery Cliff "was located immediately above the new lighthouse at Point Loma." More specifically, this battery was strategically situated at the peak of the near-vertical cliff face at the southern tip of Point Loma as seen on 18 April 1969 (Figure 6-38), with the new Coast Guard lighthouse down below at far left. The Harbor Entry Control Post is seen atop Battery Humphreys directly behind the location of AMTB Battery Cliff, with the old Spanish lighthouse just visible along the ridgeline at extreme right.

Figure 6-38. AMTB Battery Cliff was so named because of its location at the cliff edge (up arrow) on the southern tip of Point Loma (1969 US Navy photo courtesy SSC Pacific). Note Battery Humphreys directly behind AMTB Battery Cliff, the old Spanish lighthouse on the ridgeline at far right, and the new Coast Guard lighthouse at far left.

AMTB Battery Cliff was immediately adjacent to the south end of the wooden silo shelter for WWI-era Searchlight 4, as seen in the 1945 aerial photo presented in Figure 6-39. Battery Humphreys is under camouflage just to its left (north), with its associated barracks structure seen behind it. The *SCR-296A* fire-control radar for Battery Humphreys is disguised as a water tower at bottom center, just left of the powerhouse for Searchlights 3 and 4, which were not used in WWII. A later-model *AN/MPG-1* radar is seen at this location in Figure 6-38 above, circa 1969.

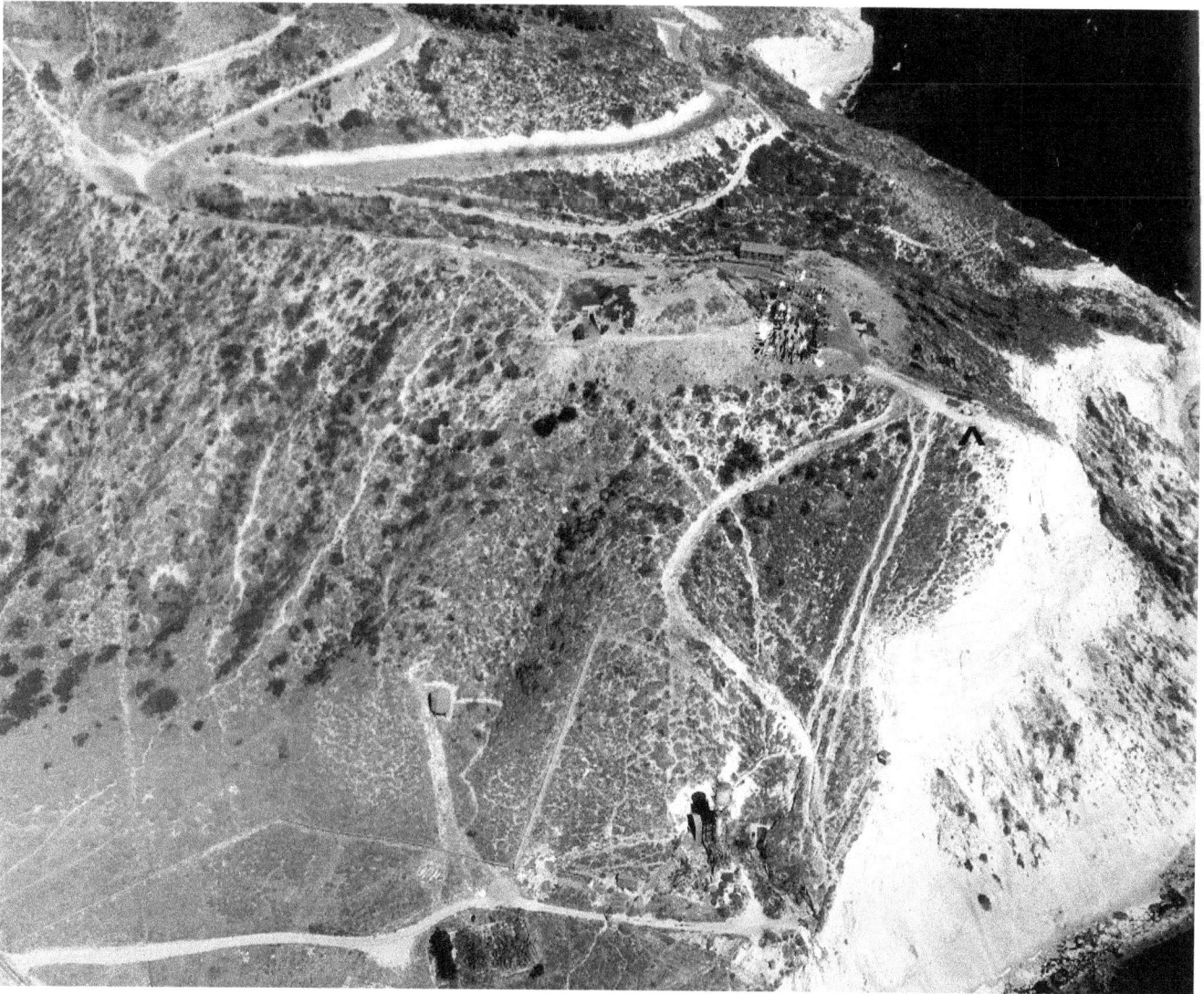

Figure 6-39. AMTB Battery Cliff is at the cliff edge (black up arrow right of center), just south (right) of the Searchlight 4 silo (dark rectangle). Note camouflage on Gun 2 at Battery Humphreys just north of the searchlight silo (1945 US Army photo courtesy SSC Pacific).

In the 1954 aerial photo shown in Figure 6-40, Battery Humphreys (with its 6-inch guns now removed) is at top right, serving as a post-war research facility for the US Navy Electronics Laboratory, previously known as the Navy Radio and Sound Laboratory during WWII. Note the post-war Harbor Entry Command Post on top of the earth-covered battery. The dirt road heading south from the former Gun 2 position at photo center leads directly to the abandoned Searchlight 4 silo, and immediately adjacent to the south end of the silo is the concrete perimeter of AMTB Battery Cliff. North Island is seen across the harbor entrance in the background.

Figure 6-40. Looking northeast in 1954 with North Island in the background, Battery Cliff is on top of the ridge at the cliff edge (far right), immediately to the right of the Searchlight 4 silo roof, with Battery Humphreys at upper left (US Navy photo courtesy SSC Pacific). The Gun 2 mount (top center) was paved over after the explosion in WWII.

During WWII, the wooden silo shelter for WWI-era Searchlight 4 was repurposed to support AMTB Battery Cliff, having been stripped of its light and associated lift for use on Searchlight 15 (Keniston, 1995). As reported in the *Cultural Resource Inventory* by Flower and Roth (1982):

"Adjacent to the south end of the searchlight shelter is a concrete-lined depression, rectangular in shape; dimensions are 22 by 14 feet; length north to south. According to employees at Battery Humphrey (sic), the feature is a World War II gun emplacement. The same employees stated that as late as 1962 the gun mounting bolts were still in place."

Their report correctly concluded this depression was the remains of AMTB Battery Cliff (Figure 6-41).

Figure 6-41. The former site of AMTB Battery Cliff is seen in the foreground of this 1958 photo looking north, having apparently been filled in with dirt. The roof of the adjacent wooden silo for Searchlight 4 is just right of the truck, with Battery Humphreys in the background (US Navy photo courtesy SSC Pacific).

This "concrete-lined depression" was no longer exposed when the Keniston survey was conducted 13 years later, but a bit of reconnaissance and some exploratory digging revealed traces about 50 feet south of the Searchlight 4 silo (Keniston, 1995):

"Excavation resulted in finding a concrete-lined earthen bank forming the southern edge of a dug-in pit. An asphalt floor was uncovered about 4 ft. below the surface; also found were asphalt "sand bags" for lining part of what appears to be a wall segment, complete with a recessed wooden storage-shelf assembly. It may well be that what was found is the site of the director for the 37-mm guns, which would have been located somewhat behind and between the gun emplacements. If this supposition is correct, at least one of the gun emplacements should be found nearby, on the adjacent easterly slope."

AMTB Battery Channel

The fourth improvised AMTB installation, Battery Channel, was reportedly built adjacent to the Coast Guard lighthouse on Ballast Point, but no detailed pictures of this troop-built installation have yet been found. The 27 November 1937 aerial photo presented in Figure 6-42 below shows the future location of this battery, just southwest of the lighthouse station at upper right. Like AMTB Batteries Bluff and Cliff, AMTB Battery Channel was equipped with two 37-millimeter *M3E1* guns, each topped with a pair of water-cooled 0.50-caliber *M2* machine guns (Thompson, 1991; Ruhge, 2017a).

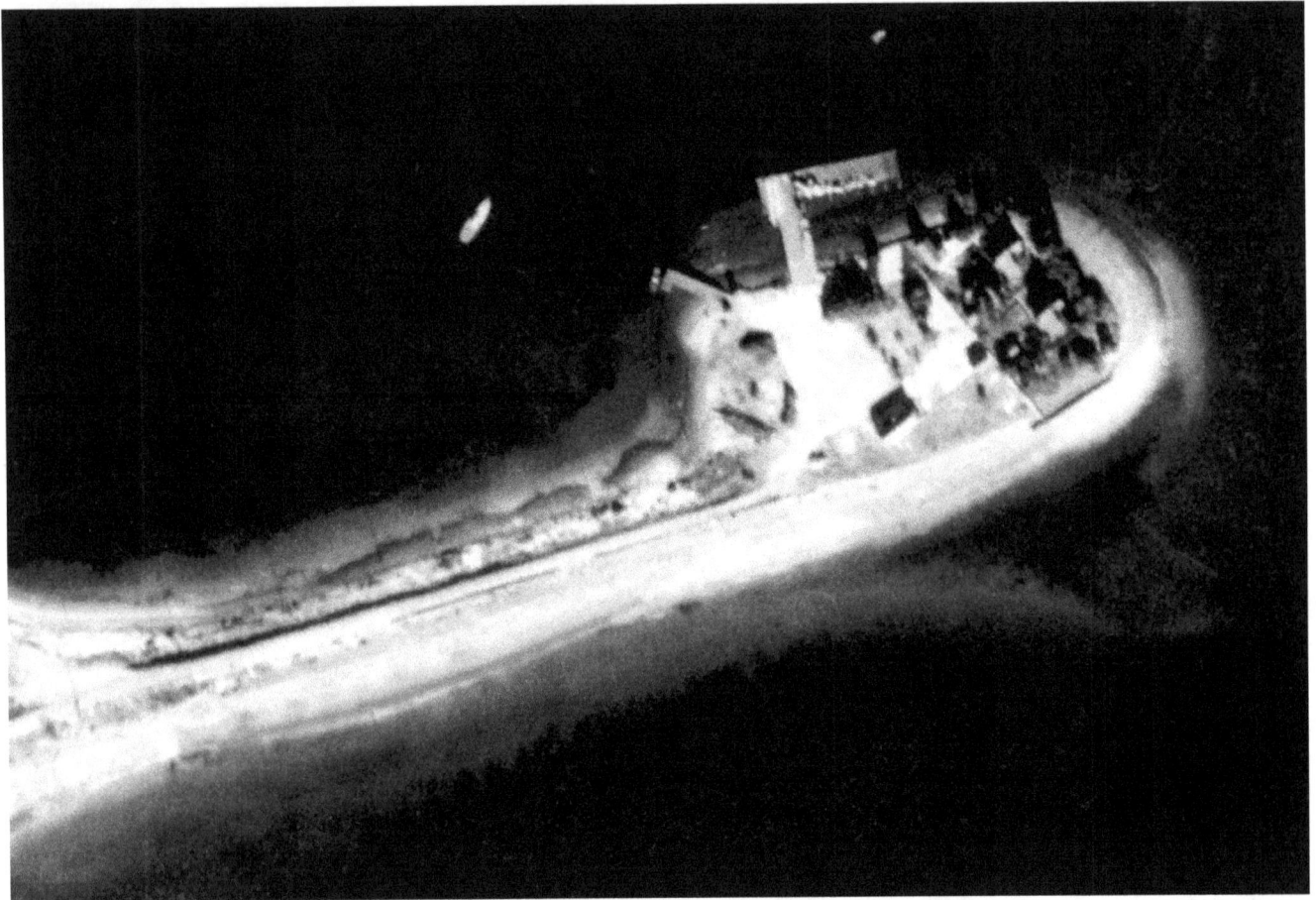

Figure 6-42. Battery Channel would be constructed in 1943 along the thin strip of land in the lower-left quadrant of this prewar aerial photo, just southwest of the Coast Guard station on the northeast end of Ballast Point (1937 US Army photo courtesy SSC Pacific). The lighthouse compound at upper right is shown in more detail in Figure 6-43.

Corporal Eldon Foster of Battery B, who was reassigned from AMTB Battery Fetterman to nearby AMTB Battery Channel in 1943, provides the following remembrance of these weapons (Overton, 1993):

"They were fine. They could chew up a target in about 5 seconds. If you hit it. We never were able to solve the problem of leaking water on the water jackets. They always leaked. It was kind of a lonely thing, you know. They had about five or six guys with you. Half of you had to be awake all the time. It got kind of monotonous."

Figure 6-43. Built in 1890 (Engel, 1960), the old Coast Guard lighthouse overlooking the entrance to San Diego Harbor from the tip of Ballast Point was torn down in 1960 (undated USCG photo courtesy Cabrillo National Monument).

The undated wartime aerial photo seen in Figure 6-44, looking east, shows the outer submarine net across the entrance to San Diego Bay in its open position, probably to have allowed the small craft seen at right to exit. Note the newly constructed barracks at bottom left, and WWI-era Battery Calef-Wilkeson at bottom right. Close examination of the necked-down portion of Ballast Point just right of the Coast Guard lighthouse station reveals what appears to be one of the two gun emplacements of AMTB Battery Channel, with perhaps the second emplacement adjacent to the northwest.

Figure 6-44. This undated wartime photo of the submarine net across the harbor entrance appears to show one of the fixed emplacements for AMTB Battery Channel, on the near side of the road in the narrow portion of Ballast Point at center left (up arrow). Note the new barracks in the lower-left quadrant, and Battery Calef-Wilkeson at bottom right (US Army photo courtesy Cabrillo National Monument.

Taken just after the war ended, Figure 6-45 reveals another fortuitous capture of the presumed AMTB Battery Channel location at image left. Just behind the road there appears to be a concrete bunker with a rectangular opening facing to the right, covered by an earthen berm (up arrow), presumably used for ammunition storage. The Gun 2 position for the battery would be just to its right. Note the "Welcome Home" sign for returning sailors left of the Coast Guard Station, and the newly constructed barracks buildings on the extended land mass west of Ballast Point in background, which can also be seen in Figure 6-44 above.

Figure 6-45. The Gun 2 position of AMTB Battery Channel is left of center in this post-war 1945 photo of the tip of Ballast Point, with an adjacent earth-covered bunker to its left (up arrow). Note similar bunker presumably for Gun 1 at far left, and Coast Guard sign welcoming sailors returning from the war (US Navy photo courtesy Cabrillo National Monument).

According to Gaines (1993), Battery Channel was manned by Battery B of the 19th Coast Artillery Regiment from January to May of 1943. It is not clear what happened after that, as no official documentation or photos have been found. This battery was not addressed in either the Flower and Roth (1982) or Keniston (1996) cultural surveys, since it was no longer extant.

Machine Gun Batteries

When the US entered WWII in 1941 following the 7 December attack on Pearl Harbor, the consequent perceived threat on the west coast was the Japanese Navy. Continuing news of Japan's seemingly unstoppable rampage across Asia and the Pacific (Scott, 2015), along with the terrible losses suffered by our own Navy, painted a dark picture of California's increasing vulnerability. As the primary mission for the AMTB batteries was defending against fast moving surface combatants trying to enter San Diego Bay, these emplacements had been sited accordingly along both sides of the harbor entrance, as already discussed.

On the other hand, the perceived targets for a Japanese carrier attack were the major gun batteries at Fort Rosecrans and Fort Emory, hence additional antiaircraft defense specific to their locations took on similar priority. Figure 6-46 shows the effective 0.50-caliber and 40-millimeter antiaircraft coverage areas associated with the 6-inch, 8-inch, and 12-inch batteries in the Harbor Defenses of San Diego, as well as the three 90-millimeter AMTB batteries. There were numerous other machine-gun emplacements throughout Fort Rosecrans for ground defense, some of which have briefly been mentioned in earlier chapters.

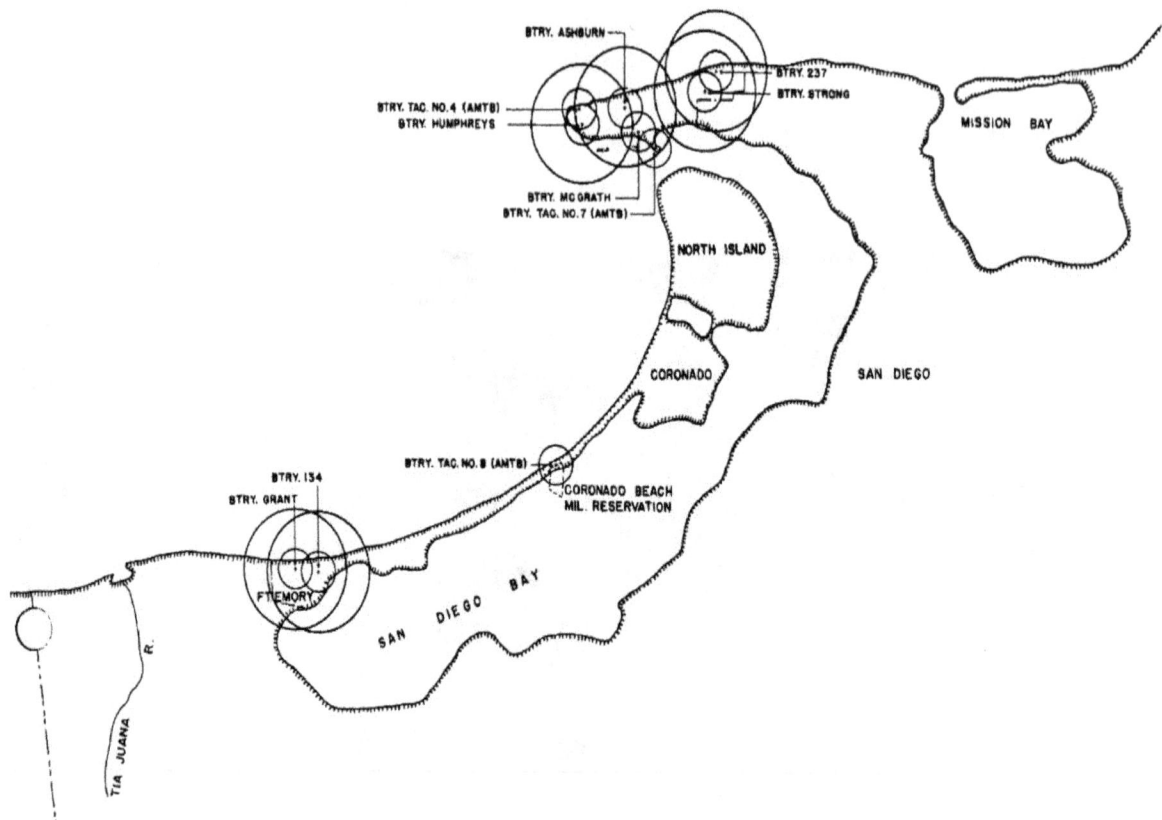

Figure 6-46. This center portion of the "Local Defense Antiaircraft Armament" drawing dated 1 July 1945 shows the effective HDSD coverage zones for 0.50-caliber (small circles, 500 yards) and 40-millimeter (large circles, 1500 yards) antiaircraft emplacements.

The 1946 *Annex A to the Harbor Defenses of San Diego* described the antiaircraft defense as follows (Annex, 1946):

"There is no antiaircraft defense authorized, except that provided by the automatic weapons with which each of the seacoast batteries is provided, and that provided by the secondary mission of the AMTB batteries, of which there are three."

In *The Guns of San Diego*, Thompson (1991) reports Group 3 was responsible for automatic antiaircraft weapons.[16]

The following excerpts from "Notes on the Harbor Defenses of San Diego" identify several positions manned by Batteries of the 19th Coast Artillery Regiment, including four numbered anti-aircraft batteries (Gaines, 1993):

- On 7 December 1941, Battery H mans its battle positions at AA Batteries 4 and 5 and Battery McGrath.[17]

- On 20 April, 1942, Battery D assumes manning responsibility from Battery I for AA Batteries 2 and 3 at Fort Rosecrans.

- On 19 August 1942, Battery A was temporarily assigned to the "Beach Defenses" of Battery Strong, manning a number of machine gun positions.

---

16 The four Groups were replaced by three Battalions in 1945.

17 Battery McGrath, a repurposed Endicott-era structure on the lower level of Fort Rosecrans just south of Battery Calef-Wilkeson (Figure 6-46) was rearmed with a pair of 0.50-caliber machine guns during WWII.

One can presume from the above numbering scheme that at some point the Harbor Defenses of San Diego also included an AA Battery 1. Keep in mind that none of the modernization batteries had yet been completed as of the early dates cited above.

The "1946 Annex" to the *Harbor Defense Project, Harbor Defenses of San Diego* provides a summary of authorized automatic weapons for antiaircraft defense at existing AMTB and larger batteries at the end of the war:

(1) Local defense:

        (a) 30 – 50 cal. machine guns.
        (b) 16 – 40 mm antiaircraft guns.

(2) AMTB – 8 – 37 mm antiaircraft guns.

Except for the eight 37-millimeter AMTB antiaircraft guns, which were not addressed, the above guns were listed as shown in Table 6-1 below (Annex, 1946). The real significance here is that these antiaircraft and machine-gun allocations are broken out by the known gun batteries in the top row, thus providing a strong clue as to their location.

Battery	Woodward	Strong	Ashburn	Cabrillo	Humphreys	McGrath	Fetterman	Cortez	Gatchell	Grant	Spares
Dir. M5	2	2	4		2				4	2	4
40-mm	2	2	4		2				4	2	
50-Cal.	2	2	4	4	2	2	4	4	4	2	

Table 6-1. Allocation of HDSD directors, 40-millimeter antiaircraft guns, and 0.50-caliber machine guns as of 1 July 1945 (Annex, 1946).

Some of these inferred sites are in fact still extant. Starting at the northern fence line for Fort Rosecrans, the distinctive remains of three machine-gun bunkers at the top of Point Loma above Battery North/Gillespie can still be seen today from Catalina Boulevard / Cabrillo Memorial Drive, looking west. These earth-covered wooden structures appear to have been situated for protecting the northern flank of the fort from ground assault (Figure 6-47). Their location was reported in the *Cultural Resource Inventory* by Flower and Roth (1982) as "260 degrees, 300 feet from NOSC Building A36," which is shown later in Figure 6-48.[18]

Figure 6-47. From right to left, the remains of machine-gun bunkers 1, 2 and 3 (up arrows), looking southwest from Catalina Boulevard / Cabrillo Memorial Drive on Point Loma, circa 2007. For reference, the San Pedro radio tower A-45 is seen at upper left (see also map in Figure 6-48).

---

18 NOSC stands for Naval Ocean Systems Center, one of many former names for the Navy lab where I worked on Point Loma (see Appendix).

Flower and Roth (1982) described this area as follows:

"July 1941, four positions of .30-caliber machine guns consisting of four guns in each emplacement were installed as anti-aircraft guns. This type of installation was for AA defense against dive-bomber attack, as well as beach defense (Fulmer, 1982). This location appears to be one of those four positions."

While the above claim of antiaircraft defense is certainly true for some of the other emplacements at Fort Rosecrans, it is somewhat questionable for this particular site.

Mounted in semi-underground bunkers covered over with dirt from the excavation, these guns would have been incapable of pointing upward with 360-degree views. A reexamination of Figure 6-46 shows the only significant antiaircraft defense for the northern portion of Fort Rosecrans was in the vicinity of Batteries Strong and Woodward, some ways to the south from the northern fence line of the fort. Due to their construction, the three bunkers shown in Figure 6-47 would only be able to pick off low flyers coming into their vertically-limited field of fire. Their primary mission was presumably ground defense, which suggests 0.50-caliber versus 0.30-caliber machine guns (see again Figure 6-46).

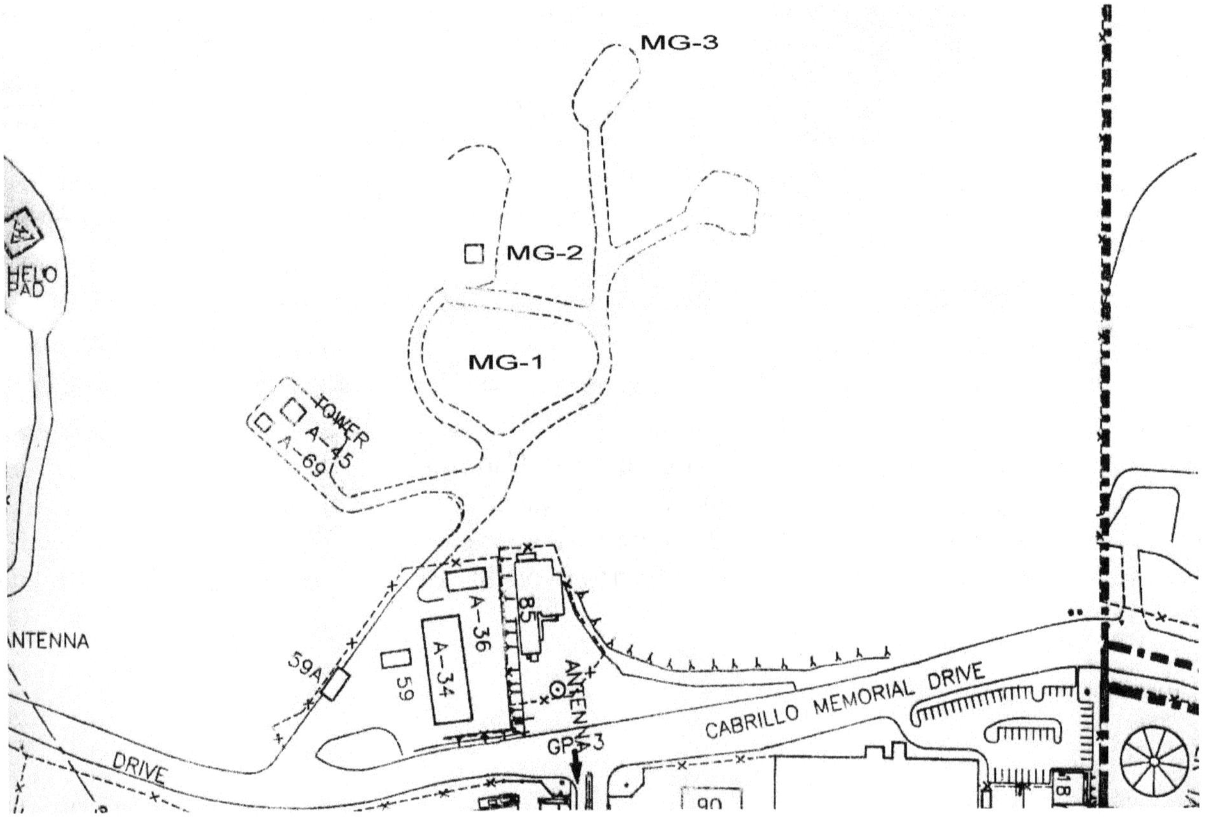

Figure 6-48. Southeast corner of "Basemap 1" (1991), showing Building A-34 at bottom center, with the San Pedro tower A-45 above it and to the left. WWII machine-gun bunker locations *MG-1*, *MG-2*, and *MG-3* (callouts added) are in an east-west line at the end of Hardware Road (not labeled) at center (US Navy drawing courtesy SSC Pacific).

A zoomed-in portion of a 1959 aerial photo of this area is presented in Figure 6-49, with Cabrillo Memorial Drive running north-south at bottom, and the San Pedro radio tower A-45 in foreground left. Building A-34 is the larger structure with the two garage doors to the right of center, in front of which Hardware Road is seen running northwest off of Catalina Boulevard / Cabrillo Memorial Drive. Machine-gun bunkers *MG-1* and *MG-2* are in the turn-around loop at the west end of Hardware Road at left center, directly behind (slightly northwest) of the San Pedro radio tower. Bunker *MG-3* is just out of the image to the left.

Figure 6-49. Looking northwest in 1959 with the intersection of Catalina Boulevard / Cabrillo Memorial Drive at right and Hardware Road branching off to the west. Building A-34 is the larger structure at upper right, with machine-gun emplacements *MG-1* and *MG-2* directly behind the San Pedro Tower A-45 at far left (US Navy photo LRO1888 3-14-59 courtesy SSC Pacific).

The Flower and Roth (1982) field-check results regarding these bunkers reads as follows:

"Two structures still intact, facing north. Field of fire covers Site 1. Both structures are wood and contain gun platforms and two bunks each. The western structure measures 12 by 16 feet ceiling to floor. The eastern structure measures 9 feet by 6 feet ceiling to floor."

Lee Guidry, while assisting in the later Keniston (1995) survey, found not two but three machine-gun bunkers at this location, which he documented in incredible detail. These bunkers were accessed via what is now Hardware Road, which runs northwest off Catalina Boulevard / Cabrillo Memorial Drive in the bottom-right corner of Figure 6-50. The "cast iron Army mine anchor" annotated along the edge of Woodward Road (unlabeled) at far left was previously discussed in the beginning of Chapter 2.

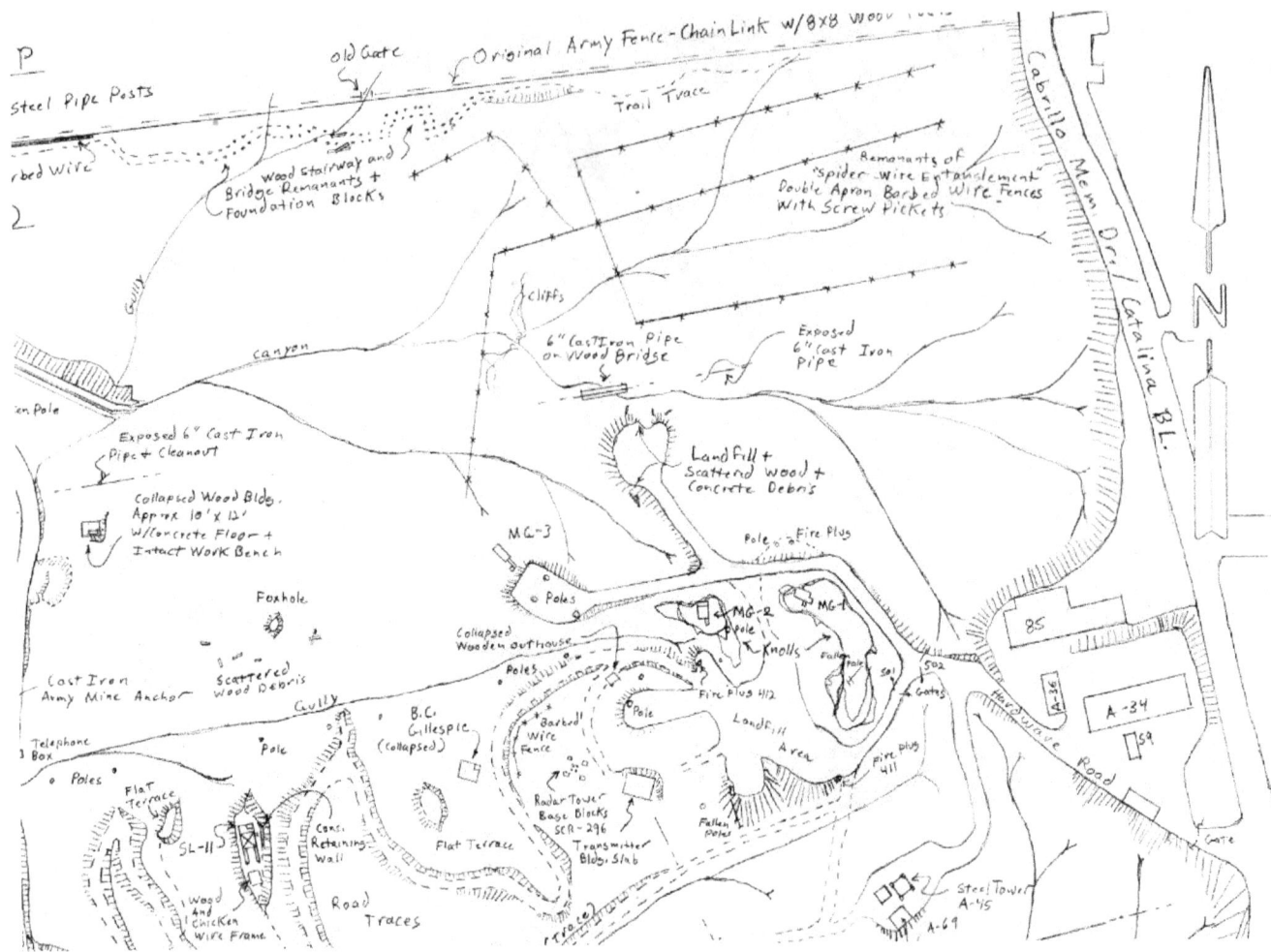

Figure 6-50. The three machine-gun bunkers just below image center are north of the *SCR-296-A* radar-transmitter foundation slab at bottom center, with the San Pedro Tower A-45 in bottom-right corner, accessed via Hardware Road from Catalina Boulevard / Cabrillo Memorial Drive (1995 drawing by Lee Guidry).

Note the foundation pillars for the *SCR-296-A* fire-control radar for Battery Woodward and the slab for its transmitter/receiver building just below bottom center in Guidry's map above. Protecting those high-value assets from a ground assault approaching from the north or the west was perhaps the primary mission of these machine-gun emplacements. The collapsed battery commander station for Battery Gillespie (Chapter 4) is just to the left (west) of this radar foundation and east of the underground-silo shelter for Searchlight 11 (lower-left corner).

The low-angle aerial photo presented in Figure 6-51 shows this area looking southeast in 1969, with the San Pedro tower at image center and Building A-34 to its left at the intersection of Catalina Boulevard / Cabrillo Memorial Drive and Hardware Road. Emplacement *MG-1* is in the large clay mound at the western end of Hardware Road (leftmost up arrow), with *MG-2* in the smaller mound further to the west. Emplacement *MG-3* was situated at the end of the dirt road running west (right), just below and to the right of the pair of power poles. The northern fence line for Fort Rosecrans is seen left of bottom center.

Figure 6-51. In this 1969 photo looking southeast, *MG-1* is in the large clay mound at photo center (left up arrow), with *MG-2* in the smaller clay mound just to its right (second up arrow). *MG-3* is further to the right at the end of the dirt road (third up arrow), just past the pair of power poles. The fourth up arrow (far right) denotes one of the four *SCR-296-A* tower-foundation pillars. Note World War II barracks at top center.

Referring again to Figure 6-51 above, a salvaged section of the *SCR-296-A* radar tower lays horizontally on the ground between the two power poles (above third arrow), albeit hard to see.[19] One of the tower's four concrete foundation pillars can barely be seen a bit further to the south, as indicated by the rightmost up arrow. Figure 6-52 below shows this lone pillar is clearly visible from Woodward Road, which runs parallel to the Pacific Ocean coast line, out of image below. Note the pair of power poles at top center, previously seen near emplacement *MG-3* in Figure 6-51 above.

Figure 6-52. Machine-gun bunker *MG-3* was situated just to the left of the pair of adjacent power poles at top center, approximately 100 yards north of the *SCR-296-A* tower foundation to the right. One of the tower's four concrete foundation pillars (up arrow) is clearly visible from the photographer's position down on Woodward Road (circa 2006).

19 These two power poles are in the cleared area slightly above the third up arrow from the left.

Note Guidry's annotation of "Remnants of Spider-Wire Entanglement, Double Apron Barbed Wire Fences with Screw Pickets" at upper right in his earlier field sketch (see again Figure 6-50). These defensive measures, several of which are still extant in this area as seen in Figure 6-53, were presumably intended to impede the advance of ground forces advancing towards the radar site. Their presence further reinforces the theory that the primary mission for these machine-gun bunkers was ground as opposed to antiaircraft defense.

Figure 6-53. Seen here looking northwest towards the north end of Woodward Road (top of left photo), extensive spider-wire entanglements supported by "screw-pickets" are still extant below the former *SCR-296-A* radar site for Battery Woodward, circa 2006.

The section view presented in Figure 6-54 below shows construction details for the northernmost bunker *MG-1*, for which the firing platform opened to both the west and north (as opposed to just north as reported by Flower and Roth). As annotated by Guidry, this view is through the bunker entrance and not the western face (see later plan view in Figure 6-55), and therefore does not show the firing platform. Instead it shows the two bunk beds against the rear (east) wall of the bunker. Note the local terrain level comes up about halfway to the overhead on the entrance (right) side of the bunker.

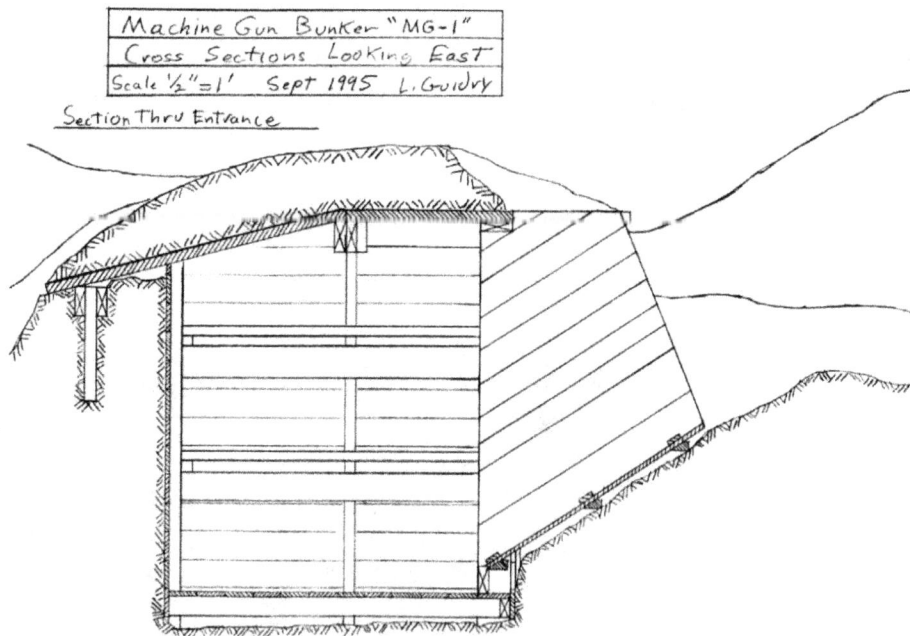

Figure 6-54. This 1995 field sketch provided by Lee Guidry shows a cross section of machine-gun bunker *MG-1*, looking east towards Catalina Boulevard / Cabrillo Memorial Drive, with descending entrance stairs on right. Note the two bunk beds on the east (far) side of the bunker.

A plan view of bunker *MG-1* is shown in Figure 6-55 below. The bunk beds previously seen in Figure 6-54 are on the east side of the layout at right, with the firing platform in the upper left corner and the entrance stairs at bottom center. As previously mentioned, the firing platform at the northwest corner has both northerly and westerly fields of fire, as this is the westernmost of the three bunkers.

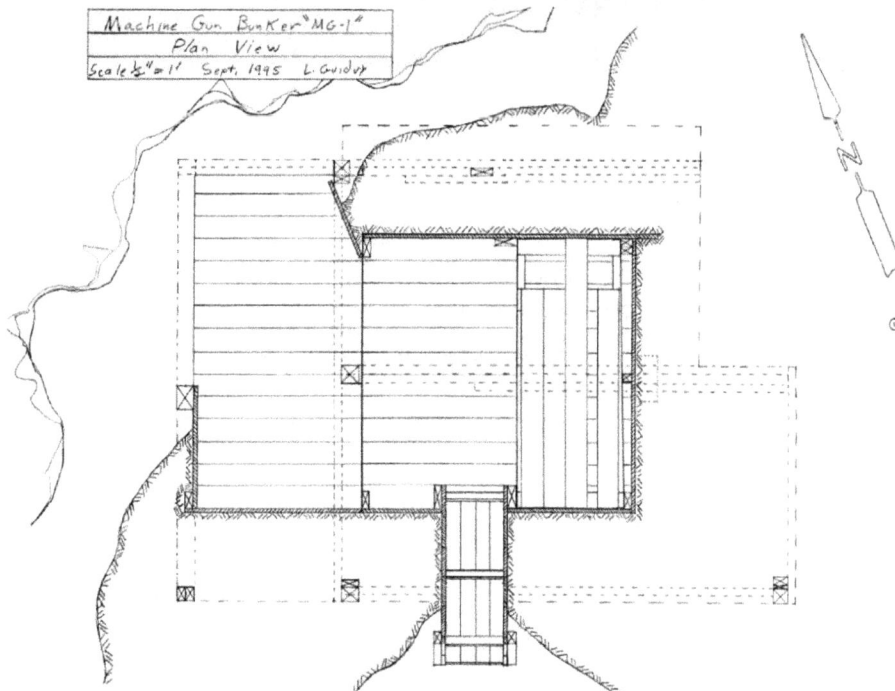

Figure 6-55. Plan view of machine-gun bunker *MG-1* sketched by Lee Guidry. Note the firing platform at upper left is open both to the north and to the west.

The remains of machine-gun bunker *MG-1* as seen looking north in March 2007 (some 12 years after Guidry's 1995 site survey) are shown below in Figure 6-56, with debris from the firing platform in the top-left quadrant. The three depressions seen across the top of the earthen mound suggest the wooden structure has for the most part collapsed, as the dirt piled on the roof during construction has suffered some 60 years of erosion. Accordingly, no interior entry was deemed possible or prudent.

Figure 6-56. The remains of machine-gun bunker *MG-1* looking north just west of Catalina Boulevard / Cabrillo Memorial Drive, circa March 2007.

Machine-gun bunker *MG-2* was located just to the south of bunker *MG-1*, between Catalina Boulevard / Cabrillo Memorial Drive and the Pacific Ocean. This large but well-hidden bunker was similarly constructed partially above the terrain level and then covered with excavated dirt (Figure 6-57). Note the hinged entry door in the upper-left corner, stairs leading down to the structure floor, and sleeping arrangements for two men just left of the gun platform at foreground right. The north-facing gun port on the right side is the opening between the two horizontal 2x12s.

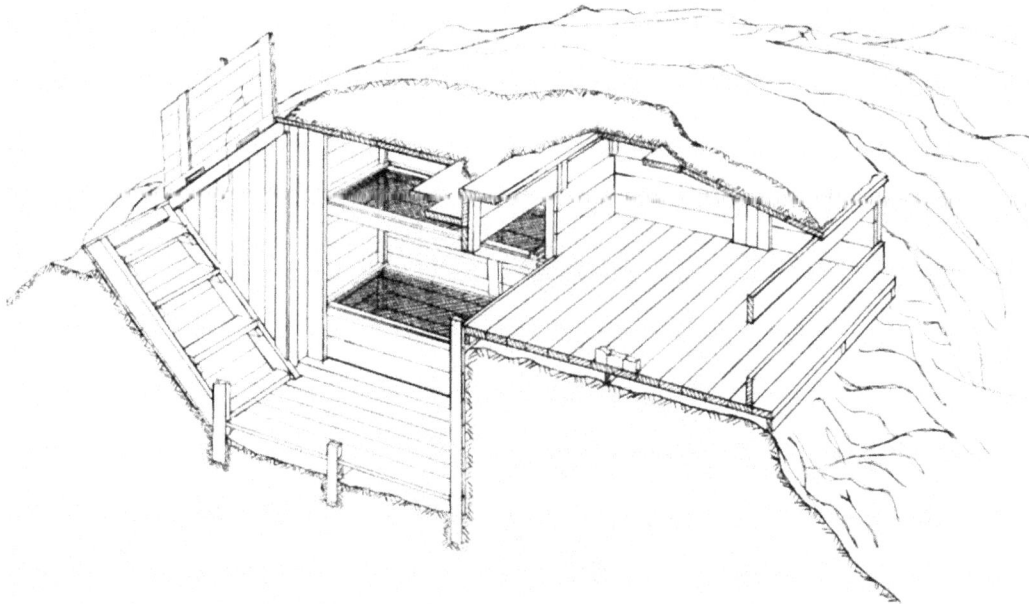

Figure 6-57. This December 1995 field sketch provided by Lee Guidry shows an isometric view of the semi-underground machine-gun bunker *MG-2*, looking southwest. The firing platform is in foreground right, with the horizontal opening facing to the north.

Guidry's 1995 plan view of the *MG-2* bunker is shown in Figure 6-58 below. Note the descending entryway stairs on the south (right) side and the elevated gun platform facing north on the left. The double-rack bunk area is in the southwest corner of the earth-covered structure.

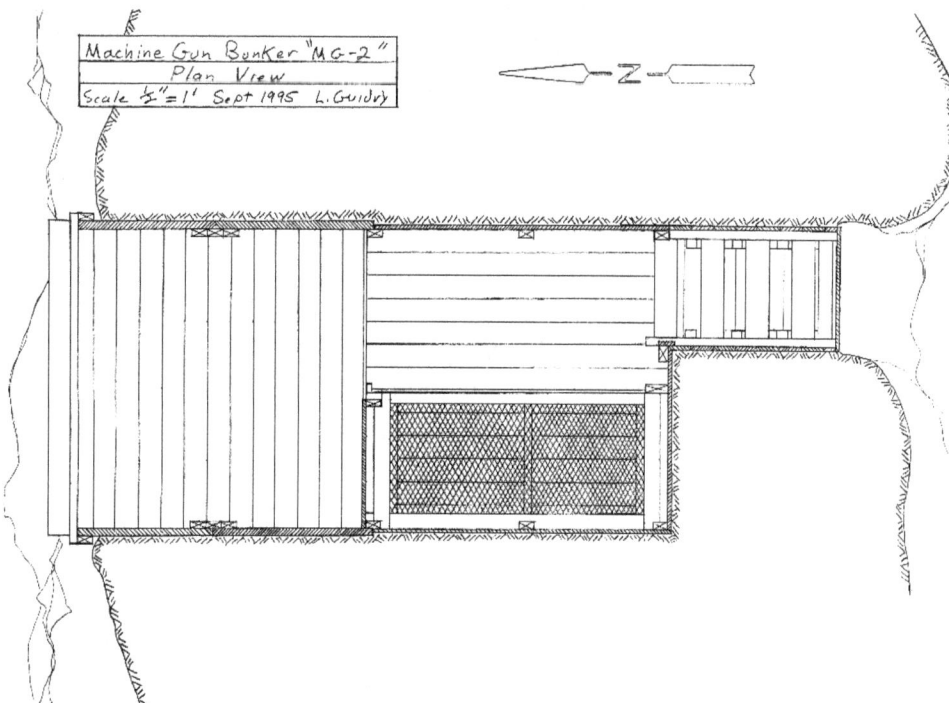

Figure 6-58. Plan view of *MG-2*, showing the south entryway at right and slightly elevated firing platform facing north at left (see again Figure 6-57 above for perspective).

Still mostly covered with dirt, a small portion of machine-gun bunker *MG-2* is seen in Figure 6-59, circa 2007. While it appeared from the surface that a good portion of this bunker might still be intact, no entry was attempted for safety reasons, especially since the warmer weather had increased the number of rattlesnake sightings on Point Loma. Due to the cave in, which filled the visible portion of the interior with spillage, it was somewhat unclear which portion of the sizeable layout shown in Figure 6-58 was found partially exposed in the photo below.

Figure 6-59. Remains of machine-gun bunker *MG-2*, circa March 2007. As this was a rather large bunker and there were no obvious signs of collapse, a considerable portion likely remained somewhat intact underground.

Figure 6-60 provides plan and section views of emplacement *MG-3*, showing the entrance at lower right to a tunnel leading to the bunk area, and the firing platform at image left facing the Pacific Ocean. This was probably the emplacement missed by Flower and Roth (1982) when they reported only two bunkers were located, as there was less erosion evident relative to *MG-1* and *MG-2*. *MG-4*, the fourth machine-gun emplacement reportedly at this site, has yet to be found, but may show up after continuing erosion takes its toll.

Machine Gun Bunker "MG-3"
Overall Plan
Scale 3/8" = 1' Sept 1995 L. Guidry

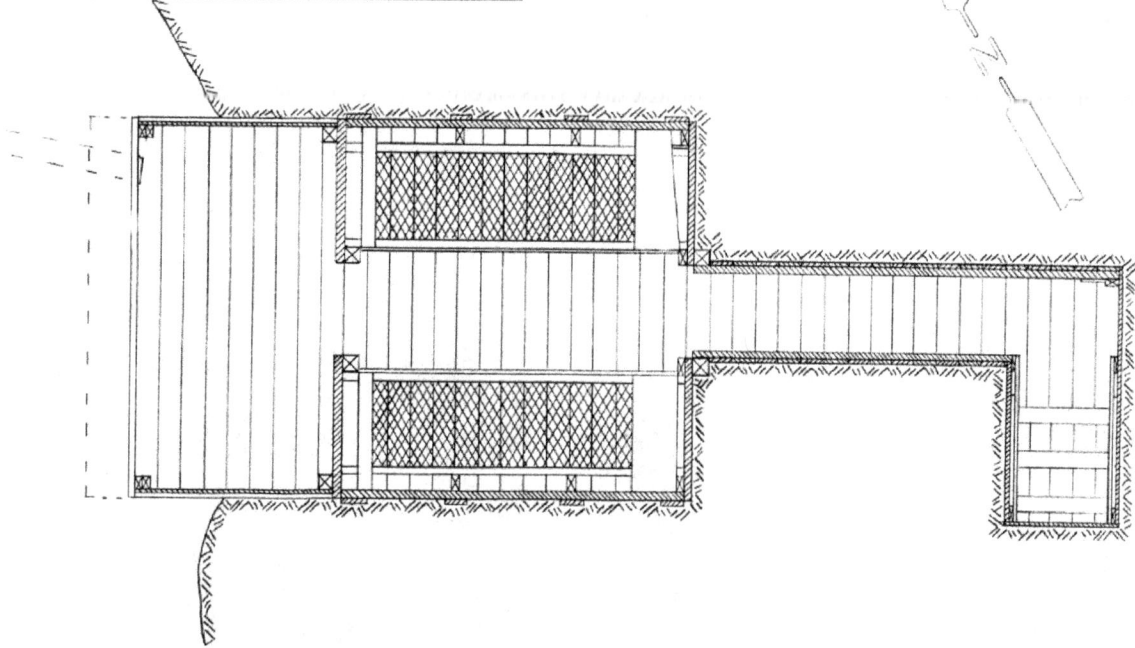

a)

Machine Gun Bunker "MG-3"
Top: Section Looking Northwest
Bottom: Section Looking Northeast
Scale ½" = 1' Sept 1995 L. Guidry

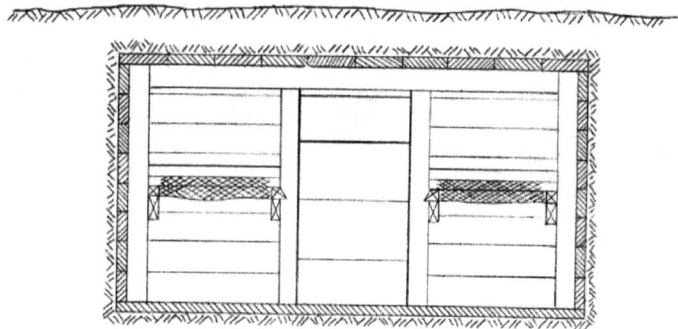

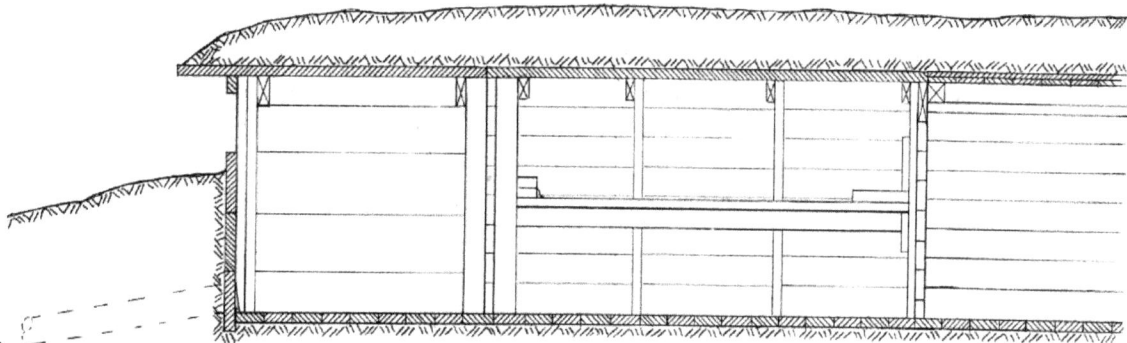

b)

Figure 6-60. **a)** Plan view of machine-gun emplacement *MG-3*, showing the entrance at lower right with L-shaped tunnel leading to the bunk area, and the firing platform at image left facing the Pacific Ocean. **b)** Section views of *MG-3* showing the firing platform at left facing northwest (drawings by Lee Guidry).

As seen in Figure 6-61a, *MG-3* was located to the northwest of the pavement at the west end of Hardware Road, just past the bare clay area at bottom right. The well-preserved remains of a semicircular defensive wall can be seen on the far side of the clay, looking northwest. The exposed debris of the northwest-facing firing platform, which is slightly upslope from this wall, is seen in Figure 6-61b, looking northeast in March 2007. As the interior of this bunker was mostly buried in dirt with the wooden structure in an obvious state of decay, no attempt at entry was deemed practicable.

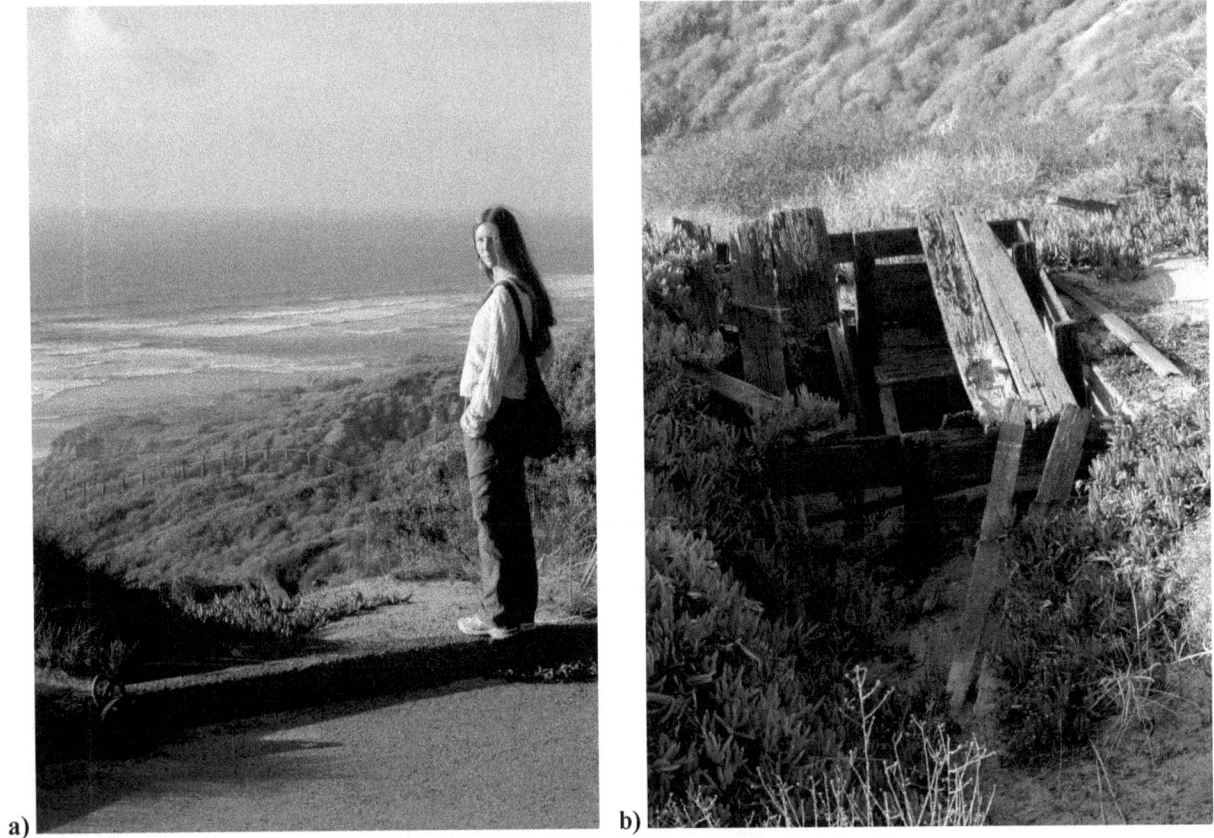

Figure 6-61. a) Northwest view towards curved defensive wall below machine-gun bunker *MG-3*, located just past the bare area beyond the asphalt curb, with the northern fence line of Fort Rosecrans running down to the coastline in the background. **b)** Remains of the collapsed *MG-3* firing platform, circa March 2007.

There were a number of observation posts and machine-gun emplacements situated along the cliff edge below these three bunkers to the west. The field sketch by Lee Guidry in Figure 6-62 shows a 0.30-caliber machine-gun pit west of Battery Gillespie, overlooking the Pacific Ocean down below. This site was perhaps part of the second of the four groups of machine-gun emplacements claimed by Flower and Roth (1982). In 1995, the associated observation post for this position was located approximately 25 feet to the west, but has since fallen into the surf below due to cliff erosion (see again Chapter 4).

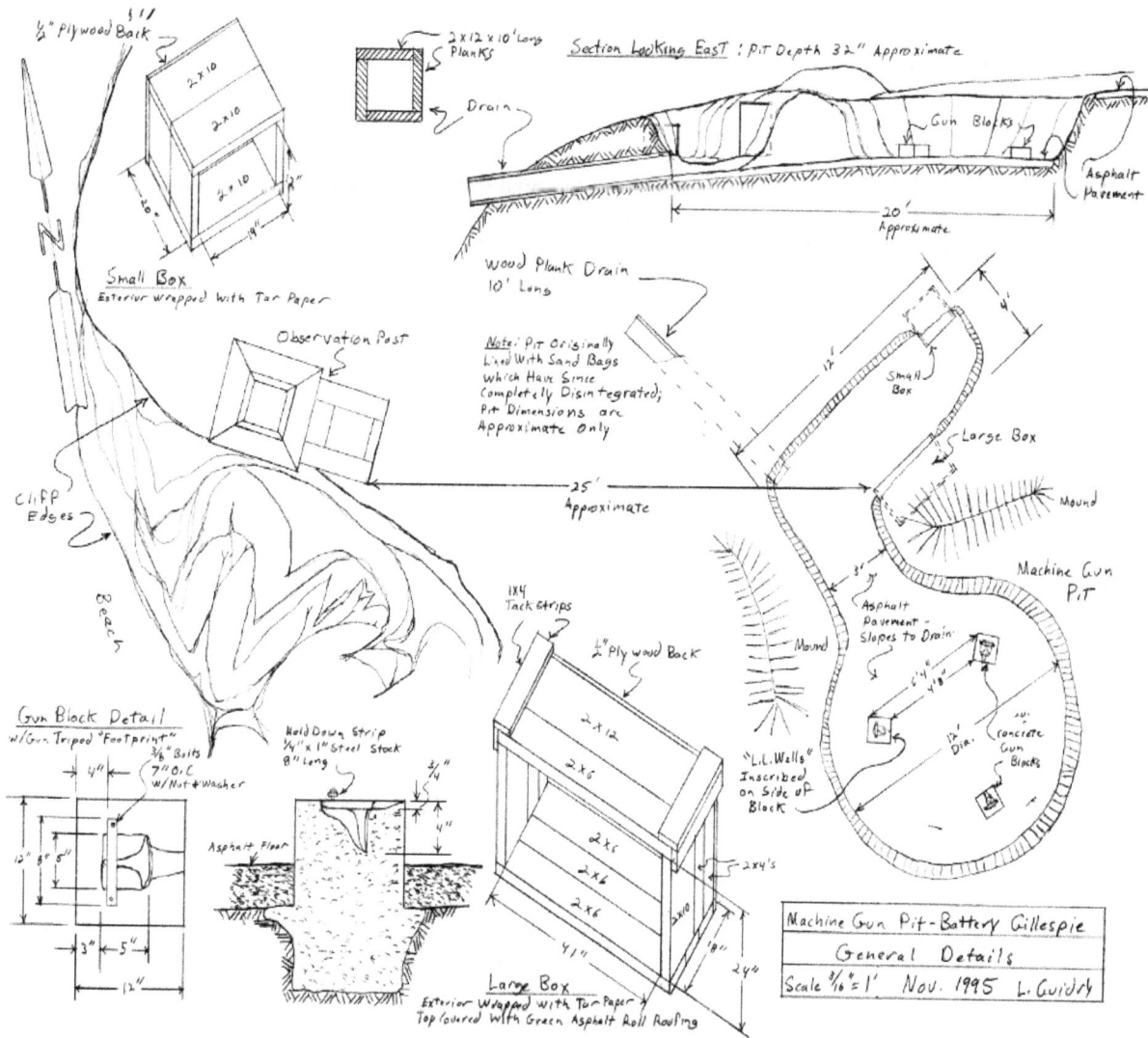

Figure 6-62. This 1995 field sketch by Lee Guidry shows a 0.30-caliber machine-gun pit above the beach just below Battery Gillespie. Note observation post on the cliff edge 25 feet to the west at left center.

Figure 6-63a shows the remains of this 32-inch-deep machine-gun emplacement as seen during the Keniston site survey in 1995, while Figure 6-63b shows this same location some 13 years later in 2008. Due to erosion, the edge of the cliff now appears just 6 feet to the west, versus 25 feet as Guidry noted in his 1995 field sketch, and the emplacement remains are now level with the trail. No traces were found of the wooden storage boxes or the 10-foot wooden pit drain. The descending ravine just north of this location provided an easy approach from the beach below to Battery Gillespie and the *SCR-296-A* radar site upslope to the east.

Figure 6-63. a) The 32-inch deep 0.30-caliber machine-gun emplacement as seen in 1995 (photo courtesy Lee Guidry). Note beach below in the upper-right corner. **b)** Lee Guidry points out the westernmost of three concrete supports for the machine-gun tripod, looking northwest in April 2008.

Figure 6-64 below shows what appears to be a line of three machine-gun emplacements just west of Battery Zeilin, which is seen west of Woodward Road below the pistol range at image center. Note the former auxiliary range on the cliff edge overlooking the Pacific in the bottom-right corner, the future site of the 6-inch Battery Woodward (237). These three machine-gun emplacements, later associated with Battery Woodward (see again Chapter 5), may have been part of the third group claimed by Flower and Roth (1982).

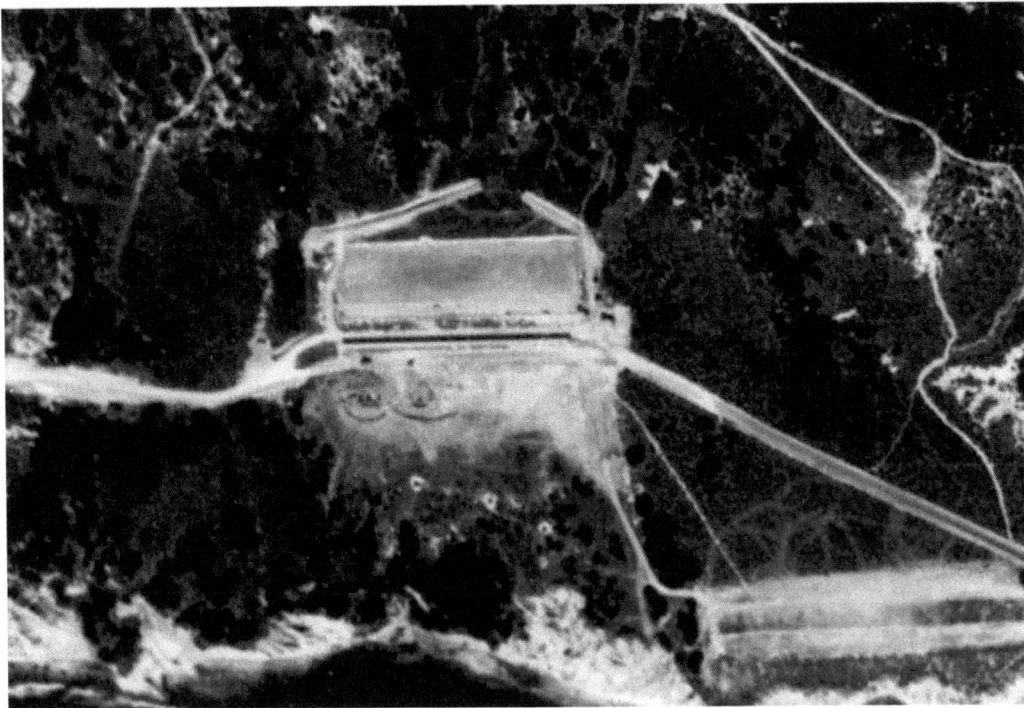

Figure 6-64. Three machine-gun emplacements are west of the 7-inch guns of Battery Zeilin, which was directly across Woodward Road from the pistol range. Construction of Battery Woodward in the auxiliary-range area at bottom right would not begin until March 1943 (undated US Army photo courtesy SSC Pacific).

Located just south of Gate 5 at the intersection of Cabrillo Memorial Drive and Woodward Road, the antiaircraft complex seen in Figure 6-65 featured positions for a 40-millimeter gun, its associated director, and a height finder (Guidry, 2008). Note the ammunition-storage bunker and access stairs to the elevated position at lower right. The steel tower in background is a more recent construction. Woodward Road is seen descending down towards the clifftops at upper left, after which it turns north on the way to Battery Woodward (out of image). This AA site was probably associated with nearby Battery Strong (see again Table 6-1 and map presented in Figure 6-46).

Figure 6-65. No longer extant, this extensive WWII 40-millimeter anti-aircraft bunker complex was located just south of Gate 5 at the intersection of Cabrillo Memorial Drive (upper right) and Woodward Road (upper left), circa 1954 (US Navy photo courtesy SSC Pacific).

One Saturday in December 2007, my son Todd and I found the remains of another (possibly the fourth) group of machine-gun emplacements east of Gatchell Road, just below the shelter for Searchlight 13. The first of these bunkers was just downslope from an attention-grabbing footbridge made from a pair of power poles, as seen in Figure 6-66. This improvised WWII structure spanned the ravine just below the buried Multi-Plate-Arch shelter for Searchlight 13, which will be further discussed in Chapter 8.

Figure 6-66. Todd Everett crosses the remains of a wooden footbridge over the ravine descending down from the shelter for Searchlight 13, circa 2007.

While climbing up the south side of the ravine to the searchlight-shelter location, our second discovery that day was little more than a bare patch of dirt covered over with the remains of chicken-wire camouflage netting and 2x4 support posts (Figure 6-67a). The out-of-place object about halfway down to Gatchell Road in the background of this picture soon caught our attention, which upon inspection turned out to be a drain-culvert inlet on the west side of a distinctive road trace. My son wrestled it out of the bushes for the close-up photo of Figure 6-67b.

a)

b)

Figure 6-67. a) Remains of a camouflaged WWII machine-gun position looking west towards a gravel-road trace, with a section of a corrugated road drain in the brush just above photo center, and Gatchell Road seen in the background. **b)** Todd Everett holds the corrugated road drain inlet for closer inspection, circa 2007.

In April of 2008, I returned to this site with Lee Guidry to further explore the bunker debris in the ravine by the footbridge, but we unfortunately found it too scattered and decayed to yield any details of the original layout (Figure 6-68). We did get the sense that the structure had originally been partially below ground and covered with back fill like the three bunkers guarding the Woodward radar (*MG-1*, *MG-2*, and *MG-3*). This dirt cover had completely washed away over the years due to rainfall runoff flowing down the ravine. Part of the improvised footbridge seen earlier in Figure 6-66 is visible at bottom left in the photo below, with the ravine running downslope to the right towards Gatchell Road.

Figure 6-68. Lee Guidry explores the scattered remains of a machine-gun bunker located downslope from and immediately adjacent to the southern end of the improvised footbridge crossing the ravine, circa 2008.

Another field conclusion we reached was that these machine-gun emplacements were not built so much for defending the nearby searchlight shelter, but rather to impede invading troops attempting to access the upper level of Fort Rosecrans via the ravine. The large tree seen at the very top of the view presented in Figure 6-69 is on the edge of Fort Rosecrans National Cemetery, just west of Cabrillo Memorial Drive on the south side of the HDCP/HECP (see also Chapter 7). Defending this critical command-and-control installation from ground assault would have been a high wartime priority.

Figure 6-69. Easterly view of the ravine, which would have provided invading troops easy access to the top of Point Loma. Note debris from the machine-gun bunker next to north end of the footbridge at bottom left. The large tree at the top of the ravine is on the corner of Fort Rosecrans National Cemetery (Figure 6-70).

The cemetery expansion on the west side of Cabrillo Memorial Drive seen in Figure 6-70, circa 1966, was added sometime after the war. The concrete upper observation level of the HDCP/HECP is visible midway between the radar reflector and the cemetery in the bottom-left corner, directly in front of the leftmost dark tree. The pathway running south from the dirt road coming upslope from Gatchell Road at upper right leads to the local searchlight operator's hut just right of top center. The road trace and drain inlet we found were in the curved section of the dirt road just below the searchlight shelter.

Figure 6-70. The underground shelter for Searchlight 13 was situated on the south side of the ravine descending from the southwest corner of Fort Rosecrans National Cemetery (lower left). Note the top of a large radar dish in the extreme bottom-left corner (9 September 1966 US Navy photo courtesy SSC Pacific).

The previously mentioned radar dish is seen just north of the cemetery fence in Figure 6-71, looking south. Building 557 is at the center of this photo, just left of the radar. The entrance to the subterranean section of the HDCP/HECP is immediately behind this building, at the west end of the short driveway. The upper and lower observation levels of the HDCP/HECP are at the end of the underground connecting tunnel as labeled. Midway between Building 557 and the cemetery fence is a circular defensive wall (up arrow), which suggests a probable machine-gun emplacement, no longer extant.

Figure 6-71. Looking south in 1966, the entrance to the HDCP/HECP is right behind the southeast corner of Building 557 at photo center. Note the circular defensive wall (up arrow) halfway between this building and the cemetery fence, which appears to have been a machine gun emplacement.

As mentioned by Ranger Brett Jones in a 5 December 1984 interview with Master Sargent George H. McGlothlin (Overton, 1993), there was an improvised machine-gun emplacement (not shown) situated along the cliff edge below the 155-millimeter *GPF* emplacements for Battery Point Loma, which is seen in Figure 6-72. An extensive north-south line of defensive positions can be seen left of and parallel to Humphreys Road in the right half of this 1962 aerial photo. This may have been the fourth of the four groups of 0.30-caliber machine guns alluded to by Flower and Roth (1982). Note Sylvester Road (now the Bayside Trail) running east from Humphreys Road at upper right.

Figure 6-72. Remains of the four 155-millimeter emplacements for Battery Point Loma are between Cabrillo Road and the trace of old Gatchell Road at left (circa 1962). Note evidence of an extensive line of defensive positions running north of Battery Humphreys at lower right, just west of and parallel to Humphreys Road (US Navy photo courtesy SSC Pacific).

This line of defensive positions very likely included the four machine guns associated with Battery Humphreys in Table 6-1. Lee Guidry's 1995 field sketch of this area presented in Figure 6-73 does in fact identify four circular pits that likely served as 0.30-caliber machine-gun emplacements. This site appears to have had vehicle access via a short dirt road off Humphreys Road (note trace seen at upper left), and was also served by a wooden stairway down to a guardhouse on Humphreys Road at upper right. Note also the bank of telephone booths depicted in the bottom-right corner.

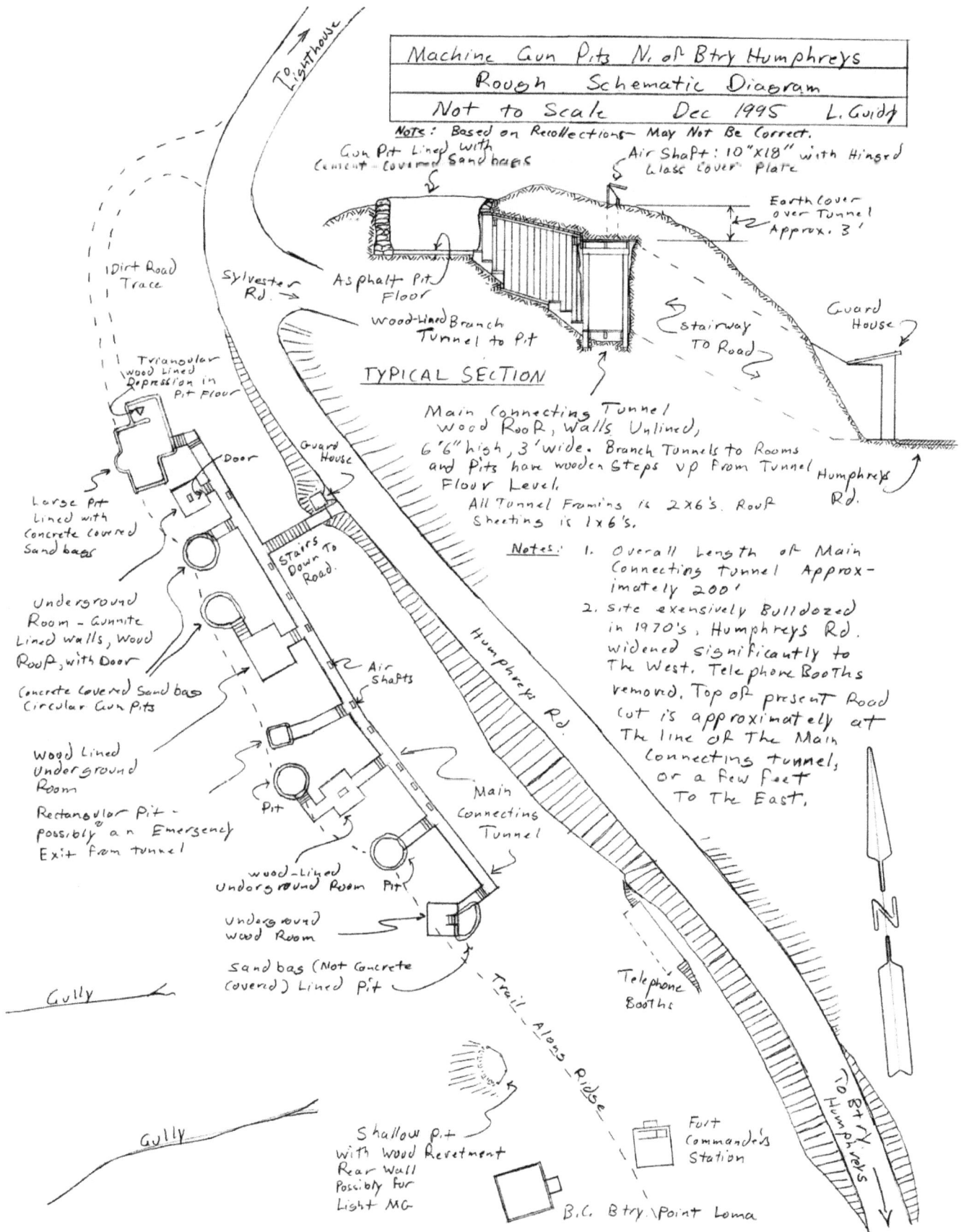

Figure 6-73. Field sketch drawn by Lee Guidry of the group of four defensive machine-gun pits previously shown north of Battery Humphreys in the post-war aerial photo of Figure 6-72. As noted on the drawing (Note 2), this site was extensively bulldozed in the 1970s, when Humphreys Road was widened to the west.

There were numerous other improvised gun emplacements constructed by troop labor throughout Fort Rosecrans shortly before and during the war. According to Wolfgast (1943), there was a 37-millimeter antiaircraft position next to the old Spanish lighthouse at the top of the point. Wolfgast further reports that an antiaircraft machine-gun position located by the temporary 155-millimeter guns at Battery Humphreys was camouflaged by the 19th Coast Artillery. As both these areas have been extensively redeveloped in the post-war era, no evidence remains of either installation.

7

Fire Control

> "Never in the field of human conflict was so much owed by so many to so few."
>
> *Sir Winston Churchill*

The Harbor Defenses of San Diego (HDSD) batteries required a distributed network of base end stations for effective fire control, and several were installed from the Mexican border all the way north to Solano Beach, CA. Each bunker (or level in multi-level bunkers) was dedicated to a specific battery and identified by a special alphanumeric designation. All batteries were assigned a "tactical number" as shown in Table 7-1. These tactical numbers were ordered north to south down the west side, then south to north up the east side of the Point, continuing back down north to south along Coronado and the Silver Strand.

Battery	T.N.	Guns	Site	Location
Woodward	1	2 – 6-inch	5	N. Rosecrans
Strong	2	2 – 8-inch	5	N. Rosecrans
Ashburn	3	2 – 16-inch	6	W. Rosecrans
Cabrillo	4	4 – 90-mm	8	S. Rosecrans
Humphreys	5	2 – 6-inch	8	S. Rosecrans
Bluff	5A	2 – 37-mm	8	S. Rosecrans
McGrath	6	2 – 3-inch	9	E. Rosecrans
Fetterman	7	4 – 90-mm	9	E. Rosecrans
Cortez	8	4 – 90-mm	11	Strand
Gatchell	9	2 – 16-inch	12	Fort Emory
Grant	10	2 – 6-inch	12	Fort Emory

Table 7-1. Tactical number designations for the HDSD gun batteries during the latter part of the war (Thompson, 1991).

The batteries listed in Table 7-1 above were initially organized into four groups, each with its own command post (Thompson, 1991):

- Group 1 – Batteries Ashburn and Gatchell
- Group 2 – Batteries Woodward and Strong
- Group 3 – Automatic antiaircraft weapons
- Group 4 – Batteries Humphreys and Grant

These four groups were replaced in 1945 by three battalions, with two at Fort Rosecrans and one at Fort Emory.

Supporting each of the batteries were three types of fire-control stations, abbreviated as follows:

- BC – Battery commander station
- B – Base end station
- S – Spotting station

Base end stations would track the position of enemy ships, while spotting stations reported the azimuth of the shot/shell splash. An alphanumeric nomenclature appended the tactical number of the associated battery as a subscript to the standardized abbreviation of fire-control functions listed above.

By way of example, the designation BC_1 indicated the structure was the battery commander station for Battery Woodward (Tac. No. 1). While each battery had but a single battery commander station, it was served by multiple base end and spotting stations, so a superscript was added to identify the individual structures of each type. B^5_1, for example, would indicate base end station number 5 for Battery Woodward, while B^1_1 indicated base end station number 1 for the same battery. Base end stations and spotting stations were often combined in a single structure or level, so B^1_2, S^1_2 would identify the combined base end and spotting station number 1 for Battery Strong (Tac. No. 2).

Additional spotting could also be provided by airborne observers (Annex, 1946):

"Coordination with Air Forces for observation of fire on naval targets by long range batteries will be as follows:

(1) The Harbor Defense Commander will make a request to the Commanding General, Southern California Sector, WDC,[1] for aerial spotting when needed.

(2) Communication between the battery firing and the observing aircraft will be as prescribed in TM 1-465, 'Air-Ground Communication.'

(3) Radio frequencies and call signs to be used will be those designated SOI."[2]

Target range and bearing were calculated through geometric triangulation, for which there were three distinct types (Smith, 2004):

- Horizontal baseline – the classic angle-side-angle triangle solution involving two base end stations at either end of a surveyed baseline of known orientation and length, with the angles measured by an optical instrument in each station.

- Vertical baseline – an angle-side-angle triangle solution involving a single elevated station, which measured the "depression angle" to the observed target. The second angle was the 90-degree relationship of the known vertical side of the triangle (station elevation above mean sea level) with respect to the surface of the ocean.

- Self-contained – An independent sighting instrument such as the stereoscopic rangefinder or coincidence rangefinder.

The "Type B" azimuth scope (AS) and the "Type A" depression position finder (DPF) were the most commonly employed observation instruments during WWII, as the self-contained systems were not sufficiently accurate at the longer ranges achieved by the new modernization batteries.

The *Azimuth Instrument M1910* shown in Figure 7-1 was used under the horizontal-triangulation scheme to visually determine the bearing to an observed enemy ship, relative to the baseline between the pair of base end stations. The depression position finder (DPF), introduced in the late 1800s, was utilized under the vertical triangulation scheme to determine both the azimuth and the horizontal range to target. An elevated observation post well above sea level was required for DPF instruments, which were sufficiently accurate out to 1,000 yards for each 10 feet of installation height above mean low water (Hines & Ward, 1910).

---

1   WDC – Western Defense Command.
2   SOI – presumably Standard Operating Information.

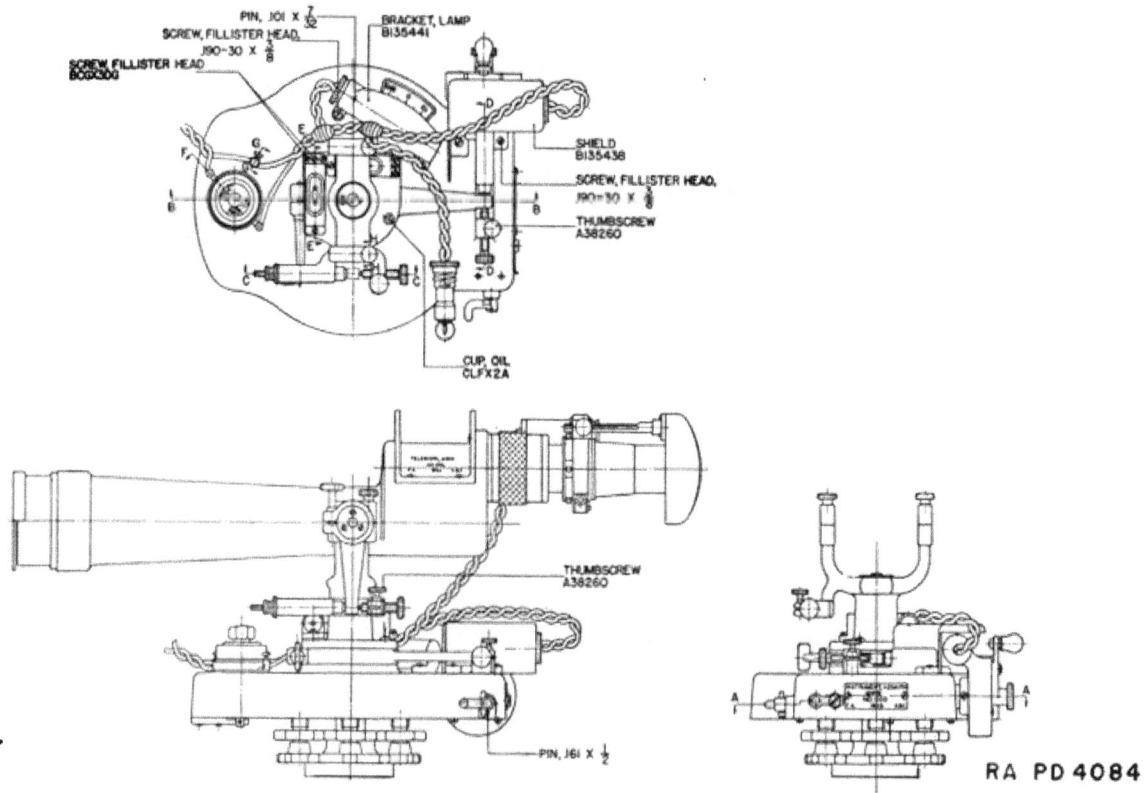

Figure 7-1. Plan and elevation views of *Azimuth Instrument M1910* (US Army diagram adapted from *TM 9-1675* (1941).

The undated WWII-era photo presented in Figure 7-2 below shows a member of the Coast Artillery at image right manning a tripod-mounted *Azimuth Instrument M1910*, with his right hand on the azimuth crank that rotated the housing. Course azimuth readings were viewed through a window on the azimuth scale, which was graduated in 1-degree intervals, while fine readings were indicated on the crank micrometer, which was graduated in 0.01-degree intervals (TM 9-1675, 1941). Three small electric lamps provided adjustable illumination for the telescope reticle, the azimuth scale, and the azimuth micrometer.

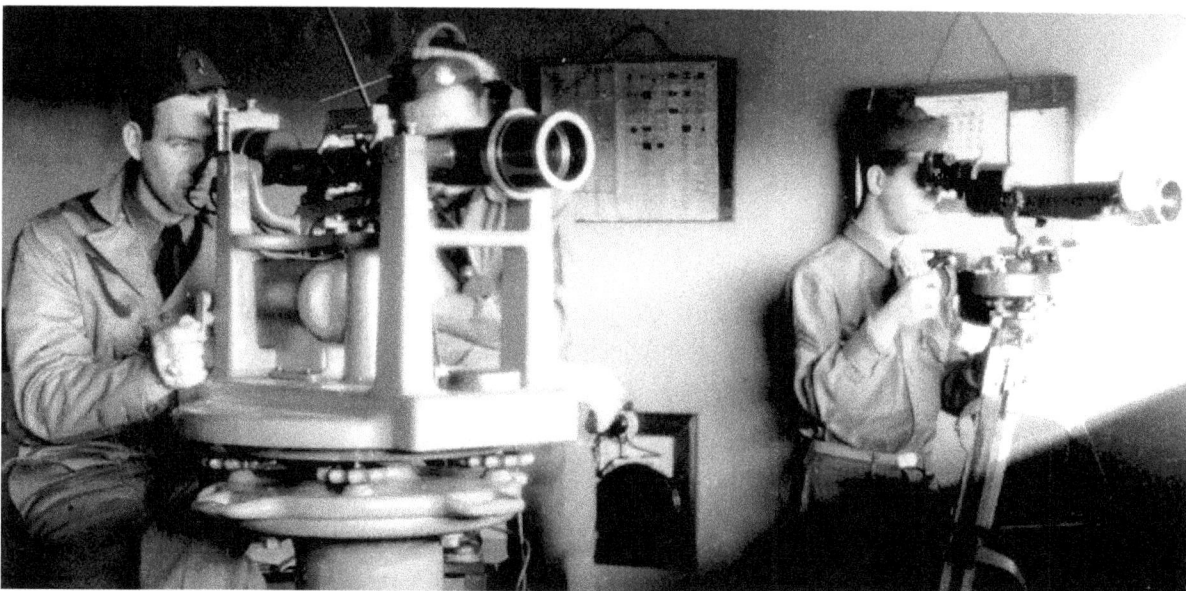

Figure 7-2. A tripod-mounted "Type B" azimuth scope is seen at image right, with a "Type A" Swasey depression position finder (discussed next) on the left (US Army photo courtesy Cabrillo National Monument).

The *M1907* depression position finder (DPF), employed during both world wars, is described in *TM 9-1685* (1941):

"The instrument consists essentially of a telescope, mount, and pedestal. The mount carries the telescope and contains mechanisms for directing it. The mount is accurately leveled on the pedestal by means of four leveling screws. When the telescope is directed on the waterline of a target the depression angle is measured mechanically and the corresponding horizontal range,[3] corrected for curvature of the earth and normal atmospheric refraction, is read directly on a range scale graduated in yards. Internal correction adjustments compensate for tidal variations, abnormal atmospheric refraction, and for heights of instrument other than those for which the range scale is calibrated."

Additional corrections had to be made for the rotation of the earth and prevailing weather.

As reported in *The Guns of San Diego* (Thompson, 1991):

"In 1917 the Chief of Ordnance notified the Coast Artillery that fourteen Warner and Swasey azimuth instruments, Model 1910, would be sent to Fort Rosecrans for use in the new fire control stations. The position finder instrument in use at that time was the Swasey Depression Range Finder, Type A, Model 1910..."

Another period photo of a Swasey DPF manned by an observer and reader is shown in Figure 7-3.

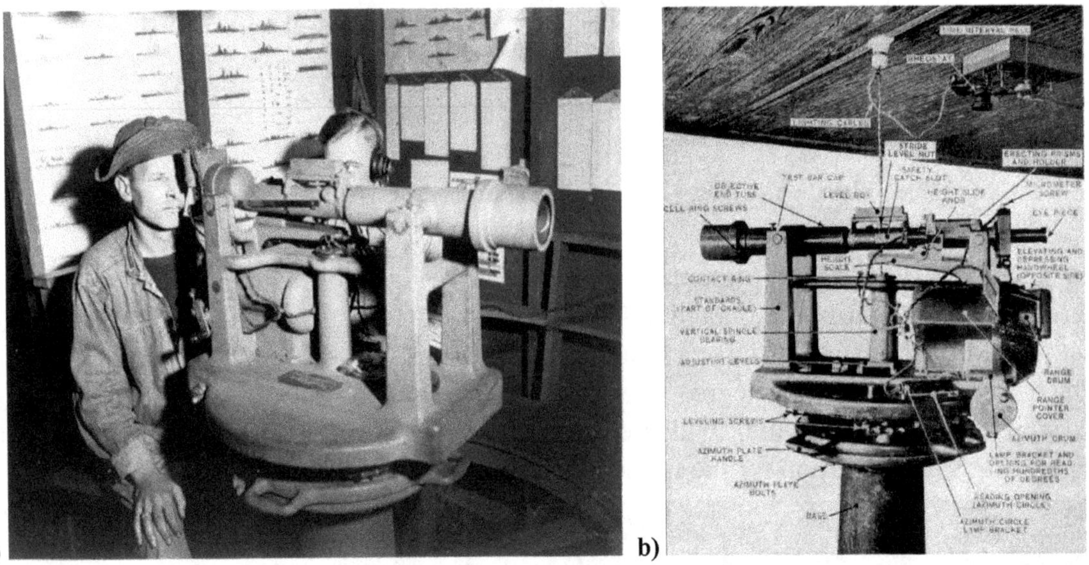

Figure 7-3. a) A Swasey *Type A-1* depression position finder (US Army photo courtesy Cabrillo National Monument). Note the ship-recognition posters on the back wall. **b)** Component breakout of the Swasey *Type A-1* Depression Position Finder (US Army diagram courtesy Wikipedia, 2019).

Some 495 pages of extensive detail regarding this complicated process can be found in *Seacoast Artillery: Fire Control and Position Finding* (FM 4-15, 1943),[4] which provides the following summary:

"In all the standard systems that employ the plotting board, relocation is performed mechanically on that instrument. It is accomplished by establishing the position of the directing point in the proper relation to that of the other points on the board and providing means for reading the azimuth and the range from the directing point to the target.

The words "from the directing point to the target" in the last sentence are a bit misleading, in that the moving target would actually no longer be at the observed location when the projectile arrived. Hines and

---

3   The horizontal crosshair was placed upon the waterline, with the vertical crosshair simultaneously placed upon the ship's stack.

4   A far more concise yet thorough treatment is provided by Coast Defense Study Group member Bolling W. Smith (2004) in *American Seacoast Defenses: A Reference Guide*, edited by Mark Berhow (2004).

Ward (1910) define "relocation" of a target as "any process whereby having the location of a target from one point, its range and azimuth from some other point may be determined without further observation." In other words, the fire-control objective was to derive (from the base-end observations) the observed target's set-forward range and bearing with respect to the directing point of the engaging battery.[5]

The battery "directing point" is discussed in *FM 4-15* (1943):

"In some of the older fixed two-gun batteries, the guns are fairly close together and the directing point has been taken as the point midway between the guns. The more modern batteries, however, have the guns spaced at a considerable distance so that it is necessary to make a displacement correction. In this case, No. 1 gun is usually taken as a directing point and a displacement correction is made for No. 2 gun."

The key elements of the fire-control problem are graphically depicted in Figure 7-4. The pivot points for the B[1] and B[2] arms seen at bottom center were located on station blocks attached to the baseline arm (not shown) that ran horizontally across the bottom of the plotting board. These blocks were carefully positioned to precisely replicate on the plotting board the relative positions of the base end stations out in the field. The plotting board thus produced a scaled-down representation of the real-world triangle defined by apices B[1], the target, and B[2], as seen in the upper half of the diagram below.

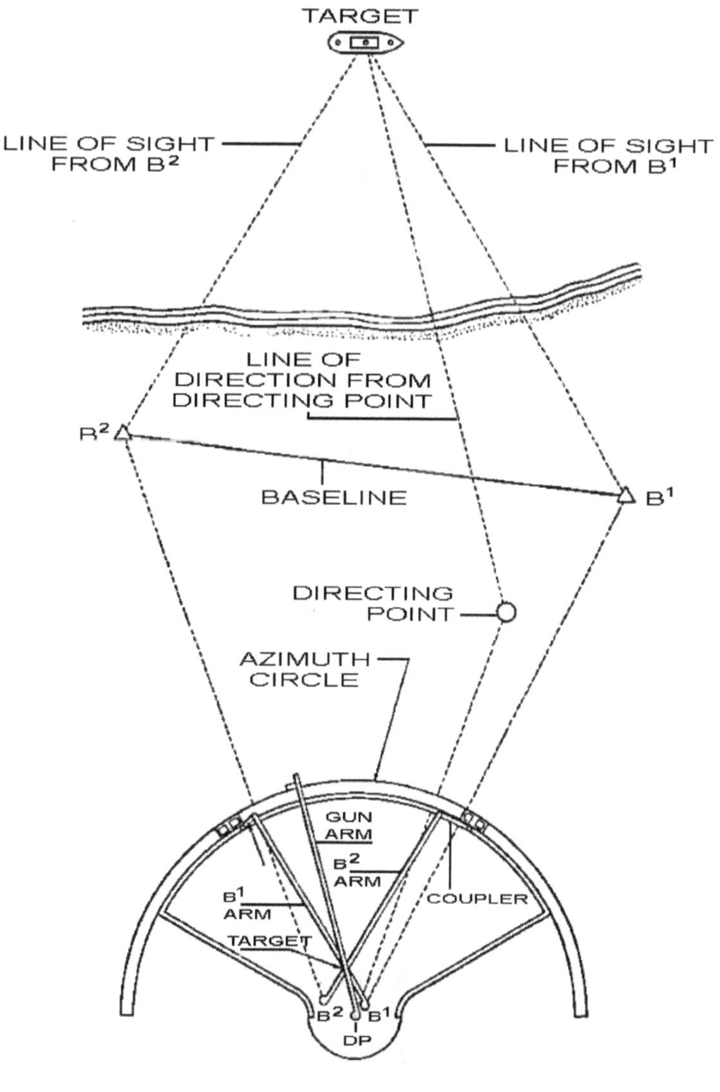

Figure 7-4. The graduated azimuth circle was oriented so that azimuth readings from base end stations B[1] (primary) and B[2] (secondary) would recreate on the plotting board a scaled-down representation of the real-world triangle defined by apices B[1], target, and B[2] at top center (relabeled from *FM 4-15*, 1943).

5   This set-forward point on the plotted target track is based upon the projectile time of flight.

The Whistler-Hearn Plotting Board *M1904* seen in Figure 7-5 could accommodate horizontal-baseline separations from 900 to 7,000 yards (Hines & Ward, 1910):

> "The station blocks are mounted on the baseline arm. One block, known as the primary block, is attached to the baseline arm and fits over the main pintle center. The other block, known as the secondary block, is attached to the baseline arm, but is arranged so it will move laterally along the baseline arm and may be set for any length of baseline on either side of the main center or primary station block. Each block is provided with a pivot or center, over which the primary arm and auxiliary arm are pivoted."

The standard scale of the Whistler-Hearn Board was 300 yards per inch, with a 450-yards-per-inch option (Baldwin, 2004).

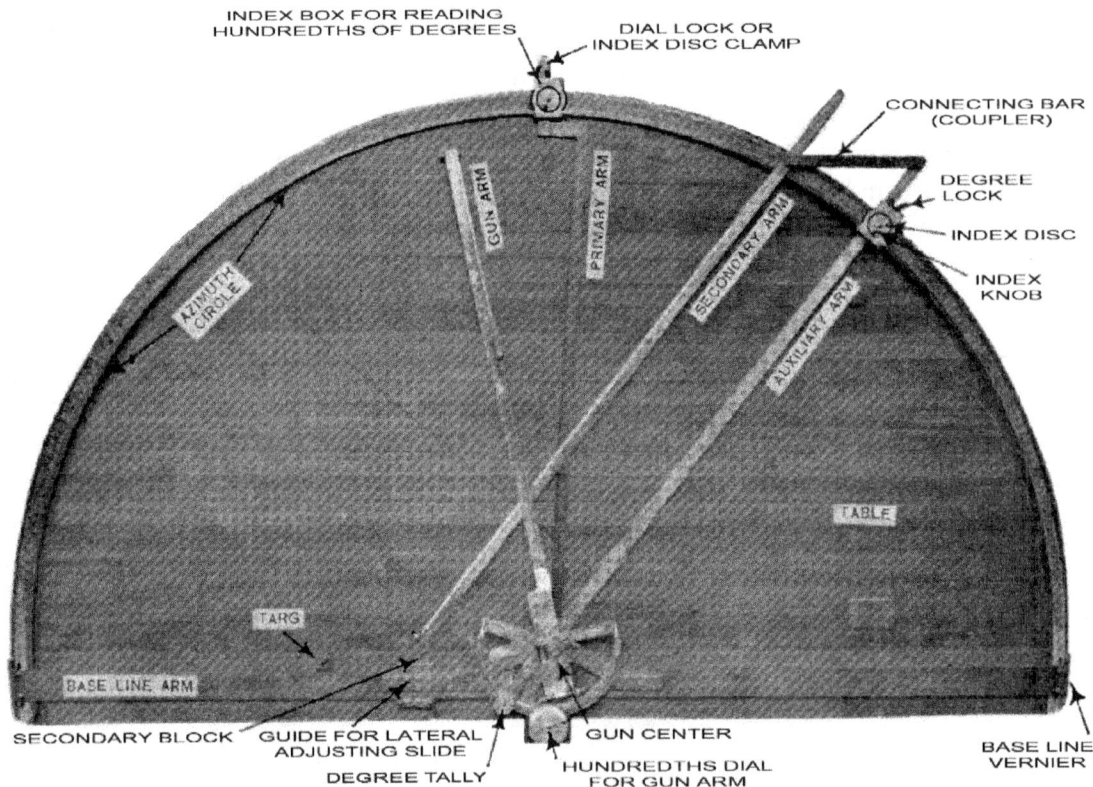

Figure 7-5. The *Whistler-Hearn Plotting Board M1904* was made from 2-inch-thick well-seasoned lumber. The bronze azimuth circle was attached only at its two extremities to avoid distortion from expansion or contraction of the wood (partially relabeled from Plate XXV, Hines & Ward, 1910).

As 7,000 yards is only about 4 miles, this plotting board was eventually replaced by newer models that could handle the extended baselines required by longer-range modernization batteries. By way of example, HDSD base end stations were situated from Solana Beach, CA, down to the Mexican border, a distance of some 37 miles.[6] The *Cloke Plotting and Relocating Board M1923* featured scales of 300, 600, 750, and 1500 yards per inch, while the *110-degree Board M3 and M4* models were specifically designed for certain long-range installations (Baldwin, 2004).

To track the target, the base end stations recorded their observed ranges and/or bearings at predetermined time intervals (typically 15 to 30 seconds) and passed this information by telephone to the plotting room. A simple but effective system of time-interval bells, which simultaneously rang at all base end stations and in the plotting room, was employed to ensure all readers reported simultaneous data at the proper time. Each time-interval event typically consisted of three short rings every prescribed number of seconds (Baldwin, 2004):

---

6   By 1944, most major-caliber batteries had six to ten different baseline options (Smith, 2004).

"At the first ring, the reader read the azimuth in whole degrees (i.e., '242 point') to the arm setter, who then moved his arm to that setting. In the meantime, the observer continued to track the target. However, at the last stroke of the three rings, he would momentarily stop tracking, the reader would read the hundredths of the azimuth from the hundredths dial and repeat this to the arm setter, who would set this on his arm index. At this time he would report 'set' to the plotter."

As Dwight Smith, who served in Battery D at Fort Rosecrans, explained to Howard Overton (1993):

"The observer stops it, the other guy reads it and puts it on the phone. So it goes right into the plotting section. Where these arms, gun arm down the middle, and the two observing stations on either side. The one on the right side goes up here on a cross... The guys have a little triangular block of steel with a point out on the end of it."

The "cross" mentioned by Smith was likely the coupler between the B² secondary arm and the auxiliary arm (Figure 7-6).

Figure 7-6. The coupler between the B² secondary arm and the auxiliary arm on this WWI-era Whistler-Hearn plotting board is just above the azimuth circle in front of the auxiliary-arm setter in lower-left corner (US Army photo courtesy Coast Defense Study Group). See also upper-right corner of Figure 7-5.

The "triangular block of steel," known as a "targ" (target), had a sharp point that was pressed into the paper to progressively mark the target track, as explained in paragraph 24 of *TM 9-1570* (1942):

"Set the primary arm and the secondary arm to the azimuth of the target. Accurately place the targ at the intersection of the arms and mark the position of this point on the plotting board. At least two, and preferably three, points must be plotted before a prediction may be made and a set-forward point located."

Note the "Targ" annotation on the left side of the baseline arm previously shown in Figure 7-5.

As further explained by Major General George S. Pugh (1988):

"The plotter then marks that point and a metal arm, using the gun position as the fulcrum, is swung out to where the plotter has anticipated the target location. The range and azimuth to that position are then transmitted to the guns and at the appropriate bell signal they are fired.

Of course, the data to the guns must take into account the difference in the height of the site from the sea-level target, the atmospheric pressure and air temperature, powder temperature, and most of all, the distance the target will travel during the time of flight of the projectile. To the maximum extent, these calculations were reduced to graphical or mechanical means."

The network linking all these fire-control stations to the various plotting rooms was the Army Tactical Communications Cable, which was classified "Secret." The southern leg of this buried cable was made up of segments *SCD-923*, *SCD-929*, and *SCD-968*, with the northern leg designated *SCD-936* (Roper, 1952b). Towards the end of the war, however, target tracking through visual observation was largely superseded by radar (as discussed in chapter 9), which worked day or night in good or bad weather, with the traditional plotting boards replaced by gun computers (CDSG, 2019).

A representative number of HDSD fire-control stations, arranged according to site number, are discussed in the following sections.

Site 1 (Santa Fe)

Base end and spotting station B^5_3, S^5_3 at Site 1 was a one-room installation assigned to the 16-inch Battery Ashburn at Fort Rosecrans (Figure 7-7). According to Thompson (1991), who cites no reference, the originally planned location for this structure had been at Site 13 by the Mexican Border:

"B^5_3 S^5_3 shifted from Mexican Border to Solano Beach (Santa Fe) along with other redesignations later in the war."

This change in siting may explain why *Army Tactical Cable SCD-936* connected all fire control installations north of Fort Rosecrans with the sole exception of Santa Fe.

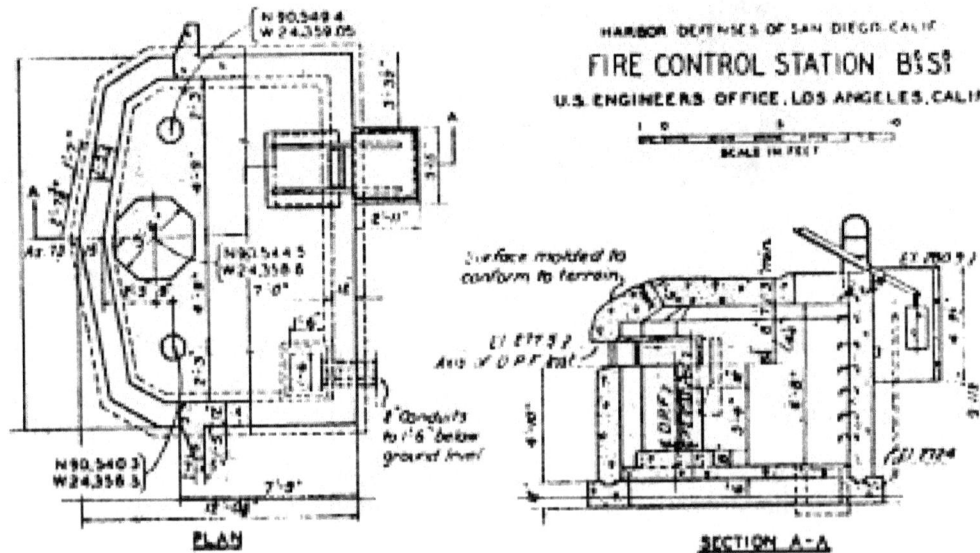

Figure 7-7. Plan and section views for base end and spotting station B^5_3, S^5_3 at Site 1 in Solana Beach, CA, which served the 100-Series 16-inch Battery Ashburn located on Fort Rosecrans.

The single-level bunker was located approximately midway between Marvista and Marview Drives, just south of Canyon Drive in Solana Beach, CA, with a detached concrete 10-foot by 12-foot back-up-generator building nearby. A 1991 photo of this fire-control station, taken by Coast Defense Study Group member Al Grobmeier, is presented in Figure 7-8. The 19th Coast Artillery soldiers manning this site were reportedly berthed in local barracks, one of which is still extant and now serves as the Fletcher Cove Community Center on Pacific Avenue.

Figure 7-8. The base end and spotting station B^5_3 S^5_3 for Battery Ashburn as it appeared in January 1991 (photo courtesy Al Grobmeier). Note steel rebar loops in concrete roof for attaching camouflage netting.

Site 2 (Soledad)

Located in Site 2 on Mount Soledad in La Jolla, CA, base end and spotting stations B^3_3 S^3_3 – B^5_9 S^5_9 were assigned to the 100-Series Batteries Ashburn at Fort Rosecrans and Gatchell at Fort Emory. The greater ranges achievable by these 16-inch batteries required a longer baseline between their respective base end stations, hence their remote locations both to the north and south of Fort Rosecrans. As indicated in the plan view of Figure 7-9, the upper level of this structure was assigned to Battery Ashburn, while the lower level served Battery Gatchell.

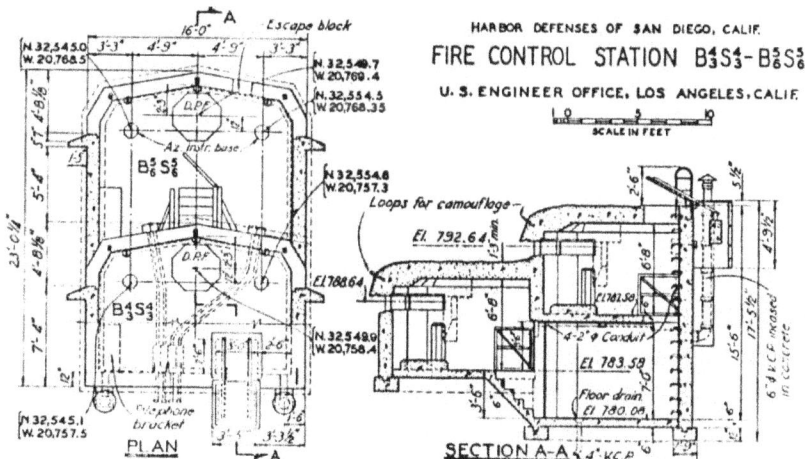

Figure 7-9. Plan and section views of base end stations B^3_3 S^3_3 – B^5_9 S^5_9 (originally designated B^4_3 S^4_3 – B^5_6 S^5_6) at Site 2 on Mount Soledad in La Jolla, CA (adapted from the Report of Completed Works). The high elevation of this site (795 feet) allowed both levels to be equipped with depression position finders (DPFs).

The location of this two-level structure on the summit of Mt. Soledad is within the elongated rectangle on the right side of the 1 July 1945 "Harbor Defenses of San Diego, Fire Control Stations at Soledad, Site 2" map presented in Figure 7-10 below. This fenced-in Army compound on Pueblo Lot 1264 was accessed from the intersection of La Jolla Drive and La Jolla Rancho Road at upper left. The base end stations were at the northwest end of the compound, the rest of which contained the Air Warning Service (AWS) receiver building, which provided electrical power.

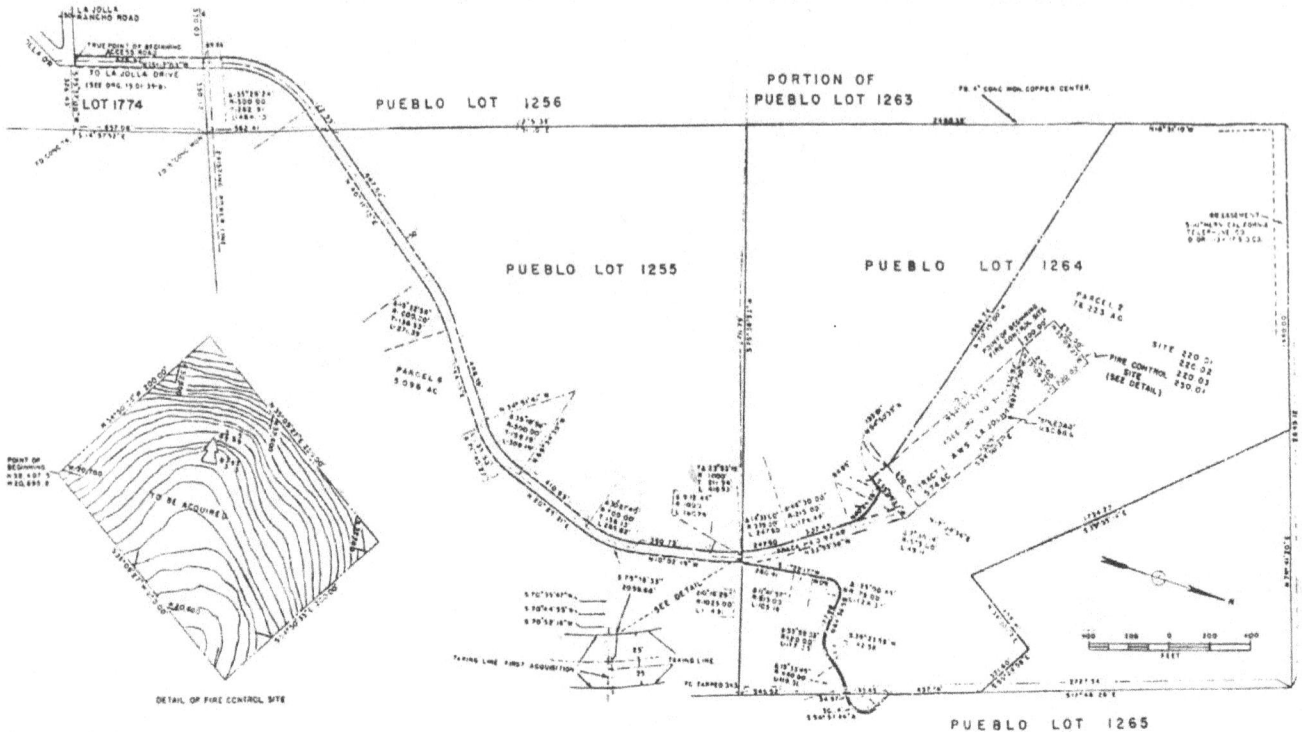

Figure 7-10. This "Harbor Defenses of San Diego, Fire Control Stations at Soledad, Site 2" map dated 1 July 1945, shows the access road to the fenced-in compound atop Mt. Soledad in La Jolla, CA. The fire-control site was at the northwest end of the compound as shown at upper right (see detail view at lower left).

Richard Markham, who served at Fort Rosecrans in Battery D of the 19th Coast Artillery, recounts his experiences at the Mount Soledad base end station in his oral history interview with Howard Overton (1993):

> "When I first went out there they just had a wooden shack out there, and then they built some nice cement-bunker-type station out there. I heard somebody built a house out there and that was in their basement."

In an editorial remark, Overton indicated the private residence of which Markham spoke was rumored to belong to the renowned author Theodore Seuss Geisel, better known to the public as Dr. Seuss.

Markham, who was assigned to B$^3_3$ S$^3_3$ for Battery Ashburn, further recalls additional supporting infrastructure developed in this area over the course of the war (Overton, 1993):

> "They built a barracks, quite a complex, out there on the point of La Jolla. ... Didn't drive back and forth after they got it built up out there. They had a mess hall. We had a radar also out there. We had two-way radio in our station in case the telephone lines, something happened to them. Because everything was synchronized by the bell. When the bell rang, you read to the plotting room and they set their arms and plotted their course."

This nearby support complex described by Markham is seen in the 1949 aerial photo presented in Figure 7-11 below.

Figure 7-11. Some 16 reasonably sized structures can be seen in the barracks area described by Markham, shown left of center in this portion of 1949 aerial survey *AXN 1F 19*. Note footpaths leading to the base end stations (circled), and the AWS La Jolla[7] compound (up arrow) to the southeast at lower right (photo courtesy San Diego County Annex).

A valuable clue as to the location and type of radar at AWS Site B-5 in La Jolla was found in Chapter 8 of the *History of the Western Defense Command* (HWDC, 1945b). The itinerary for an 11 March 1942 inspection tour of US radar facilities for a visiting Mexican delegation included a "radar installation at La Jolla (S D Area)." Since the presumed intent of this tour was to view working equipment of the type to be installed in Mexico, it is reasonable to conclude that a mobile *SCR-270* was operational in La Jolla shortly after Pearl Harbor. [8]

As the maximum detection range of this early-warning air-defense radar was proportional to antenna height, it makes sense the equipment would have been situated on top of Mount Soledad. With an elevation of 822 feet, this was the highest peak along the San Diego coast, and its existing road access plus ready availability of commercial power would have expedited installation after Pearl Harbor. Electrical power was derived from the "V.H.F. Receiver Bldg." located within the AWS La Jolla area adjacent to the southeast boundary of the Site 2 compound (Figure 7-12).

---

7   AWS – Air Warning Service.

8   The only other radar possibility would have been Battery Ashburn's *SCR-296-A* fire-control radar, which was not on Mt. Soledad, but in the nearby Hermosa (Site 3) neighborhood of La Jolla. The tower-mounted *SCR-296-A* fire-control radars didn't need further elevation due to the limited range of the gun batteries.

Figure 7-12. Base end stations B^3_3, S^3_3 – B^5_9, S^5_9 are right of center as photographed on 2 May 1953 in aerial survey AXN 7M 188, with the VHF Receiver Building southeast of these bunkers at far right (photo courtesy San Diego County Annex). Note many of the former barracks at upper left have already been razed.

The US Navy Electronics Laboratory (NEL) acquired this AWS La Jolla site from the Army after the war. The nearby VHS Receiver building in the former compound today houses a VHF relay connecting the Southern California Offshore Range (SCORE) on San Clemente Island to Naval Air Station North Island (Sturgeon, 2007). A close-up view of the two-level observation bunker is seen in the 1977 photo presented in Figure 7-13. As alluded to earlier by Markham, the B^3_3, S^3_3 – B^5_9, S^5_9 bunker was incorporated into a private-residence construction in the late 1970s (Pierson, 2005). Further investigation supports the rumor reported by Overton that the homeowner was in fact Theodore Geisel.

Figure 7-13. The Mount Soledad base end and spotting stations B^3_3, S^3_3 – B^5_9, S^5_9 are seen looking northeast in 1977 (photo courtesy Al Grobmeier).

"Seuss House" was built on and around this three-level bunker, which was accessible from the outside via a retrofitted doorway cut into the right-front (southwest) face of the partially exposed lower observation level. Whether access could be gained from inside the house via the top trapdoor entry is unclear. The steel shutters adjacent to the new doorway were replaced with conventional sliding glass windows (Pierson, 2005). Both the house and the base end stations were demolished by a new owner in 1999 to facilitate construction of a larger residence (Figure 7-14).

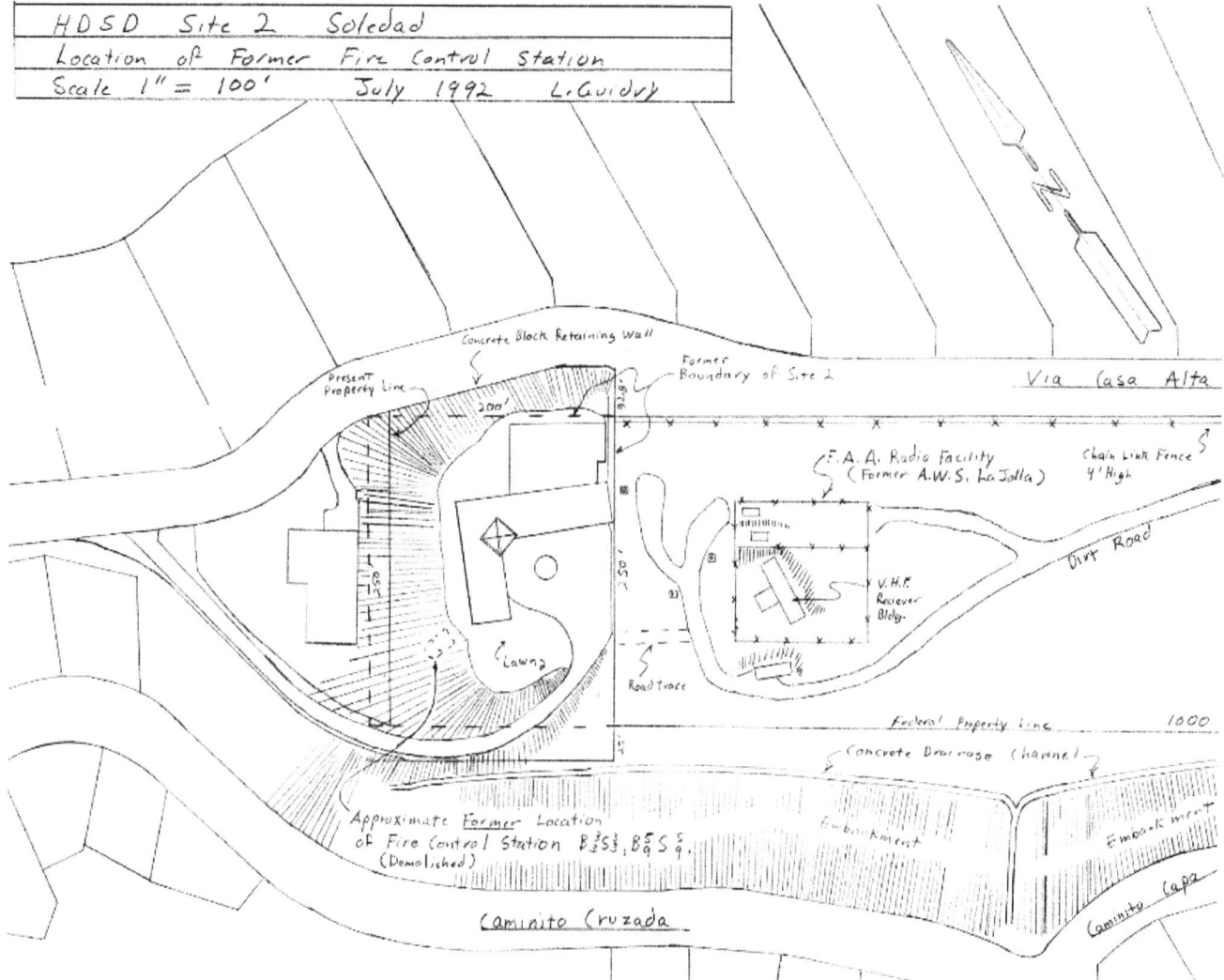

Figure 7-14. Demolished in 1999, base end and spotting station $B^3_3 S^3_3 - B^5_9 S^5_9$ was originally located near the southwest corner of Site 2 (left of center), south of what is now Via Casa Alta (1992 field sketch by Lee Guidry).

Site 3 (Hermosa)

Three base end and spotting stations were constructed at Site 3 (Hermosa): the combined structure $B^5_1 S^5_1 - B^5_3 S^5_3$, which serviced Batteries Woodward and Humphreys, plus $B^5_2 S^5_2$, which serviced Battery Strong (Figure 7-15a). Note the *SCR-296-A* fire-control radar for Battery Ashburn immediately behind $B^5_2 S^5_2$ in the detail map at lower left in Figure 7-15b below. The radar antenna was mounted inside a wooden shelter on top of a 100-foot Signal Corps tower (Type *TR-18*), with the entire configuration disguised as a water tank. The transmitter/receiver equipment and a 25-KVA emergency generator, which also provided power for the base end stations, were housed in wooden shelters that looked like beach cottages.

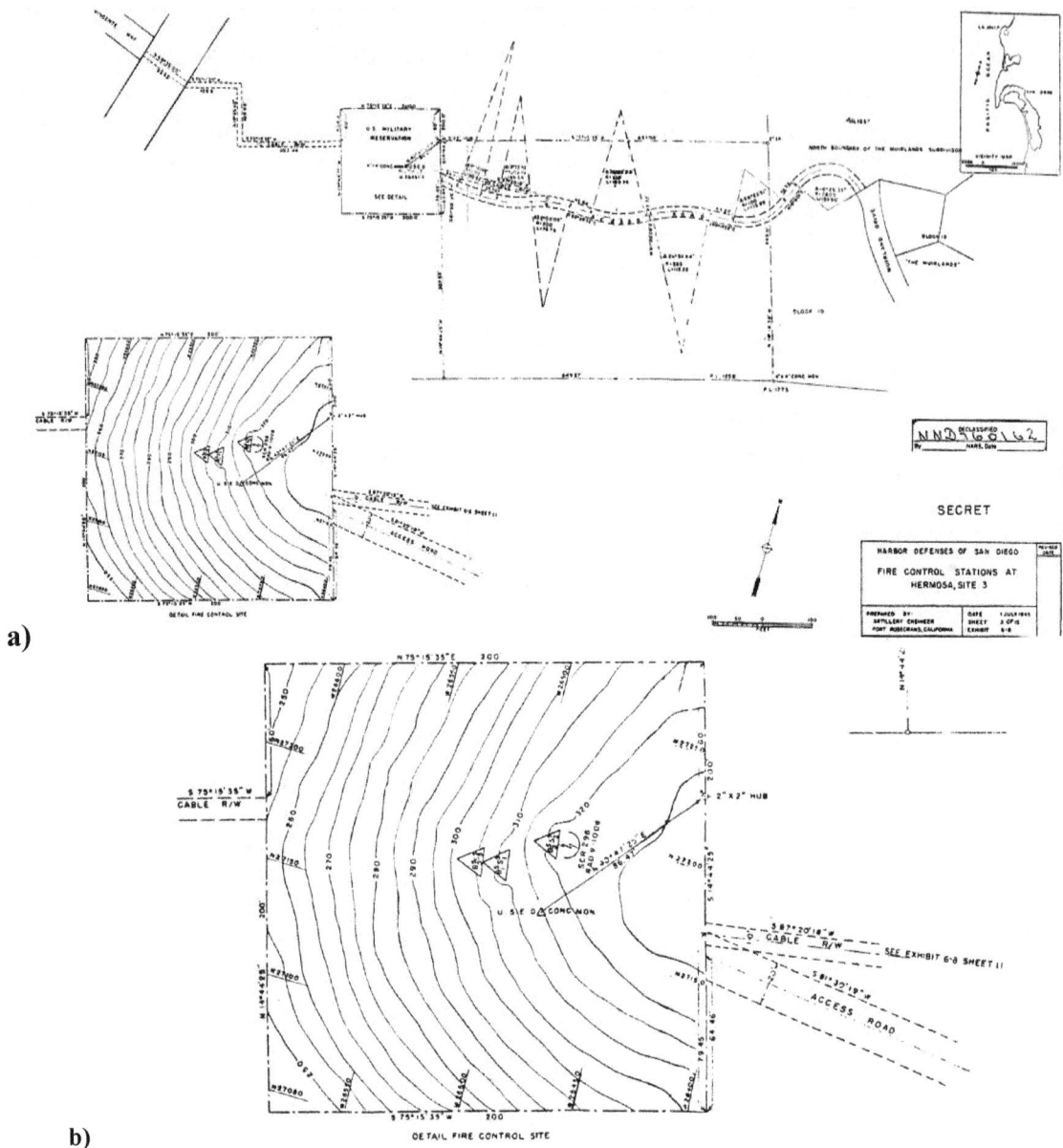

SECRET

Figure 7-15. a) Map of "Harbor Defenses of San Diego, Fire Control Stations at Hermosa, Site 3," showing new access road beginning at the termination of "Muirland Drive" (sic) in upper-right corner.[9] Note section of the Army Tactical Cable entering the square fenced compound from Vincente Way at extreme upper left. **b)** The expanded detail map shows an *SCR-296-A* radar for Battery Ashburn immediately behind $B^5_2 S^5_2$.

Plan and section views of base end and spotting stations $B^5_1 S^5_1 - B^5_4 S^5_4$ for Batteries Woodward (237, Tac. No. 1) and Humphreys (238, Tac. No. 4) are shown in Figure 7-16a.[10] The combined structure for these two bunkers was identical to that shown earlier for $B^3_3 S^3_3 - B^5_9 S^5_9$ in Site 2 (Soledad). Situated at a reference elevation of 312 feet, both levels were equipped with azimuth instruments and depression position finders. Figure 7-16b shows plan and section views of $B^5_2 S^5_2$ for Battery Strong (Tac. No. 2), which was also equipped with both types of observation instruments.

---

9   The correct spelling is "Muirlands."

10   Table 7-1, tactical number designations for the HDSD gun batteries during the latter part of the war (Thompson, 1991), lists AMTB Battery Cabrillo as Tac. No. 4 towards the end of the war, but the earlier RCW for this construction has a penciled annotation of "238" (Battery Humphreys) next to $B^5_4 S^5_4$.

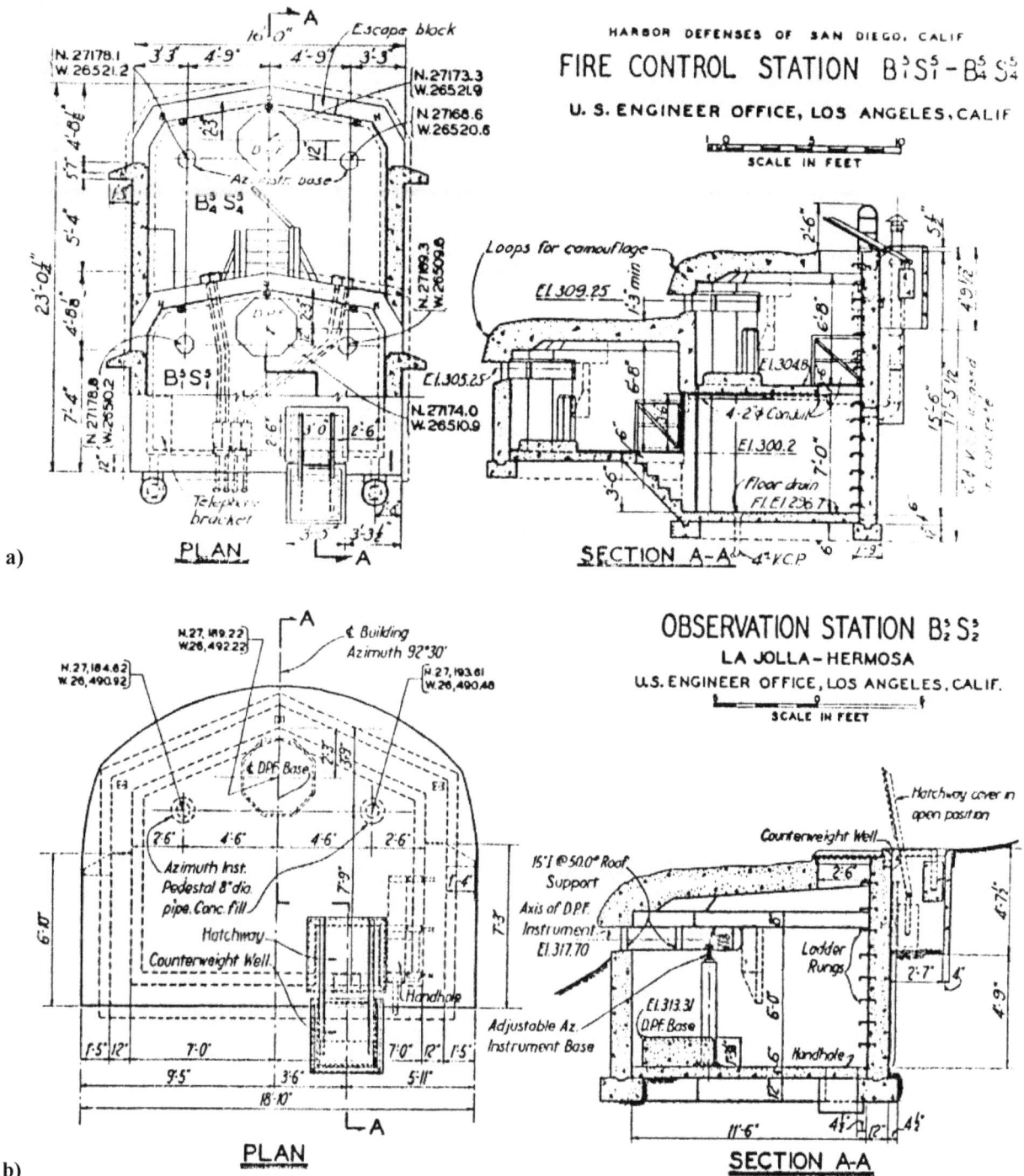

Figure 7-16. **a)** Plan and section views of $B^5_1 S^5_1 - B^5_3 S^5_3$ at Site 3 (Hermosa) for Batteries Woodward and Ashburn. **b)** Plan and section views of $B^5_2 S^5_2$ for Battery Strong (NARA).

The location of this Site 3 compound was just left of the termination of Muirlands Drive as seen in the 1946 La Jolla street map provided below by Coast Defense Study Group member Al Grobmeier (Figure 7-17). Vincente Way, shown earlier in the upper-left corner of Figure 7-15a, is marked by the up arrow to the west. Site 3 was served by portable Searchlights 7 and 8 in adjacent Site 3A (Neptune) further to the west, where the street Neptune Place is seen running north-south along the coastline.

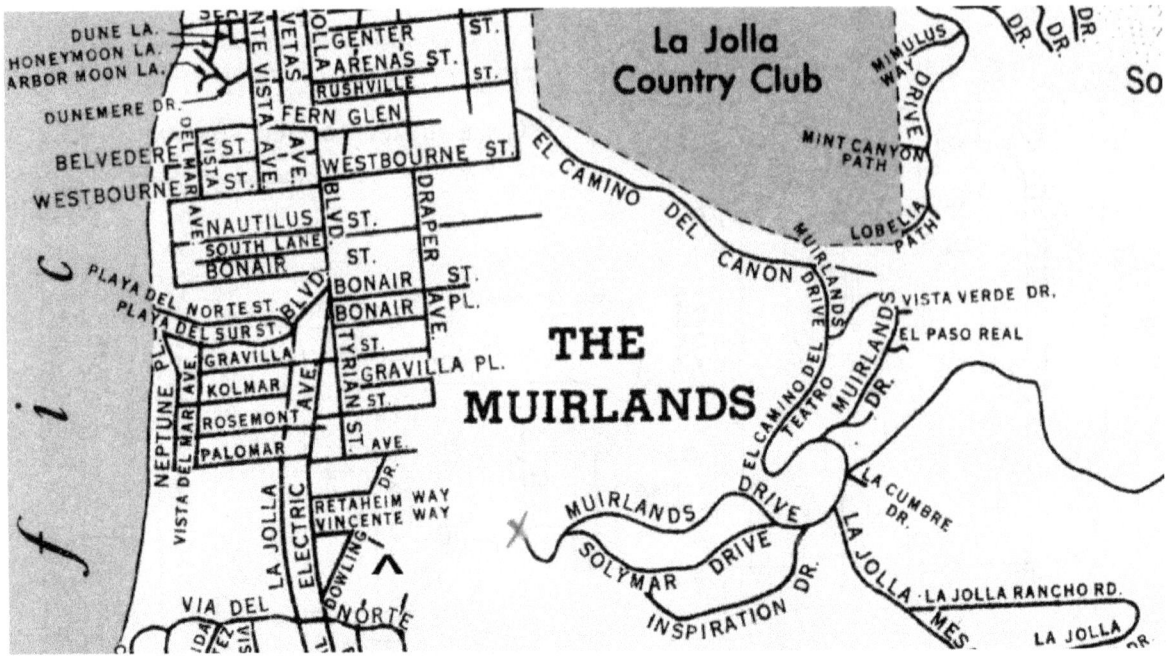

Figure 7-17. . The Site 3 compound was situated west of the termination of Muirlands Drive, just left of bottom center on this 1946 La Jolla street map (courtesy Al Grobmeier). Note the intersection of Dowling Drive and Vincente Way a bit further to the west (up arrow), and Neptune Place paralleling the coastline.

The square fenced compound for the base end stations at Site 3 can be seen just left of center in the 1949 photo from aerial survey *AXN 1F 19* presented in Figure 7-18 below. Cable access for Searchlights 7 and 8 in Site 3A entered from the intersection of Dowling Drive and Vincente Way, just west of the upper-left corner of the compound. The concrete observation stations can be seen in the center of the compound, with what appears to be a few small buildings to their right. Muirlands Drive is the wider light-colored road snaking in from the east in the upper-right corner.

Figure 7-18. A few small buildings are seen east of the Site 3 base-end stations, which are at the center of the square fenced compound just left of center in this cropped portion of 1949 aerial survey *AXN 1F 19* (photo courtesy SD County Annex).

This Site 3 compound was turned over to the Navy sometime after the war (Houston, 1955):

"The SCD 936 cable line extends from the north boundary of Fort Rosecrans approximately 9.6 miles to a point where it divides, with one segment running approximately 1.52 miles to the former La Jolla-Hermosa Fire Control Station and the other segment running approximately .51 mile to the former La Jolla AWS Station, both of which installations are now under the jurisdiction of the Department of the Navy."

Figure 7-19 shows residential encroachment in the lower-left corner of the former Site 3 compound during construction of Muirlands Vista Way in May 1953. Still extant today, the three fire-control bunkers are beneath a private home on present-day Avenida Manana. Although visible from the air, the lower level of $B^5_1 S^5_1 - B^5_3 S^5_3$ is obscured from street view by the dense foliage. The upper level and $B^5_2 S^5_2$ are completely hidden beneath the residence.

Figure 7-19. Remains of the Site 3 compound (just left of center) as seen in aerial survey *AXN 7M 188* on 2 May 1953 during the construction of Muirlands Vista Way, the black-top road seen right of center (photo courtesy San Diego County Annex).

Site 4 (Sunset)

Three two-level concrete bunkers for a total of six base end and spotting stations are found in Site 4 (Sunset), just north of Fort Rosecrans:

- $B^2_3 S^2_3 - B^1_1 S^1_1$ for Batteries Woodward (Tac. No. 1) and Ashburn (Tac. No. 3).

- $B^2_4 S^2_4 - B^4_2 S^4_2$ for Batteries AMTB Cabrillo (Tac. No. 4) and Strong (Tac. No. 2).

- $B^4_9 S^4_9 - B^5_{10} S^5_{10}$ for Batteries Gatchell (Tac. No. 9) and Grant (Tac. No. 10).

The plan and section views from the Report of Completed Works for one of these observation structures, $B^2_4 S^2_4 - B^4_2 S^4_2$, are presented in Figure 7-20 below.

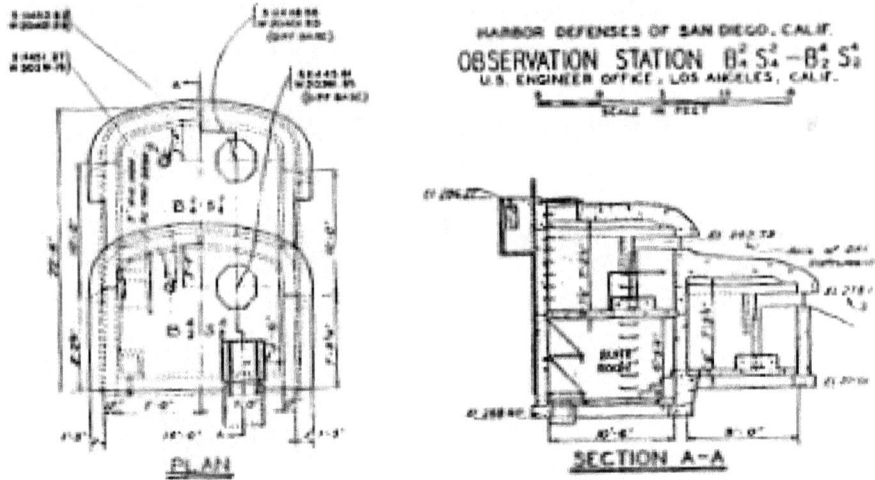

Figure 7-20. Plan and section views from the Report of Completed Works for base end stations $B^2_4 S^2_4 - B^4_2 S^4_2$ for Batteries AMTB Cabrillo (Tac. No. 4) and Strong (Tac. No. 2).

These three fire-control bunkers were collocated on either side of present-day Stafford Place in the Sunset Cliffs area, just to the north of Point Loma Nazarene University. The 1999 field sketch prepared by Lee Guidry seen in Figure 7-21 indicates the current Loma Mar subdivision boundaries are identical to those of the former Site 4 compound. Structures $B^2_3 S^2_3 - B^1_1 S^1_1$ for Batteries Woodward and Ashburn and $B^4_9 S^4_9 - B^5_{10} S^5_{10}$ for Batteries Gatchell and Grant are located beneath private residences west of Stafford Place, while $B^2_4 S^2_4 - B^4_2 S^4_2$ for Batteries Cabrillo and Strong are upslope to the east in dense brush.

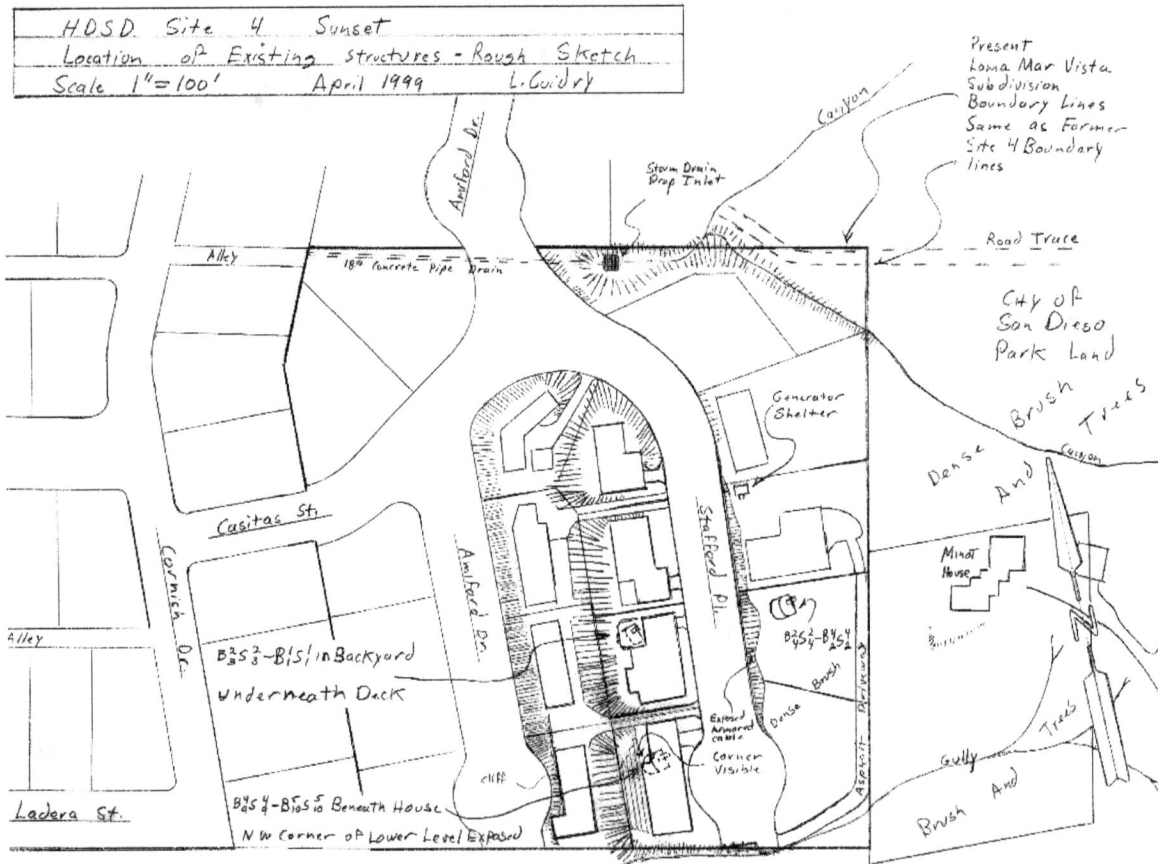

Figure 7-21. The locations of the three Site 4 fire-control bunkers in the Loma Mar Vista subdivision are seen on either side of Stafford Place in the bottom right corner of this 1999 field sketch (courtesy Lee Guidry). Note generator shelter right of center, just east of Stafford Place.

In the 1962 aerial view of the Loma Mar Vista subdivision shown in Figure 7-22, base end and spotting stations $B^2_3 S^2_3 - B^1_1 S^1_1$ are not yet covered over by the homeowner's deck (as Guidry later reported to be the case in 1999) and can still be seen (upper-left up arrow). The northwest corner of $B^4_9 S^4_9 - B^5_{10} S^5_{10}$ is barely visible directly below to the south, while $B^2_4 S^2_4 - B^4_2 S^4_2$ is fully exposed upslope on the east side of Stafford Place (right up arrow). In the upper-left corner of the property lot immediately above this last bunker and just east of Stafford Place are the remains (small white dot) of the WWII generator shelter that provided back-up power (see also Figure 7-21).

Figure 7-22. This November 1962 aerial photograph, rotated into alignment with Figure 7-21, shows the three Site 4 base end bunkers (up arrows) and their back-up-generator shelter on either side of Stafford Place, just north of Fort Rosecrans (US Navy photo courtesy SSC Pacific).

The easternmost (right up arrow in Figure 7-22 above) of the three base end and spotting stations, $B^2_4 S^2_4 - B^4_2 S^4_2$, is shown in Figure 7-23 below, just east of Stafford Place, circa April 1979. In a later photo taken by Al Grobmeier of the trapdoor entrance atop this structure (not shown), looking west towards Stafford Place in October 1990, the surrounding brush had grown to where the bunker was completely hidden from a street-level view. The exposed concrete roof sections were easily located from above on Google Earth in 2018, however.

Figure 7-23. The eastern-most base end and spotting station $B^2_4 S^2_4 - B^4_2 S^4_2$ is seen just east of Stafford Place, looking east in April 1979 (photo courtesy Al Grobmeier).

Site 5 (North Rosecrans)

The numerous Site 5 fire-control structures are depicted in the 1945 map excerpt seen in Figure 7-24. The plotting and spotting room (PR_2) for Battery Strong is southeast of the battery in the lower left corner. The battery commander station BC_1 and base end and spotting station $B^3_1 S^3_1$ for Battery Woodward are east of that battery just left of top center. The combined battery commander station BC_2 and base end station $B^1_2 S^1_2$ for Battery Strong is seen at map center, with the Battalion 2 Command Post ($BNCP_2$) below and to the right, west of Catalina Boulevard. The 'Unassigned" structure west of the *SCR-296-A* radar in the upper-right corner previously served as the battery commander station for Battery Gillespie.

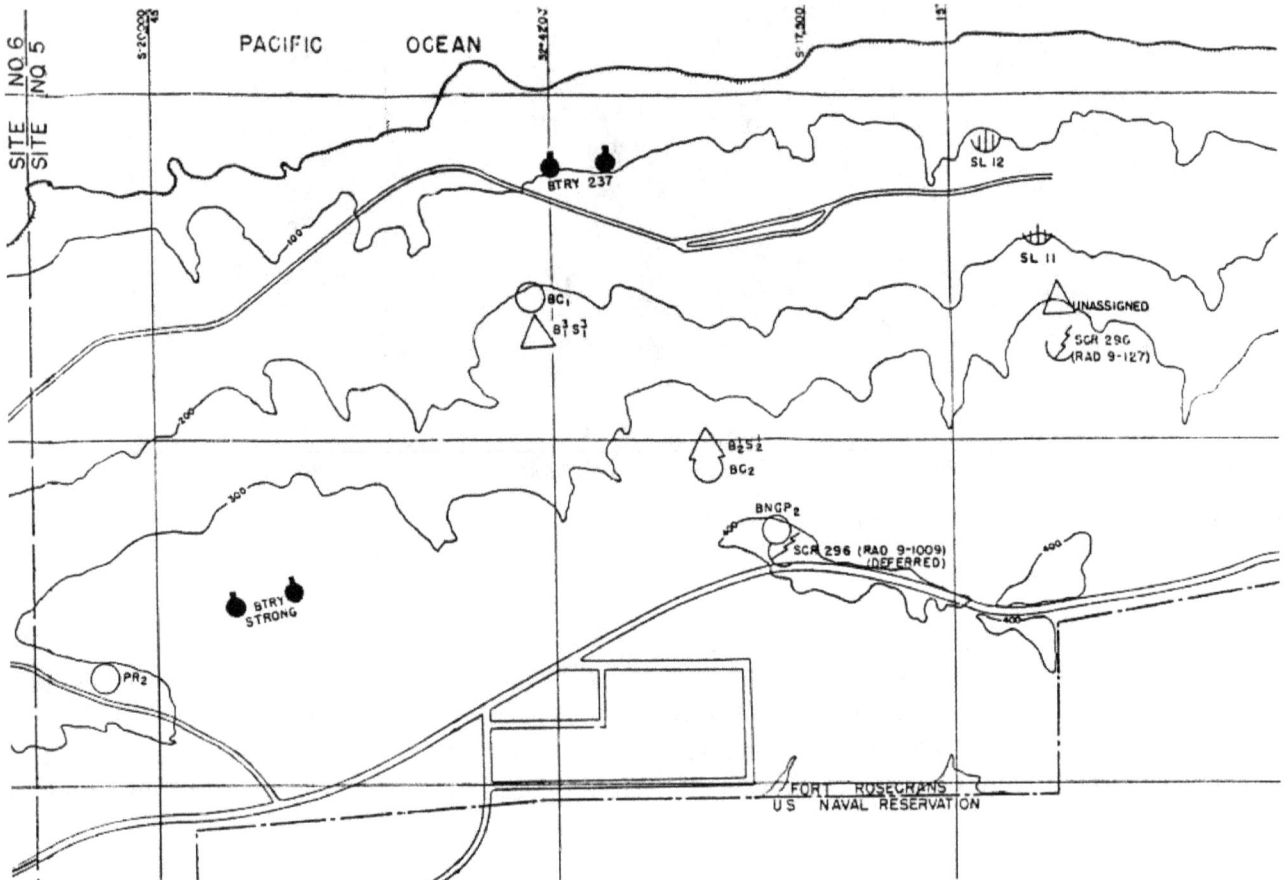

Figure 7-24. Zoomed in portion of 1 July 1945 map entitled "Fire Control Stations at North Fort Rosecrans, Sites 5, 6, 9." PR_2 on the northwest side of Woodward Road at lower left is the Plotting and Spotting Room for Battery Strong.

Battalion 2 Command Post / Standby Harbor Defense Command Post

Both these dug-in bunkers are situated on top of a small knoll on the west side of Catalina Boulevard, across from Battery Whistler and just south of the present-day TRANSDEC[11] pool (Figure 7-25). A 1 July 1945 "Harbor Defenses of San Diego, Main Cable Routings" blueprint (sheet 2 of 4) shows two structures in this area: "BN CP-2" (formerly the Group 2 Command Post), and an adjacent eastern structure identified as "CH-2" (Cable Hut 2). The same map indicates this was the proposed site of a deferred *SCR-296-A* fire-control radar for the unfinished 16-inch Battery Gatchell at Fort Emory.

---

11 TRANSDEC – Transducer Evaluation Center.

Figure 7-25. The Battalion 2 Command Post (up arrow) is just left of center on the west (lower) side of Catalina Boulevard, just south of the TRANSDEC pool at lower left, circa 1984. Battery Whistler (later Arctic Submarine Lab) is the large white structure across the street at upper-left (US Navy photo courtesy SSC Pacific).

Figure 7-26 shows the exposed concrete roofs for these two conjoined bunkers on the west side of Catalina Boulevard in 1946, with Battery Whistler just out of image to the right on the east side of the road. The northwest corner of the upper cantonment area is seen in the foreground, with the topside base chapel in the upper left corner. The extended barracks-roof overhangs allowed the windows to remain open for cross ventilation during the rainy season. The two sets of wide stairways leading from the barracks area to Catalina Boulevard might suggest more people go to church during wartime.

Figure 7-26. The Battalion 2 Command Post (left up arrow) and the Harbor Command Post (right up arrow) are seen on top of the knoll on the far side of Catalina Boulevard at upper right, circa September 1946 (US Navy photo courtesy SSC Pacific).

The single-level Battalion 2 Command Post was transferred to the Coast Artillery on 9 May 1944, with a total cost to that date of $6,806.77. The Report of Completed Works, updated to 17 November 1944, indicates the rear Operations Room was equipped with Guth "Mazelight" lighting, which the plan-view seen in Figure 7-27 describes as 40-watt two-lamp fixtures, an early type of fluorescent lights. During our site visit in May 2006, however, my son and I instead found Wheeler reflectors for incandescent bulbs installed on the ceiling in each corner of both the Operations Room and the Observation Room.

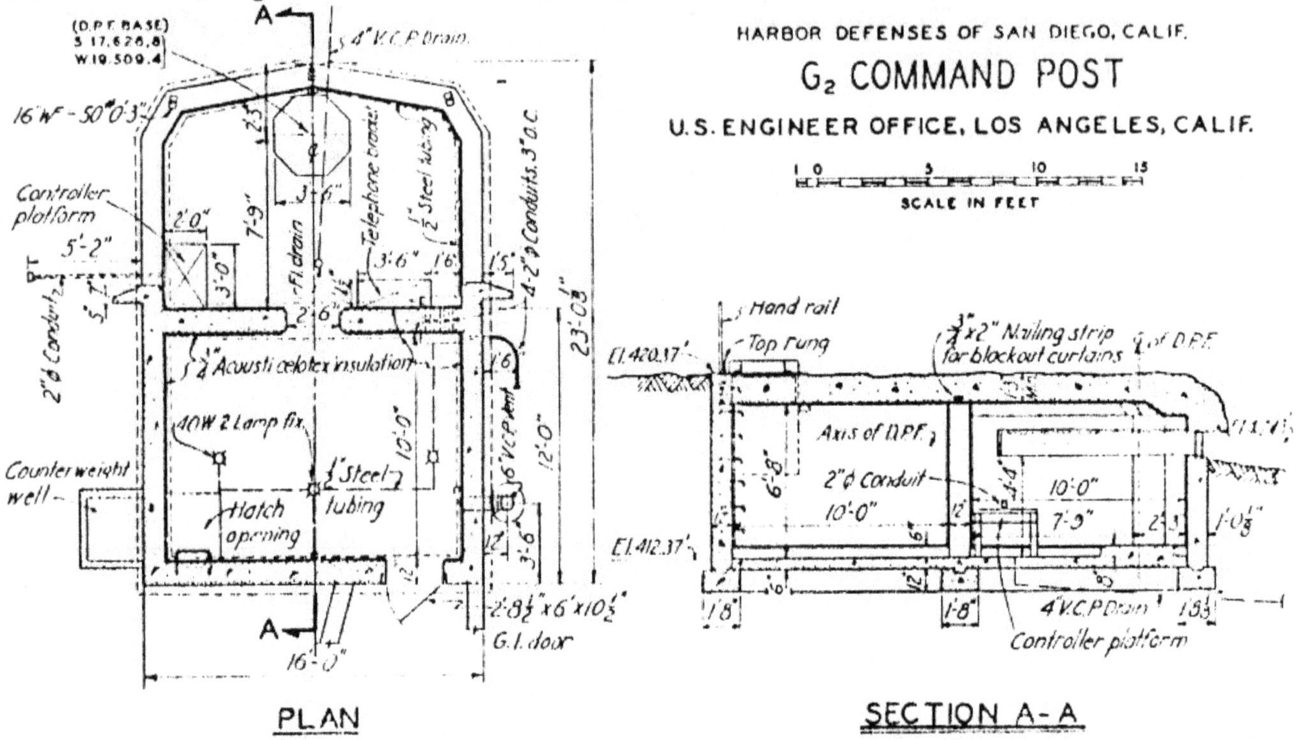

Figure 7-27. Plan and section views of the WWII Group 2 Command Post, later the Battalion 2 Command Post,[12] reproduced from the Report of Completed Works corrected to 17 November 1944. As shown in the plan view, there were supposedly three two-lamp fluorescent-light fixtures in the rear Operations Room (NARA).

An interesting detail in the plan-view drawing above is the "G.I. door" (presumably galvanized iron door) shown in the lower-right corner of the rear Operations Room. Since this was an underground bunker, I was initially confused as to the purpose of this door, which appears to go nowhere. Upon entering the bunker in 2006, the door was found to open into a connecting passageway leading to an adjacent bunker (Figure 7-28), which during WWI had served as Primary Fire Command Station F'1 for Battery Whistler (Flower & Roth, 1982).[13] It was enlarged during WWII with the rear addition shown in Figure 7-27, and repurposed as the Standby Harbor Defense Command Post.

---

12 Recall the Groups were redesignated Battalions in 1945.

13 F'1 for Battery Whistler was transferred to the Coast Artillery on 21 February 1920 (Willard, 1920).

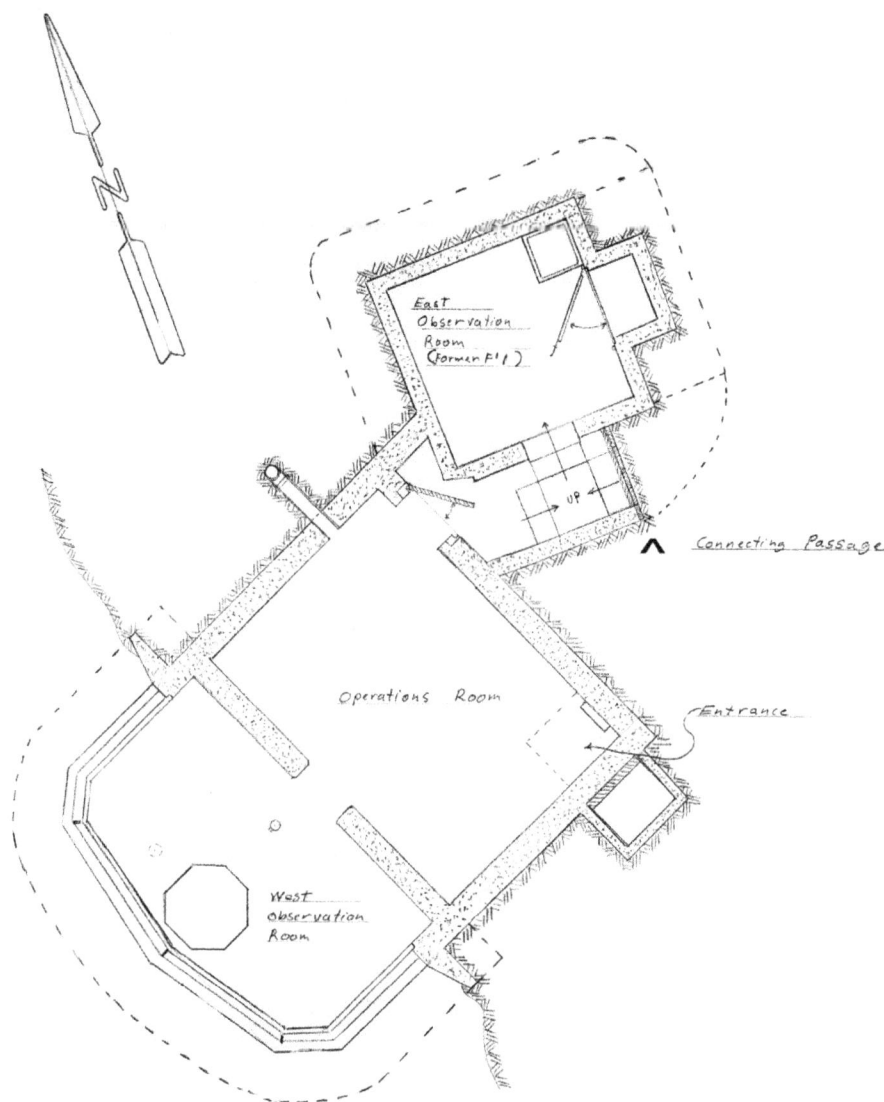

Figure 7-28. Plan view of the combined structure for the Battalion 2 Command Post (bottom) and the Standby Harbor Defense Command Post (top). The east-end wall (up arrow right of center) of the underground passage joining the two structures is made of wood (1995 field sketch by Lee Guidry).

Figure 7-29a shows the West Observation Room of the Battalion 2 Command Post, which for a number of reasons is one of the most intriguing fire-control bunkers at Fort Rosecrans, the first being it was a WWII expansion of a WWI structure. Secondly, note the steps leading down to a very unusual wooden wall at the eastern end of the connecting passageway between the original bunker and the new addition to the southwest (see again Figure 7-28 above). Some thoughts on a possible purpose for this wooden wall are presented at the end of this section.

The most obvious oddity, however, is a large octagonal foundation on the west roof (Figure 7-29b), which Apple et al. (1995) describe as a "poured concrete gun mount that appears to have been added during the Second World War." This seems most unlikely, as the added section seems too thin, with insufficient mounting bolts of inadequate diameter for such an application, and placing a gun mount on top of a manned camouflaged observation bunker makes little sense. Furthermore, in a 1962 US Navy aerial photo (LSF22-11-62, not shown), this foundation appears much newer than the bunker roof.

Figure 7-29. a) The West Observation Room of the Battalion 2 Command Post is west of Catalina Boulevard at an elevation of 420 feet, with the adjoining WWI-era bunker behind the brush to the right (circa May 2006). **b)** Close up of the post-war foundation seen on top of the bunker roof in Figure 7-29a.

The Battalion 2 Command Post received commercial 110-volt-AC power from the Battery Whistler switchboard across Catalina Boulevard, with backup power provided by same. Figure 7-30a shows the vertical steel-rung access ladder beneath the trap-door entry, which was found unlocked. Note the plywood kick shield behind the ladder rungs, presumably to protect the acoustic insulation tiles on the wall from being damaged by a climber's shoes. The concrete counterweight for the heavy steel hatch is seen inside the adjacent counterweight well in Figure 7-30b. A 1933 Geodetic marker was found near the northwest edge this bunker during our site visit in May 2006.

Figure 7-30. a) The steel cover for the access hatch is in the raised position, out of sight to the right, revealing the access ladder to the Battalion 2 Command Post. **b)** The suspended concrete weight that counterbalances the heavy hatch cover is seen inside the counterweight well on the right side of the hatch opening.

Figure 7-31 shows the northeast corner of the Operations Room directly below the entry hatch, with the padlock presumably used to secure the hatch found hanging from one of the steel ladder rungs embedded in the concrete wall. The wall switch for the overhead lighting is just left of the entry ladder, with a Wheeler reflector attached to the ceiling-mounted fixture directly above. A good many of the acoustic tiles have fallen from the ceiling and walls, with considerable debris present on the floor. Several sections of metal pipe had at some point been stored in his room by our Facilities Department.

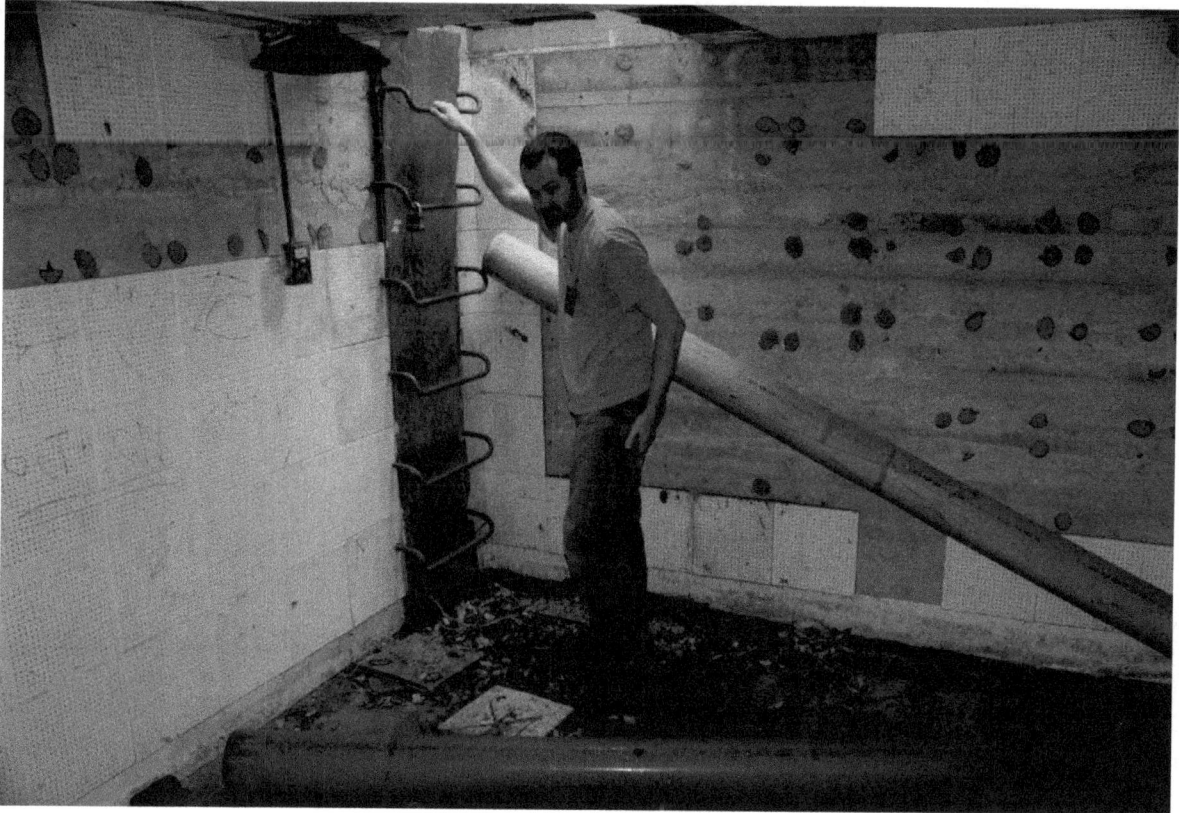

Figure 7-31. Todd Everett stands inside the cluttered Operations Room of the Battalion 2 Command Post after following me down the ladder from the access hatch above, circa May 2006. Note the Wheeler reflector on the incandescent light fixture attached to the overhead, and the associated wall switch left of the ladder.

Figure 7-32 shows the connecting doorway leading to the West Observation Room, looking northwest. The twin rods for two sets of blackout curtains above either side of the top of the doorway formed a light trap to shield the darkened Observation Room from the illuminated Operations Room. Another Wheeler reflector is seen attached to the overhead above the wall-mounted fuse panel and the electrical junction boxes seen in the background at image right. Altogether there were eight such light fixtures, one in each corner of both rooms, as opposed to fluorescent lights.

Figure 7-32. View looking through the interior doorway to the West Observation Room of the Battalion 2 Command Post. Note the pair of light-trap curtain rods above the doorway, and the Wheeler reflector on the ceiling-mounted light fixture above the electrical panel in the upper right corner of the photo.

The West Observation Room is seen in Figure 7-33a, with the steel shutters dropped down in the open position and the octagonal concrete foundation for a depression position finder at foreground center. The vertical black stripes seen through the left observation slot are from the front grill of my Jeep Wrangler, washed out by the bright sunlight. The acoustical insulation here was found to be more intact than that seen in the Operations Room, and equally marked up with post-war graffiti. A strange "For Official Use Only" sign is stenciled just above the surface-mount electrical conduit for the convenience outlets in the lower-right corner.

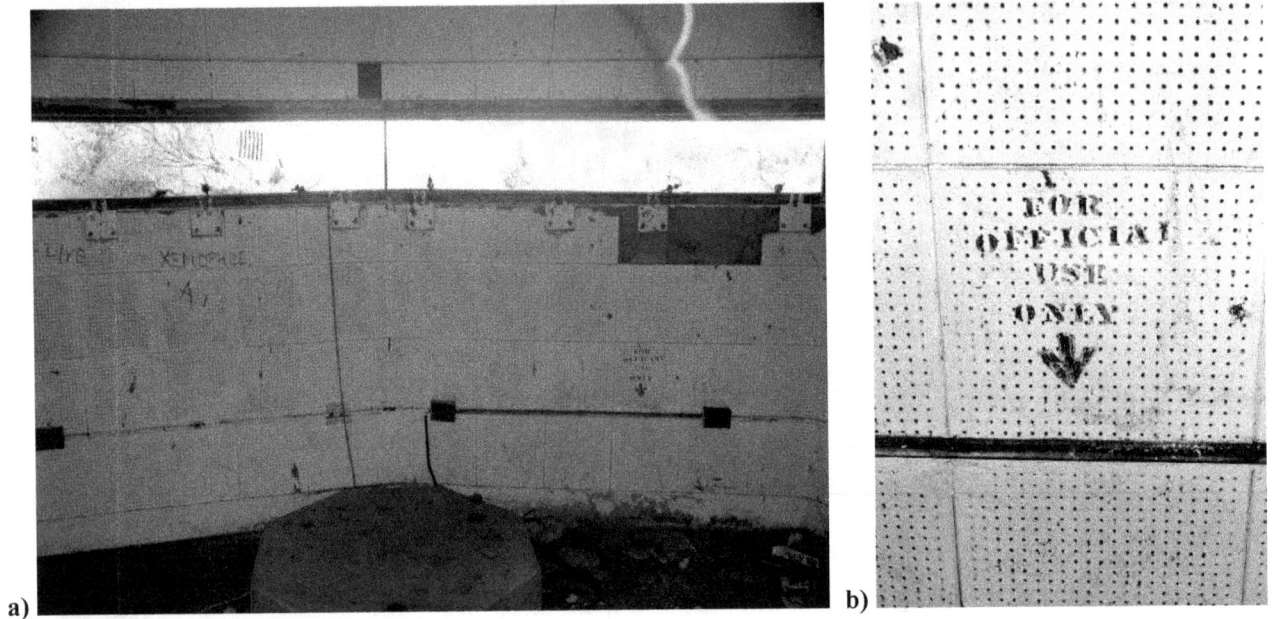

Figure 7-33. a) Concrete foundation for a DPF in the Observation Room of the Battalion 2 Command Post, with "For Official Use Only" stenciled above the post-war electrical conduit in the lower right corner (Figure 7-33b). **b)** Close up of what is probably a post-war crayon stencil above the electrical conduit.

The Report of Completed Works indicates the west Operations Room was lined with ¼-inch acoustic Celotex insulation. Lee Guidry, however, found 1-inch-thick tiles in both the Observation and Operations Rooms during his 1995 site survey. This upgrade, the new electrical outlets, and the octagonal concrete foundation on top of the structure suggest this bunker served in some new capacity in the late 50s or early 60s timeframe, perhaps with a surface-search radar installed nearby. The elevated location and heavy brush surroundings currently preclude visual sighting from nearby Catalina Boulevard.

Figure 7-34a shows the concrete roof of the adjoining WWI-era bunker on the top of the knoll, just east of the previously discussed WWII addition, with its steel shutters in the open raised position. This square bunker was originally equipped during WWI with four observation slots for 360-degree viewing, but only the east-facing window was retained in WWII, as the adjoining new addition covered the remainder. The concrete foundation seen behind and to the right of the bunker is a post-war installation for a cylindrical pedestal that supported a Navy *F-4 Phantom* aircraft display, which can be seen in Figure 7-34b.

Figure 7-34. a) The WWI-era bunker faces east towards San Diego Bay, with the Pacific Ocean visible in background center and the concrete foundation for the no-longer-extant *F-4 Phantom* pedestal in upper right corner. **b)** The *F-4 Phantom* aircraft display is just north (right) of the bunkers at lower left (undated US Navy photo courtesy SSC Pacific).[14]

Figure 7-35 shows a section view of the East Observation Room, looking south towards the doorway to the connecting passage leading to the WWII addition, as previously shown in the plan view of Figure 7-28. During this new construction, a layer of protective concrete was poured on top of the original steel-plate roof of the WWI structure, sealing off its overhead entry hatch and three of the four observation slots. A subterranean concrete chamber was also added on the north side of the structure for Cable Hut 2, which provided a communication connection to the Army Tactical Cable.

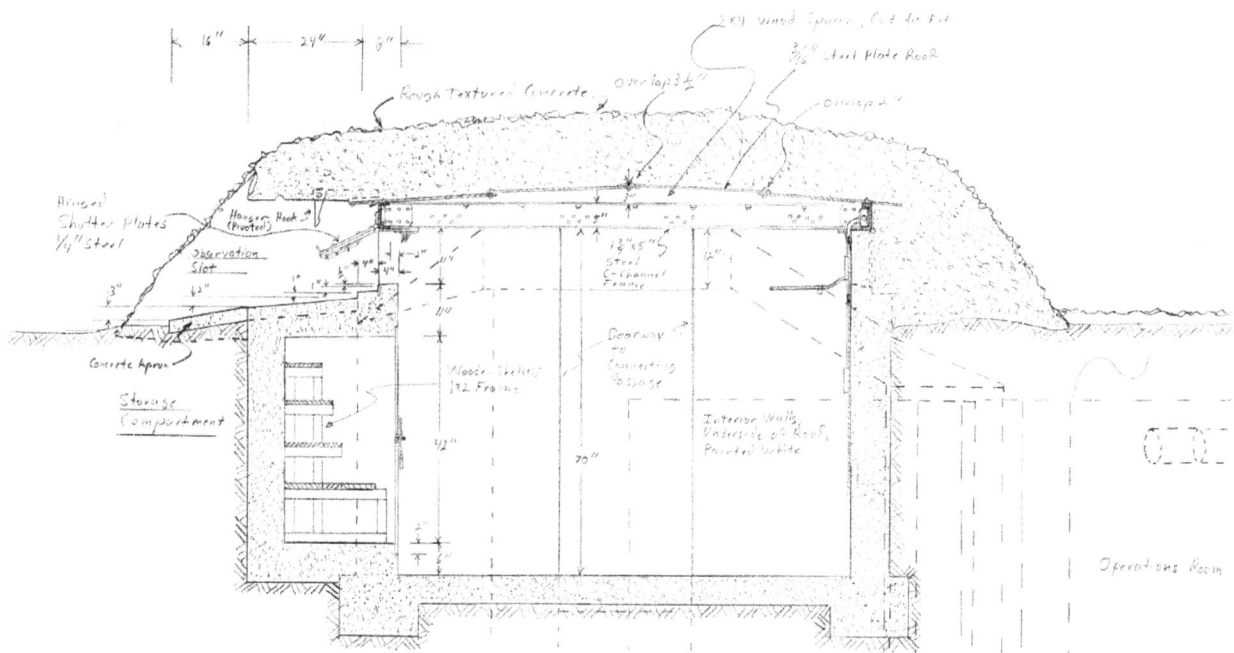

Figure 7-35. Section view of the East Observation Room, looking south. Note the unusual storage compartment at lower left, which extends outward beneath the eastern observation slot, and the rough-textured concrete roof poured on top of the original steel-plate roof (1995 field sketch by Lee Guidry).

14 The WWII pistol range, and the exposed Gun 1 emplacement for Battery Zeilin seen on the far side of Woodward Road in the upper-left corner, date this photo somewhere in the late 1960s to early 1970s timeframe.

At the east end of the connecting passage there was a five-step ascending stairway leading up to the East Observation Room, each of these steps a different height for some reason, for a total gain in elevation of 42 inches. The first two steps led to the west side of a small landing, from which the next three continued on from the north side of the landing into the repurposed WWI bunker (Figure 7-36). Interestingly, on the east side of this landing, another step led down to the previously mentioned wooden end wall (up arrow, Figure 7-28). There may have been plans for a temporarily entrance at this location, perhaps to bring in equipment that wouldn't fit through the overhead trapdoor in the Operations Room.

Figure 7-36. Interior of the East Observation Room of the Standby Harbor Defense Command Post, looking southwest towards the doorway to the connecting passage from the Operations Room. Note Todd Everett's flashlight barely visible behind the right side of the doorframe as he ascends the first two steps.

There was no such large equipment found in the Observation Room, however, and the wooden wall, comprised of vertical 2x12 lumber with horizontal 2x4 cross bracing, appeared undisturbed as seen in Figure 7-37. Note the improvised rifle rack nailed to the inside of this wall, and the outside rock and clay visible through the crack between its left side and the concrete bunker. There was considerable evidence of soil spillage beneath this crack, adding to the mystery of what led to the incorporation of this unusual feature in the first place.

Figure 7-37. An improvised rifle rack is mounted on the unusual wooden wall made of vertical 2x12s at the east end of the connecting passage, with rock and clay from outside fill visible through the crack on the left. For perspective, the inverted plastic bucket at bottom left was previously shown at lower left in Figure 7-36.

Battery Commander Station for Batteries North and Gillespie

At the time of the Japanese attack on Pearl Harbor, there were no permanent base end stations for the 155-millimeter coast-defense batteries (HSD, 1945). Battery Gillespie was initially served by a temporary wooden structure that had originally been built for Battery North. Located upslope and behind the gun emplacements (Figure 7-38), it was approximately 950 feet west of where Building A36 is now located Topside, according to the Flower and Roth (1982) survey, which describes conditions found at the time:

"A base end station built of wood covered with tar paper. Dimensions: 9 feet, 8 inches, by 14 feet. Floor to ceiling, 6 feet, 10 inches. Structure contains 3 bunks against back wall. Octagon base and map table for "A" instrument and rifle rack very similar to that in Gillespie. The station appears to never have had a mount for a "B" instrument. Approximately 15 feet due west up the slope from the structure are concrete pillar radar mounts. Anti-Japanese graffiti dated 1945 is scratched in the top bunkheads."

Figure 7-38. As seen looking southeast from Battery Gillespie in February 2008, the remains of the temporary battery commander station are seen on the ridgeline (right up arrow), just down slope from the foundation pillars of Battery Woodward's *SCR-296A* fire-control-radar tower (left up arrow).

This wooden structure was not a base end station for Battery North as claimed, but the battery commander station for Battery North/Gillespie. In addition, the concrete foundation pillars for Battery Woodward's *SCR-296-A* fire-control-radar tower are due east of the site as seen previously in Figure 7-38, not west as reported above. The previously mentioned Type A instrument for Battery Gillespie was located further down slope in a small wooden observation station just south of the Gun 3 emplacement, as seen Figure 7-39 below.

Figure 7-39. The azimuth instrument for Battery Gillespie was located inside the wooden observation station just left of the power pole (up arrow), as previously discussed in Chapter 4. The former Gun 2 and Gun 3 emplacements for the battery are left of center immediately behind the dirt road (1954 US Navy photo).

Plan and section views of the original battery commander's station, reproduced from field sketches prepared by Lee Guidry during the Keniston (1993) survey, are shown in Figure 7-40. Note the observation slot on the west side extends around a bit along both the north and south walls for a broader field of view, with a Type A depression position finder (DPF) foundation in the southwest corner to better observe the harbor approaches. The entry hatch is on the northeast corner to the left of the bunks.

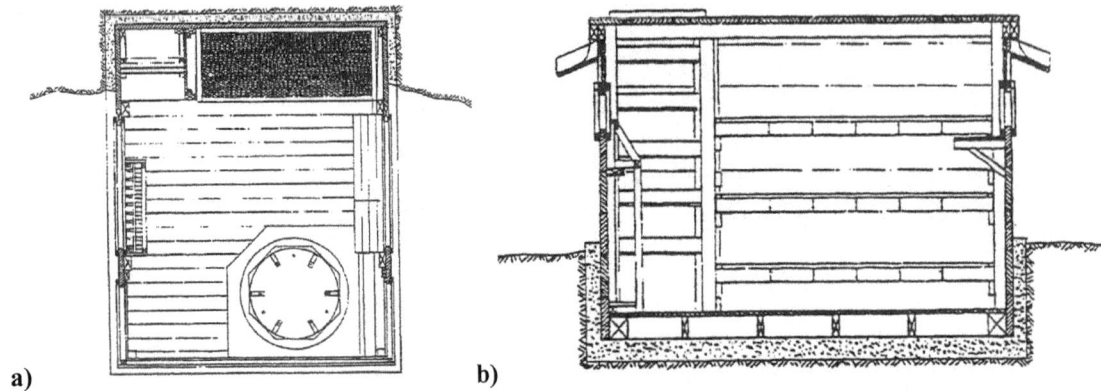

a)   b)

Figure 7-40 a) Plan view of the battery commander station showing (clockwise from upper left) ladder, bunks along east wall, shelf, DPF base, and gun rack. **b)** Section view looking south showing entry ladder at left with the three bunks on the right (1995 field sketches by Lee Guidry).

Figure 7-41a shows the temporary battery commander station looking north in 1971, with the rolled-roofing material still partially covered by residual camouflage debris. A close-up shot of the roof is seen in Figure 7-41b, with the access hatch above the inclined entry ladder seen on the northeast corner, and waves breaking along the shoreline of the Pacific Ocean in the background. The wooden hatch cover has fallen down into the structure interior, as shown later in Figure 7-43b.

a)   b)

Figure 7-41. a) The temporary battery commander station behind and upslope from Batteries North/Gillespie, looking north in 1971. **b)** The trapdoor entry on the northeast corner of the roof, looking southwest (photos courtesy Lee Guidry).

Guidry's 1971 photo seen in Figure 7-42a shows the still-extant lower rungs of the wooden ladder beneath the trapdoor on the northeast corner of the roof (see again Figure 7-41b). Although the upper ladder rungs were missing, the rest of the interior was found in remarkably good shape. The temporary wooden structure later collapsed as seen in Figure 7-42b, with indications that it had been intentionally chain sawed found by Guidry in 1973. The mattress supports in the bunk frames appear to be made from fence wire only, with no chicken-wire overlay as seen in the later berthing shelters constructed for Battery Point Loma.

Figure 7-42. a) Remains of the inclined access ladder in the northeast corner, with the end of the bunks just visible on east wall at far right, circa 1971. **b)** In August 2004, two of the three wooden bunks along the east wall were found pancaked in the rubble following the structure's collapse (photos courtesy Lee Guidry).

The interior was found in reasonably good shape by Guidry in 1971, as evidenced by the concrete DPF base and wooden bench seen in Figure 7-43a. The Keniston (1995) site survey reported the wooden structure had largely collapsed, with one bunk frame found outside. The August 2004 photo presented in Figure 7-43b shows the west wall has fallen in upon the DPF station, which is surrounded by a pile of rubble. Note the wooden trapdoor cover in the northwest corner (bottom right) leaning against the curved DPF bench.

Figure 7-43. a) The well-preserved DPF station is seen in the southwest corner, circa 1971. **b)** The west wall was found collapsed inward upon the DPF station in August 2004 (photos courtesy Lee Guidry).

Figure 7-44 shows a makeshift wooden rifle rack near the midpoint of the north wall, circa August 2004, similar in construction to those encountered at Battery North/Gillespie and in the nearby Searchlight 12 silo. Traces of yellow paint on the horizontal wall boards suggest the wood was salvaged from some other location for reuse in this temporary construction. As seen at far left, the westernmost section of this wall has also collapsed inward, allowing the sandy soil to spill into the interior.

Figure 7-44. The remains of a makeshift wooden rifle rack (see again plan view in Figure 7-40a) was found reasonably intact along the north wall of the structure in August 2004. Note soil entry from collapsed wall at left (photo courtesy Lee Guidry).

Battery Commander and Base End Station for Battery Strong

Prior to WWII, this underground structure had originally served as B^1_3 S^1_3 – BC_3 for Battery Strong (Figure 7-45), but was redesignated under the Modernization Program in 1941 as B^1_2 S^1_2 – BC_2, per the 18 September 1942 Report of Completed Works. Situated some 1650 feet northwest of the battery, this dug-in bunker was transferred to the 19th Coast Artillery Regiment on 20 March 1941, with a total cost of construction of $3,086. Each level contained an adjustable base for an azimuth instrument, with the lower station further equipped with an octagonal concrete base for a depression position finder.

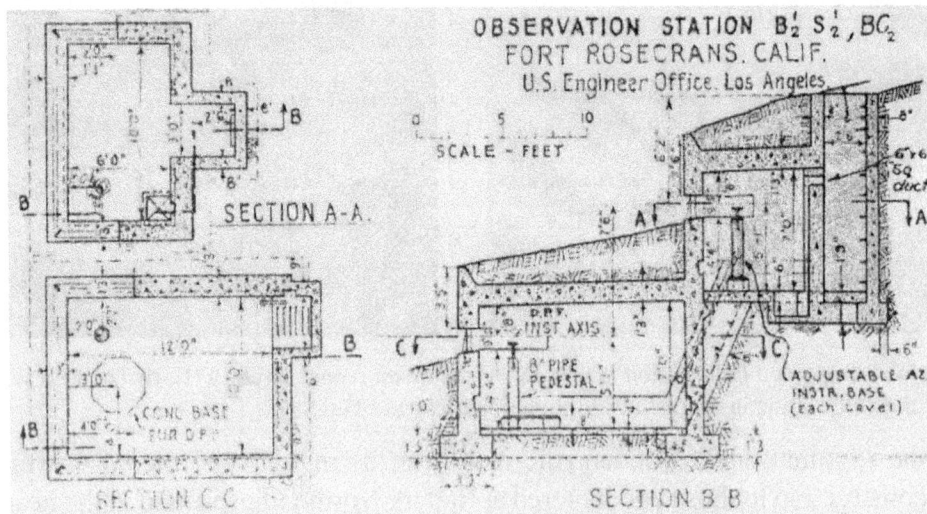

Figure 7-45. The upper level of this structure served as the battery commander station BC_2, while the lower level was base end station B^1_2 S^1_2 for Battery Strong.

The location of this two-level bunker is marked by the up arrow at top center in the 1962 aerial photograph presented in Figure 7-46 below, directly upslope from Battery Woodward. Note the unidentified small wooden structure below this bunker (question mark), where the dirt road running from the upper level down to Woodward Road intersects both a footpath to the north and another dirt road leading south. The construction evident in the foreground of this photo involved the tunneling of a large-diameter underground pipeline beneath the Point Loma Ridge for the new San Diego sewer plant to the south.

Figure 7-46. The battery commander and base end / spotting station for Battery Strong (up arrow top center) were built near the top of the Point, directly above Battery Woodward, for which the Gun 1 and Gun 2 emplacements are labeled at bottom center. Note the unknown wooden structure just right of the question mark (1962 US Navy photo courtesy SSC Pacific).

Figure 7-47 shows this combined battery commander and base end station as it appeared in December 2007, looking northeast. The steel shutters for the observation slots were locked in the closed position to keep out rodents, but the steel trapdoor on top of the upper level was found unlocked. Note the fill dirt on top of and around the bunker depicted in the section view of Figure 7-45 has been completely washed away by rain over the past 60 years. A 9 September 1975 Naval Electronics Laboratory Center "General Development Map" (sheet 3 of 12) shows this structure designated as "F-22U."

Figure 7-47. The combined battery commander and base end and spotting station for Battery Strong, situated below (west of) what is now the Safety Office (Building 311), looking northeast in August 2007. Note the San Pedro Radio Tower to the north at background left.

Access to this upper level of this massive concrete structure was via the overhead steel trapdoor seen at far right in Figure 7-48 below. The upper level served as battery commander station BC_2 for Battery Strong, while the lower level was the battery's base end and spotting station $B^1_2 S^1_2$. Evidence of the original terrain level on the south side of the structure can be seen in the concrete scarring above and to the right of the rusted steel shutter at bottom center. Several buildings overlooking the ocean in the Unmanned Systems compound are visible on the far side of Woodward Road at lower left.

Figure 7-48. Fellow roboticist Rachel TenWolde points out the elevated trapdoor entry on the upper level, which served as battery commander station BC_2 for Battery Strong. Part of our Unmanned Systems compound is visible on the far side of Woodward Road at lower left, circa November 2007.

There was no DPF foundation in the upper BC_2 level, but an 8-inch steel pedestal for the azimuth instrument was bolted to a small hexagonal concrete foundation in the southwest corner as shown in Figure 7-49a. The steel shutters, hinged at the bottom so as to hang down outside the window when open, were held shut by wingnuts at the end of long eyebolts passed through slotted brackets bolted to the concrete wall. A steel-rung ladder cast into the sloped concrete wall leads westward down to the lower-level base end station (Figure 7-49b).

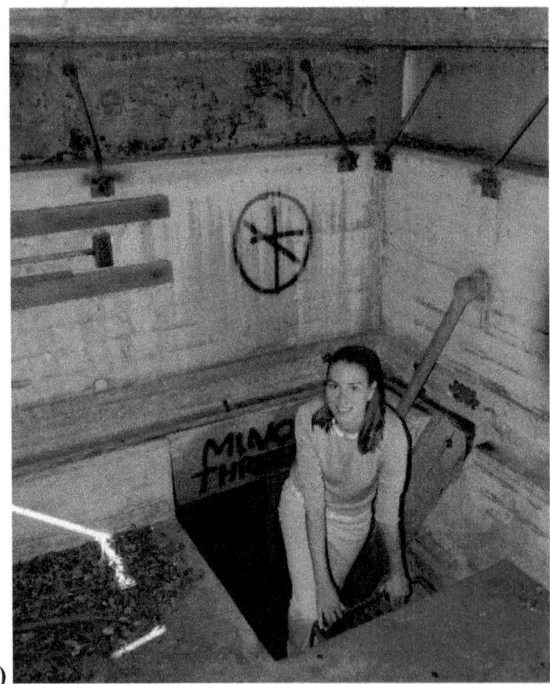

a) b)

Figure 7-49. **a)** The 8-inch-diameter steel pedestal for the BC_2 azimuth instrument in the southwest corner of the upper level. **b)** Rachel TenWolde carefully descends the embedded steel ladder in the northwest corner down to base end and spotting station $B^1_2 S^1_2$ for Battery Strong on the lower level (circa November 2007).

For safety purposes, we set up a portable electrical generator on the roof and ran an extension cord down to a pair of halogen work lights, which also provided better interior illumination for photography. The embedded steel ladder leading down from the upper level is seen in Figure 7-50a; note the adjacent handrail and the metal conduit providing upper-level telephone service on the north wall. Figure 7-50b shows the well-preserved field-telephone distribution panel on the east wall of the lower level, just to the right of the ladder.

a)   b)

Figure 7-50. a) The extension cord running down the ladder in the northeast corner of the lower level provided power from a portable generator for temporary lighting. **b)** The remains of the original field-telephone terminal panel and associated cabling are seen on the east wall to the right of the ladder (circa November 2007).

The partial remains of a rotating wooden bench attached to the octagonal concrete foundation of a depression position finder were found in the southwest corner of the lower level (Figure 7-51a). The L-shaped wooden brackets attached to the concrete overhead provided temporary storage for removable glass windows that could be put in place when the steel shutters were open. Figure 7-51b shows the 8-inch steel pipe that supported the azimuth instrument in the northwest corner. Note the wooden field-telephone mounting cleats below the steel shutters along the west wall in both photos.

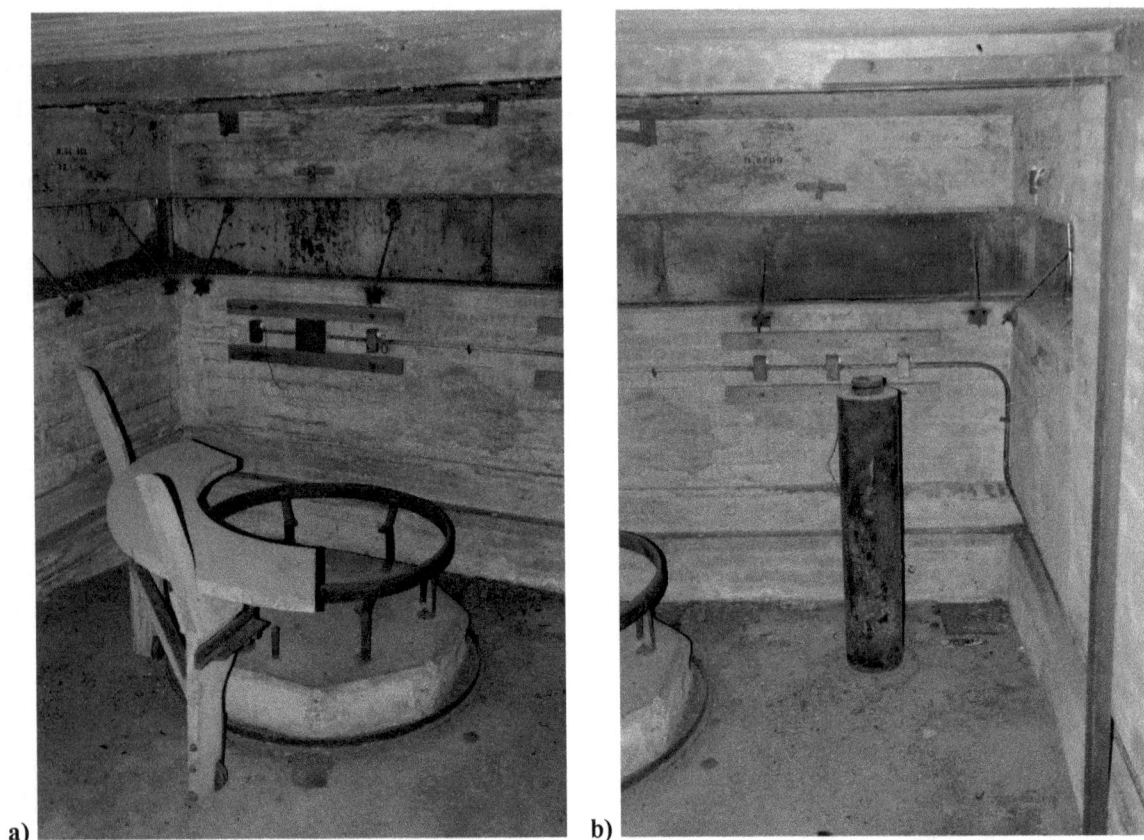

Figure 7-51. a) Remains of the rotating wooden chair and hexagonal concrete foundation for the depression position finder in the southwest corner of the B^1_2 S^1_2 lower level. **b)** The concrete-filled 8-inch steel post for the azimuth instrument is located in the northwest corner of the lower level (circa November 2007).

Plotting and Spotting Room for Battery Strong

Stephen R. Van Wormer, the evaluator for the 1995 "Historic and Archeological Resources Protection Plan" for the Point Loma Naval Complex, described the Plotting and Spotting Room (PSR) for Battery Strong as follows (Apple, Van Wormer, Cleland, 1995):

"This structure consists of a subterranean bunker built into the bank of a southeasterly trending slope. Surface manifestations consists (sic) of two separate entrances approximately 100 feet apart. The westerly entrance consists of a concrete façade approximately 50 feet wide and 20 feet high. It exhibits two large riveted steel doors. Remains of a sand bag retaining wall are also present around the doors. The easterly entrance is about 50 feet high and 18 feet wide. It is recessed into the bank approximately 8 feet. This entrance has double riveted steel doors and two additional riveted steel panels over smaller openings."

It would appear Van Wormer mistook the battery's combined Rest Room, First Aid Station, Latrine, and Radio Room, to be discussed in the next section, for this assumed "easterly entrance" to the PSR,[15] although these two bunkers are clearly more than 100 feet apart (see lower-left quadrant of Figure 7-52). Battery Strong is seen in the upper-left corner of the map, with its PSR to the south on the northwest side of Woodward Road, now Building 549 at the Center. The Rest Room, First Aid Station, Latrine, and Radio Room (Building 559) is to its north. Note the abandoned course of Woodward Road to the east of its current location.

---

15 The PSR is actually the southerly versus easterly of the two bunkers, as seen in Figure 7-52. Battery Strong is in the upper-left corner of the map, due west from the intersection of Woodward Road and Catalina Boulevard / Cabrillo Memorial Drive at Gate 5 (1993 field drawing by Lee Guidry). Its associated PSR is just left of bottom center, south of the Rest Room, First Aid Station, Latrine, and Radio Room..

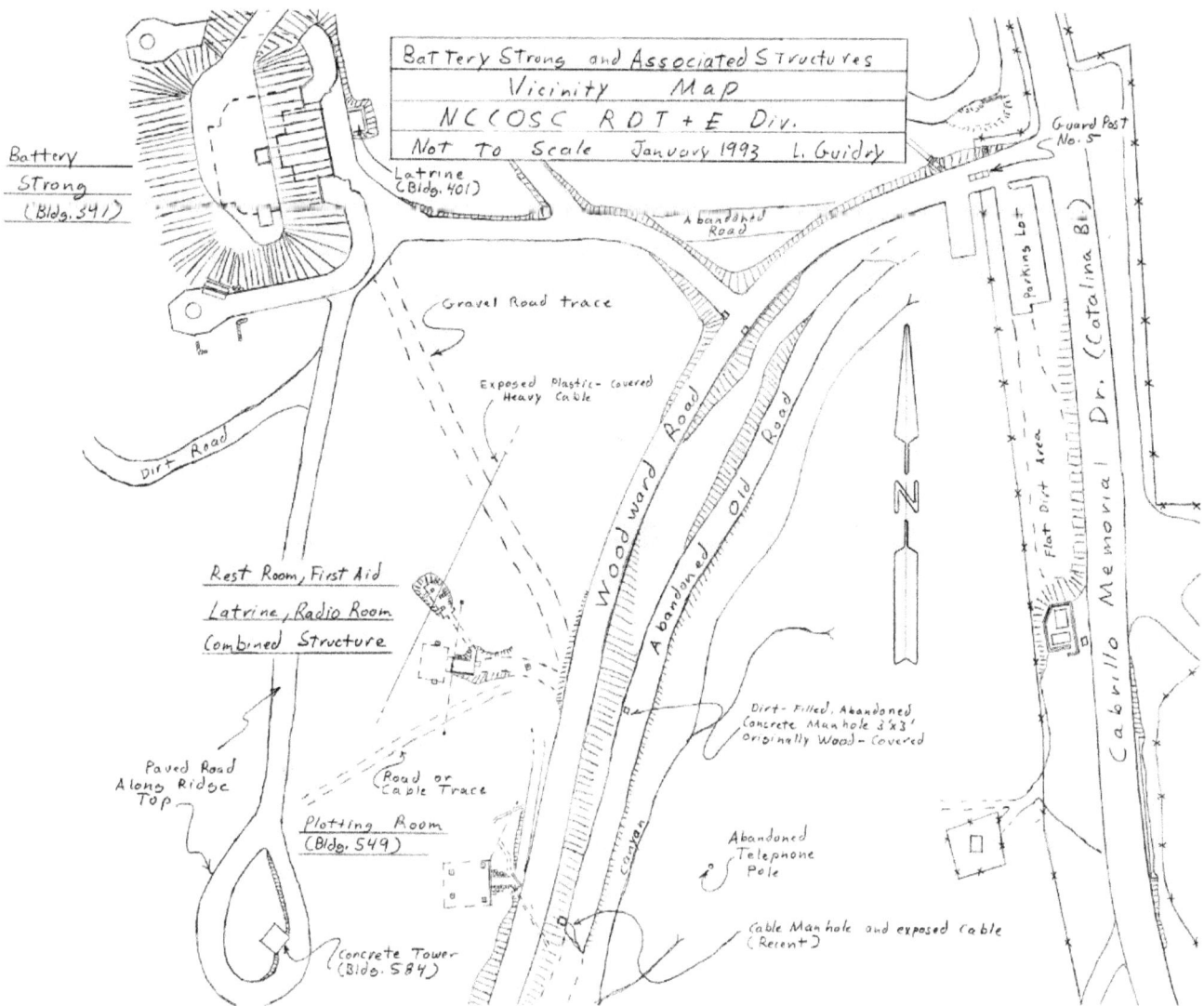

Figure 7-52. Battery Strong is in the upper-left corner of the map, due west from the intersection of Woodward Road and Catalina Boulevard / Cabrillo Memorial Drive at Gate 5 (1993 field drawing by Lee Guidry). Its associated PSR is just left of bottom center, south of the Rest Room, First Aid Station, Latrine, and Radio Room.

The PSR is seen downslope from the post-war microwave tower (Building 584) in Figure 7-53, depicted earlier in the lower-left corner of Figure 7-52. The open-frame tower construction and vintage car to its left suggest this undated photo was taken in the late 1950s timeframe. [16] Two of the concrete-encased air vents for the underground PSR bunker can be seen behind and to the right of its entrance (up arrow), and several camouflage-support posts are still extant. The original course of Woodward Road, which ran more or less north/south in front of the bunker as shown in Guidry's drawing above, is occluded from view by the elevated underbrush and dirt in the foreground.

---

16 This open-frame steel tower was later upgraded with a concrete exterior, as seen in the 1968 photo presented in Figure 7-55. The PSR for Battery Strong is seen on the west side of the original route of Woodward Road, which is just visible at bottom center, just down the slope from the post-war Building 584 microwave tower at upper left, circa 1968. Note the concrete cable pit in foreground between the PSR entrance and the original location of Woodward Road (US Navy photo courtesy SSC Pacific)..

Figure 7-53. The entrance (up arrow) to the PSR for Battery Strong is downslope from the post-war Building 584 microwave tower at top center (undated US Navy photo courtesy SSC Pacific). Note the two concrete-encased air-vent shafts above and to the right of the entrance, and the camouflage-net support poles in front.

Figure 7-54 shows the plan and section views of the PSR from the Report of Completed Works. Note the 4-foot-thick concrete walls around the main gas-tight section on the right, which is divided into two rooms by a cement-tile wall with a 36-inch-wide doorway. The larger of these two rooms has a pair of frosted windows that open to the vestibule for ventilation, each equipped with gas-tight steel shutters. The overhead for the gas-tight area was 3.5-feet thick, above which was an 18-inch-thick concrete burster slab covered with dirt. The structure was transferred on 12 October 1942 with a total cost to that date of $19,145.63.

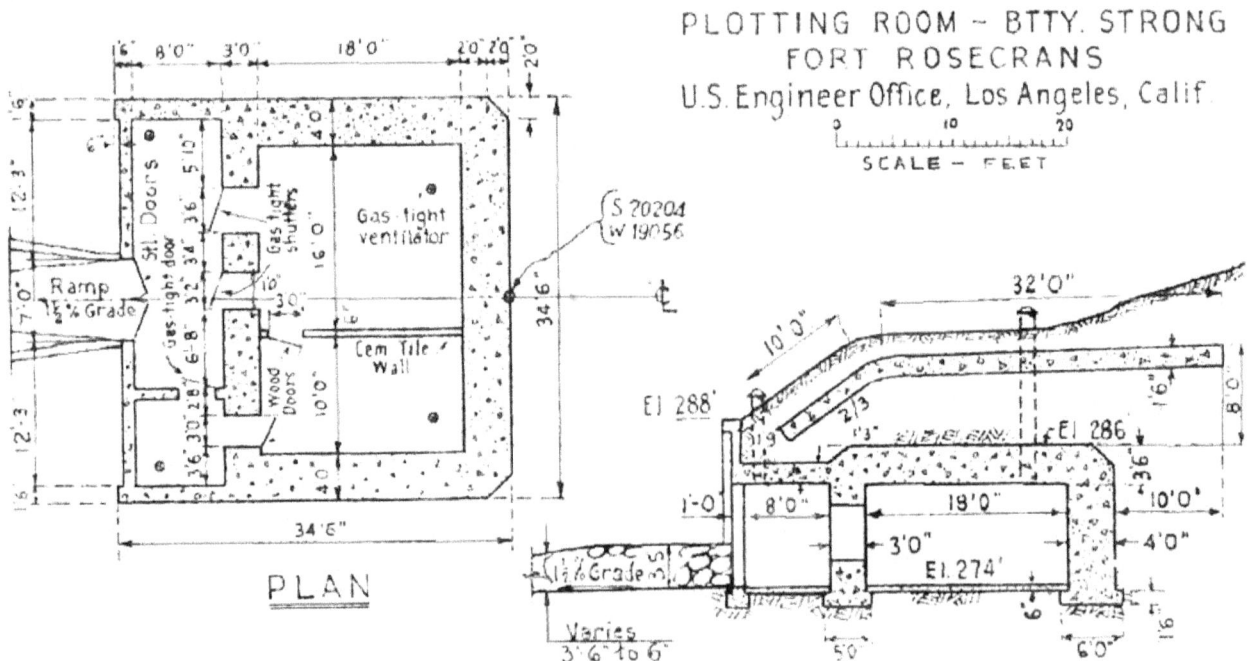

Figure 7-54. Plan and section views of the PSR for Battery Strong (NARA). There was no dedicated Power Room, as the PSR was connected to commercial power and two standby generators in the magazine for nearby Battery Strong.

The 1968 aerial photo seen in Figure 7-55 provides a closer look at the PSR entrance, which is still reasonably well hidden in this view from a Woodward-Road perspective, even with the overhead camouflaged nets long since removed. All four of the concrete-encased vent shafts can be seen above the subterranean structure, as well as the still-extant camouflage-net support poles out in front. The newly enclosed Building 584 microwave tower is on the ridgeline at upper left. Note the discarded lumber scattered down slope below the tower, which appears to be WWII-vintage.

Figure 7-55. The PSR for Battery Strong is seen on the west side of the original route of Woodward Road, which is just visible at bottom center, just down the slope from the post-war Building 584 microwave tower at upper left, circa 1968. Note the concrete cable pit in foreground between the PSR entrance and the original location of Woodward Road (US Navy photo courtesy SSC Pacific).

As seen in Figure 7-56 below, circa May 2006, the PSR for Battery Strong is now situated immediately adjacent to Woodward Road, which has been rerouted between the cable pit and bunker entrance seen in the foreground of the above 1968 photo (Figure 7-55). The steel pipe running horizontally across the top of the front side probably served as an attachment point for the overhead camouflage netting. The metal sign above the door indicates this massive underground bunker was repurposed as a fallout shelter at some point during the Cold War. While gas tight, this subterranean structure was not equipped with a Chemical Warfare Service (CWS) air scrubber.

Figure 7-56. Todd Everett examines the concrete ramp leading to the Battery Strong PSR entrance, circa May 2006. This rerouted section of Woodward Road can be seen turning right at background left to rejoin the original route, which parallels the coastline up to the northern fence line of Fort Rosecrans.

Figure 7-57 below shows the concrete ramp leading down to the PSR entrance. Note the distinctive impressions in the retaining wall to the left of the outer doors, made by stacked sandbags used to create a form for the concrete pour. A similar retaining wall is hidden behind the shrubbery on the right side of the ramp. The previously mentioned fallout-shelter sign is screwed to the concrete above the steel doors, which now reveal the more recent function of the bunker with the painted inscription: "Storeroom Code 84."

Figure 7-57. The entrance to the PSR for Battery Strong, as seen looking west in May 2006. The painted sign on the steel doors says "Storeroom Code 84" (see also Figure 7-59).

The antechamber just inside the steel outer doors of the PSR is shown in Figure 7-58 (see again plan view in Figure 7-54). The gas-tight door seen in the open position at far right leads to a small airlock for the two-room gas-proof rear section of the bunker. The gas-tight shutter for the northern frosted window is in the open position, with the southern window out of sight to the left. Note the two additional fall-out-shelter signs near either end of the armored electrical cable on the overhead.

Figure 7-58. As seen from just inside the steel outer doors, the gas-tight door at far right opens to a small airlock leading back to the main gas-proof section of the bunker, circa June 2006. The northern gas-tight window shutter at image center is seen in the open position.

The plotting room shown in Figure 7-59, the larger of the two rooms in the gas-free section, was originally equipped with a *110-degree Plotting Board M-3*. The telephone mounting cleats and cable conduits for the various field lines from the base end stations assigned to the battery are seen along the west wall at background center. The object hanging from the ceiling at upper left appears to be a rotating steel shaft with two large discs attached at the bottom, perhaps the remains of a heavy-duty ceiling fan. The gas-tight ventilation-duct opening is nearby to the right, midway between the two light fixtures.

Figure 7-59. This interior view shows the former plotting room on the west side of the gas-free area, repurposed as a Code 84 storeroom, circa June 2006. Note the mounting cleats and cable conduits for the field telephones along the back wall and the distinctive WWII-era acoustical tile on the overhead.

Also repurposed for storage, the interior of the smaller east room inside the gas-free area of the bunker is shown in Figure 7-60, looking east towards Woodward Road. Note the air-vent pipe in the upper-left corner above the wood-paneled door leading back to the airlock. The wooden door at far right on the cement-tile divider wall leads to the larger south room that served as the plotting room for Battery Strong during WWII (see again plan view of Figure 7-54). The abandoned junk stored in this room appears to date back to the late 50s – early 70s timeframe.

Figure 7-60. Interior of the smaller room on the east side of the gas-free area, looking east. The wooden door on the left leads back to the airlock, while the door on the right leads to the larger plotting room on the south side of the bunker, as seen earlier in Figure 7-59.

Rest Room, First Aid Station, Latrine, Radio Room for Battery Strong

As seen in Figure 7-61, the combined Rest Room, First Aid Station, Latrine, and Radio Room for Battery Strong, now Building 559 at the Center, is on the west side of Woodward Road just inside Gate 5 (Guard Post 12), about 60 yards north of the previously discussed Plotting and Spotting Room. In WWII Army parlance, the term "rest room" was literally just that, a place to rest, while the term "latrine" was used for what today we more commonly call a rest room. Like the associated PSR, this underground concrete structure was later used as a civil-defense shelter during the Cold War.

Figure 7-61. The entrance to the abandoned Building 559 as seen in 2006. The wooden debris and door on the right side are from the WWII addition, which expanded the entryway and provided tunnel access to a large underground berthing shelter, as shown later in Figure 7-63b.

Construction details of Battery Strong's combined Rest Room, First Aid Station, Latrine, and Radio Room are presented in the plan and section views of Figure 7-62 from the Report of Completed Works. An interesting detail is the handhole seen at bottom center in the plan view, which leads to an unidentified 3-foot-deep compartment below the bunker floor, as seen in the section view. This void was probably the fuel tank for the two generators in the nearby power room, with the handhole providing access to the fill pipe. The three-holer "latrine" is shown in the bottom-left corner of the plan view.

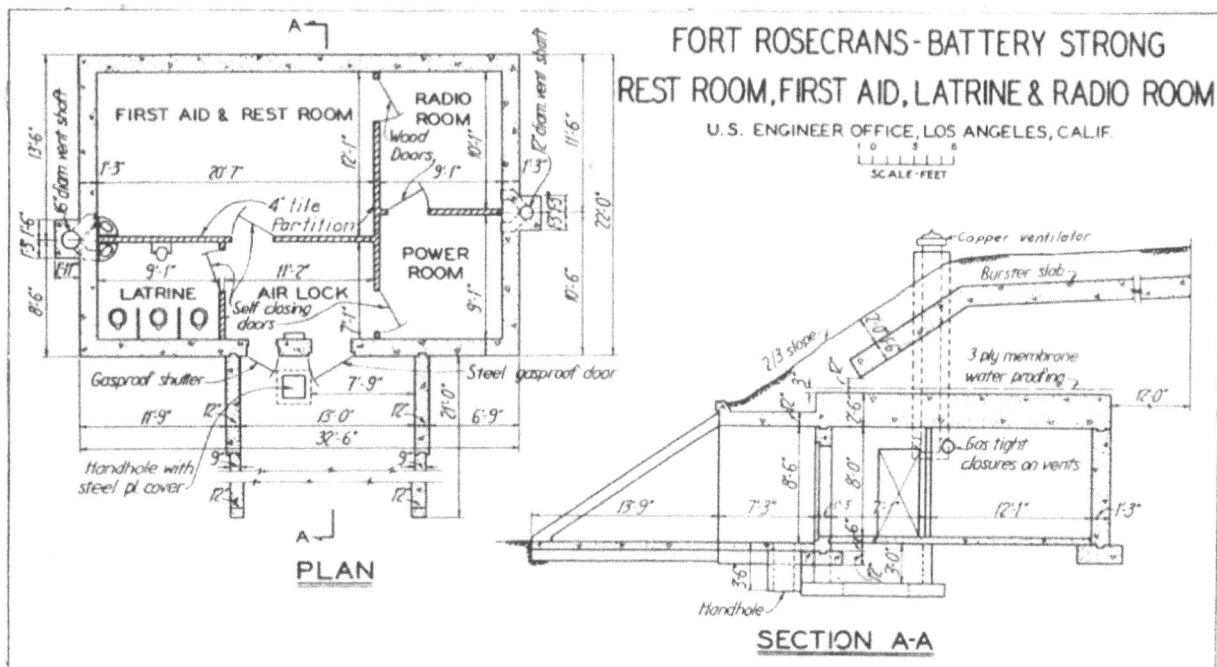

Figure 7-62. Plan and section views of the combined Rest Room, First Aid Station, Latrine, and Radio Room for Battery Strong (NARA). Note the handhole outside the gas-tight entry doors, which appears connected to the 3-foot-deep compartment seen below the bunker floor in the section view. A troop-built wooden expansion was added to the structure during WWII, as shown in Figure 7-63b.

The spring-actuated self-closing mechanism on the gas-tight front door to the dug-in bunker is seen at far right in Figure 7-63a below. A detailed field sketch of the expanded site layout prepared by Lee Guidry during a cultural-resource site survey in 1993 is presented in Figure 7-63b. The interior partitions of the original structure in the upper-left corner are made of 4-inch cement tile. The later addition to the entrance and the underground sleeping quarters to the north, plus their interconnecting tunnel, were constructed of wood with troop labor.

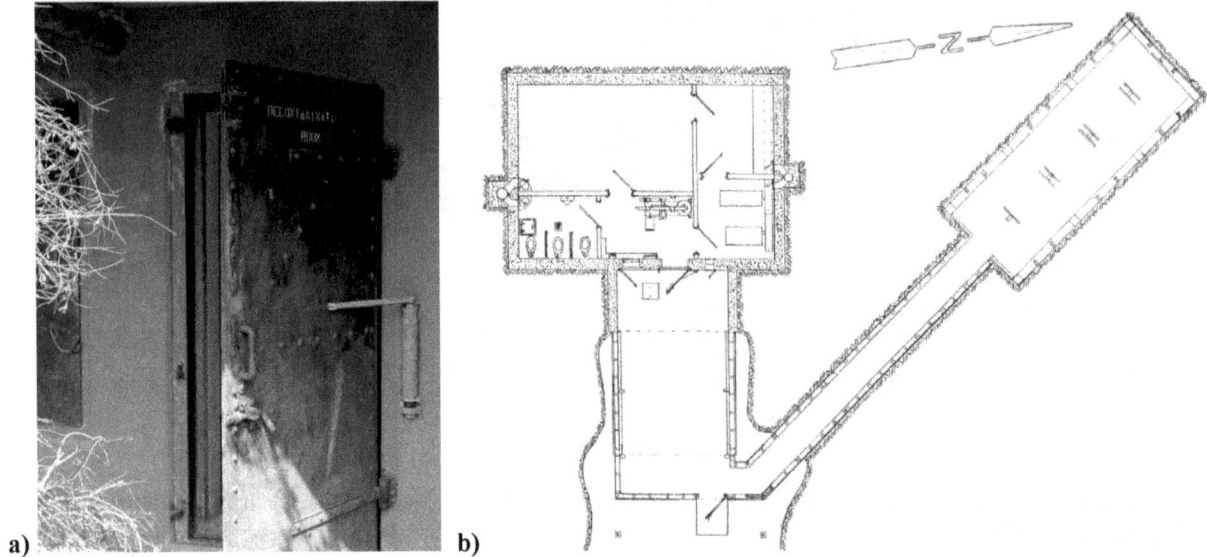

Figure 7-63. a) Note self-closing mechanism for the steel entry door to the concrete bunker. **b)** Plan view of the expanded structure, with a 37-foot tunnel connecting the new entry-room addition to a 12-by-30-foot wood-framed bunk room approximately 12 feet below grade level (1993 field sketch courtesy Lee Guidry.)

Figure 7-64a shows the doorway to the former Radio Room in the northwest corner of the bunker, with the wooden equipment bench seen against the north wall. A post-war parabolic antenna sitting on the floor in front of the open doorway to the CWS room is seen in Figure 7-64b, with the latrine doorway just barely visible at lower left. Figure 7-64c shows a collapsed shelf assembly laying across the two generator foundations in the Power Room in the northeast corner of the bunker, looking north. The scattered floor debris apparently entered through the exhaust-discharge port in the upper-left corner of the outer wall (not shown).

Figure 7-64. a) View from the First Aid and Rest Room looking north into the Radio Room at the equipment bench along the far wall. **b)** View from the First Aid and Rest Room looking southeast, with the open doorway to the latrine barely visible just inside the air lock. **c)** Concrete generator foundations in the Power Room.

While the previously discussed PSR for Battery Strong was only gas tight, the combined Rest Room, First Aid Station, Latrine, and Radio Room was further equipped with an active Chemical Warfare Service (CWS) air scrubber as shown in Figure 7-65a. This equipment was located inside the airlock opposite the front entrance as previously depicted in Guidry's field sketch presented in Figure 7-63b. The intake and output diverter valves could be simultaneous actuated to engage the system by stomping on the chain-linked kickplate illustrated in Figure 7-65b.

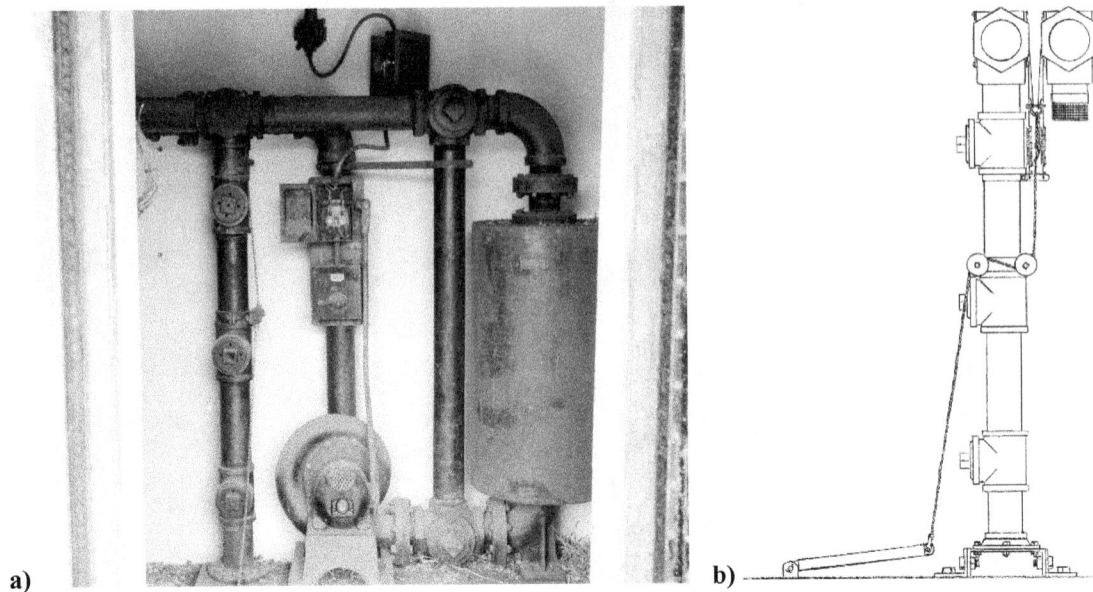

Figure 7-65. The brass plate on the CWS canister reads: "Chemical Warfare Service U.S.A., Collective Protector Canister, Capacity 200 Cu. Ft. per Minute, E.A. Serial Number 385, Lot 2695-1, Edgewood Arsenal Maryland." **b)** Kickplate-actuation scheme for the diverter valves (1993 field sketch by Lee Guidry).

Oliver Unruh of Battery D, who served as a gun mechanic on Battery Strong during the war, helped build the underground sleeping quarters shown in Figure 7-66 before being transferred to Battery Ashburn (see again Chapter 5). As he explained to Howard Overton (1993) in his June 1992 interview at Cabrillo National Monument, the tool room he oversaw at Battery Strong was relocated to the new wooden addition to the Rest Room, First Aid Station, Latrine, and Radio Room shown earlier in Figure 7-63b. It had previously been housed in Battery Strong's powder galleys behind the guns.

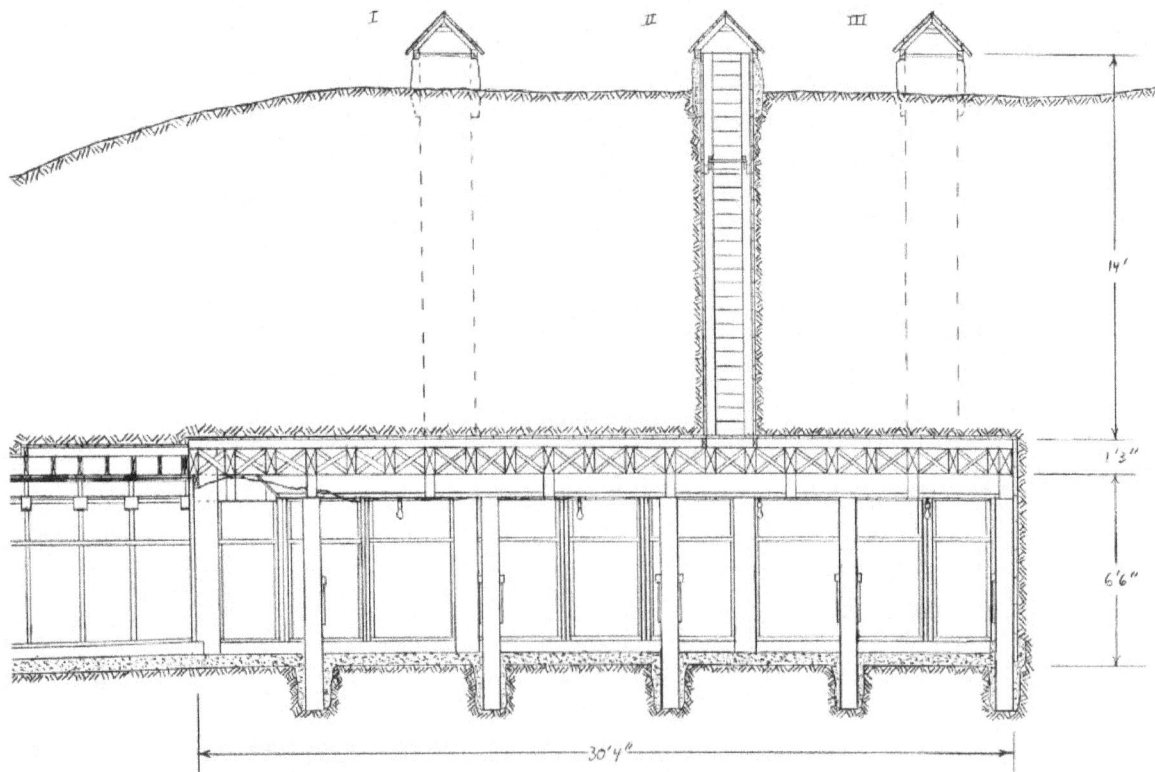

Figure 7-66. Longitudinal section view of the troop-built 12-foot by 30-foot underground wooden structure that served as a berthing shelter, showing three ladder-equipped escape/air shafts to the surface some 14 feet above (1993 field sketch courtesy Lee Guidry). This area collapsed sometime prior to my site visit in 2006.

The above-ground entrance to the 37-foot-long troop-built connecting tunnel leading to this underground berthing structure is seen in Figure 7-67a below, with its interior now filled with several feet of spillage. Lee Guidry found there were four slightly different cross-sectional designs encountered during his 1993 site visit, mostly dealing with the width and spacing of the ceiling joists and the associated 2x4 triangular bracing. Figure 7-67b, taken from the northwest end of the tunnel looking back to the southeast, shows the collapsed tunnel midsection illuminated by sunlight from above.

Figure 7-67. a) The above-ground entrance to the 37-foot connecting tunnel depicted in Figure 7-63b, looking northwest in May 2006 at remnants of horizontal 1x6 sheeting and alternating 2x4 and 4x4 posts covered with ½" gypsum drywall. **b)** View looking to the southeast towards the collapsed midsection of the tunnel.

Battery Commander and Base End Stations for Battery Woodward

The battery commander and base end and spotting stations repurposed during WWII for Battery Woodward are located west of and directly downslope from the present-day radar-calibration target at the Center as seen in Figure 7-68. A 1 July 1945 "Harbor Defenses of San Diego, Main Cable Routings" blueprint (sheet 2 of 4) identifies the upper structure as $B^3_1 S^3_1$ and the lower structure as BC_1. The Flower and Roth (1982) cultural resource survey states these fire-control structures were originally built for the WWI-era 12-inch mortar batteries White and Whistler.

Figure 7-68. Located just below the Center's radar-calibration target on top of the Point Loma ridge, as seen from Woodward Road looking east, circa May 2006, these two fire-control stations were initially built to serve the 12-inch mortar Batteries White and Whistler.

The upper structure, at an elevation of 250 feet, later served as the base end and spotting station for Battery Woodward during WWII. The Report of Completed Works, corrected to 11 September 1942, shows an original transfer date of 16 January 1920 at a cost of just $733.59. This early history explains why both structures have 8-inch hexagonal concrete pillars for azimuth-instrument mounts as shown in Figure 7-69, versus the steel-pipe columns more typical of WWII construction. Both were also equipped with octagonal concrete foundations for depression position finders.

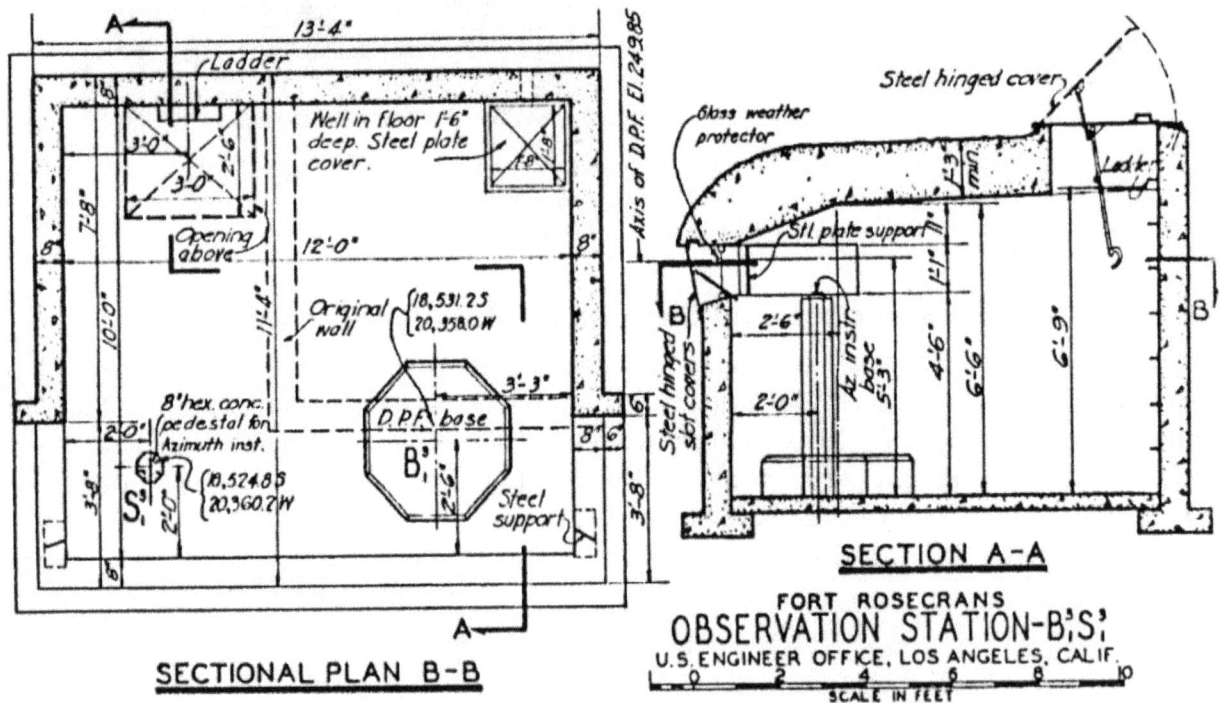

Figure 7-69. Plan and section views of base end and spotting station $B^3_1 S^3_1$ for Battery Woodward, which was equipped with an 8-inch hexagonal concrete pillar for an azimuth instrument and an octagonal concrete base for a depression position finder (NARA).

At an elevation of 250 feet, the upper bunker served as the base end and spotting station B^3_1, S^3_1 for Battery Woodward during WWII (Figure 7-70). A 9 September 1975 Naval Electronics Laboratory Center "General Development Map" (sheet 3 of 12) shows this structure designated as F-21U. A sizeable layer of soil and brush covering the trap-door entry on the northeast corner of the roof precluded easy access to the interior during our site visit in December 2007. Since it was an identical design to the lower structure, we elected not to climb through the unsecured viewing slot seen at foreground center.

Figure 7-70. Todd Everett catches his breath on the roof of the upper bunker after ascending the steep slope from Woodward Road. Access to the interior of this structure was hampered by soil and brush seen in background center covering the trap door, but the southwest viewing slot just to his right was found open.

After Battery Woodward replaced Battery Zeilin in August 1944 under the HDSD Modernization Program, the lower structure became the battery commander station (BC$_1$) for Battery Woodward (Tac. No. 1). Referring now to Figure 7-71, this WWI-era bunker is identical to that previously shown for the battery's base end and spotting station in Figure 7-69. Note the "glass weather protection" callout at upper left in the section view, which refers to wood-framed glass windowpanes that could be inserted into the viewing slots in the event of bad weather. An example of such is seen later in Figure 7-74c.

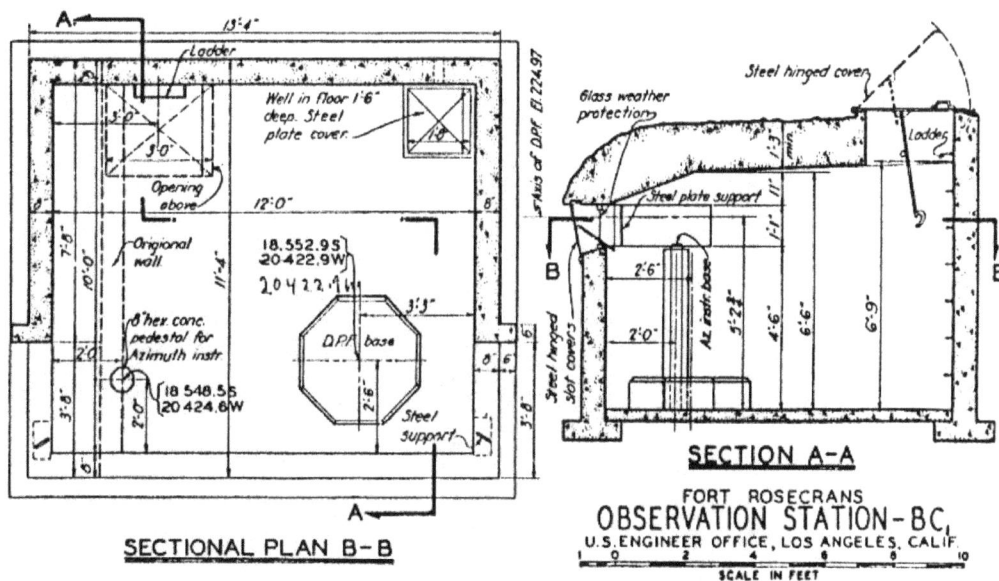

Figure 7-71. The battery commander station (BC$_1$) for Battery Woodward was also equipped with an 8-inch hexagonal concrete post for an azimuth instrument and an octagonal concrete base for a DPF (NARA).

Figure 7-72a shows the lower bunker for the battery commander station at an elevation of 175 feet, with the upper structure just below the radar target in the top-right corner. The loose boulders seen along the south side at lower right once helped camouflage the concrete roof, but were probably dislodged when the soil covering washed away over the years. The rusted angle-iron in the upper-left background was found to be protruding from a small unidentified concrete vault as seen in Figure 7-72b. Equipped with three hinges for a missing lid, the purpose of this nearby structure is unknown.

Figure 7-72. a) At an elevation 175 feet, the lower WWI-era bunker served as BC_1 for Battery Woodward in WWII. Note the dislodged boulders along the south side that once helped camouflage the concrete roof.   **b)** Close up view of the small concrete vault situated northeast of BC_1, with our Unmanned Systems Compound overlooking the Pacific Ocean on the far side of Woodward Road in background left.

Figure 7-73 shows the interior of the battery commander station (BC_1) for Battery Woodward, looking in through the northwest observation slot at the hexagonal azimuth instrument post and octagonal DPF foundation. The field-telephone mounts and conduit can be seen in front of the DPF and azimuth positions along the front bulkhead of the bunker in the lower-right corner, with a third location on the south side next to the wall-mounted worktable. A 9 September 1975 Naval Electronics Laboratory Center "General Development Map" (sheet 3 of 12) shows this lower structure designated as F-20U.

Figure 7-73. Interior view of the lower bunker (base end station for Battery Woodward) looking through the north-side observation slot. The tall concrete pedestal in the foreground was the post (pier) for an azimuth instrument. On the floor behind it is the octagonal concrete foundation for a depression position finder.

The southwest interior of the battery commander station is seen in Figure 7-74a below, with a large wooden cable spool resting on the DPF foundation. The concrete pier for the azimuth instrument in the northwest corner is shown in Figure 7-74b. The wooden twist latches seen above the observation slots in both these photos were for securing wood-framed glass windowpanes in the event of rain or cold weather. One such windowpane insert is still extant in its overhead storage rack in the upper right corner of Figure 7-74c.

a) b) c)

Figure 7-74. a) The interior of the lower structure, looking southwest, with the southern steel shutter blocked by a small boulder. **b)** Interior looking northwest, with the hexagonal concrete azimuth pier on left. **c)** View of the telephone junction panel in the southeast corner, with windowpane rack on the overhead at upper right.

Site 6 (West Rosecrans)

Harbor Defense Command Post / Harbor Entrance Control Post

Army and Navy representatives met at Fort Rosecrans in early 1941 to select a suitable location for the Harbor Entrance Control Post (HECP), which required a full and unobstructed view of the harbor approaches (Thompson, 1991). One potential candidate was the old Spanish lighthouse on Cabrillo National Monument, which certainly had the view but lacked sufficient space. In September 1941, the Western Defense Command proposed placing the HECP in the Harbor Defense Command Post (HDCP), with the lighthouse serving as a supporting watchtower and signal station (Thompson, 1991).

The combined HDCP/HECP became operational later that year (Figure 7-75), located just west of Catalina Boulevard / Cabrillo Memorial Drive near the northwest corner of what is now Fort Rosecrans National Cemetery (Thompson, 1991):

> "The large bomb-proof structure was buried under a covering of earth with two large concrete-and-steel observation stations facing the sea. As first constructed, it contained the harbor defense operation post and command post, the harbor defense intelligence center, Group 1 command post,[17] Group 3 command post, and the antiaircraft intelligence center."

This was the highest elevation on Point Loma, which placed the depression position finders in the two observation posts at 421 and 427 feet above mean low water (RCW, 1943).

---

17 Note the upper observation post in the original HDCP/HECP blueprint shown in Figure 7-76 is labeled "G_2". This was changed to "G_1" in the Report of Completed Works corrected to 20 January 1943.

Figure 7-75. Observation levels of the combined Harbor Defense Command Post / Harbor Entrance Control Post as seen from the northwest corner of Fort Rosecrans National Cemetery (undated Army photo courtesy Cabrillo National Monument). Each level contained a depression position finder (DPF).

The observation levels were connected by a 40-foot descending tunnel to the multi-room subterranean bunker seen in the plan and section views presented in Figure 7-76 below. This bomb-proof structure was protected by a 2-foot-thick concrete burster course covered by 2 feet of earth, with between 4 and 7 more feet of earth between the burster course and the bunker's concrete roof, which itself was 2.5-feet thick. An 8-kilowatt AC generator in the power room provided back-up electrical service. Outside structures included a concrete pump house and the open-air Group 3 Command Post.

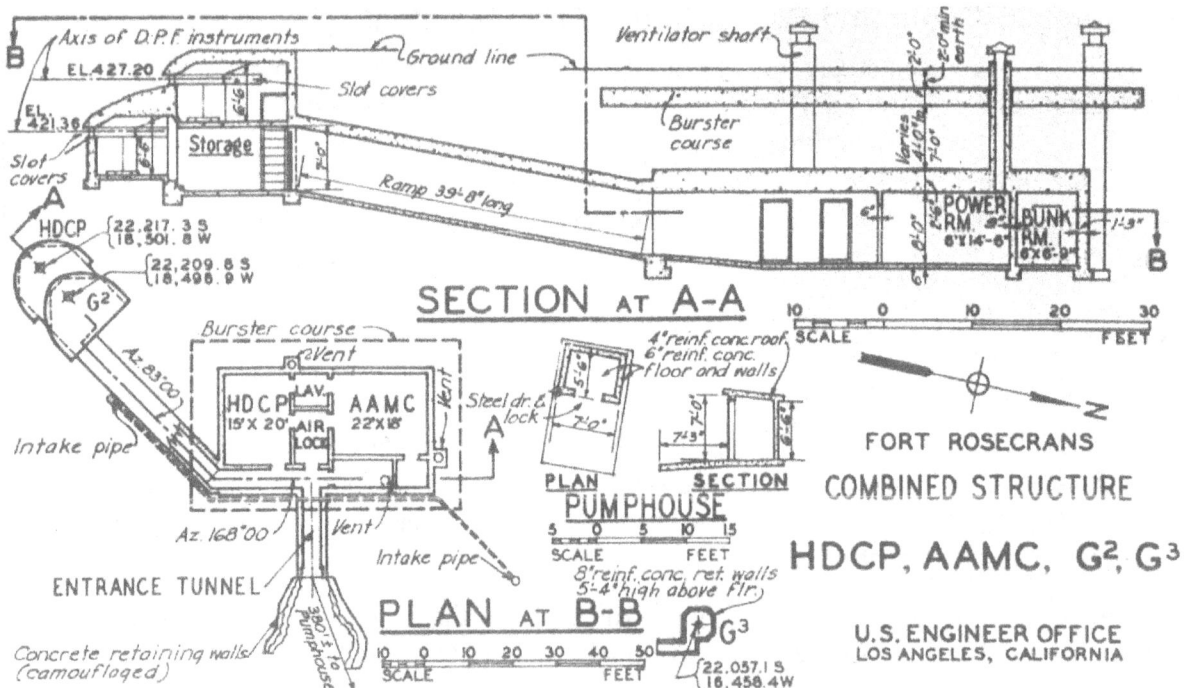

Figure 7-76. Original layout, prior to additions, of the HDCP/HECP structure from the 1942 Report of Completed Works, which also housed the Antiaircraft Message Center (AAMC) and the Group 2 (later Group 1) Command Post. The open-air Group 3 (G³) Command Post (bottom center) was about 25 yards to the northeast.

The old Spanish lighthouse on top of Point Loma was repurposed as a temporary HECP signal station just prior to the US entering WWII, as reported by Colonel George Ruhlen (2016), USA (Ret.):

> "On May 17 1941, the Secretary of the Interior issued a special-use permit to the War Department that turned over Cabrillo National Monument for military use. The Los Angeles District Engineer designed a signal mast in connection with the conversion of the old lighthouse into an HECP signal station.[18] The lighthouse now became the signal station and was equipped with a signaling searchlight, an observation instrument, and telephone communication to Battery McGrath, the alert battery,[19] and the alert searchlight. A temporary signal mast was installed, along with a set of flags for visual signaling to surface craft."

In his May 1992 oral-history interview at Cabrillo National Monument, Navy Signalman Harris Wright recalled his WWII service at the old Point Loma lighthouse (Overton, 1993):

> "Our main function out here was to challenge any ship or craft coming into this area. We did that with the international challenge sign, by INT sign, Item, Nan, Tare.[20] And they in turn would reply with what we called the recognition signal for that period of time, day or hours. And we in turn would report that to what we called HECP, down here across from the cemetery where you go back underground. Ships would request permission to enter the harbor."

According to Wright, authorized vessels were given permission to enter the harbor by signal light (Overton, 1993):

> "We had a blinker gun that you aimed that was about 4 inches in diameter. We had a 12-inch light and then we had a 24-inch carbon-arc light, the big one, which of course we only used in emergencies."

Alternatively, the interrogation and/or permission-to-enter message could be given by flag hoist via the signal mast.

---

18 The old Spanish lighthouse on Point Loma originally saw service from 1855 to 1891 (Thompson, 1991).

19 Recall from Chapter 2 that the 3-inch Battery McGrath was at the southwest end of Ballast Point, just northeast of Battery Calef-Wilkeson.

20 Wright is referring to the Navy's phonetic alphabet, a list of words (since revised) used to identify the letters assigned to flags: "Item" for I, "Nan" for N, "Tare" for T. "INT" is short for interrogative, which identifies the subsequent message as a question.

REPORT OF COMPLETED WORKS — SEACOAST FORTIFICATIONS.   COAST DEFENSES OF SAN DIEGO, CALIF.
        (Fire control or Torpedo Structures)   FORT ROSECRANS.
                                                                STRUCTURE FORT SIGNAL MAST.

Form 2.   Corrected to   June 16, 1920

Location	Crest of Point Loma near Old Spanish Light.	Type of data transmission	Telephone.
Date of transfer	June 16, 1920	Date of transfer	
Cost to that date	$ 1483.97	Cost of data transmission equipment	
Type of construction	Wooden (Oregon fir) Mast.	For tide stations give description of tide gauge	
a) Roof			
b) Remainder of bldg.			
How concealed	Not concealed.	For datum points give Forts from which visible	
How protected	Not protected.	For dormitories give stations served	
Height above concealment	80 feet.	For cable hut give S. C. type	
Height above protection	80 feet.		
Conspicuous at 25,000 yards	14 statute miles.		
Source of electric current	Power plant, Bty. White (in station)		
Kilowatts required	.32 K.W. (in station)		
Type of lighting fixtures	Water Tight (in station)		
How heated	Coal Stove (in station)		
Connected to water mains	No.		
Connected to sewer	No.		
Type of latrine	None.		
Permanent or temporary installation	Permanent.		
Present condition	Good.		
Reference of site	420 feet.		
Reference of instrumental axis	None.		
Type of observing inst.	Field glasses.		
Type of plotting board	None.		
Type and capacity of crane	None.		
Max. dimensions of reel handled	None.		

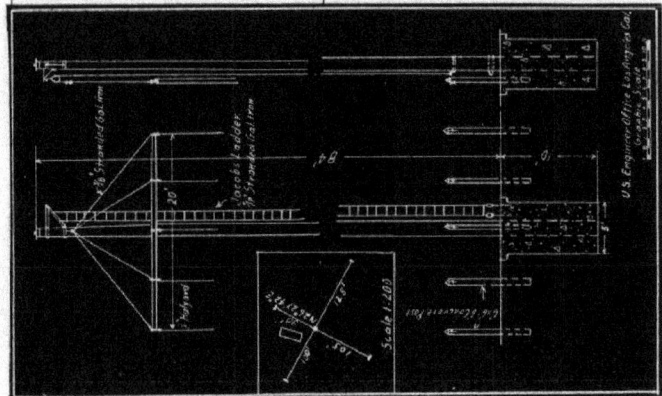

Figure 7-77. The first Fort Rosecrans signal-mast was installed in 1920, adjacent to the old Spanish lighthouse on the crest of Point Loma at an elevation of 420 feet.

According to the Report of Completed Works shown in Figure 7-77 above, the first such signal mast at Fort Rosecrans had been erected adjacent to the old Spanish lighthouse back in 1920, to take advantage of the excellent all-around visibility. The period post card reproduced in Figure 7-78 below seemingly depicts this early installation, but there are a number of issues that raise questions regarding its authenticity. For starters, the four signal-flag halyards look way too thick, and the rightmost pair appears to be strangely anchored to the roof edge, while the leftmost pair seems to descend to the ground beyond the parked trucks. All the above suggest the signal mast was poorly retouched by an artist, perhaps when the original presumably black-and-white photo was colorized.

Figure 7-78. An undated postcard depicting the first (1920) Fort Rosecrans signal mast. The actual mast location was just southeast of the lighthouse, as shown in Figure 7-79a below (courtesy Cabrillo National Monument.

A photograph that better shows the relative position of this early signal mast with respect to the lighthouse is presented in Figure 7-79a, with a similar-vintage truck seen in foreground right. The halyards on the signal mast at upper right appear thinner than those previously depicted in the presumably doctored post card, even though this mast is much closer than that seen in Figure 7-78 above. For perspective, note the significant difference in the ratio of the halyard thickness to the support-pole diameter in both images. This 1920 signal mast was no longer extant when WWII broke out.

Figure 7-79. **a)**. The first signal mast was erected just southeast of the old lighthouse in 1920. **b)** This early WWII-era photo shows a temporary signal mast atop the lighthouse, with Army and Navy personnel seen relieving the watch at bottom right (undated US Army photos courtesy Cabrillo National Monument).

Part of the WWII harbor-entry-control procedures involved opening and closing the anti-submarine nets guarding the harbor, as explained by Navy Signalman Shorty Beights (Overton, 1993):

"We had control of two anti-submarine nets. The first one was here at Ballast Point that extended across the harbor entrance. Then perhaps half a mile further inland was the second anti-submarine net. They had barges at each end with powerful winches to raise and lower these submarine nets. We had telephone circuits hooked up here from the old lighthouse down to the section base and to both of these anti-submarine-net barges. The outer submarine net had a crash boat or a ready boat tied up alongside."[21]

This outer anti-submarine net at Ballast Point is seen in Figure 7-80.

Figure 7-80. The two winch barges for the outer submarine net are seen midway between Ballast Point in the foreground and North Island at upper left, with the net still in the lowered position following passage of the vessel leaving the harbor at far right (undated US Army photo courtesy SSC Pacific).

The temporary HECP signal station was relocated in 1944 from the old Spanish lighthouse to a purpose-built wooden tower constructed above the new Group 4 Command Post in Site 7,[22] as recalled by Navy veteran Pete Cutri (Overton, 1993):

"Because of the Army, we built a new tower beyond the lighthouse, south of it. The point goes north and south. I used to be a navigator in the tuna fleet and I can remember some of these things. Immediately below the four legs of the tower was an entrance to an underground... what do you call it? It was a command post, sort of, that's where the radar was..."

The HDCP was enlarged in early 1942 after Fort Rosecrans commanding officer Colonel Ottosen complained it was too small (Figure 7-81), with the Army and Navy radio operators having to compete with all the other activities going on in the same room (Thompson, 1991):

---

21 The crash boat was dispatched by the signal-station crew anytime an Army aircraft went into the water while attempting to take off or land from Rockwell Field, now North Island Naval Air Station.

22 The Group 4 Command Post is discussed later in the Site 7 section of this chapter entitled "Battalion 1 Command Post and Signal Station."

"As a result of his letter, a radio room was added to the HDCP and a concrete dugout signal station was constructed south of the lighthouse.[23] Another enlargement of the command post occurred in 1943 when General DeWitt, Western Defense Command, paid a visit to San Diego. At the HDCP he was unhappy that there was no room wherein senior Army and Navy officers could hold private conferences. The Chief of Engineers promptly made $11,600 available for the additional bomb-proof room."

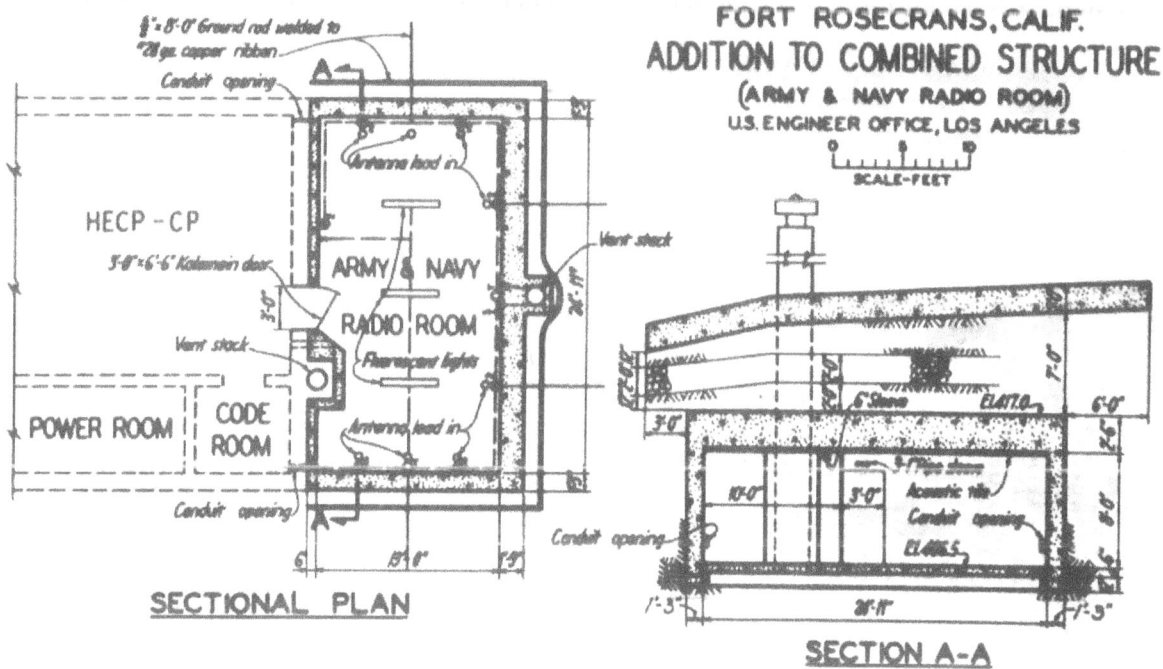

Figure 7-81. Plan and section drawings of the 1942 expansion of the underground HDCP/HECP structure (left side) to the Army and Navy Radio Room at right (now Room 107 of Building 557 at the Center). Entry was via a 36-inch Kalamein (fireproof) door on the north side of the original structure.

This second addition (now Room 101) is seen in the post-war floorplan for Building 557 presented in Figure 7-82 below.

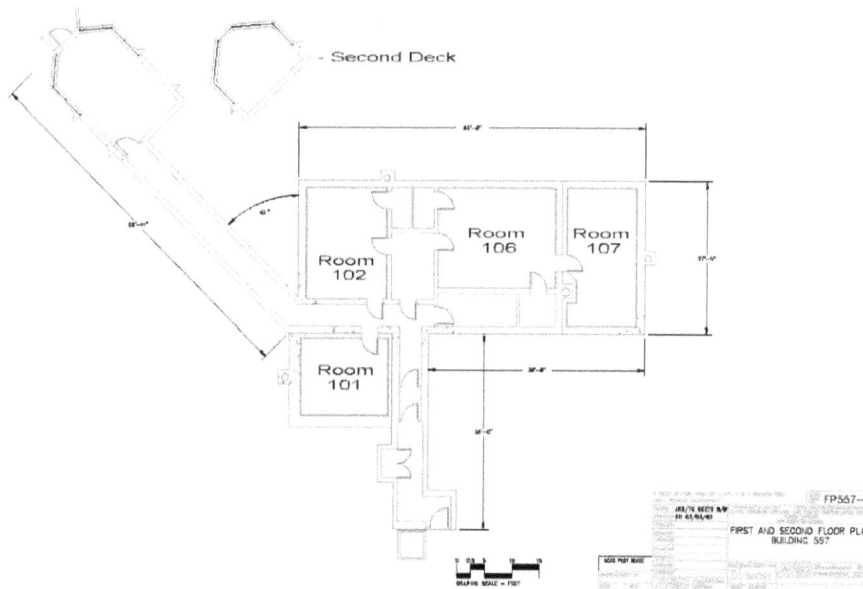

Figure 7-82. The floorplan of the former HDCP/HECP, now Building 557 at SSC Pacific. The Army-Navy radio-room is now Room 107. The second addition to this structure, Room 101 at lower left, was used as a conference room for senior Army and Navy officers in WWII (US Navy drawing courtesy SSC Pacific).

23 The new signal tower was directly above the Battalion 1 Command Post and Signal Station in Site 7.

The map seen in Figure 7-83 below, which was cropped and slightly recomposed for better presentation, shows the permitted lease for a post-war Navy RACON (radar beacon) installation on the east side of the old HDCP/HECP. The RACON concept involved a transponder attached to certain structures or potential obstacles in maritime environments, so as to ensure their radar detection and positive identification. The idea likely evolved from the introduction of IFF (Identification Friend or Foe) radar transponders to identify friendly aircraft during WWII. Note the long underground tunnel leading from the HDCP/HECP to the covered drive-in parking garage on Catalina Boulevard / Cabrillo Memorial Drive at far right.

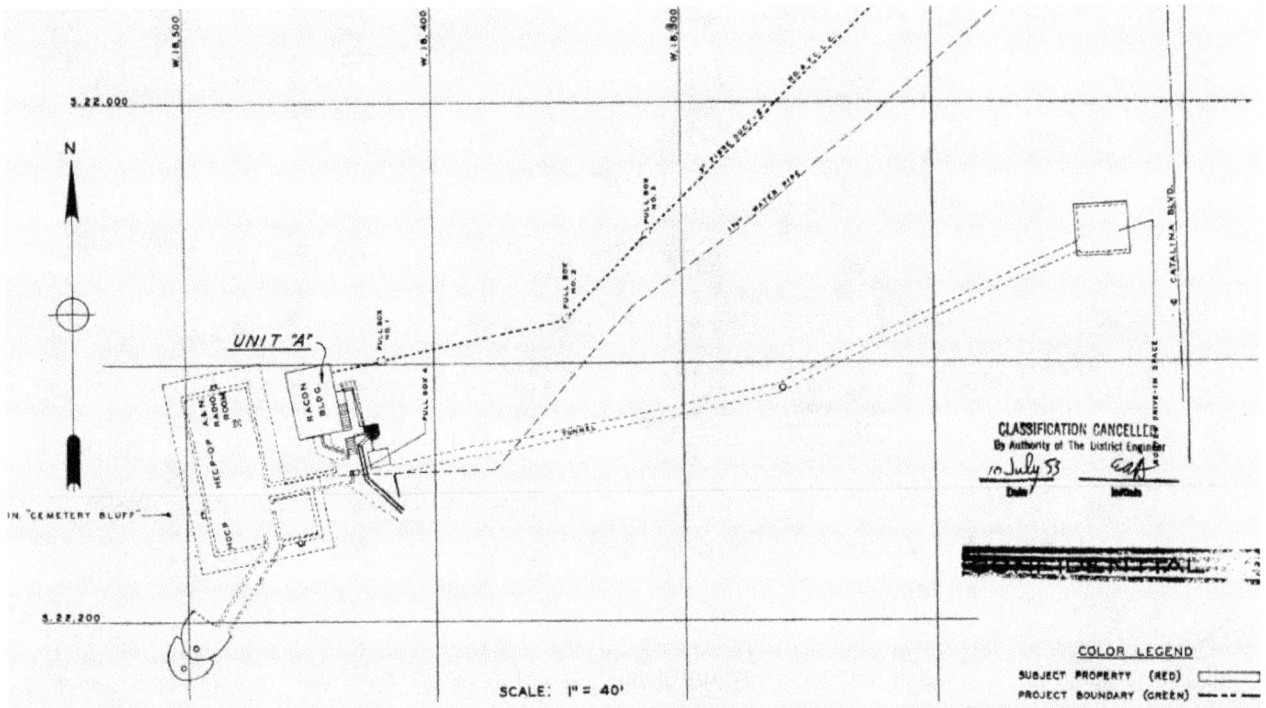

Figure 7-83. The tunnel from the covered drive-in garage (upper right) on Catalina Boulevard is shown in this zoomed-in portion of a more recently updated 1946 drawing detailing the location of the RACON "Unit A" addition (Building 567) next to Building 557, the former HDCP/HECP (US Navy drawing courtesy SSC Pacific).

The roof of this drive-in parking garage on the west side of Catalina Boulevard / Cabrillo Memorial Drive is barely visible in the bottom left quadrant (left up arrow) of the undated post-war aerial photo presented in Figure 7-84 below.[24] Note the dirt access road leading to the HDCP/HECP entrance (bottom-right up arrow), and the observation levels to the southwest (top-right up arrow). Trace evidence of the underground entry tunnel can still be seen just beyond this dirt road, starting at the southwest corner of the garage (left up arrow).

---

24 This photo was taken prior to 1965, when Fort Rosecrans National Cemetery expanded west of the road, as seen in the 1966 photo presented later in Figure 7-86.

Figure 7-84. The HDCP/HECP garage (left up arrow), above-ground entryway (middle up arrow, lower right), and observation levels (right up arrow) are seen in this post-war aerial photo looking southeast. The octagonal structure at extreme lower left is fresh-water Reservoir 423 (undated US Navy photo courtesy SSC Pacific).

The undated post-war photo presented in Figure 7-85 shows the construction of a paved driveway for Buildings 557 and 567, looking east with San Diego Bay and North Island just visible through the fog in background. This upgrade destroyed the underground tunnel to the old drive-in parking garage, the roof of which can be seen on the west side of Catalina Boulevard / Cabrillo Memorial Drive at right center (up arrow). On the far side of the newly graded driveway shoulder, slightly below and to the right of the up arrow, is an exposed portion of the remaining tunnel section leading back to the garage.

Figure 7-85. The connecting tunnel between the HDCP/HECP and its covered garage (up arrow) was destroyed during installation of a new concrete driveway. An exposed portion of the severed garage connection is visible beyond the driveway shoulder below the up arrow (undated US Navy photo courtesy SSC Pacific).

This new driveway for Buildings 557 and 567 is seen at bottom right in Figure 7-86 below, with the observation levels of the former HDCP/HECP indicated by the up arrow at lower left. The concrete roofs for these two bunkers are now exposed due to erosion of their earth covering since the war. The covered parking garage for this site, if still extant, would have been just out of the image to the south on the west side of Catalina Boulevard / Cabrillo Memorial Drive at bottom right. Note Gatchell Road in upper left corner, which intersects Woodward Road at top center, and octagonal fresh-water Reservoir 423 at bottom right.

Figure 7-86. The HDCP/HECP observation levels (up arrow) are adjacent to the western expansion of Fort Rosecrans National Cemetery at lower left (undated, but post 1965, US Navy photo courtesy SSC Pacific). The above-ground entrance to the structure is behind the white vehicle parked on the dirt-covered asphalt this side of Building 567.

Figure 7-87 shows a close-up view of the above-ground HDCP/HECP entrance (Building 557) adjacent to the RACON building (567), circa September 1966, with the northwest expansion of Fort Rosecrans National Cemetery at upper left. Just to the right of the northern cemetery fence line are the exposed observation levels (top up arrow) of the HDCP/HECP. The antenna tower adjacent to the top-level bunker suggests the underground structure has been repurposed for some cold-war function. Note the remains of a defensive machine-gun emplacement (bottom up arrow) midway between Building 567 and the cemetery fence.

Figure 7-87. a) Looking west in September 1966, the remains of a defensive emplacement (lower up arrow) are seen left of the HDCP/HECP entrance (Building 557). The observation levels (top up arrow) are north (right) of the cemetery (US Navy photo courtesy SSC Pacific). **b)** The ground-level entrance circa October 2007, with Building 567 in upper background beyond Building 557.

Figure 7-88 shows a new radome installation behind the dug-in bunker, looking northeast during our site visit in October 2007. Note also the post-war addition of a large exterior door on the center portion of the lower-level HDCP, the purpose of which is unknown, as the steep terrain on this side of the structure was rather difficult to traverse. It's a bit wider than necessary for an emergency exit, which suggests it may have served as an access for some craned-in piece of post-war equipment.

Figure 7-88. View of the HDCP/HECP observation levels as seen during our site visit in October 2007, showing the post-war retrofit of a large exterior door on the center face of the lower HDCP level. The white radome above and behind the structure, not present in Figure 7-87c, was added sometime after 1966.

The short hallway inside Building 557 leading to the connecting tunnel for the HDCP/HECP observation levels is seen in Figure 7-89a, with the former conference room (now Room 101) on the left side and the HDCP (now Room 102) on the right. Figure 7-89b shows the 40-foot ascending tunnel leading to the bottom-level storage room of the observation bunker (see again section view of Figure 7-76). Note the fluorescent lighting fixtures on the concrete overhead and the cable channel running along the top of the right-hand wall.

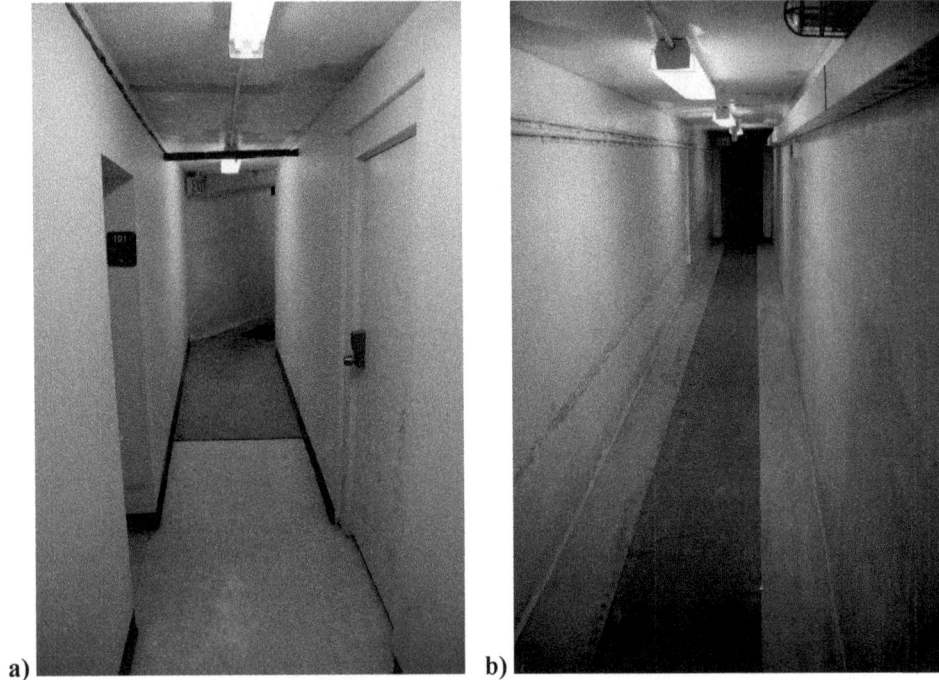

a)   b)

Figure 7-89. **a)** Looking south in the hallway towards the connecting tunnel leading to the HDCP/G[1] observation levels, with the entrance to Room 101 (former conference room) on the left and the closed doorway to Room 102 (former HDCP) on the right. **b)** The ascending 40-foot tunnel is seen looking southwest.

Figure 7-90a shows the interior of the bottom-level storage room directly below the Group 1 Observation Room (see again upper-left corner of Figure 7-76), with what appears to be a post-war door upgrade. The access ladder to the upper G[1] level is just out of image to the left. The telephone-circuit terminal box is right of the doorway on the southeast wall, with a close-up view presented in Figure 7-90b. Note the severed feed from the Army Tactical Cable near the top of the wall at upper left, and the newer cable above it coming down from a hole in the overhead.

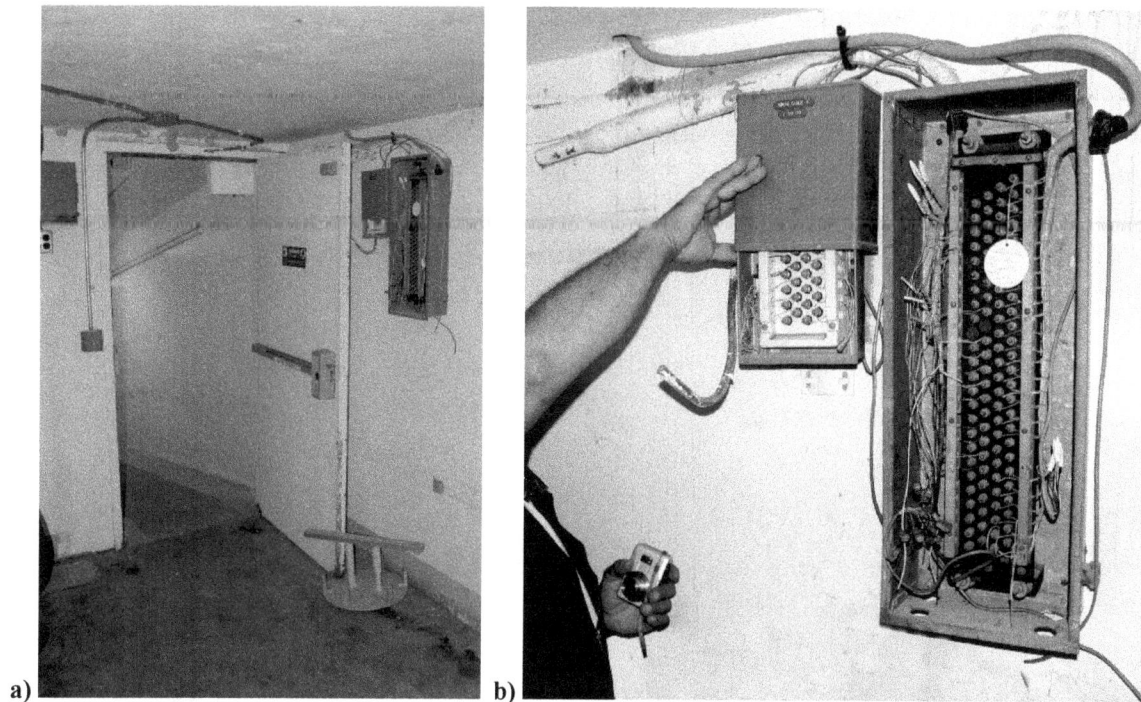

Figure 7-90. a) The storage room on the bottom level behind the HDCP Observation Post, with the telephone-circuit terminal box on the wall to right of the tunnel-door entry. **b)** Close-up of the original terminal panel (right), with a more recent terminal box to its left labeled "House Cable 4: 1 to 10."

Figure 7-91a, looking south from the storage room, shows how the concrete stairs that led up into the HDCP observation room have been removed and the floor cut down to accommodate the new exterior doorway. A post-war exit sign is attached to the sloped overhead in front of this door, with what appears to be a new fluorescent light fixture at top center. Made from a section of ½-inch pipe, the curtain rod mounted above the interior doorway was for the WWII-era light-trap curtains. Figure 7-91b shows the access ladder leading up to the northwest corner of the Group 1 observation level.

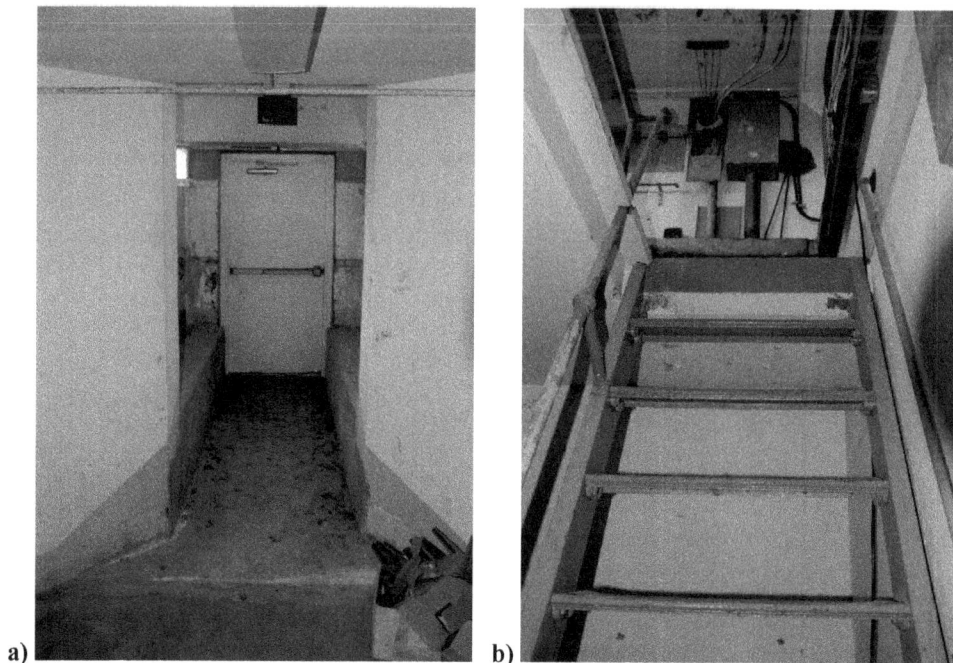

Figure 7-91. a) View from the storage room looking southwest, showing the concrete steps removed and the floor of the higher HDCP observation level cut down to accommodate the new exterior door. **b)** View looking southwest up the ladder to the Group 1 observation level above the storage room.

Figure 7-92 shows the interior of the upper-level Group 1 observation room, looking west. A number of changes have been made over the years to accommodate post-war tenants, such as the compressed-air supply on the west wall at photo center. The steel shutters have been replaced by glass windows, with a ventilation-fan installation in the west viewing slot. The top cable-entry port in the upper-right corner has eight truncated steel electrical conduits that presumably once provided power to outside equipment. The middle and bottom ports appear to be the source of electrical power into the structure.

Figure 7-92. Interior of the Group 1 observation level, looking west over the ladder from the storage room. All the original steel shutters in this room have been replaced by glass windows except the one at top center, which has an aluminum insert supporting a ventilation fan. Note the fluorescent-light fixtures at upper right.

Site 7 (Cabrillo)

The fire-control structures located in Site 7 discussed in this section include: 1) the combined battery commander station BC_3 and base end and spotting station B^1_3 S^1_3 for Battery Ashburn; 2) base end and spotting stations for Batteries Humphreys, Grant, and Woodward; and, 3) the combined Battalion 1 Command Post, Signal Station, and *SCR-682* harbor-surveillance radar.

Battery Commander and Base End Stations for Battery Ashburn

The combined battery commander station BC_3 and base end and spotting station B^1_3 S^1_3 for Battery Ashburn (Tac. No. 3) is located northwest of the old Spanish lighthouse at Cabrillo National Monument, as seen in the upper right corner of Figure 7-93 below. Connected to commercial power with a backup 3-KVA generator, this two-level dug-in bunker was transferred to the Coast Artillery on 27 June 1944, with a final construction cost of $12,713.05. The steel-roofed Fort Rosecrans Meteorological Station, which had previously been the Army Radio Station, is just upslope to the east, 300 feet from the old Spanish lighthouse.

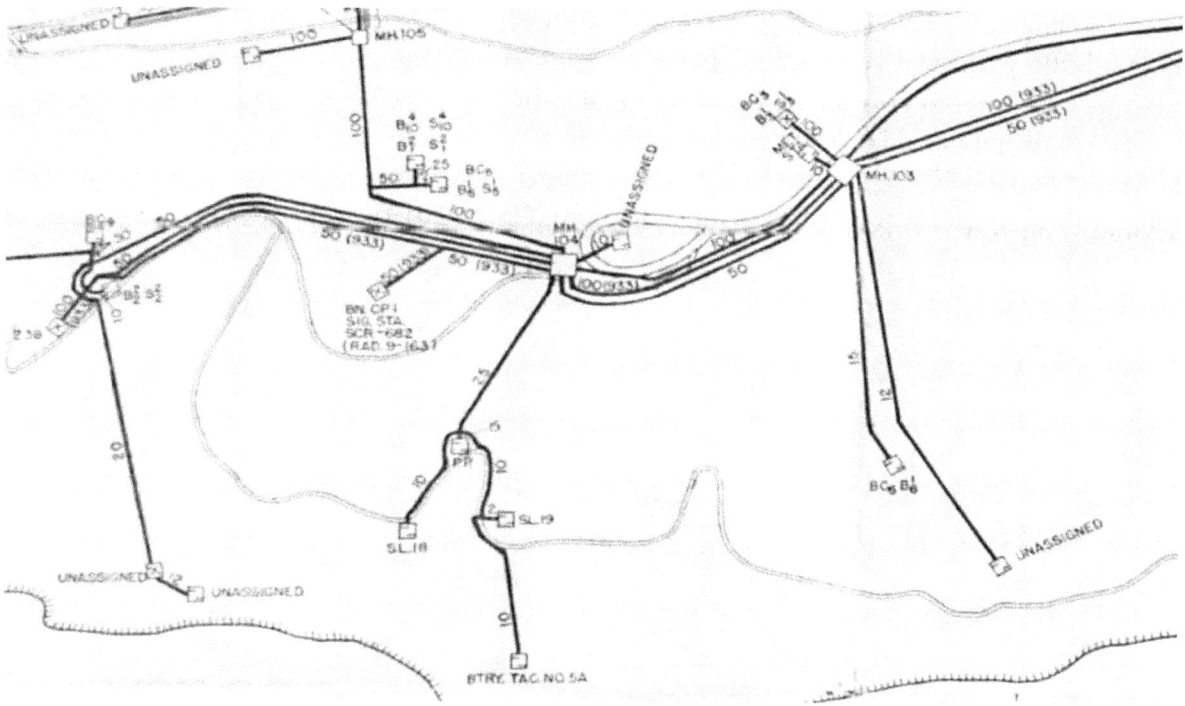

Figure 7-93. The battery commander station BC$_3$ and base end station B$^1_3$S$^1_3$ for Battery Ashburn (upper right corner) were located northwest of the old Spanish lighthouse ("unassigned" at top center) at an elevation of 408 feet. The Fort Rosecrans Meteorological Station is seen adjacent to Manhole 103 in the upper-right corner (US Navy blueprint courtesy SSC Pacific).

This somewhat unusual bunker was accessed from above via a concrete stairwell leading to the upper observation level, inside of which was a ladder down to the bunk-room level in the very bottom as seen in Figure 7-94. From the bunk room, a short set of stairs led back up to the lower-level B$^1_3$S$^1_3$ observation room, which was equipped with a depression position finder and two *M1910A1* azimuth instruments. BC$_3$ in the upper observation room had an azimuth instrument only. The dug-in structure had no provision for heat and a single 4-inch porcelain light fixture in the bunk room.

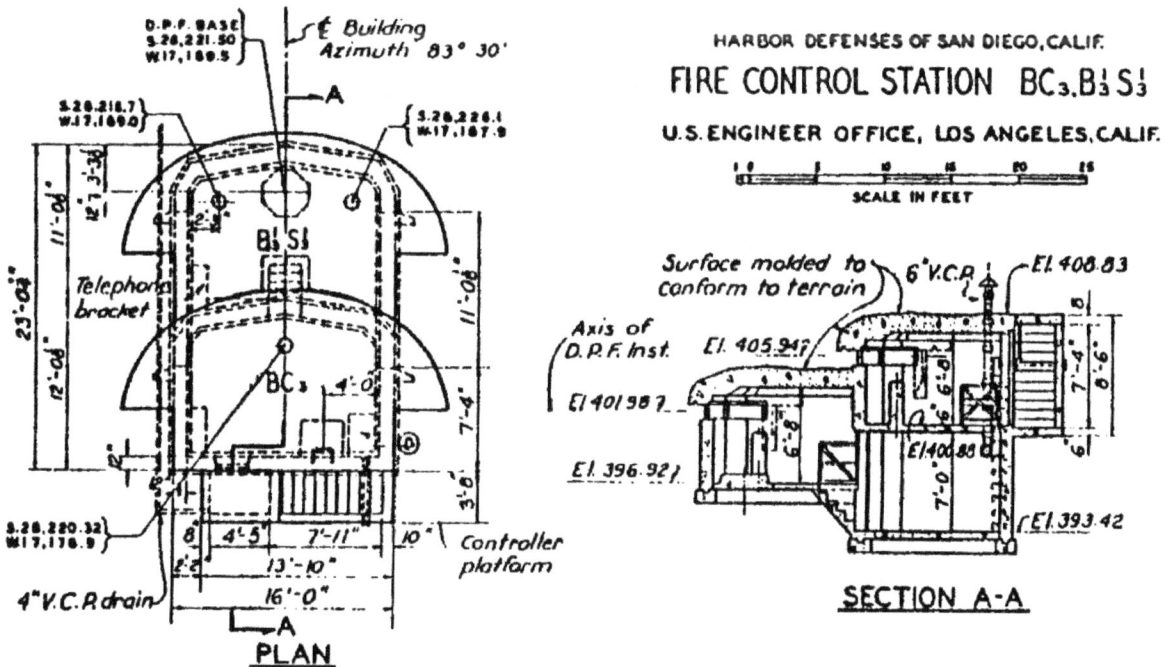

Figure 7-94. Plan and section views of battery commander station BC$_3$ and base end and spotting station B$^1_3$S$^1_3$ for Battery Ashburn (NARA). The bunk room is the lowest level seen in the section view on the right.

The access stairwell leading down to the upper-level observation room of the battery commander station BC$_3$ for Battery Ashburn is seen shrouded in the early morning marine-layer mist in Figure 7-95. The protective pipe railing on top was more recently added by the Cabrillo National Monument for visitor safety. Open to the public, this bunker has been fully restored to showcase its original WWII appearance, which includes two *M1910A1* azimuth instruments, a target-recognition poster, period telephone sets, and three vintage double-rack WWII bunks in the bunkroom (CMN, 2019).

Figure 7-95. The stairwell for the combined battery commander station BC$_3$ and base end and spotting station B$^1_3$S$^1_3$ below the Army Meteorological Station (now the Military History Museum at Cabrillo National Monument), looking southwest. A 3-inch-high antenna foundation with four ½-inch bolts is behind the sign.

Figure 7-96 below shows a section view of the entry stairwell to the bunker, drawn by Lee Guidry in 1992, looking west towards the Pacific Ocean. Note the previously mentioned 3-inch-high antenna foundation on top of the rough-textured concrete roof. The louvered steel door just left of image center is portrayed in the partially open position to reveal the BC$_3$ observation slot. The upper portion of the 6-inch ventilation duct on the west side of the entrance to BC$_3$ has been sealed off, and the vent opening in the northwest corner of the lower-level bunk room plugged with concrete, presumably for safety reasons.

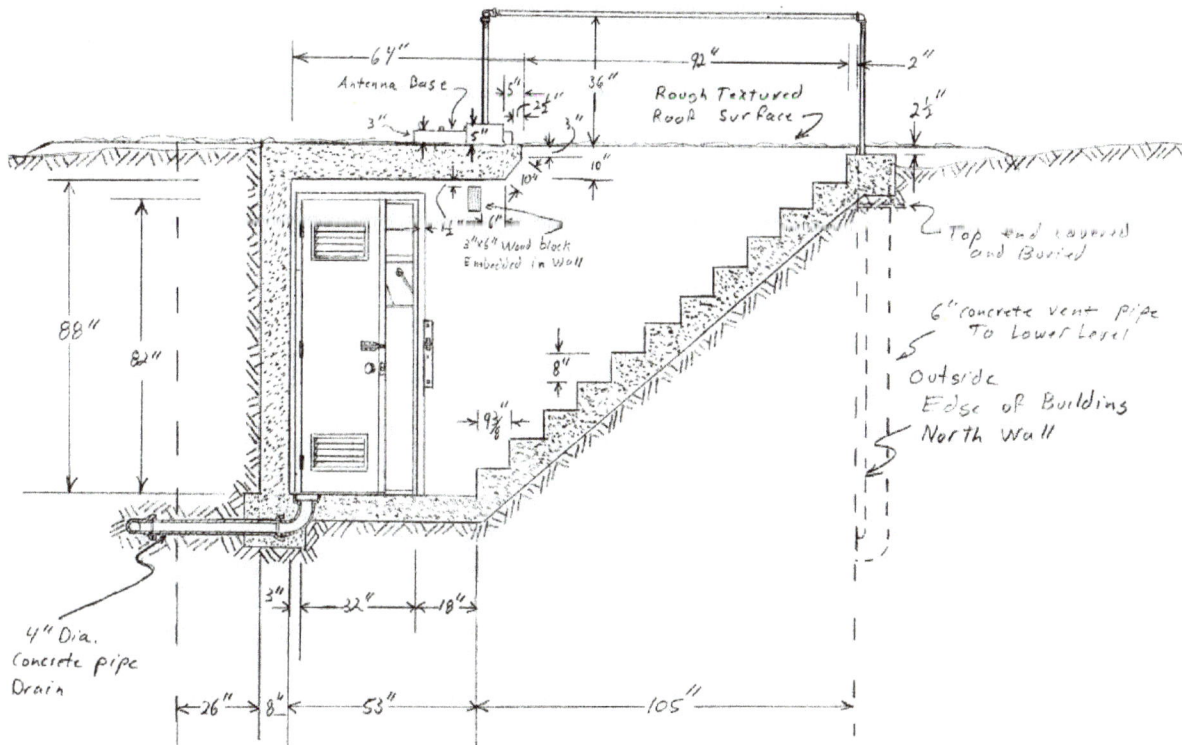

Figure 7-96. Section view looking west of the descending stairwell entry to the upper-level BC$_3$ observation room, which was behind the louvered steel door just left of center (1992 field sketch by Lee Guidry).

The rough-textured concrete roofs of the battery commander and base end and spotting stations B$^1_3$ S$^1_3$ for Battery Ashburn are seen in Figure 7-97. The U-shaped loops of steel re-bar embedded in the concrete were for attaching the camouflage netting. The former Marine Mammal Lab of the Naval Undersea Center (NUC) was once located in the small compound seen along the coast behind Cabrillo Road (upper) and Gatchell Road (lower) in the background at top center. The Coast Guard lighthouse is out of the image to the south at upper left.

Figure 7-97. View looking towards the tide pools beyond Cabrillo and Gatchell Roads on the southwest tip of Point Loma. The upper-level of the bunker is the battery commander station BC$_3$ for the 16-inch Battery Ashburn, while the lower level is the battery's base end and spotting station B$^1_3$ S$^1_3$.

The foreground of Figure 7-98 provides a close-up of the small boulders embedded in the rough-textured concrete roof of the lower base end and spotting station B^1_3 S^1_3. Note the sandbag impressions beneath the camouflage-net retaining loops along the roof perimeter for the upper-level BC_3 observation room. The steel shutters for the observation ports were all locked or rusted shut in the closed position during my site visit in May 2006, but the northwest view port at far left has since been opened after bunker restoration by the Cabrillo National Monument.

Figure 7-98. Right-front view of the upper-level observation room of battery commander station BC_3 for Battery Ashburn, with its steel shutters locked in the closed position in May 2006. Following the recent restoration of the bunker, the far-left shutter has since been reopened for park visitors.

Base End Stations for Batteries Humphreys, Grant, and Woodward

These two fire-control bunkers are located in Site 7 on the west side of Point Loma, approximately 580 feet south and slightly west of the old Spanish lighthouse in the turnaround loop below image center in Figure 7-99, just below the Whale Overlook at Cabrillo National Monument. For perspective, the 16-inch Battery Ashburn (Tac. No. 3) is seen at upper right in Site 6, with the 6-inch Battery Humphreys (Tac. No. 5) in Site 8 at left center. AMTB Battery Cabrillo (Tac. No. 4), also in Site 8, is in the upper-left corner to the west of Battery Humphreys.

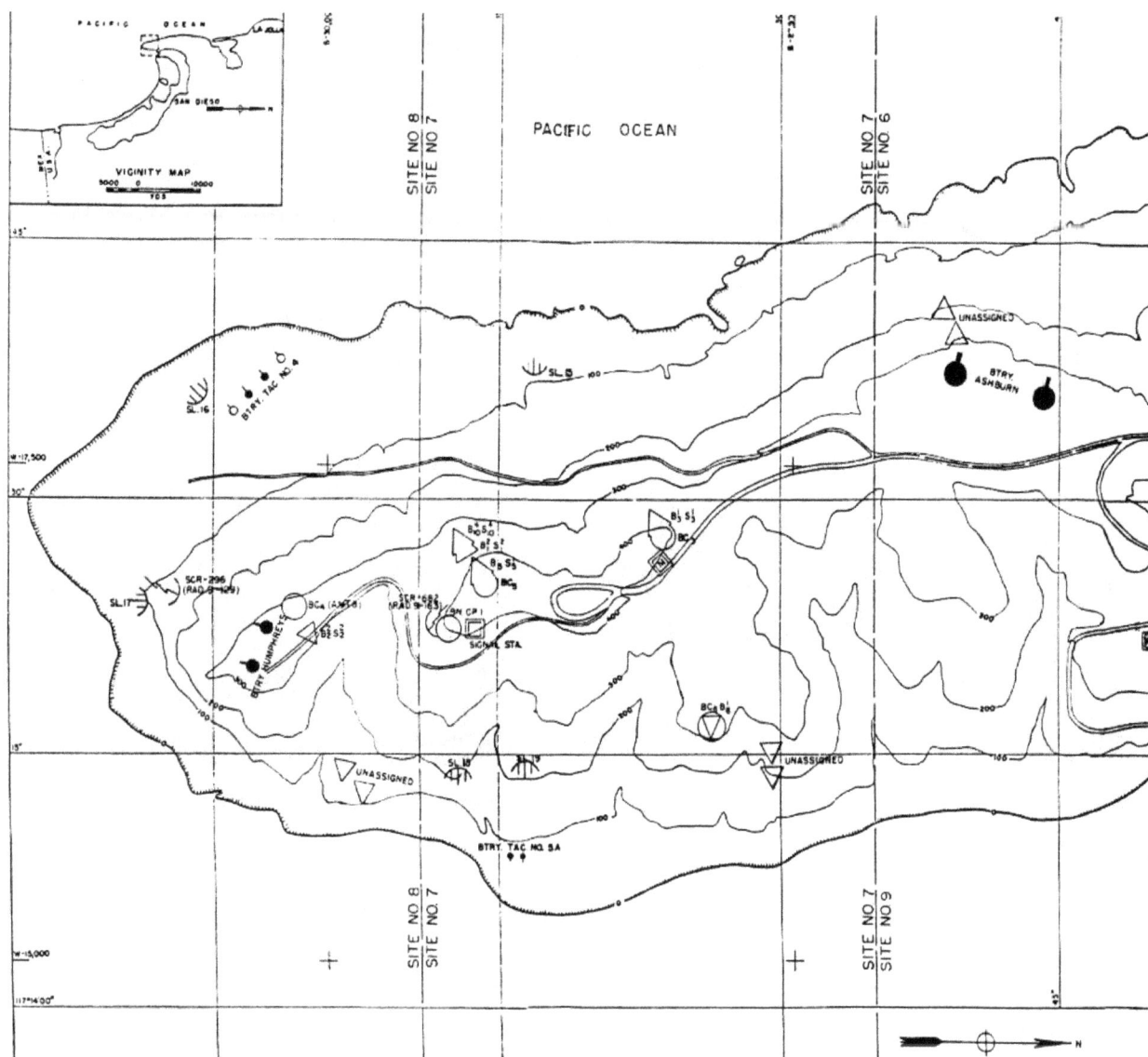

Figure 7-99. 1 July 1945 map of fire-control stations in Sites 6, 7, 8, and 9 at Fort Rosecrans. The base end stations for Batteries Humphreys, Grant, and Woodward are left of image center near the southern boundary of Site 7 (US Army map courtesy CDSG).

The upper structure seen in the foreground of Figure 7-100 served as the battery commander station BC_5 and base end and spotting station $B^1_5 S^1_5$ for the nearby 6-inch Battery Humphreys at the southern tip of the point. The lower bunker at background left comprised the base end and spotting stations $B^2_1 S^2_1$ for Battery Woodward and $B^4_{10} S^4_{10}$ for Battery Grant. The small compound seen in the upper-left corner is the former location of the Naval Undersea Center's Marine Mammal Lab. Note the external ladder rung next to the entry hatch, which served as a safety handhold when transitioning to or from the vertical ladder inside.

Figure 7-100. The structure in the foreground served as the battery commander station (upper level) and base end station (lower level) for Battery Humphreys, circa 2005. The second structure just below and to the left comprised base end stations for the 6-inch Batteries Woodward (upper level) and Grant (lower level).

Figure 7-101 shows a section view of the lower structure, which was transferred to the Coast Artillery on 1 February 1943 at a final cost of $2,688.21. Each observation level was equipped with an octagonal concrete foundation for a depression position finder, with the steel support ring for the semicircular wooden benches still intact as shown in Lee Guidry's 1992 field sketch. While the Report of Completed Works originally indicated there was no electrical service, Guidry found electrical and telephone feeds entering through steel conduits on the northeast wall on the left side of the drawing, with the fuse box and telephone distribution panel in the lower-level bunkroom.

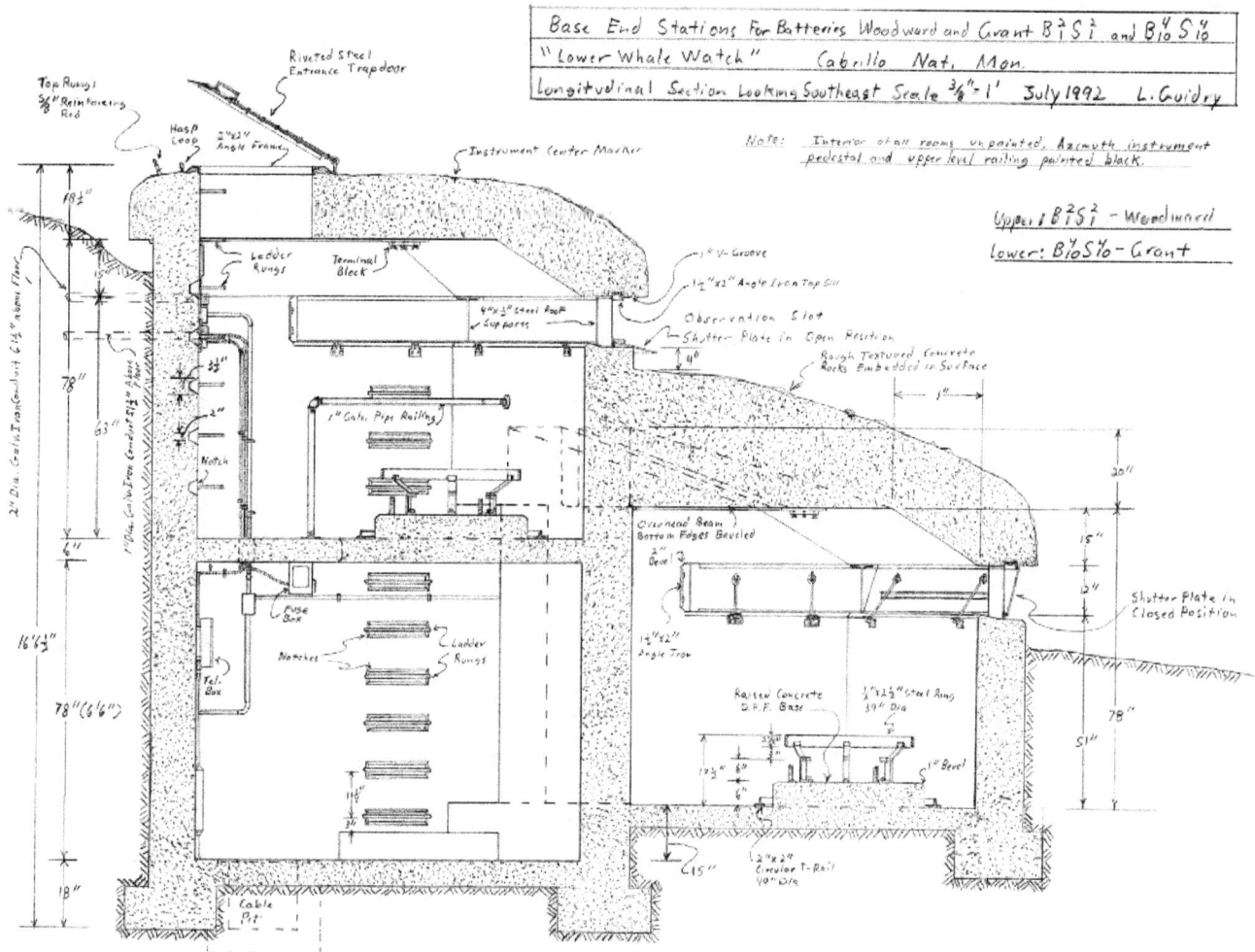

Figure 7-101. Section view of the base end and spotting stations B^2_1 S^2_1 for Battery Woodward (upper level) and B^4_{10} S^4_{10} for Battery Grant (lower level) below the Whale Watch at Cabrillo National Monument (1992 field sketch courtesy Lee Guidry). Note the DPF foundations in both observation rooms.

The top level of the dug-in bunker depicted in Figure 7-102 served as battery commander station BC_5 for Battery Humphreys.[25] This installation was structurally identical to that shown earlier in Figure 7-101 above, but with a few minor differences regarding interior details. The telephone distribution panel was in the upper-level observation room versus the bunk room, and there was a DPF foundation in the lower-level observation room with azimuth-instruments in both levels. A 4-inch-conduit entry was found on the southeast wall of the upper-level observation room, with the electrical fuse box in the bunk room below.

---

25 This RCW image shows Battery Humphreys as Tac. No. 4, which was changed to Tac. No 5 when AMTB Battery Cabrillo was designated Tac. No. 4 in 1943.

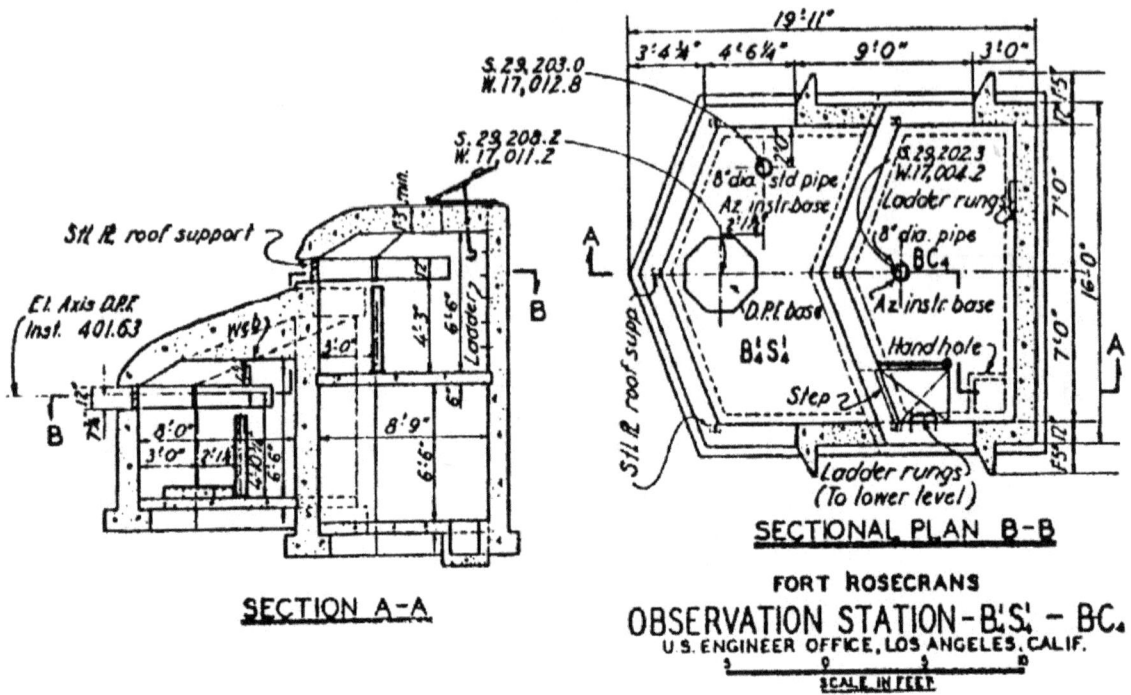

Figure 7-102. Plan and section views of the upper bunker's base end and battery commander stations for Battery Humphreys from the Report of Completed Works corrected to 18 February 1943 (NARA). When AMTB Battery Cabrillo came online as Tac. No. 4 in August 1943, Battery Humphreys became Tac. No. 5.

Figure 7-103 below shows the top-level battery commander station BC_5, with a present-day radio tower at upper left on top of the Battalion 1 Command Post and Signal Station to the southeast. Situated at an elevation of 400 feet, this bunker was transferred to the Coast Artillery on 1 February 1943, with a final construction cost of $2,688.22. Both this upper structure and the previously discussed one below it were found to be well maintained by Cabrillo National Monument, but with shutters and access hatches locked to where no entry was possible during my site visit in May 2006.

Figure 7-103. This is a view of the northwest side of the top level of the upper structure, which served as battery commander station BC_5 for Battery Humphreys. The post-war antenna tower in background left is adjacent to the Battalion 1 Command Post and Signal Station discussed in the next section.

Battalion 1 Command Post, Signal Station, and SCR-682 Radar

The combined Battalion 1 Command Post (BNCP1, previously the G_4 Command Post) and HECP Signal Station are located on top of the Point just south of the old Spanish lighthouse at Cabrillo National Monument (Figure 7-104). Flower and Roth (1982) claim this structure was built between 1916 and 1920 and describe it as follows:

"It is an octagonal concrete building approximately 20 feet across. The roof is concrete, some 4 feet thick, and is covered with boulders. The structure is built into a bank on its north side with glass windows on the south, east, and west. A large underground bunker lies directly to the north of the structure."

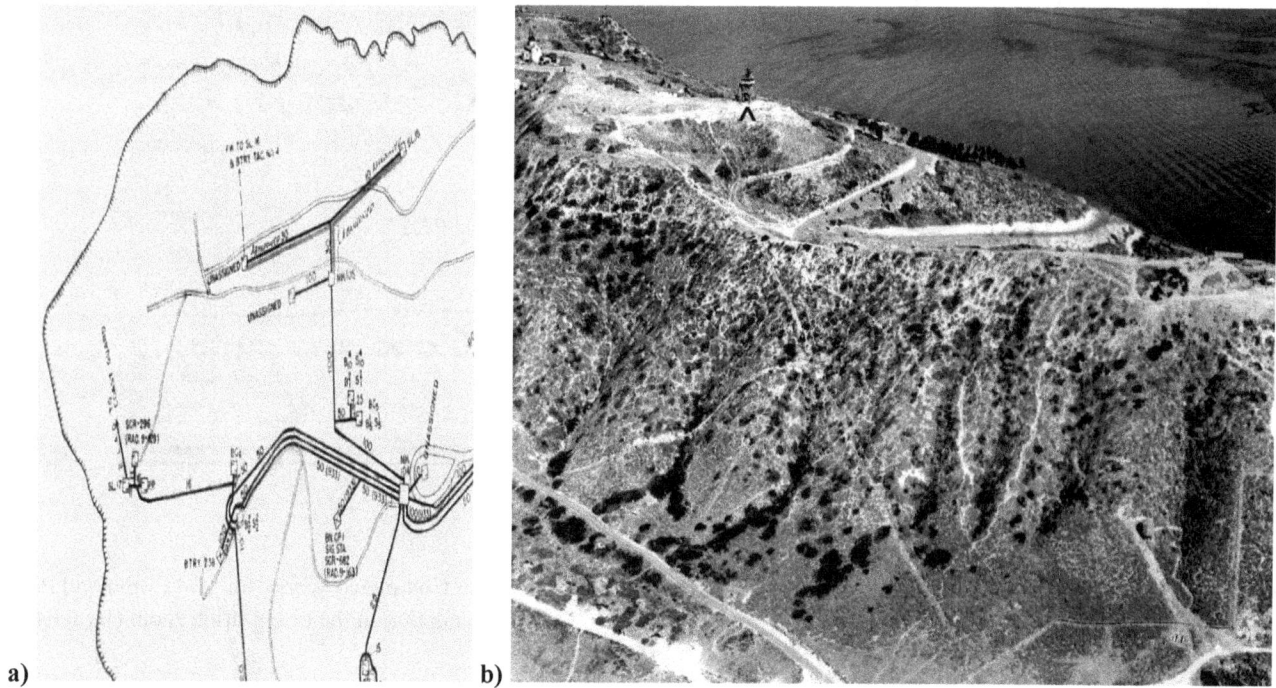

Figure 7-104. a) The WWII Battalion 1 Command Post and Signal Station were just south (left) of the old Spanish lighthouse (unassigned manhole in turn-around loop at lower right). **b)** The new Signal Station (up arrow) is south of the lighthouse, which is seen in the extreme upper-left corner, circa December 1945. Note the previously discussed signal tower near the lighthouse (Figure 7-79c) has been removed.

The "octagonal concrete building" and the "underground bunker" are both part of the same structure, which the Report of Completed Works shows to have actually been constructed during WWII and turned over to the Coast Artillery on 19 December 1944. In addition, the observation room is not octagonal, but instead retangular, with eight connected facets forming a semicircular southeast face as seen in the plan and section views of Figure 7-105.

The *SCR-682* harbor-surveillance radar equipment was located in the Army Operations room at the southeast end of the bunker, with the antenna at the very top of the new signal tower above. The southern entry hatch was situated on the east side of the bunker, midway between the Army and the Navy Operations rooms, with the northern hatch on the west side directly below the tower. To aid in concealment, the 12-inch-thick concrete roof of the subterranean structure was covered over with small boulders and earth as seen in the section view.

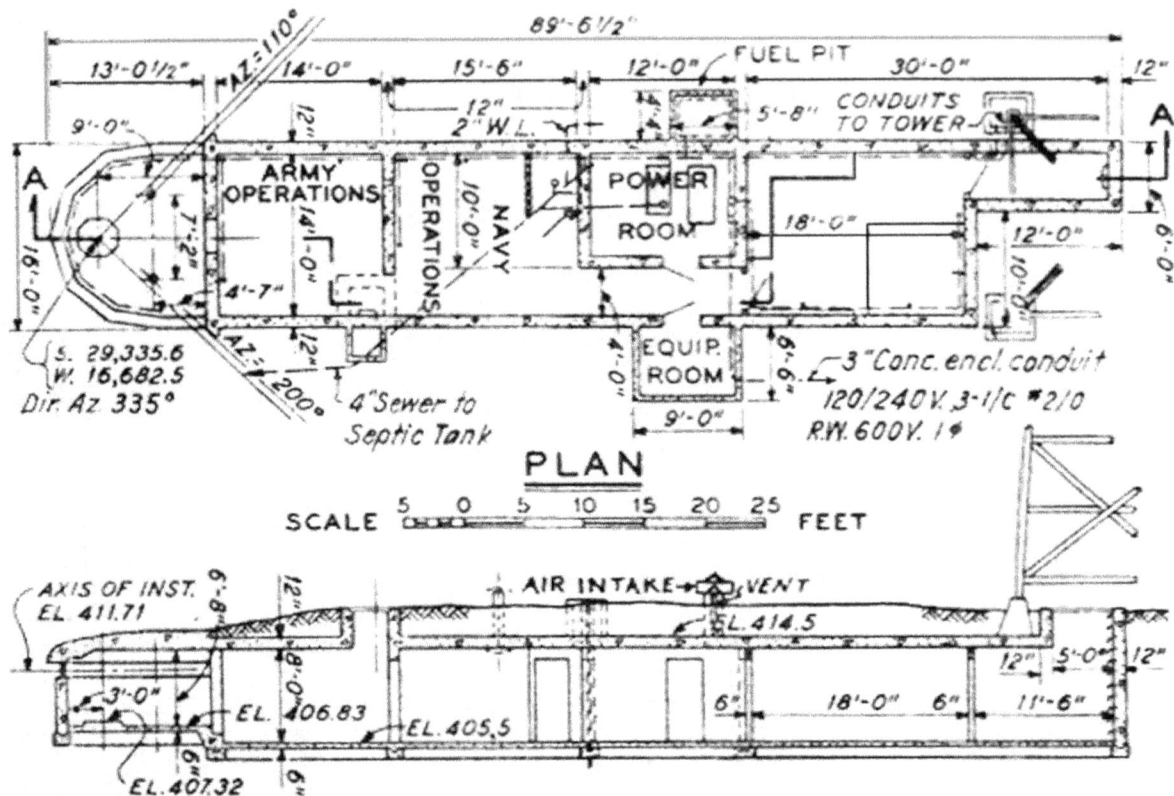

Figure 7-105. Plan and section views of the bunker, extracted from the Report of Completed Works for "G4 Command Post – Harbor Defense Signal Station," corrected to 16 November 1945. Note DPF foundation in the observation room (far left) and the new signal tower directly above the northwest hatch (far right).

In addition to manning the new signal tower, the Navy also oversaw the underwater harbor defenses, as recalled by WWII Navy veteran Shorty Beights (Overton, 1993):

> "They had transmitters and receivers hooked onto the buoys out here... These sound men would try... They would go to school and listen to all these prop noises. Learn how to identify a diesel, an electric, or... I remember there was a mustang in charge.[26] A real nice guy."

This school Beights mentioned was probably the Naval Training School (Harbor Defense) on Fishers Island in Suffolk County, NY (NTS, 2019).

The "transmitters and receivers" mentioned by Beights would seemingly imply this was an active harbor-defense sonar system called *Herald* (Walding, 2019):

> "The *Herald* – Harbor Echo Ranging and Listening Device – is a special sea-bed-mounted form of the ship-borne sonar. It was developed by the Royal Navy towards the end of WWI and first used in 1939. In the USA, experiments continued at Columbia University, which led to the formation of research teams combining the US Navy, the Submarine Signal Company, General Electric and Western Electric at Nahant, MA. In the US it was called 'Herald' instead of the British name 'Harbour Defense Asdic.'"

The shore-based *Herald* operator would trigger a short 15-kHz "ping" that would theoretically reflect off a submarine and create a telltale echo (Walding, 2019).

---

26 Mustang – a naval officer with former enlisted service.

Chapter 16 of *Naval Sonar*, entitled "SOFAR, Harbor Defense, and Other Sonar Systems," states WWII-era harbor detection systems were usually composed of three lines of defense (BNP, 1953):

1. Magnetic loops, which were the most dependable and required the least operator attention.
2. Cable-connected hydrophones or radio sonobuoys, listening equipment that allowed operators to verify a contact and establish its approximate position.
3. *Herald* systems, which gave bearing and range of the target.

No substantiating HDSD records of a *Herald* installation have yet been found, however. The "Radio Room" seen in the elevation view of the signal tower presented in Figure 7-106 was actually a passive listening station for an array of hydrophones (item 2 above) installed at the harbor entrance. The electronic equipment and PPI display for the Army's *SCR-682* harbor-surveillance radar was in the underground bunker, with the rotating antenna inside the blister housing at the very top of the signal tower.

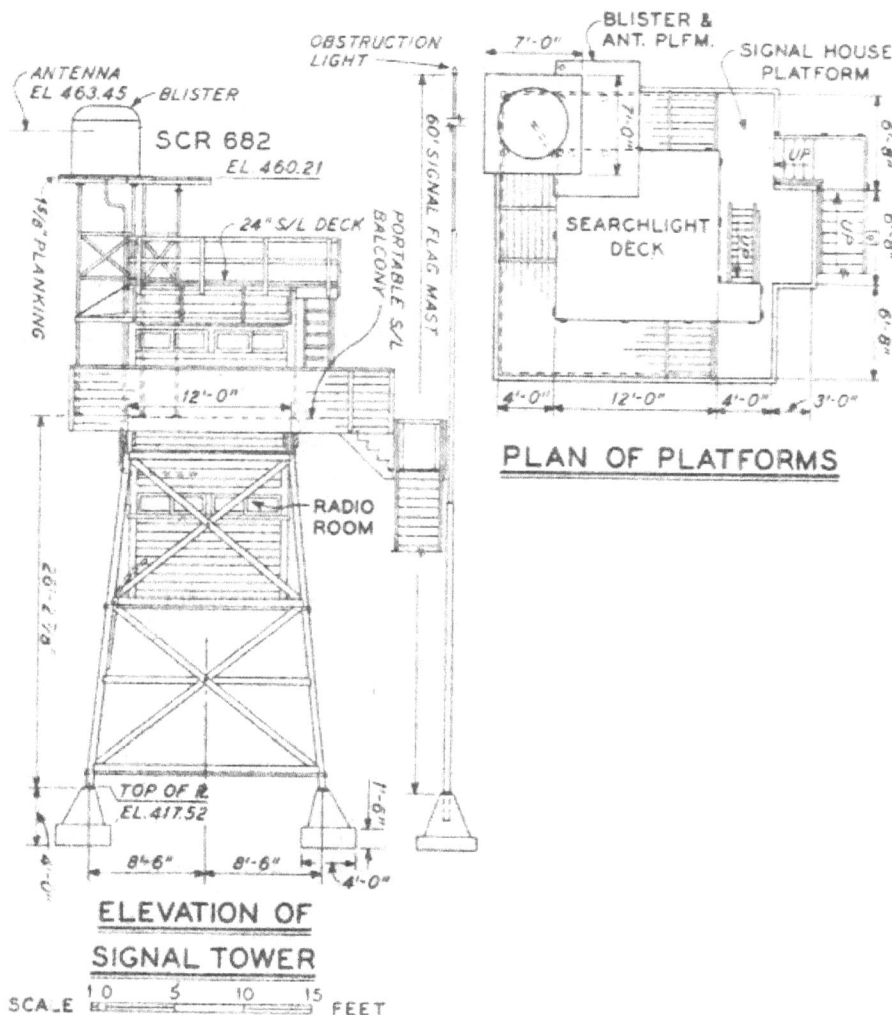

Figure 7-106. The "Radio Room" on this drawing of the new signal tower was a passive listening station for a hydrophone array installed at the harbor entrance. The *SCR-682* harbor-surveillance-radar equipment room was located in the subterranean bunker complex below the tower (NARA).

Due to its size and weight, the 24-inch carbon-arc searchlight for the HECP had originally been mounted on a raised wooden platform adjacent to the old lighthouse, accessible via temporary wooden stairs descending from the second-floor window of the south bedroom. The searchlight deck of the new signal tower, however, was built to accommodate this light as seen in Figure 7-107. The blister for the *SCR-*

682 harbor-surveillance radar has not yet been installed on the antenna platform at upper left, nor has the 60-foot-tall signal mast been erected to the structure's right.

Figure 7-107. a) The 24-inch carbon-arc searchlight on the new signal tower is visible on the southeast corner of the search-light deck, with the radar-antenna platform at upper-left. **b)** Richard Jenson is seen manning the 24-inch searchlight in 1943 (Shorty Beights' photos courtesy Cabrillo National Monument).

Figure 7-108 below shows the new signal mast just south of the signal deck, looking over the flag locker towards the old Spanish lighthouse in the background. As explained by Beights (Overton, 1993):

"We had the signal tower on the top and right below it was a hand-over-hand vertical ladder to go down into the sonar sound room, which was basically the same size as our cubicle above, but we had about a 3- or 4-foot catwalk all the way around our tower."

"But then if we wanted the head, we go all the way down, go through this trap door and go down in where the radar was. The Army manned the radar. But they had a head in there." [27]

The plan view of Figure 7-105 shows a 4-inch septic line connecting to this head in the northwest corner of the Navy Operations room.

---

27 Head – Navy term for latrine.

Figure 7-108. View from the flag locker on the signal deck of the new signal tower, with Meyler Road (later Sylvester Road, now the Bayside Trail) running north towards the old Spanish lighthouse in background center, circa 1943 (US Army photo courtesy Cabrillo National Monument).

The wartime photo seen in Figure 7-109 below shows a sailor entering the trapdoor above the Army Operations Room. According to Beights, the *SCR-682* radar operators would pass the range and bearing information of incoming vessels to the Navy sailors manning the signal tower above (Overton, 1993):

"We would turn our signal light in that direction and challenge whatever that bleep was that they had found. Quite often we'd be surprised to find a little signal light returning our message."

From the elevated searchlight deck situated at the very top of Point Loma, the 24-inch searchlight could communicate with a big ship some 35 miles out to sea (Overton, 1993).

Figure 7-109. This view from the signal tower shows a sailor entering the counter-balanced hatch at the southeast end of the underground Battalion 1 Command Post, circa 1943 (Shorty Beights' photo courtesy Cabrillo National Monument). Note power room intake and exhaust vents at bottom right.

The wooden signal tower and the *SCR-682* harbor-surveillance radar were declared surplus and removed sometime after the war. The still extant Battalion 1 Command Post is now cordoned off inside a small fenced compound on Center property inside Cabrillo National Monument, north of Battery Humphreys as seen in Figure 7-110a. The bunker was taken over by the Coast Guard in the 1980s for a harbor surveillance role, with what had been the Navy Operations Room repurposed as the Radar Room for the new radar equipment (Figure 7-110b). Note the new ground-level doorway on the south side of the observation level, presumably added by the Coast Guard for easier access.

a)

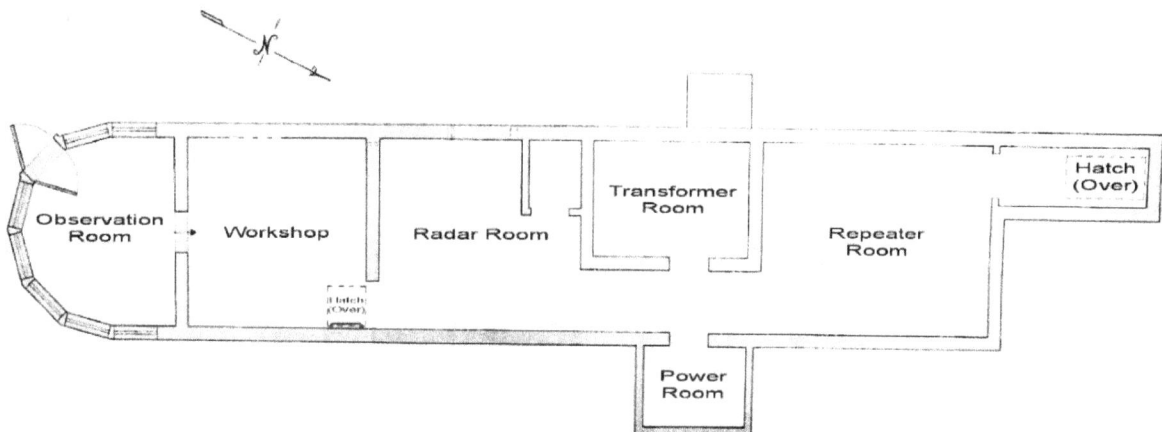

b)

Figure 7-110. a) A similar view but from a lower perspective than Figure 7-109, looking southeast across the Battalion 1 Command Post towards Battery Humphreys in July 2019. **b)** US Coast Guard "Bunker B-T-17" floorplan, circa 7 June 1983 (US Coast Guard blueprint courtesy SSC Pacific).[28]

A steel communications tower (today designated as Tower 502) was erected just west of the dug-in bunker sometime prior to 1979, as shown in Figure 7-111a below.[29] The southernmost entry hatch on the roof is seen at right center, just left of the tower behind the structure in background. Figure 7-111b shows the front of the observation room in November 2007, surrounded by several new radar and video surveillance sensors. The trailer-mounted extendable mast on the far right was part of the *Mobile Inshore Undersea Warfare* project managed by our Unmanned Systems Group and manned by Navy Reservists.

---

28   Cleaned and relabeled from Coast Guard "VHF-FM High Site Point Loma" blueprint 7111-87.
29   This is the same tower seen earlier in the background of Figure 7-103.

Figure 7-111. a) The observation room of the former Battalion 1 Command Post, showing the lower portion of a new antenna tower installation to the west, circa August 1979 (photo courtesy Al Grobmeier). **b)** Another view from the front showing this same tower and several new harbor surveillance sensors, circa November 2007.

The ground-level doorway alteration previously depicted on the Coast Guard blueprint in Figure 7-110b is seen in Figure 7-112 below, circa May 2006. A short set of stairs inside the observation-room level leads down to the five-room underground bunker extending back to the northwest. Note the impressions along the roof perimeter left by the sandbags used as a form for the concrete pour during construction, an additional steel tower on the east side just visible at background left, and the new surveillance-camera installation on the roof to its right.

Figure 7-112. View of the post-war Coast Guard entry to the former Battalion 1 Command Post, now Building T-17, looking east towards North Island in May 2006, with the jetty barely visible through the fog at background right. Note additional harbor-surveillance video camera on roof at upper right.

Figure 7-113 shows a telephone switching panel on the east wall in what is now Room 105 in Building T-17,[30] with the eight stepping relays that selected the desired phone line seen at far right. The armatures of these relays were incrementally advanced by the make-and-break action of the caller's rotary telephone dial. One such relay was incremented each time a new digit was dialed, thereby selecting one of the remaining seven relays to count the circuit interruptions representing that particular digit. This setup could therefore handle a seven-digit number, which makes it a post-war installation, but no longer in service.

Figure 7-113. All seven-digit telephone lines in the exchange were connected to the eight stepping relays on the right, which selected which pair of wires should be connected to the incoming line for proper call routing. A label on one of the terminal strips at lower left reads: "Cinema Engineering Co., Burbank, Calif."

Site 8 (Loma)

The Site 8 fire-control structures discussed in this section include the battery commander station BC_4 for AMTB Battery Cabrillo and base end and spotting station B^2_2 S^2_2 for Battery Strong, both behind Battery Humphreys left of center in Figure 7-114 below. AMTB Cabrillo is depicted as "Btry. Tac. No. 4" just north of Searchlight 16 in the upper-left corner, while Battery Strong is in Site 5, out of image to the right (north).

---

30   Room 105 is the original Power Room (Figure 7-105).

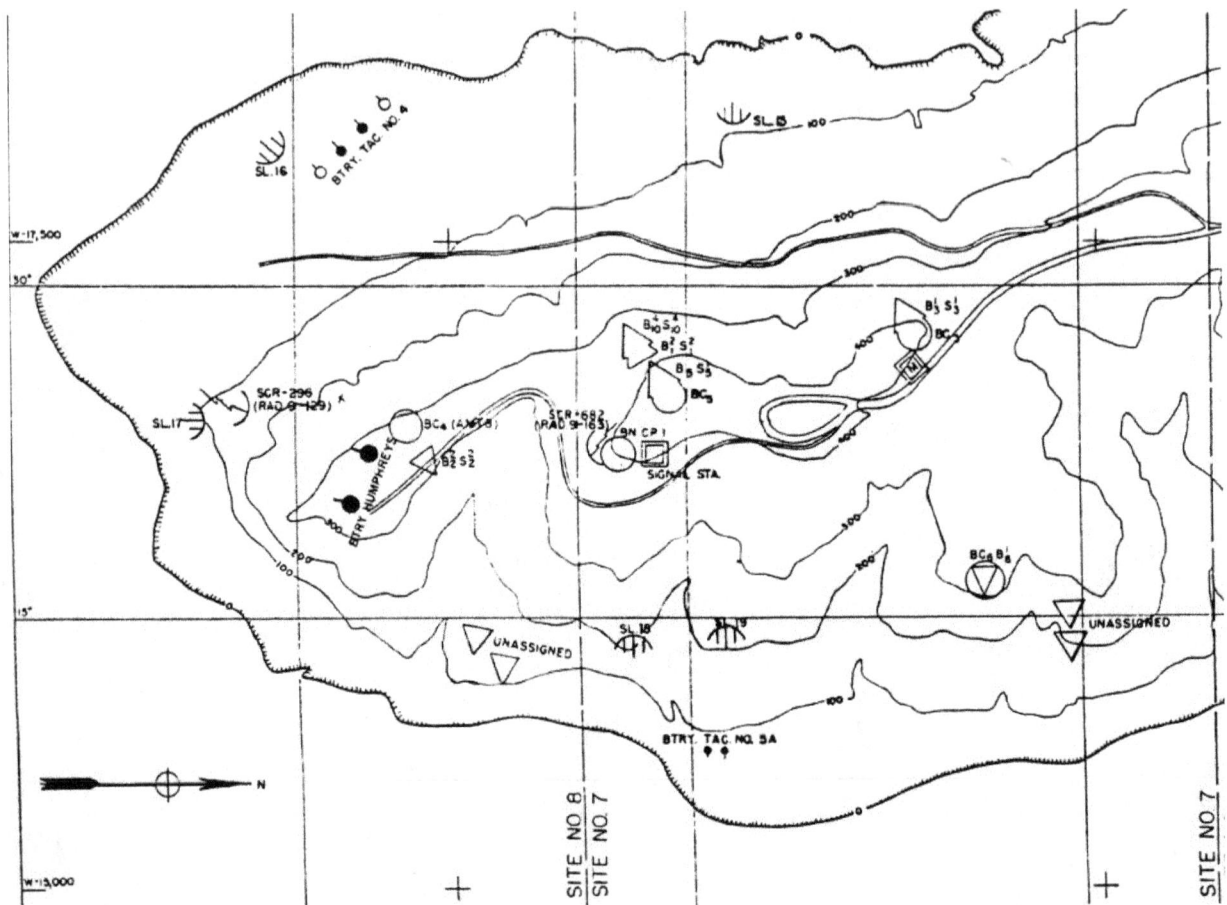

Figure 7-114. Battery commander station BC_4 for AMTB Battery Cabrillo and base end and spotting station $B^2_2 S^2_2$ for Battery Strong are at image center, slightly to the right of Battery Humphreys in this zoomed-in portion of a 1 July 1945 map entitled "Fire Control Stations at South Fort Rosecrans, Sites 6, 7, 8, 9."

Base End Station for Battery Strong

Base end and spotting station $B^2_2 S^2_2$ for Battery Strong (Tac. No. 2) was situated northwest of the battery commander station BC_4 for AMTB Battery Cabrillo, just north of Battery Humphreys (see again Figure 7-99). According to the Report of Completed Works, corrected to 19 October 1944, this bunker was an enlargement of the original fort commander station (see "Original Structure" in the upper right corner of the plan view of Figure 7-115), which had been transferred on 24 February 1920 after an initial outlay of $2,102.27. The WWII modernization cost for enlargement and splinter-proofing incurred an additional $3,719.66.

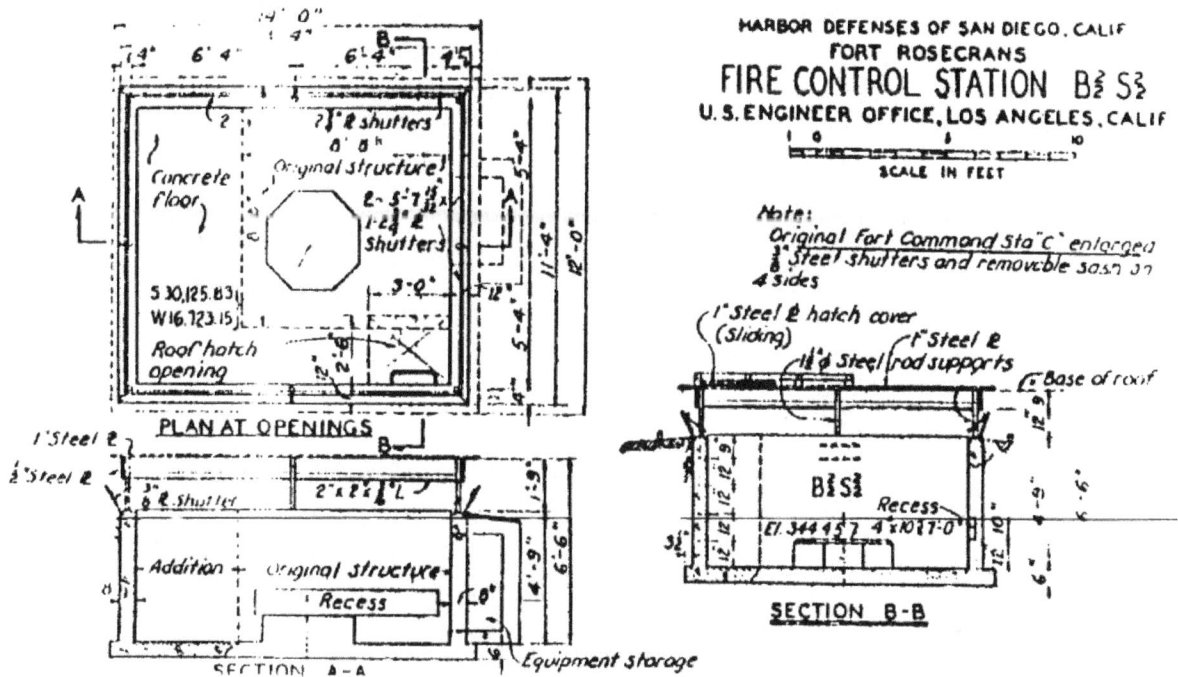

Figure 7-115. Plan and section views for base end and spotting station B^2_2 S^2_2 for Battery Strong (NARA). The 1-inch-thick steel roof was not covered with dirt or concrete, but was presumably camouflaged. Note the added concrete floor depicted on the left and bottom sides of the plan-view drawing.

The 1-inch-thick steel roof was supported by vertical steel rods in the horizontal observation slots on all four sides at an elevation of 348 feet.[31] A depression position finder was mounted on a raised octagonal concrete foundation in the center of the enlarged floorplan, which facilitated a 360-degree field of view. The underground structure (up arrow in Figure 7-116) was connected to commercial and back-up electrical power from nearby Battery Humphreys, but was not equipped with heating or lighting. Humphreys Road is seen running diagonally from the battery towards the upper-right corner of the photo.

Figure 7-116. Base end and spotting station B^2_2 S^2_2 for Battery Strong (up arrow in upper right corner) received commercial and backup electrical power from nearby Battery Humphreys to the south. The battery commander station for AMTB Battery Cabrillo, not yet constructed in this photo, was slightly above and to the left of B^2_2 S^2_2 (undated US Army photo courtesy SSC Pacific).

---

31 This vertical-rod construction technique facilitating view ports on all four sides was perhaps unsuited for supporting a heavy earth or concrete covering atop the flat steel roof.

The remains of base end station $B^2_2 S^2_2$ just north of Battery Humphreys are seen in Figure 7-117, circa May 2006, with Humphreys Road in background right. The sliding hatch cover on top of the 1-inch-thick steel roof was found to be rusted shut, precluding entry. The steel shutters were folded down in the open position, but the entire roof assembly appeared to have collapsed down upon the side walls, as if the four 1½-inch vertical support rods depicted in Section B-B in Figure 7-115 had all been purposely cut.

Figure 7-117. The badly rusted sliding hatch cover on top of the 1-inch-thick steel roof of base end station $B^2_2 S^2_2$ for Battery Strong is seen circa May 2006, with Humphreys Road in background right. The hatch was rusted shut, and the entire roof assembly appears to have collapsed upon the side walls.

Battery Commander Station for AMTB Battery Cabrillo

The battery commander station BC_4 for AMTB Battery Cabrillo (Figure 7-118) is just down the slope and slightly northwest of Battery Humphreys, as previously shown in the 1 July 1945 map of fire-control stations presented in Figure 7-99. Most of the earth covering that sloped upwards towards the slightly elevated entry hatch at upper right has been washed away over the years by rain. During WWII, AMTB Battery Cabrillo was located in the vicinity of the small compound seen near the cliff edge in background left, the former site of the Naval Undersea Center's Marine Mammal Facility.

Figure 7-118. The battery commander station BC_4 for AMTB Battery Cabrillo, just northwest of Battery Humphreys, is directly upslope to the east of the former Marine Mammal Facility seen in background left, circa May 2006.

Site 9 (East Rosecrans)

Plotting and Switchboard Room for Battery Ashburn

The combined Plotting and Spotting Room for Battery Ashburn and Switchboard Room for Fort Rosecrans was located just north of the battery and east of Catalina Boulevard / Cabrillo Memorial Drive.[32] This massive subterranean bunker is shown as PSR_3 on the relevant portion of the 1 July 1945 HDSD map entitled "Fire Control Stations at South Fort Rosecrans, Sites 6, 7, 8, 9" in Figure 7-119 below. Catalina Boulevard / Cabrillo Memorial Drive, which runs north-south across the top of the map, was the dividing line between Sites 6 and 9.

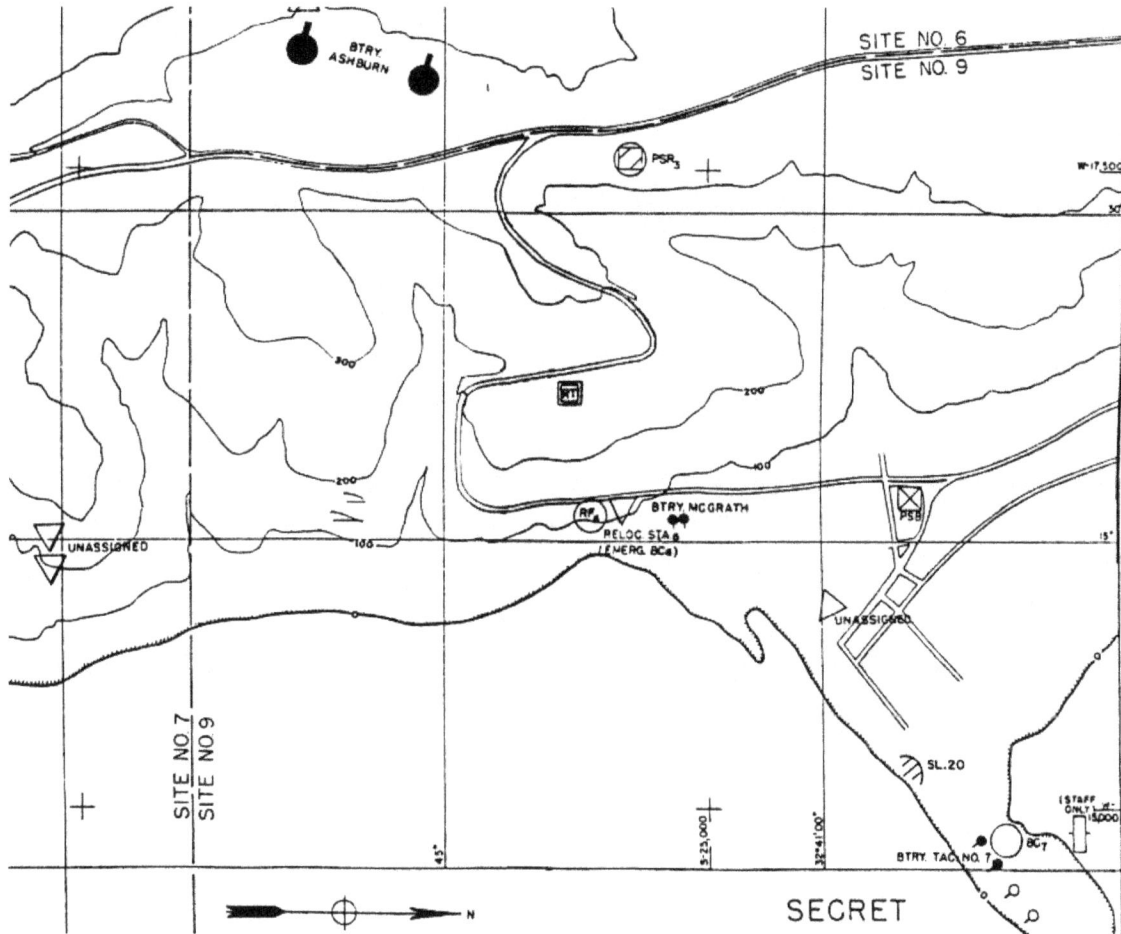

Figure 7-119. This 1 July 1945 HDSD map shows Battery Ashburn just left of top center in Site 6 (West Rosecrans), with its Plotting and Switchboard Room (PSR_3) on the opposite (east) side of Catalina Boulevard / Cabrillo Memorial Drive in Site 9 (East Rosecrans).

This underground bunker was truly massive, with 12-foot-wide concrete perimeter walls and a 20-foot-thick concrete roof, as seen in the plan and section views presented in Figure 7-120. The concrete walls on either side of the 20-foot entrance tunnel were 6-feet wide, with a pair of triangular wingwalls extending an additional 30.5 feet to the east. The telephone Switchboard Room on the right-hand side of the plan view served all of Fort Rosecrans. The combined structure was transferred to the Coast Artillery on 16 June 1944, with a total cost to that date of $168,141.72.

---

32   There is some confusion throughout the literature with regard to the abbreviation "PSR." For purposes of this text, the entire combined structure is referred to as a "Plotting and Switchboard Room," while the plotting room within said structure is referred to as the "Plotting and Spotting Room." For those installations lacking a switchboard room, "PSR" stood for "Plotting and Spotting Room."

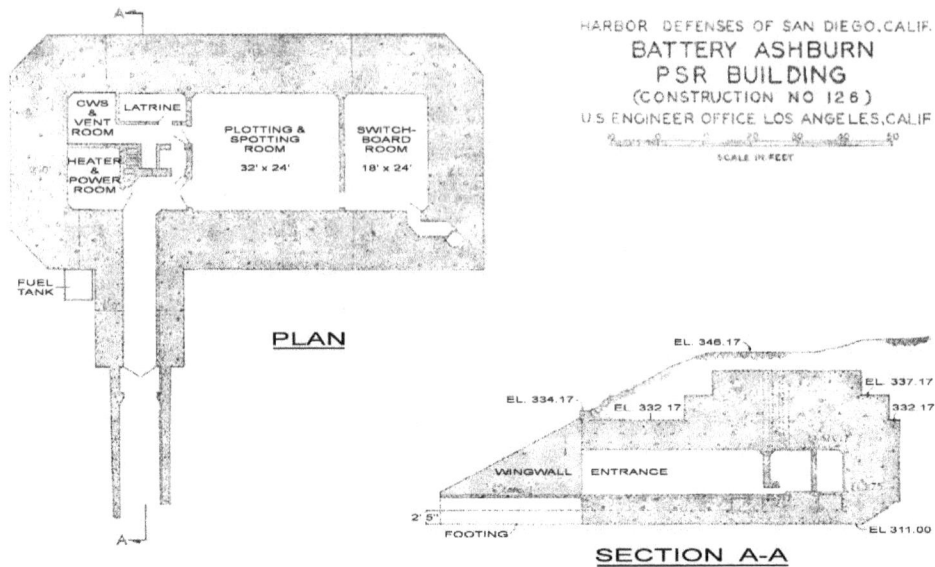

Figure 7-120. Cleaned and relabeled plan and section views of the combined Plotting and Spotting Room for Battery Ashburn and the associated Switchboard Room that served all of Fort Rosecrans. Note the vertical escape hatch to the surface in the lower-right (northeast) corner of the plan view.

The 1984 Naval Ocean Systems Center (NOSC) map of the "Battery Ashburn Area" presented in Figure 7-121 shows the PSR as NOSC Building 563 on the east side of Catalina Boulevard / Cabrillo Memorial Drive, approximately 800 feet north of the battery. The turn-around loop seen on the opposite side of this road is the southernmost portion of Fort Rosecrans National Cemetery. The entire area below the battery to the west was extensively graded and terraced in 1962 to accommodate relocation of the San Diego Sewer Treatment Plant.

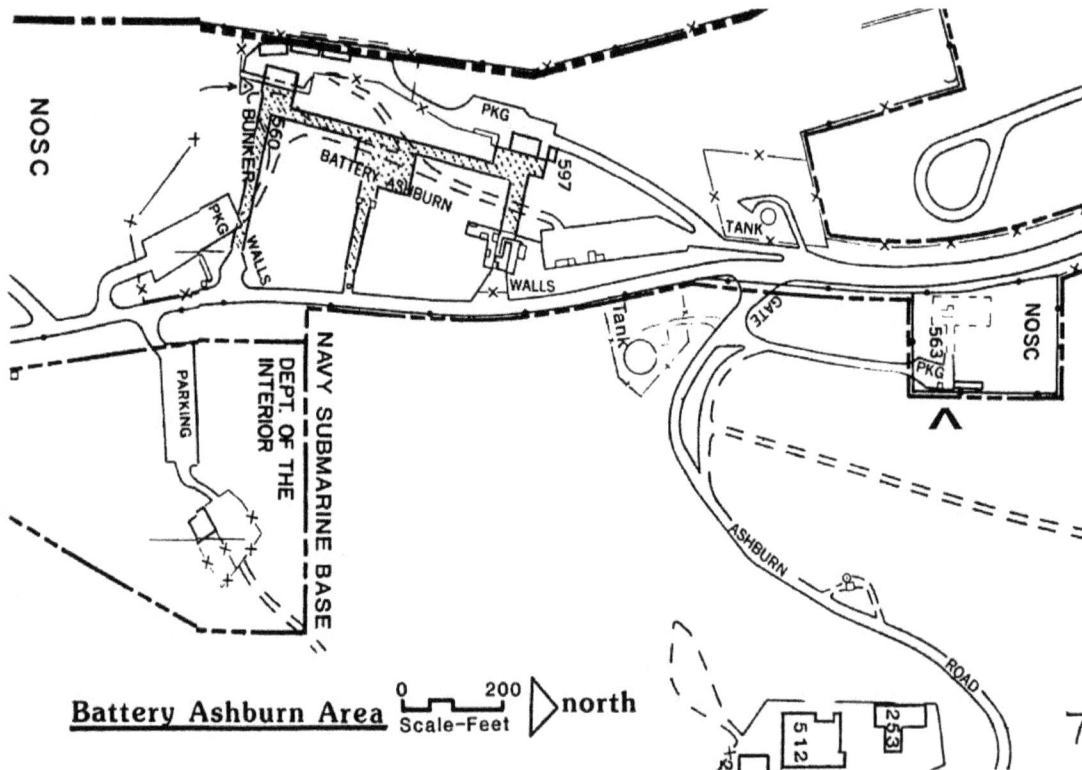

Figure 7-121. This October 1984 Naval Ocean Systems Center (NOSC) map of the Battery Ashburn Area shows the Plotting and Switchboard Room for the battery as NOSC Building 563 (up arrow) at right center on the east side of Catalina Boulevard / Cabrillo Memorial Drive (US Navy map courtesy SSC Pacific).

Battery Ashburn is seen under construction sometime in 1942 in the aerial photo shown in Figure 7-122, with the Gun 1 emplacement on the north end partially completed. The PSR is north of the battery on the east side of Catalina Boulevard / Cabrillo Memorial Drive, just west of the deactivated WWI-era Battery John White in the lower-right corner. Note the six wooden barracks buildings north of the Plotting and Switchboard Room in the upper-right corner, and the rows of parked vehicles for the construction workers along the west side of the road.

Figure 7-122. The air-conditioned Plotting and Switchboard Room for Battery Ashburn was located north of the battery on the east (near) side of Catalina Boulevard / Cabrillo Memorial Drive, just west of (above) deactivated Battery John White (four 12-inch mortars) in the lower-right corner (1942 US Army photo courtesy SSC Pacific).

As recalled by Oliver Unruh, Battery D, who was transferred as gun mechanic from Battery Strong to Battery Ashburn during WWII, there was also an underground berthing structure midway between the battery and its PSR to the north (Overton, 1993):

"I lived at the guns at Ashburn between the plotting room and the guns. I was at the command post most of the time, but that's where the gun mechanics stayed."

Probably similar to the previously discussed berthing shelters for Battery Point Loma (Chapter 4), this berthing structure likely would have been built underground by troop labor sometime after the battery construction was completed. No photos or documentation related to this shelter have yet been found.

The photo taken by Al Grobmeier presented in Figure 7-123 shows the PSR entrance looking west in 1979, with the escape-trunk exit and a vent pipe partially visible above the bunker at upper left. The new glass entry doors were obviously added some time prior to this date, perhaps when the bunker was repurposed by the Navy as a computer center in the 1960s timeframe. The open doorway, interior lighting, and packages outside suggest this facility was still in use at the time the photograph was taken by Grobmeier.

Figure 7-123. Entryway to the PSR looking west in December 1979 (photo courtesy Al Grobmeier). Note the air-vent pipe and generator exhaust housing exposed by erosion above the bunker in the upper-left corner.

Figure 7-124a shows the PSR entrance in May 2008, looking west, with the previously seen vent pipe and escape-trunk trapdoor exit now hidden by brush. The civil-defense sign on the right wingwall indicates this underground structure was also used as a fallout shelter during the Cold War. The interior of the entrance tunnel inside the glass doors is seen in Figure 7-124b, looking west towards the air lock, out of sight to the right at the far end of the tunnel. The former Heater and Power Room is out of sight to the left. The PSR still had electricity and running water at this time but was no longer occupied.

Figure 7-124. a) View from inside the concrete wingwalls leading to the PSR entrance in May 2008. The vent pipe and escape-trunk exit are hidden behind the brush at upper left. **b)** Interior of the entrance tunnel looking west, showing the lift-out floor panels for cable-run access.

Figure 7-125 shows the interior of the former PSR in May of 2008, repurposed by the Center's Safety Office as a temporary storage facility for hazardous materials pending disposal. The PSR was originally equipped with incandescent overhead lights with Wheeler industrial reflectors. The upgraded fluorescent lighting seen below was probably installed in the 1960s when the air-conditioned facility was used by the Navy laboratory to house vacuum-tube-based computer equipment. The WWII acoustic tiles were removed from the ceiling to accommodate the lighting upgrade, but are still extant along the walls.

Figure 7-125. Interior view of the former PSR, repurposed as a temporary storage facility, looking north in 2008 towards the former Switchboard Room. The original air-conditioning units seen along the upper walls were equipped with electrical strip heaters for use during winter.

The gas-tight steel doorway to the escape trunk in the northeast corner of the PSR bunker is seen in Figure 7-126a, with Figure 7-126b showing the wooden ladder ascending into the lower section of the escape trunk (see again plan view in Figure 7-120). The upward view of the vertical shaft presented in Figure 7-126c shows the midlevel offset leading to the upper-shaft section that continues to the trapdoor exit. The purpose of this offset was to (hopefully) stop a direct bomb hit on the above-ground trapdoor from penetrating all the way down to the PSR level below. It arguably served the same purpose in the event of a falling human body.

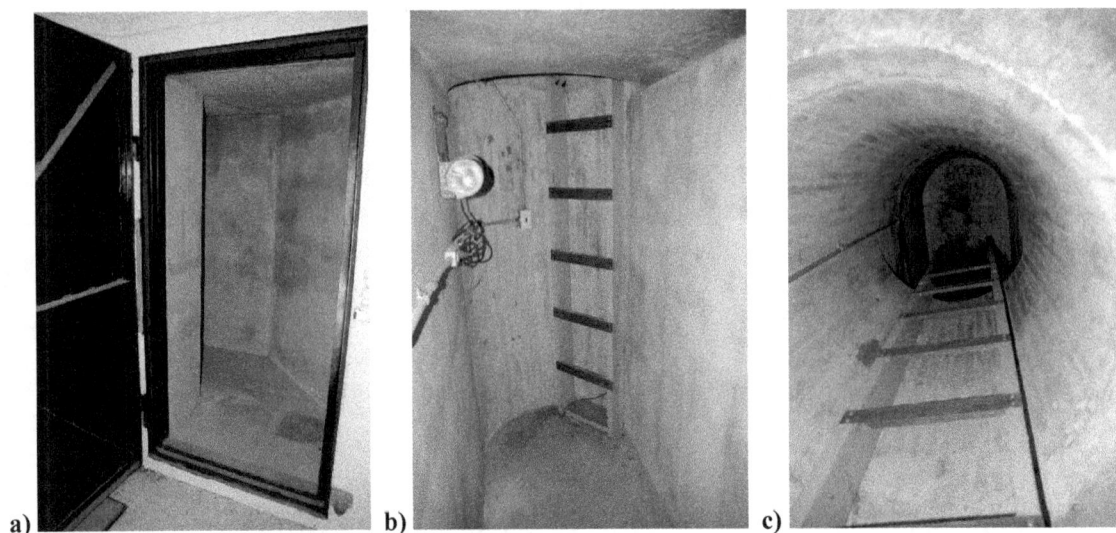

Figure 7-126. a) The open gas-tight doorway to the escape trunk, circa 2008. **b)** The ladder leading to the escape-trunk shaft. **c)** Upward view of the vertical shaft of the escape trunk showing the midlevel offset to the upper section of the shaft leading to the above-ground trapdoor exit.

Figure 7-127a shows the interior hallway looking south, with the door to the latrine at foreground right and a post-war doorway addition in background center. The electrical controller on the left hallway bulkhead enabled remote activation of the CWS system. The former CWS and Vent Room is at the far end of the hallway on the right side, directly across from the Heater and Power Room on the left side. The PSR was connected to commercial electrical service, with backup power from Battery Ashburn and an onsite auxiliary generator. Figure 7-127b shows the southwest corner of the latrine.

Figure 7-127. a) View of the hallway looking south through a post-war-doorway addition towards the former CWS and Vent Rooms in the southwest corner of the PSR bunker. **b)** Interior view of the latrine on the west side of the hallway.

Figure 7-128 shows the interior of the former Heater and Power Room (see again Figure 7-120), looking west. The metal partition on the left side of the photo is a post-war addition to accommodate a darkroom for developing X-ray films. The high-powered X-ray equipment was used in the early 1990s to look for cracks in lightweight torpedo housings manufactured in the Center's Bayside foundry (Building 188).[33] Technical-manual excerpts found taped to the walls outline the proper procedures for warming up 300-kilovolt double-focus X-ray tubes after idle periods or new installation, and the developing and fixing times in minutes for Kodak X-ray film.

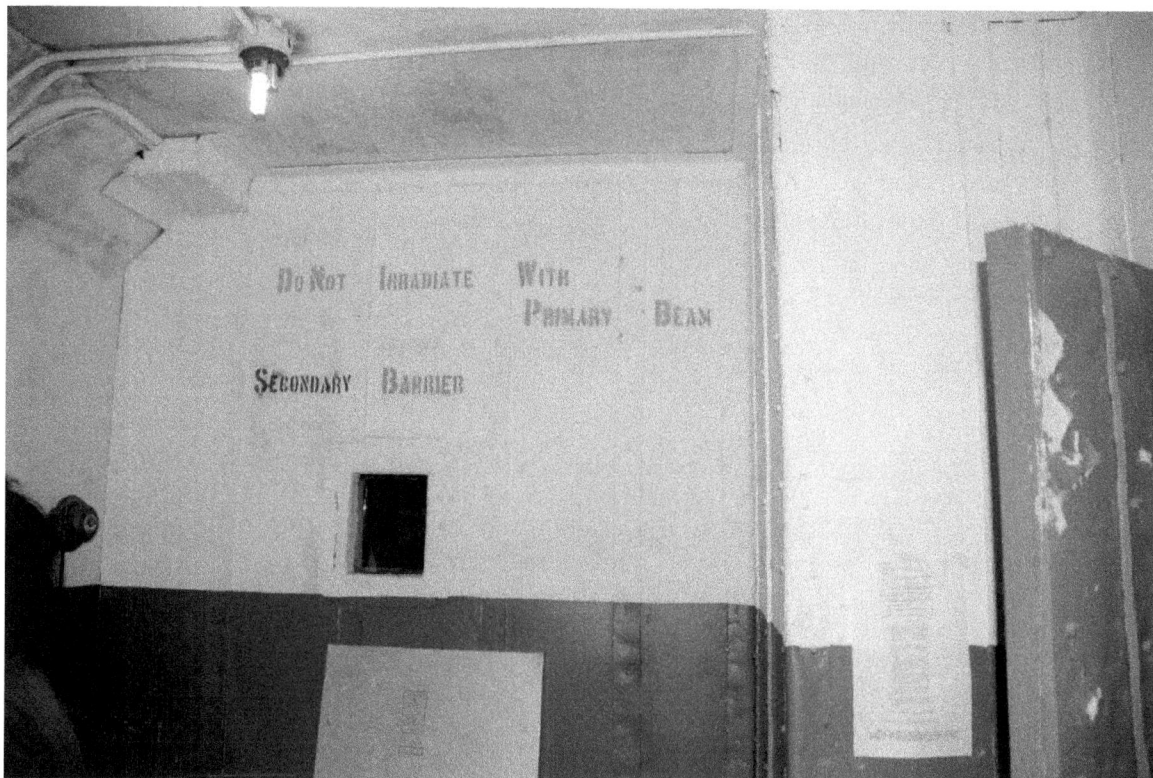

Figure 7-128. The former Heater and Power Room, repurposed in the 1990s by the Center as an X-ray facility for inspecting torpedo housings, looking west towards the darkroom from the assumed location of the X-ray machine.

---

33 This work was transferred to the east coast by the 1995 Base Realignment and Closure Commission.

Site 10 (Coronado)

As seen on this lower-right-corner excerpt of a 1 July 1945 map entitled "Fire Control Stations at North Island Site 10," the four WWI-era fire-control stations in Site 10 (Coronado) were all unassigned during WWII (Figure 7-129).

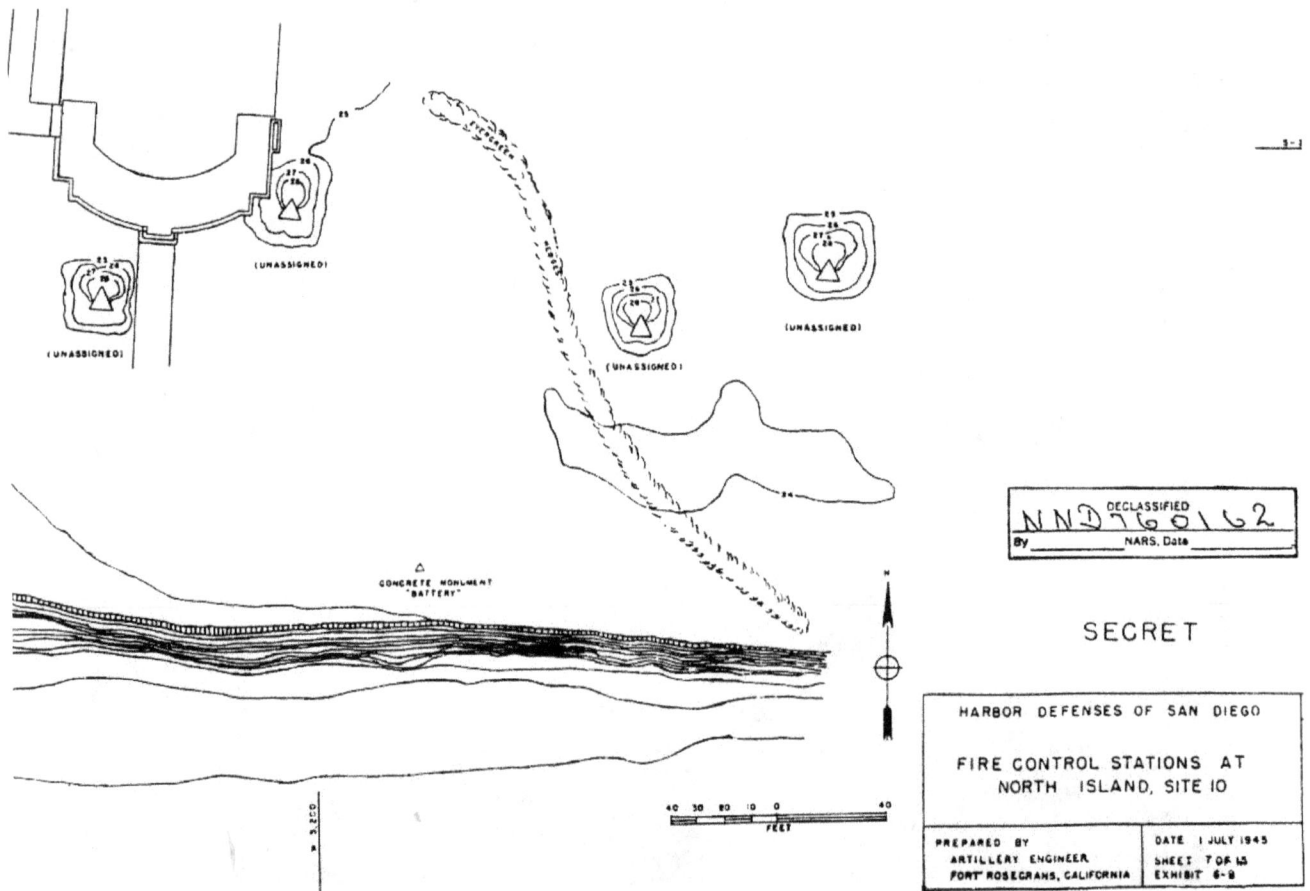

Figure 7-129. The four WWI-era fire control structures, which were all facing north towards San Diego Bay, were consequently unassigned during WWII (NARA).

Site 11 (Strand)

There were three WWII fire-control structures on the Coronado Beach Military Reservation in Site 11 (Figure 7-130):

- The combined base end and spotting stations B^3_9 S^3_9 for the 16-inch Battery Gatchell (Tac. No. 9) and B^3_{10} S^3_{10} and the 6-inch Battery Grant (Tac. No. 10).

- The combined base end and spotting stations B^3_2 S^3_2 for the 8-inch Battery Strong (Tac. No. 2) and B^3_4 S^3_4 for the 6-inch Battery Humphreys (later Tac. No. 5).

- The battery commander station BC_8 for AMTB Battery Cortez (Tac. No. 8).

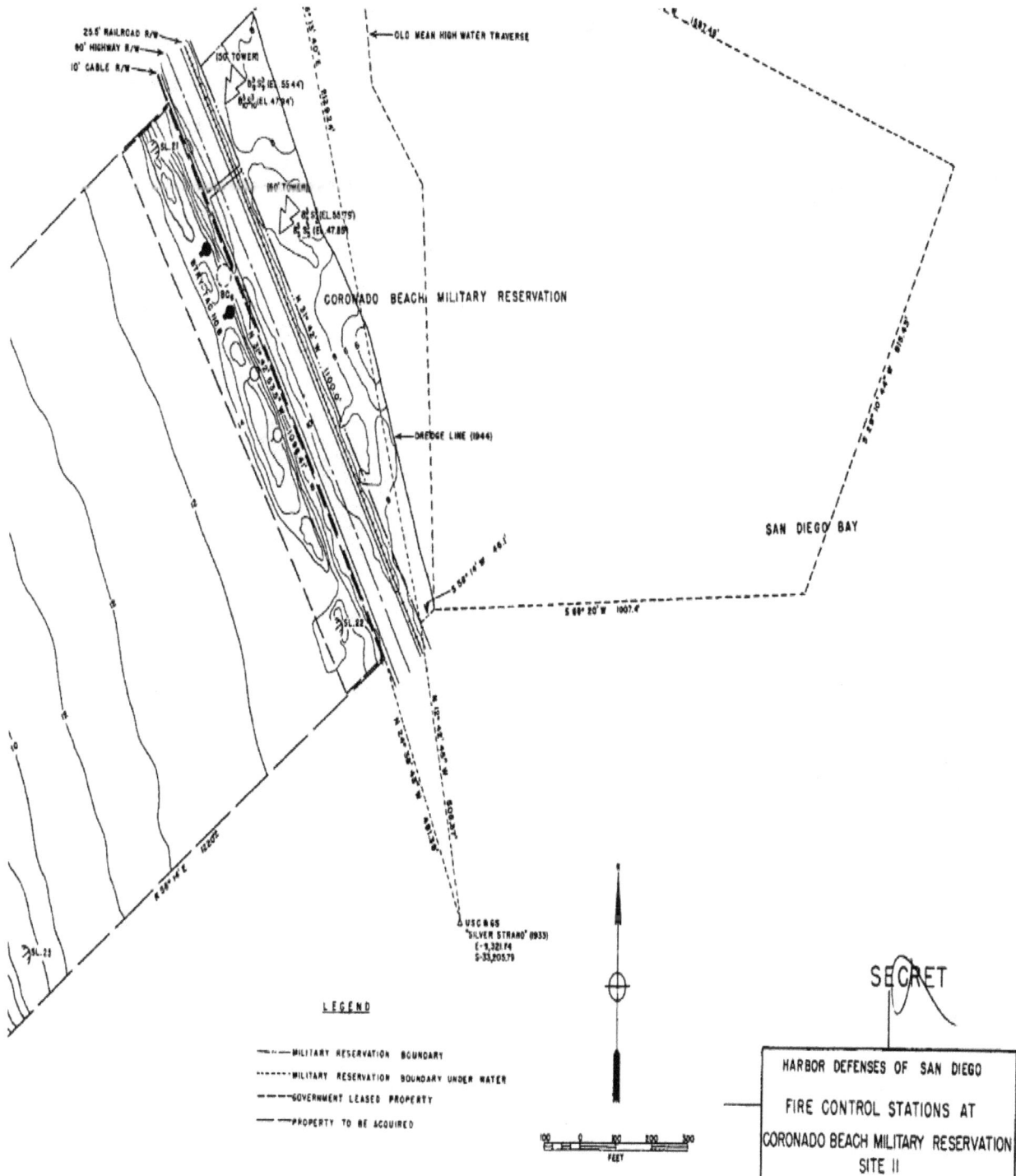

Figure 7-130. The three fire-control structures located on the Coronado Beach Military reservation in Site 11 are seen behind Searchlight 21 and AMTB Battery Cortez (Tac. No. 8) in the upper-left corner (NARA).

Base End Stations for Batteries Gatchell and Grant

The combined base end and spotting stations $B^3_9 S^3_9$ and $B^3_{10} S^3_{10}$ for the 16-inch Battery Gatchell (Tac. No. 9) and 6-inch Battery Grant (Tac. No. 10) were located atop a two-level 50-foot wooden tower disguised as a water tank at Fort Emory on the Silver Strand. The wooden substructure of the two splinter-proof observation rooms was covered with reinforced concrete, as was the roof. The Report of Completed Works, corrected to 26 October 1944, indicates the tower was transferred to the Coast Artillery on 3 May 1943 at a final cost of $9,756.48.

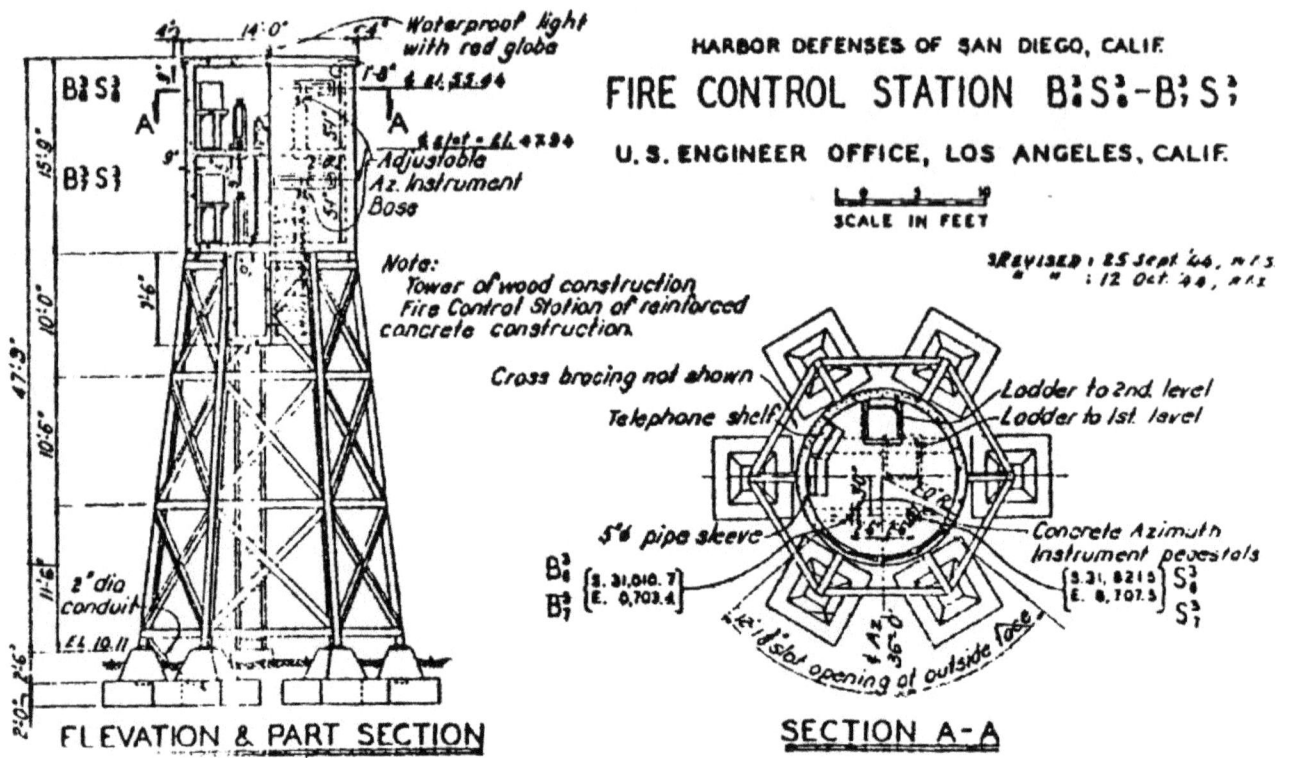

Figure 7-131. Disguised as a water tower, the base end and spotting stations for Batteries Gatchell and Grant were each equipped with two azimuth instruments featuring adjustable bases on concrete support pedestals (NARA). The elevation view on the left indicates the tower was 52 feet 3 inches high.

As indicated in the plan and section views above, there were two azimuth instruments in each level of this fire-control tower, mounted on concrete pedestals with adjustable jacks to vary the telescope height. The center of the upper observation slot was at an elevation of 55.44 feet, with the lower slot at 47.94 feet. Commercial electrical power (125 VAC) was available for a single overhead light fixture in both levels, with a backup 3-KVA auxiliary generator shared with the nearby base end stations for Batteries Strong and Humphreys, discussed in the next section. This structure is no longer extant and no period photos have yet been found.

Base End Stations for Batteries Strong and Humphreys

The base end and spotting stations $B^3_2 S^3_2$ for Battery Strong (Tac. No. 2) and $B^3_4 S^3_4$ for Battery Humphreys (later Tac. No. 5) were located atop a two-level steel tower (Figure 7-132) on Coronado Beach in Site 11 on the Silver Strand (see again Figure 7-130). The Report of Completed Works, corrected to 9 November 1944, shows this 50-foot tower was transferred to the Coast Artillery on 18 February 1943 at a cost of $12,298.42. Each observation level was equipped with a pair of azimuth instruments and lighted when necessary by a single overhead fixture. Commercial power was backed up by a 125-volt AC generator shared with base end and spotting stations $B^3_6 S^3_6$ - $B^3_7 S^3_7$ for Batteries Gatchell and Grant, as previously discussed.

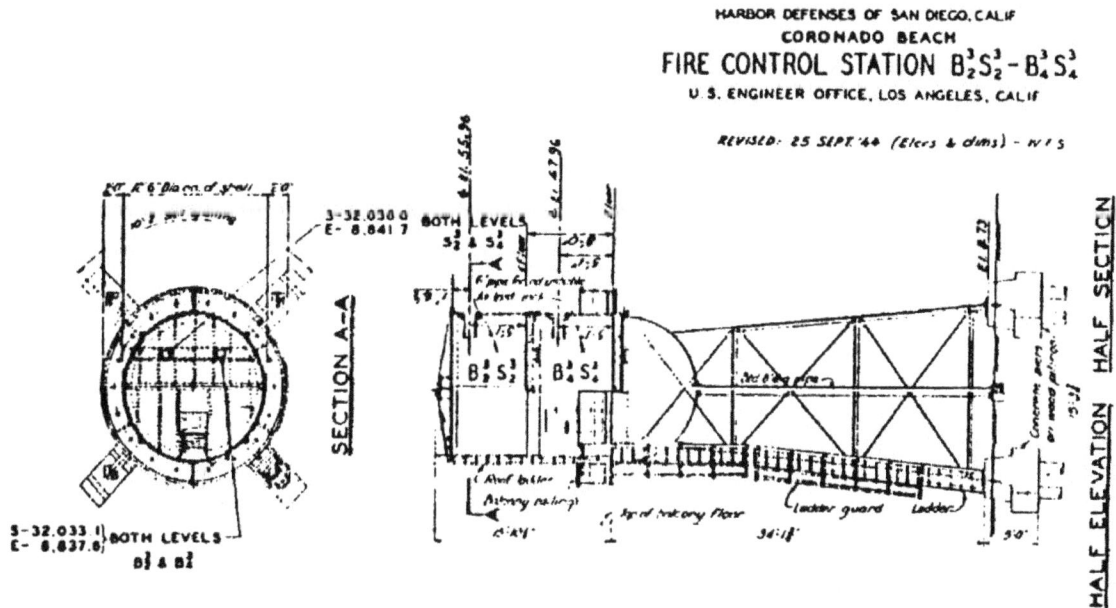

Figure 7-132. Elevation and section views of base end and spotting stations $B^3_2 S^3_2$ (upper level) for Battery Strong and $B^3_4 S^3_4$ (lower level) for Battery Humphreys from the Report of Completed Works, corrected to 9 November 1944 (NARA). This structure is no longer extant and no photos have been found.

Battery Commander Station BC$_8$ for AMTB Battery Cortez

Battery commander station BC$_8$ for AMTB Battery Cortez (Tac. No. 8) was centered between the battery's two fixed 90-millimeter gun emplacements as shown in Figure 7-133. No photographs of this structure or details of its construction have been found to date.

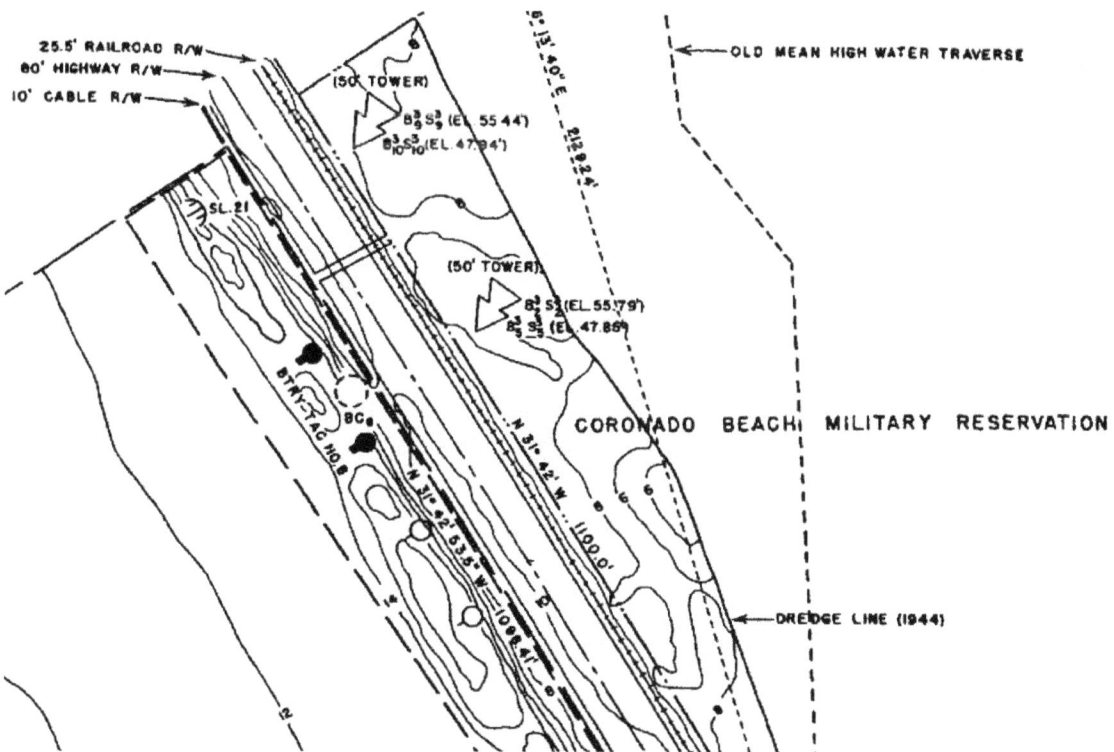

Figure 7-133. The battery commander station BC$_8$ for AMTB Battery Cortez (Tac. No. 8 just left of center) was situated directly behind the battery's two fixed 90-millimeter guns (NARA).

Site 12 (Emory)

There were three tower-mounted fire-control installations at Site 12 in Coronado Heights, arranged in a north-south line just east of Batteries Gatchell (Tac. No. 9) and Grant (Tac. No. 10), as shown in the 1 July 1945 map presented in Figure 7-134. The Plotting and Spotting Room (PSR_9) for the unfinished 16-inch Battery Gatchell is at the extreme north end in the lower-right corner of the map. Just south (left) of the PSR is a combined tower structure for BC_9 and $B^1_9 S^1_9$, with the Battalion Control Post 3 tower to its left at the center of the map image. At the southern end of this line is a combined tower structure for BC_{10}, B^1_{10} S^1_{10}, and the *SCR-296-A* radar (9-1007) for Battery Grant.

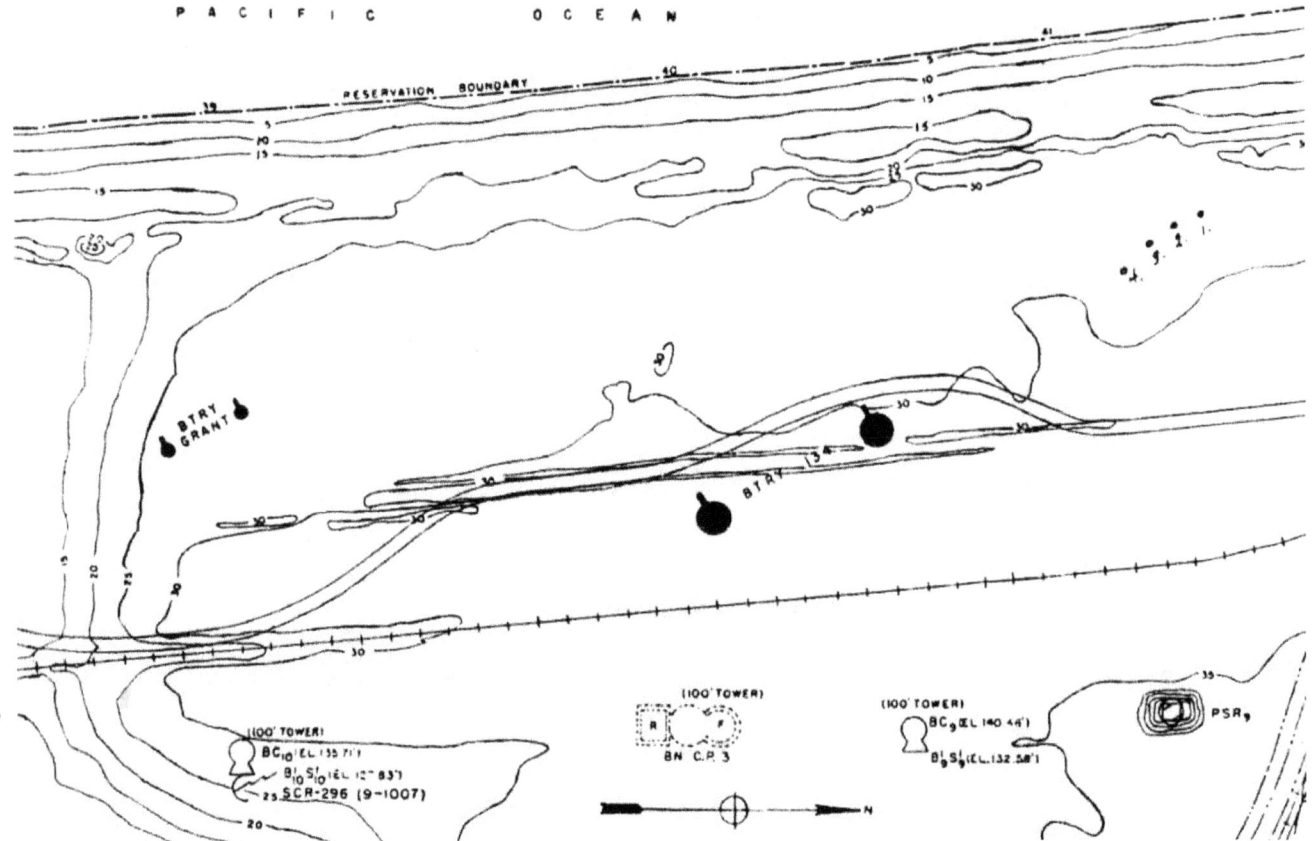

Figure 7-134. The three 100-foot tower structures east of Batteries Grant and Gatchell are seen along the lower portion of this 1 July 1945 map depicting "Fire Control Stations at Fort Emory, Site 12" (NARA). Note PSR_9 for Battery Gatchell in bottom-right corner, and the railway running north-south above it.

Plotting and Spotting Room (PSR) for Battery Gatchell

The combined Plotting and Spotting Room for Battery Gatchell (and Switchboard Room for Fort Emory) in Site 12 seen in Figure 7-135 was virtually identical to that for Battery Ashburn at Fort Rosecrans in Site 6. One key difference was the lack of a Radio Room, which was perhaps the reason there was also no air conditioning. This massive underground structure was situated northeast of the 16-inch battery as previously shown in the lower-right corner of Figure 7-134 above. Transfer to the 19th Coast Artillery Regiment was completed on 25 April 1944 with a total cost to that date of $121,384.15.

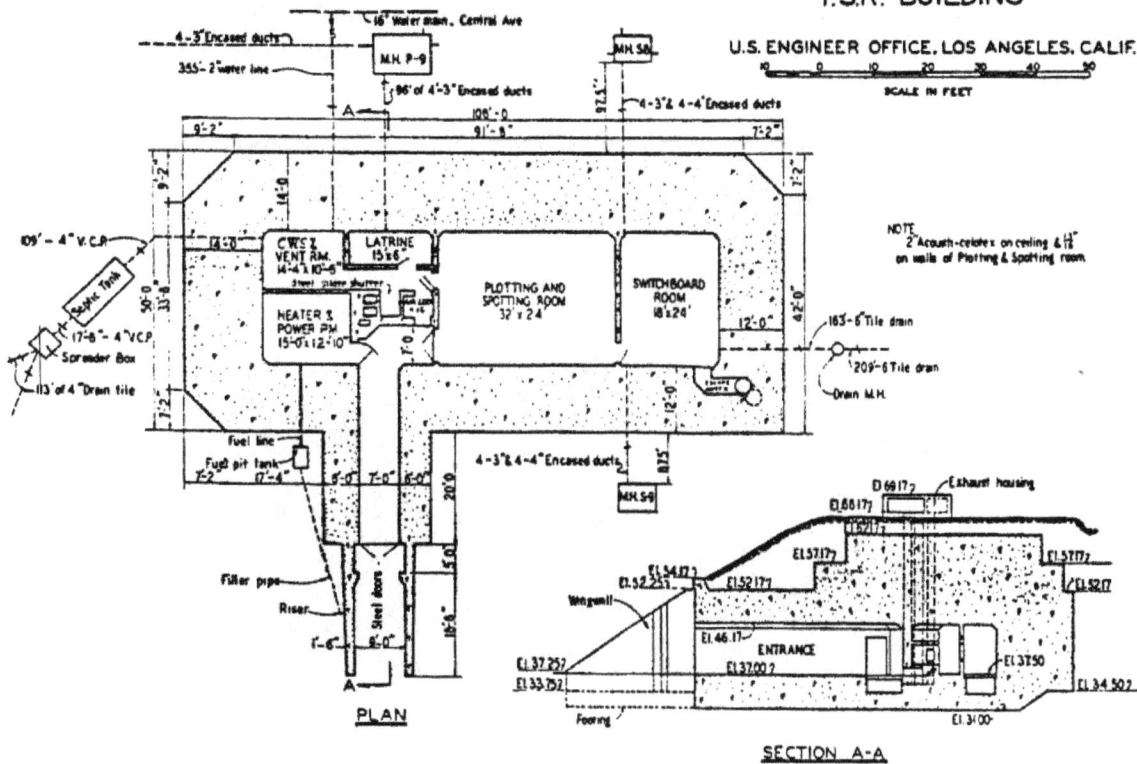

Figure 7-135. Plan and section views of the combined Plotting and Spotting Room (PSR) for Battery Gatchell (Constr. No. 134) and Switchboard Room for Fort Emory, from the RCW corrected to 26 May 1944 (NARA).

Figure 7-136a below shows a 1991 photo of the entrance to the Battery Gatchell PSR taken by Coast Defense Study Group member Mark Behrow. Note the rectangular generator-exhaust housing at top center (see again section view in Figure 7-135). The two vertical pipes in the upper-left corner were presumably for supporting the overhead camouflage netting. The fresh-air vent pipe is out of image to the left, but can be seen in the more recent 2014 photograph presented in Figure 7-136b, which was also taken by Mark Berhow.

Figure 7-136. a) The entrance to the former PSR (now Building 98) for Battery Gatchell looking west, circa 1991. **b)** The same entrance as seen in 2014 (photos courtesy Mark Berhow, CDSG).

Unlike the PSR for Battery Ashburn in Site 6, which was gutted and then refurbished for post-war use, the interior of the PSR for Battery Gatchell was found to be well preserved in its original state. Figure 7-137a shows the 1944 electrical panel for the Caterpillar *D4600* generator in the Power Room, with what appears to be a three-phase AC step-down transformer on the adjacent wall (Figure 7-137b). During another Coast Defense Study Group visit in 2015, the original Chemical Warfare Service (CWS) scrubber equipment was found intact and in excellent condition in the CWS room.

Figure 7-137. a) The original electrical panel in the Power Room of the PSR, circa September 2015. **b)** A three-phase step-down transformer is mounted on the adjoining wall above the upper-left corner of the electrical panel (Marvin Henze photos courtesy Mark Berhow).

Battery Commander and Base End Station for Battery Gatchell

Disguised as a water tank, the battery commander station BC_9 and base end and spotting station B^1_9 S^1_9 for the 16-inch Battery Gatchell were behind the battery just south of the PSR (Figure 7-138). This poor-quality image is a digital scan of a photocopy of a microfiche printout of a wartime photograph of this area from a report on camouflage activities in the Harbor Defenses of San Diego (Wolfgast, 1943). The tower positions relative to the railroad tracks and road surface in the lower-left corner indicate the view is to the south, with the photographer standing west of the PSR (see again Figure 7-134).

Figure 7-138. This wartime image shows the three 100-foot fire-control towers in Site 12, with the battery commander and base end station for Battery Gatchell at foreground center, looking south (US Army photo adapted from Wolfgast, 1943). The PSR for Gatchell would be just out of image to the right.

Plan and section views of this 100-foot fire-control tower are presented in Figure 7-139 below. The Report of Completed Works for this structure, corrected to 7 November 1944, indicates it was transferred on 26 September 1944 at a cost of $17,883.07. Situated at an elevation of 34 feet above mean lower low water, the observation rooms were equipped with a depression position finder in the lower-level base end station and an azimuth instrument in both levels. This tower is no longer extant and no additional period photos have been located.

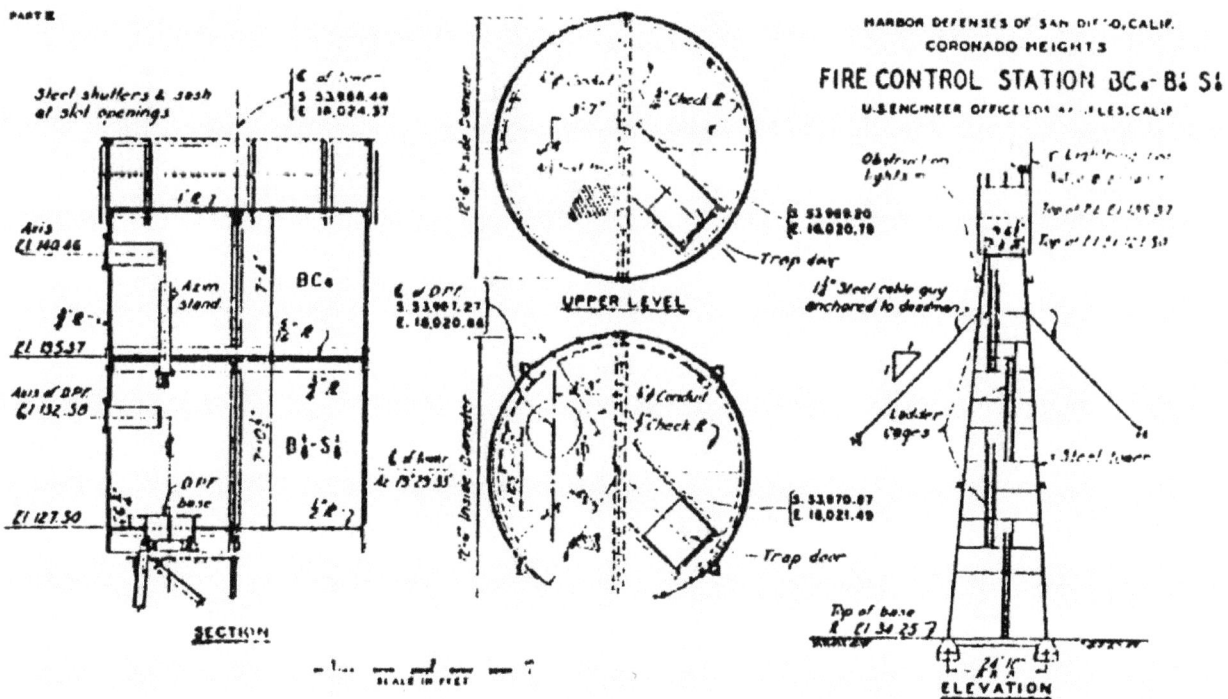

Figure 7-139. Plan and section views for battery commander station BC_9 (upper-level) and base end and spotting station B^1_9 S^1_9 (lower level) for Battery Gatchell at Fort Emory on Coronado Heights (NARA). The tower elevation diagram appears to be generic, as it doesn't match the wartime photo presented in Figure 7-138 above.

Battery Commander, Base End Stations, and Radar for Battery Grant

The battery commander station BC_{10} and base end and spotting station $B^1_{10} S^1_{10}$ for Battery Grant at Fort Emory were located atop a 100-foot steel tower disguised as a water tank (Figure 7-140). The antenna housing for the *SCR-296-A* radar was above the upper observation level for BC_{10}, which was equipped with a variable-height azimuth instrument. The lower-level observation room for $B^1_{10} S^1_{10}$ had a fixed-height azimuth instrument and a depression position finder. Nearby to the east, the transmitter and generator buildings for the radar featured standard Thermolite side panels and interlocking sheet-steel roofs.

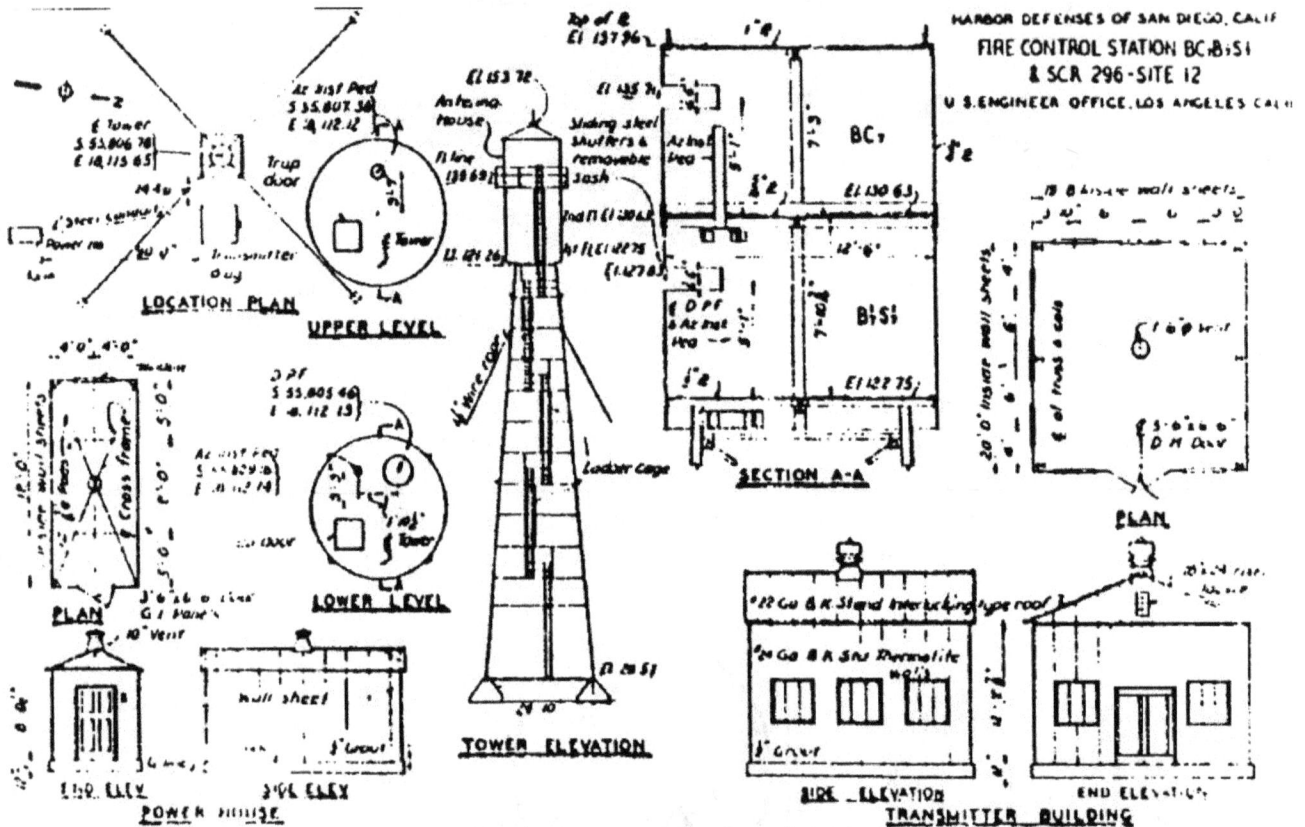

Figure 7-140. Stabilized by four guy wires, the 100-foot tower supporting battery commander station BC_{10}, base end station $B^1_{10} S^1_{10}$, and the *SCR-296-A* fire-control radar antenna for Battery Grant had an enclosed ladder access to the observation levels (NARA). The tower elevation diagram again appears to be generic.

The relevant portion of the 1945 "HDSD Main Cable Routing" blueprint seen in Figure 7-141 shows the combined battery commander station, base end and spotting stations, and fire-control radar for the 6-inch Battery Grant. It is interesting to note there is no indication of the Battalion Command Post 3 tower seen earlier in the 1 July 1945 map of Figure 7-134, which was depicted in dotted lines between the fire-control towers for Batteries Grant and Gatchell. The Report of Completed Works for this two-level tower for Battery Grant, corrected to 8 November 1944, shows a transfer date of 4 October 1944 with a cost of $29,893.52.

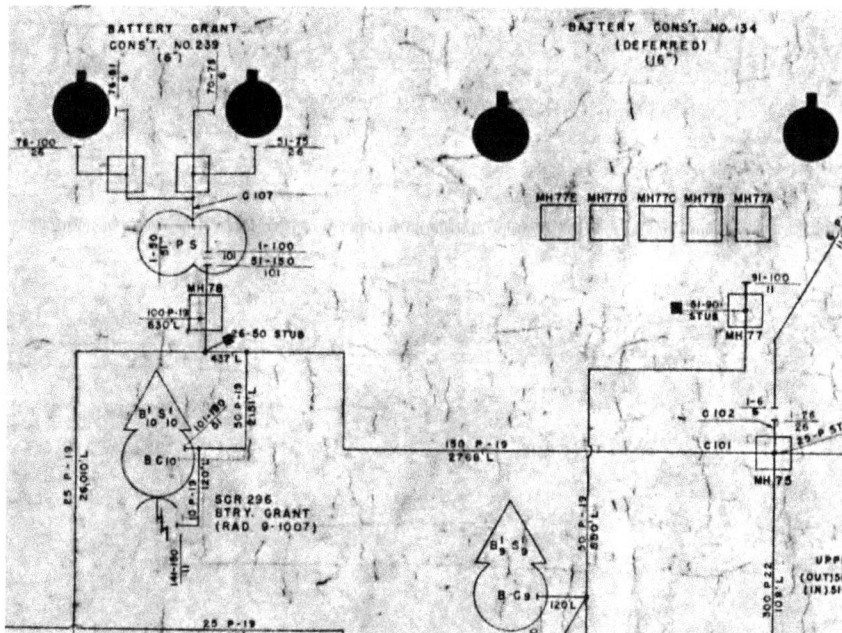

Figure 7-141. Battery Grant, the nearby tower for BC_{10}, B^1_{10} S^1_{10}, and the *SCR-296-A* radar at Fort Emory are shown on the left side of this zoomed-in portion of a 1945 "HDSD Main Cable Routing" blueprint (NARA).

Battery Commander Station for Battery Imperial

The battery commander station for the four 155-millimeter GPF guns at Battery Imperial was situated behind that emplacement, disguised as a 50-foot wooden water tower on reinforced concrete piers. The RCW indicates this wooden structure was transferred on 31 October 1942 at a cost of $7,901.02. The observation room featured a splinter-prof inner wall made of brick, with a steel-plate floor and ceiling, over which was a conical roof made of redwood sheathing. A 180-degree field of view was provided for two azimuth instruments mounted on adjustable bases embedded in 8-inch concrete-filled pipes.

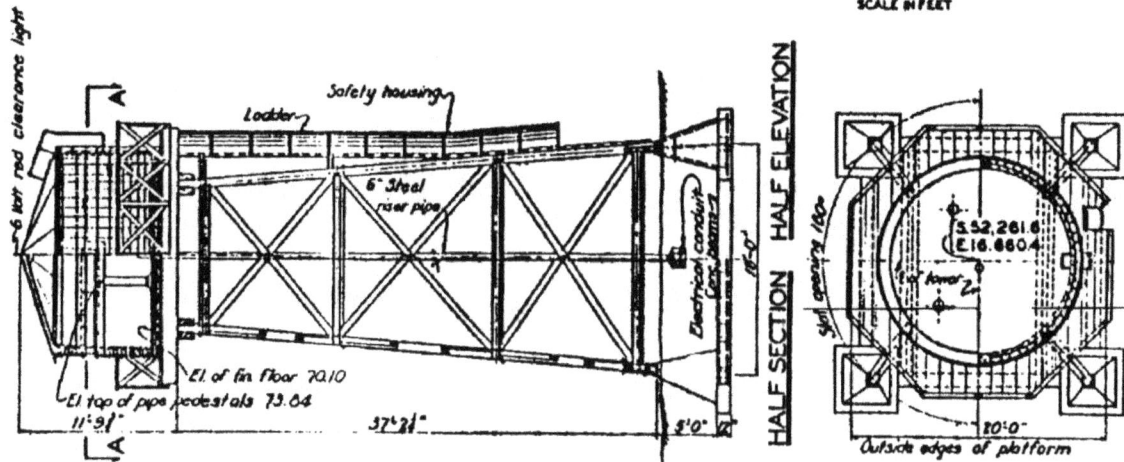

Figure 7-142. Disguised as a wooden 50-foot water tower, the battery commander station for Battery Imperial was located behind the 155-millimeter GPF battery (see again Figure 7-134). This tower is no longer extant, and no additional period photos have been found (NARA).

Site 13 (Border)

Located just north of Tijuana, Mexico, Site 13 (Border) consisted of 22.4 acres plus a permanent road easement (Roper, 1952b). As shown in Figure 7-143, there were four fire-control structures in this area:

- HDOP$_3$, B$^6_2$ S$^6_2$ – B$^2_9$ S$^2_9$ at an elevation of 350 feet.
- *SCR-296-A* radar 9-126 at an elevation of 350 feet.
- B$^4_1$ S$^4_1$ – B$^4_5$ S$^4_5$ at an elevation of 320 feet.
- B$^4_3$ S$^4_3$ – B$^2_{10}$ S$^2_{10}$ at an elevation of 320 feet.

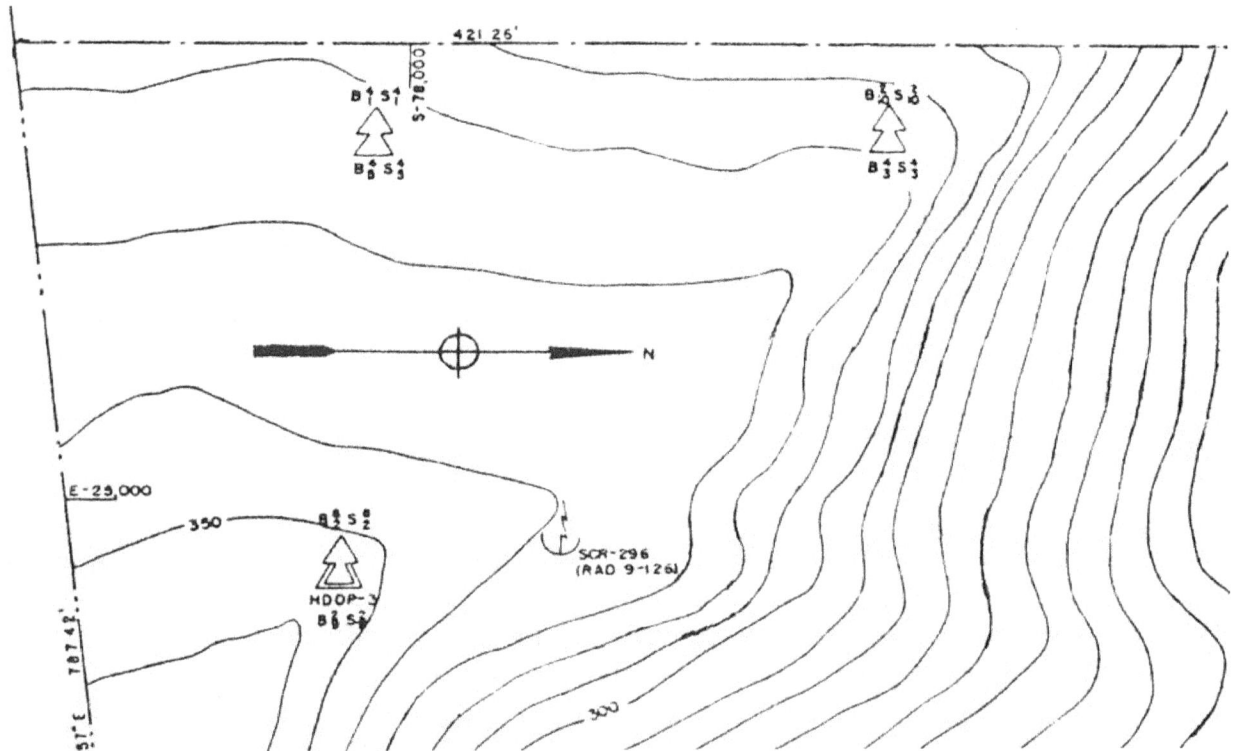

Figure 7-143. Zoomed in portion of 1 July 1945 map entitled "Report of Completed Works, Fire Control Stations at Border, Site 13" (NARA), with the international border at far left and the Pacific Ocean (not shown) approximately a mile due west (up). Note *SCR-296-A* fire-control radar for Battery Strong at bottom center.

All three base end stations at Site 13 featured the same design (Figure 7-144), but their alphanumeric designations changed several times during the course of the war as they were reassigned to different batteries. To minimize confusion, further discussion will use those designations shown in the 1 July 1945 RCW map presented earlier in Figure 7-143 above. Note the unusual vertical extension of the access ladder seen in the upper-right corner of the section view below.

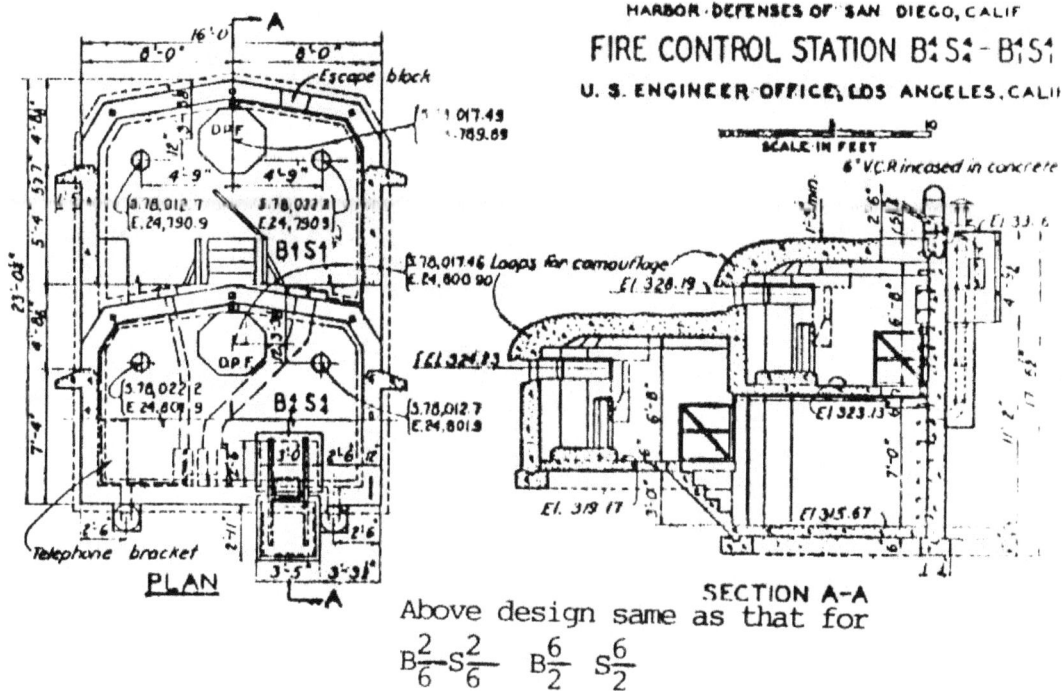

Figure 7-144. The plan and section views for all three fire-control structures at Site 13 (Border) were identical, but their alphanumeric designations changed several times over the course of the war (NARA).

The military "Class A Card" provided by Al Grobmeier seen in Figure 7-145a belonged to Henry L. Pope, who served in Battery I of the 19th Coast Artillery Regiment at Fort Rosecrans during WWII. His overnight liberty pass for Christmas Eve in 1944 shows he was assigned as an observer to the Mexican Border Detachment at Site 13. After the war, Mr. Pope settled in El Cajon, CA, where he worked for 40 years as a cabinet maker for Valley Cabinet before retiring at age 65; he passed away on 3 January 1998 at the age of 80 (UT, 1998).

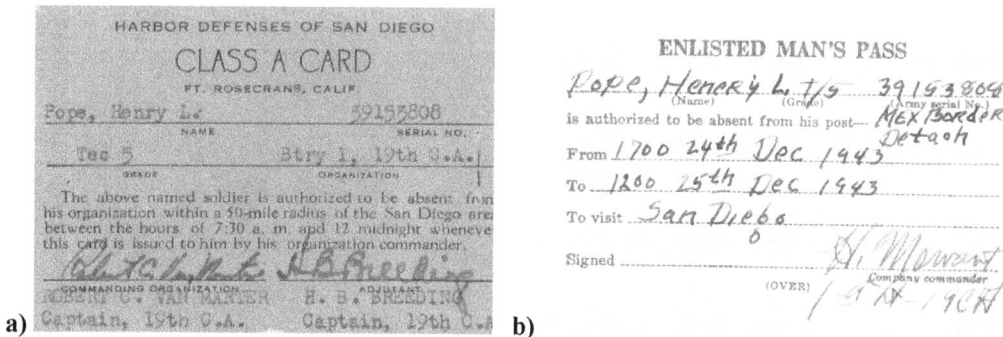

Figure 7-145. a) The "Class A Card" ID for Henry L. Pope indicates he served in Battery I of the 19th Coast Artillery at Fort Rosecrans. **b)** This overnight liberty pass for Christmas Eve, 1944, shows Pope was assigned to the Mexican Border Detachment (images courtesy of Al Grobmeier, Coast Defense Study Group).

Base End Stations for Batteries Strong and Gatchell

The Harbor Defense Observation Post 3 (HDOP$_3$) and base end and spotting stations $B^6_2 S^6_2 - B^2_9 S^2_9$ for Batteries Strong (237, Tac. No. 2) and Gatchell (134, Tac. No. 9) were situated just north of the international border (see again map in Figure 7-143). Constructed at an elevation of 357 feet, these were the highest of the three base end stations at Site 13, with a commanding view of the Pacific Ocean (Figure 7-146a). The associated RCW, corrected to 16 October 1944, shows this two-level structure was transferred on 26 September 1944 with a total cost of $10,916.34.

Figure 7-146. a) The HDOP$_3$ and base end stations B$^6_2$ S$^6_2$ – B$^2_9$ S$^2_9$ are seen looking west towards Tijuana at upper left, with B$^4_1$ S$^4_1$ – B$^4_5$ S$^4_5$ (up arrow) directly in front about halfway to the ocean. **b)** The foundation pillars for the *SCR-296-A* fire-control radar can be seen just left of background center in March 2008.

Looking north towards San Diego, Figure 7-146b above shows the four concrete foundation pillars for the *SCR-296-A* fire-control radar for Battery Strong at Fort Rosecrans. The front of B$^6_2$ S$^6_2$ – B$^2_9$ S$^2_9$ is seen in Figure 7-147 below, looking northeast. The remains of an adobe foundation that presumably dates back to the WWII timeframe or earlier were found just south of this dug-in bunker, but the origin and purpose of this small structure are unknown. Note the steel loops for securing camouflage netting embedded in the concrete roof, and the access-ladder extension above the top level.

Figure 7-147. Front view of the two-level HDOP$_3$ and base end stations B$^6_2$ S$^6_2$ - B$^2_9$ S$^2_9$, circa March 2008, showing steel loops embedded in the concrete for securing camouflage netting. The access ladder against the rear wall of the upper level is barely visible through an open steel shutter.

Base End Stations for Batteries Woodward and Humphreys

Base end and spotting stations $B^4_1 S^4_1 - B^4_5 S^4_5$ for Batteries Woodward (237, Tac. No. 1) and Humphreys (238, Tac. No. 5) at Fort Rosecrans were situated down slope from $HDOP_3$ and $B^6_2 S^6_2 - B^2_9 S^2_9$ at an elevation of 331 feet. The Report of Completed Works, corrected to 16 October 1944, shows the dug-in structure was transferred on 26 September 1944 with a total cost to that date of $10,291.95. This fire-control bunker is seen in Figure 7-148, with the Plaza Monumental de Toros (Bullring by the Sea) and lighthouse in Tijuana, Mexico, in the upper right corner.

Figure 7-148. Rear view of base end stations $B^4_1 S^4_1 - B^4_5 S^4_5$ for Batteries Woodward and Humphreys, with Tijuana's Plaza Monumental de Toros and lighthouse at background right, silhouetted against the Pacific Ocean, circa March 2008. Note access-ladder extension to the right of the counter-weight housing.

A front view of the two-level $B^4_1 S^4_1 - B^4_5 S^4_5$ bunker is seen in Figure 7-149 below, looking upslope to the northeast. Note the distinctive impressions around the roof-structure perimeters left by the stacked sandbags used as a form to contain the concrete pour during construction. All three fire-control structures in this area were surprisingly well preserved with no signs of graffiti, as they are in plain sight from the coastline down below, where the US Border Patrol maintains a consistent and obvious presence. We gave them a wave as we drove in from the north and they left us free to explore.

Figure 7-149. Front view of base end stations $B^4_1 S^4_1 - B^4_5 S^4_5$ just north of the international border, looking northeast in March of 2005.

Base End Stations for Batteries Ashburn and Grant

In his 13 June 1992 interview with Howard Overton (1993), Sergeant Major John Hennely, Battery C, recalls his service at $B^4_3 S^4_3 - B^2_{10} S^2_{10}$, which supported Batteries Ashburn (Tac. No. 3 at Fort Rosecrans) and Grant (Tac. No. 10 at Fort Emory):

> "I was mainly an observer of the guns at Fort Emory, they called it. Down on the Mexican border. They had a radar unit down there. The chief of section of radar was Pete Sagan from Chicago. ... we overlooked what is known now as Border Field. And the Navy fired their guns down there and the Marines manned the antiaircraft guns."

Figure 7-150 below shows the concrete roof and trap-door entry for base end and spotting stations $B^4_3 S^4_3 - B^2_{10} S^2_{10}$, looking northwest towards what is now Border Field State Park.

Figure 7-150. Base end stations $B^4_3 S^4_3 - B^2_{10} S^2_{10}$, looking northwest towards Border Field State Park. The southern tip of Point Loma is barely visible in the mist just right of top center. Note the apparent concrete patch, and the lack of an access-ladder extension as found on the otherwise identical bunkers in Site 13.

During WWII, Border Field was a Naval Outlying Field (NOLF) reporting to NAS San Diego, located on a 100-acre parcel of land about 2 miles south of Ream Field (Everett, 1995). Its primary function was to host an aerial free-gunnery school, for which it was equipped with five target ranges along the Pacific coastline (Figure 7-151a). Rail carts carrying sleeve targets ran along the serpentine track seen in Figure 7-151b, while Navy gunners in stand-alone turrets fired colored ammunition at them to master the concept of leading their targets (Jones, 2000).

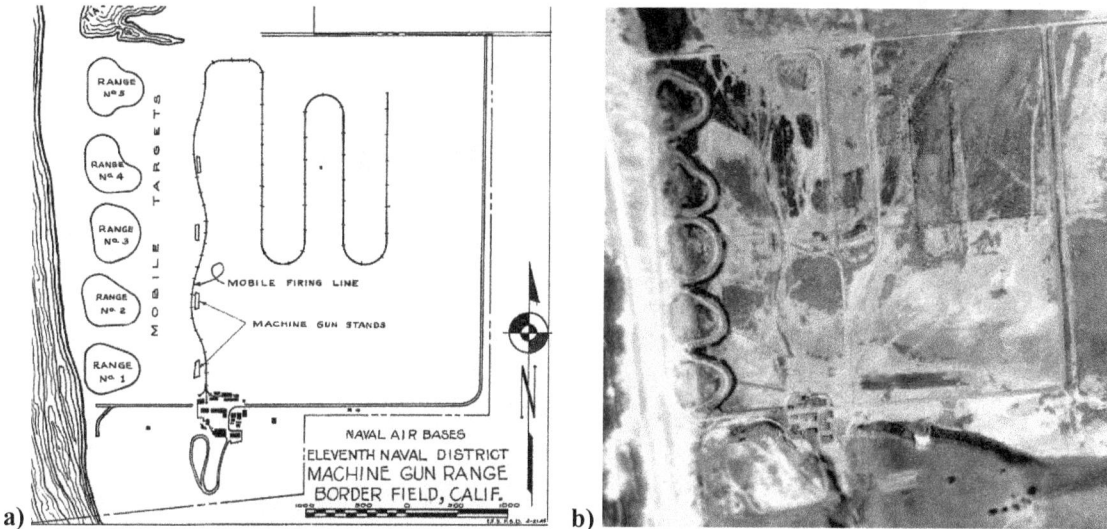

Figure 7-151. a) WWII map of the US Navy Machine Gun Range at Border Field, now Border Field Sate Park (NARA). **b)** The remains of the five sleeve-target ranges at NOLF Border Field can be seen along the coastline at image left, circa April 1945 (photo courtesy San Diego County Annex).

Figure 7-152 below provides a front view of base end and spotting stations $B^4_3 S^4_3 - B^2_{10} S^2_{10}$, looking northeast in March of 2005. One interesting observation common to all three fire-control bunkers at Site 13 was that they were missing their steel entry hatches, which were found mostly intact at other HDSD sites. One possible explanation for this anomaly might be the hatches were removed by the Border Patrol to prevent migrants from becoming entrapped inside while seeking shelter, although such is purely speculation.

Figure 7-152. Todd Everett consults his map of the surrounding area while standing in front of base end and spotting stations $B^4_3 S^4_3 - B^2_{10} S^2_{10}$, looking northeast in March 2005. Cowles Mountain in the San Carlos area of San Diego is just barely visible on the horizon at left.

Transfer of Army Tactical Cable

In the early 1950s, the Navy expressed interest in the retention of the Government-owned land at Site 13 (Roper, 1952a):

> "The Border Field Fire Control Station site is located adjacent to Border Field where the Navy Electronics Laboratory is carrying on experimental work. This site has a potential value as a location for a broadcasting transmitter, and the operation of such transmitter, if in private control, would completely nullify the usefulness of the Border Field Station operated by the Electronics Laboratory. It is considered a protective measure that this land be retained under Government control."

The Army Tactical Cable that provided the communication backbone for the Harbor Defenses of San Diego had been declared surplus in November 1948 (Houston, 1955). Transfer of the southern leg was requested by the Commandant, Eleventh Naval District, on 26 March 1952, with follow-on (30 July 1952) correspondence citing ongoing and intended usage (Kurtz, 1952):

> "**a.** The segment (Army Cable No. 4) from Fort Rosecrans (3-BD 75) to the North Boundary of Fort Emory (2-BD 75) has been requested by the Officer in Charge, Harbor Defense Unit, San Diego for use in connection with Harbor Defense telephone requirements as they arise...
>
> **b.** Cable throughout Fort Rosecrans is currently being used by the Harbor Defense Unit and the Electronics Laboratory for experimental purposes in connection with harbor defense...
>
> **c.** Cable in the Fort Emory area (now part of the Radio Station, Imperial Beach) is being used to maintain normal telephone communications with the Radio Station...
>
> **d.** Cable near the Mexican Border is being used by the Electronics Laboratory for use in connecting with their operations at Border Field..."

This southern leg from Fort Rosecrans to the Mexican border was formally transferred to the Navy in December 1953 (Fogler, 1955). A request for transfer of the north leg (*SCD-936*) was submitted by the Navy in 1955 (Houston, 1955):

> "By letter dated 21 June 1955 the Department of the Navy, Bureau of Yards and Docks, made a formal request for transfer of the entire SCD 936 cable line, with such interests as the Government may have in the cable rights of way, and stated that the requisite approvals for the transfer had been obtained."

This request was granted the following year, with control of subject property transferred to the Department of the Navy effective 13 January 1956 (Rea, 1956).

8
Searchlights

> "Anyone who has to fight, even with the most modern weapons, against an enemy in complete control of the air, fights like a savage against modern European troops, under the same handicaps and with the same chances of success."

Field Marshall Erwin Rommel

Development History

Prior to radar, all artillery was by necessity aimed through visual means, and if you couldn't see it, you likely couldn't hit it. High-intensity narrow-beam searchlights were therefore introduced to illuminate targets at night, so range and bearing could be determined for a gun-laying (fire-control) solution. Convened in 1886, the Endicott Board of Fortifications recommended installing 200 lights for US coast defense, and by 1901 the Army Board of Engineers had increased this number to 344. A variety of sizes (i.e., 24-, 30-, 36-, and 60-inch) were evaluated, but by 1902 only the 36-inch (Figure 8-1) and 60-inch options received much further attention (Suter et al., 1902).

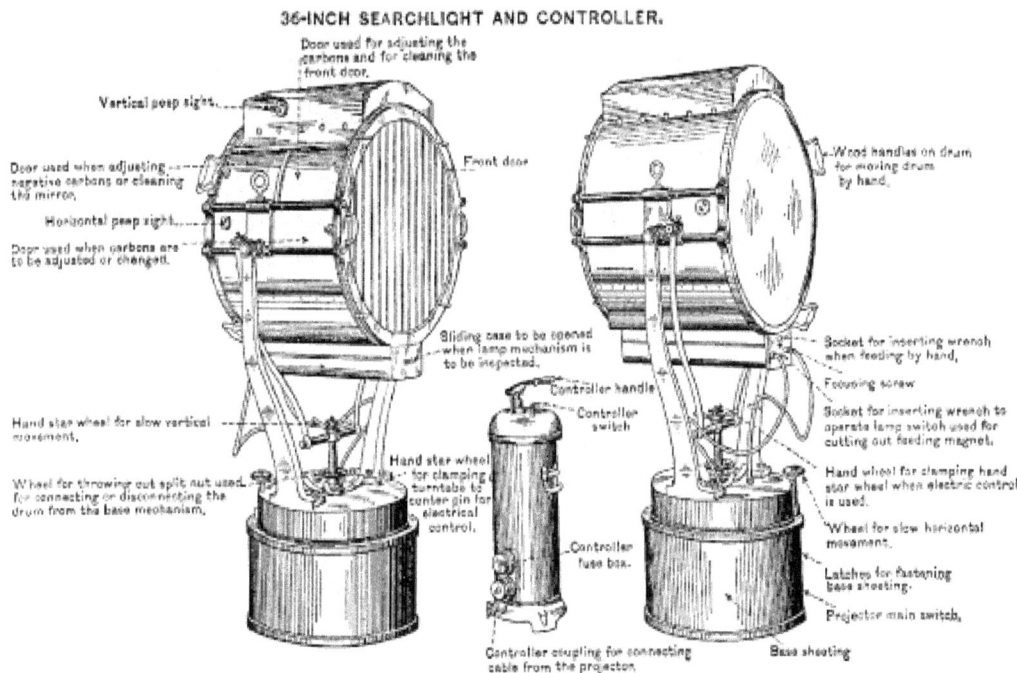

Figure 8-1. The 36-inch searchlight, which cost $1800 in 1902, drew 130 amps at 55 volts DC (reproduced from Smith, 1999, Coast Defense Study Group).

When later improvements to the gasoline engine enabled portable generation of DC power (Miller, 1917), searchlights were used to create "artificial moonlight" for nocturnal ground attacks during WWI, an offensive strategy developed by tank-warfare theorist J.F.C. Fuller (Wikipedia, 2006). Turn-of-the-century versions of the 60-inch light required 200 amps DC at 60 volts and were usually serviced by a 110-volt DC generator to account for cable losses, often powered by a 40-horsepower steam engine with a 55-horsepower boiler (Suter et al., 1902). Two DC motors in the searchlight base provided pan-and-tilt actuation. A combination of 2-to-1 gear reduction and a series resistor that halved the drive-motor rpm was used to achieve four different operating speeds for azimuth traversal.

Since aircraft had not yet been widely introduced, these early lights were intended for maritime targets, with limited elevation control, approximately 11 degrees below to 32 degrees above horizontal (Roessler, 1904). A series resistor allowed for two different speed options in elevation. Electrical remote control, not yet fully trusted in 1901 (Robert et al., 1901), was strongly endorsed just three years later. Initially limited to a maximum operator standoff of 1000 feet, a significantly improved scheme was introduced by General Electric in 1904 that could control their 30- and 60-inch projectors out to 16,000 feet (Roessler, 1904).

The development of the antiaircraft searchlight began in the 1930s in response to the unsettling combination of political unrest in Europe and the aggressive development of air power (Hill, 1993). Some effective means was required to illuminate enemy planes for nighttime engagement by defensive batteries, in an effort to increase their rather dismal effectiveness against the ever-escalating airborne threat. It is estimated, for example, that British gunners in 1940 expended an average of 20,000 rounds of ammunition for each German aircraft destroyed (Skylighters, 2006).

It is perhaps useful at this point to examine the three basic components to the evolving antiaircraft fire-control solutions that addressed this critical need: 1) locator devices that determined the elevation and azimuth of incoming aircraft; 2) searchlights to illuminate the identified targets; and 3) the antiaircraft guns and artillery that would subsequently open fire.

Early locator devices were large sound-gathering horns mechanically coupled to the operators' ears as shown in Figure 8-2. On 9 September 1940, two *M-2 Sound Locators* were received by Battery N (later redesignated Battery K) of the 19th Coast Artillery at Fort Rosecrans (Gaines, 1993). Although manpower intensive and only marginally effective, this primitive system was the best solution available at the time. Its operators would attempt to binaurally balance the engine noise of incoming aircraft by mechanically rotating the horn array in the direction of maximum intensity. To simplify the process, one man worked the elevation axis while the other concentrated on azimuth. The resulting pan-and-tilt angles were electrically coupled to the searchlight-control station through a master/slave selsyn (self-synchronizing) network, a key component of WWII-era control systems.

Figure 8-2. One operator listened to the left and right horns of this early "sound locator" to determine azimuth, while the other used the upper and lower horns to determine elevation. The Acoustic Correction Operator at left attempted to compensate for the sound delay (US Army photo courtesy *Skylighters.org*).

The selsyn, later known as the "synchro," was an early form of electrical shaft encoder used to transmit angular information from one place to another with great precision (Schwartz & Grafstein, 1971). In essence, the selsyn forms a variable-coupling transformer with an AC-excited rotor winding (primary) and two or more stator windings (secondaries) symmetrically arranged around the rotor. The effective magnetic coupling between the rotor and the surrounding stator windings thus varies as a function of rotor orientation. As a consequence, the stator outputs form a set of AC signals, the relative magnitudes of which uniquely define the rotor angle at any given point in time (Everett, 1995).

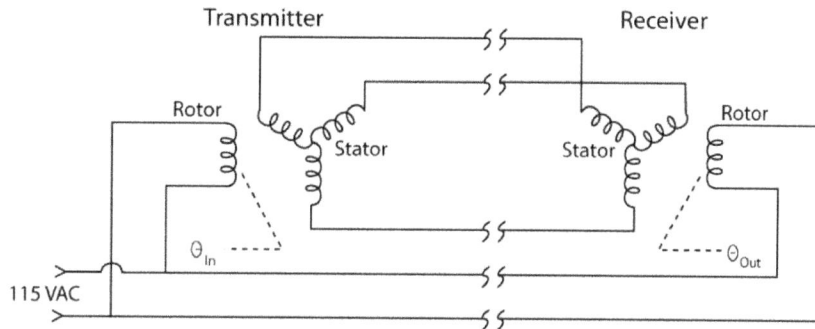

Figure 8-3. Schematic diagram of a typical remote-indicating selsyn (i.e., synchro) configuration (adapted by Lindsay Seligman from Schwartz & Graftstein, 1971).

A number of selsyn types exist, but the most widely known configuration is probably the three-wire transmitter-receiver pair commonly used for remote shaft-angle indication. The slave selsyn receiver is electrically identical to the transmitter and connected so that the three stator windings for both devices are in parallel as shown in Figure 8-3 above. The rotor windings on both the remote transmitter and the receiver ("CX" devices in the shorthand notation of synchro parlance) are excited by a common AC source supplied through slip rings (ILC, 1982). When the receiver and transmitter rotors are in identical alignment with their respective stator windings, the individual stator outputs will be equal for the two units, and consequently there will be no current flow (Everett, 1995).

If the transmitter rotor shaft is turned by some external force, this equilibrium condition is upset, and the resulting voltage differences generate current flows in both sets of stator windings. These current flows induce an identical torque in both rotors, but since the transmitter rotor is constrained, the torque on the receiver rotor acts to restore alignment and thus equilibrium (Diermengian, 1990). As a result, the receiver output shaft would precisely track any rotational displacement seen by the remotely located transmitter input shaft. While this was an effective means of moving a needle or dial to display some measured angle, there was insufficient torque available for much else, such as repositioning a bulky 60-inch searchlight assembly.

To address this problem, the selsyn control transformer, or "CT" for short, was introduced to output a two-wire signal proportional to the sine of the angular difference between two rotating devices, as shown in Figure 8-4a (ILC, 1982). The output of the control transformer could then be amplified and used to drive a much larger servomotor that could handle the desired load. Alternatively, the CT was often employed in conjunction with a zero indicator (Figure 8-4b), which provided a simplified visual display of the direction and magnitude of misalignment between the two shafts. As we shall soon see, this interim approach allowed a human operator to serve as a surrogate for the more complex component needs of the desired servo-control loop, until such time as the required technology was readily available.

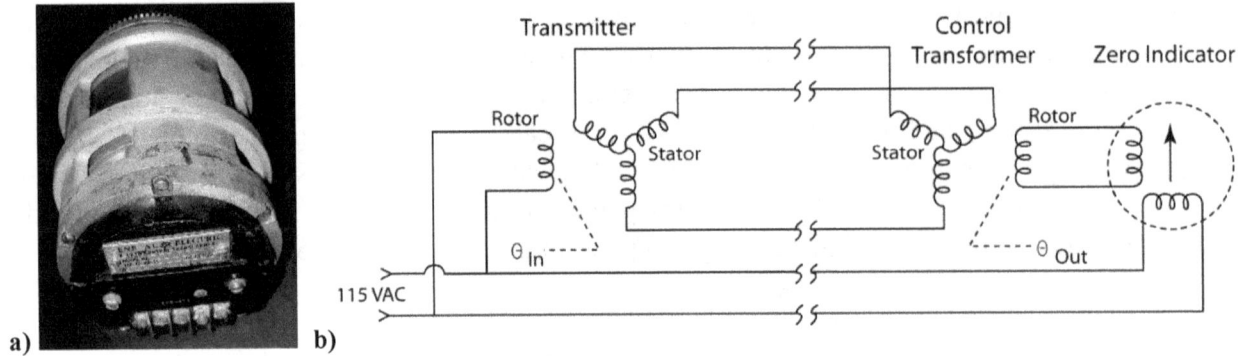

Figure 8-4. **a)** An original General Electric control transformer (CT) restored by Bob Meza, Santa Clarita, CA. **b)** The ubiquitous WWII zero-indicator (far right) was a zero-center AC voltmeter (with dual inputs for rotor and stator) that compared the amplitude and phase of the transmitter (CX) and CT rotor windings to indicate the degree and direction of shaft misalignment (drawing by Lindsay Seligman).

For remote operation of the searchlight projector, the selsyn signals from the sound locator were passed to the searchlight control station, also known as a Distant Electric Control (DEC) unit, typically manned by three operators. Two of these men turned azimuth and elevation hand cranks in response to zero indicators that displayed the selsyn mismatch between the DEC and the sound-locator device. Once a particular DEC axis was in alignment with the locator, the associated indicator would point to zero at center scale (Figure 8-5). By watching which direction and how far his needle deflected from center, each operator knew which way and how fast to turn his crank to restore alignment.

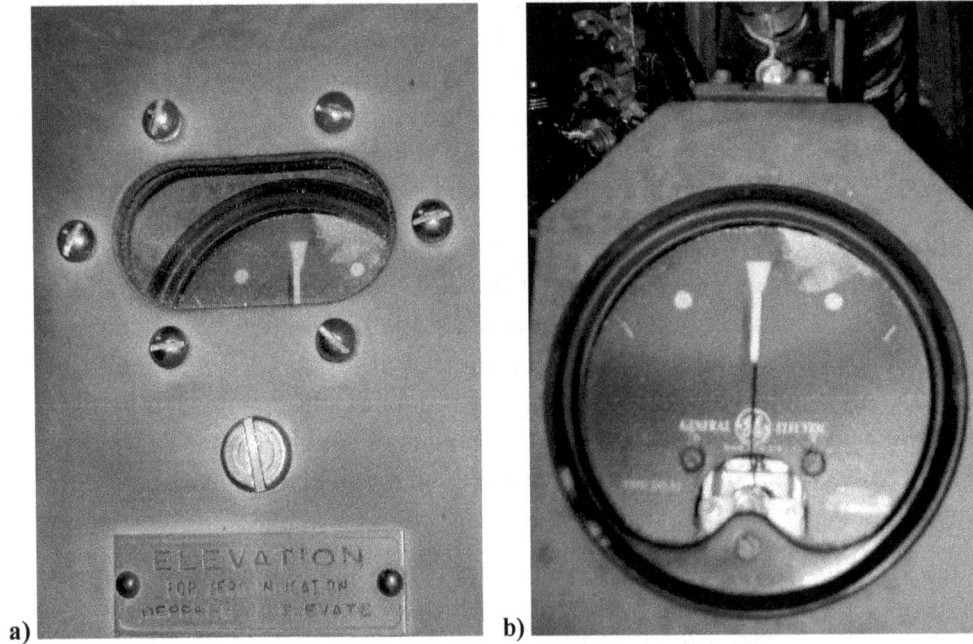

Figure 8-5. **a)** The elevation-axis zero indicator as seen through the watertight viewing port of a WWII Distant Electric Control (DEC). **b)** The zero-indicator inside the DEC unit with cover removed (photos courtesy Bob Meza).

The third operator (Figure 8-6a) looked through a pair of binoculars attached to a harness that physically followed the pan-and-tilt motions of the DEC (Figure 8-6b), and was therefore constrained to point directly at the perceived aircraft location. Once the target was visually acquired in such fashion, it could be more precisely tracked by this "observer," who manipulated a second set of hand cranks mechanically coupled to those of the other two operators.

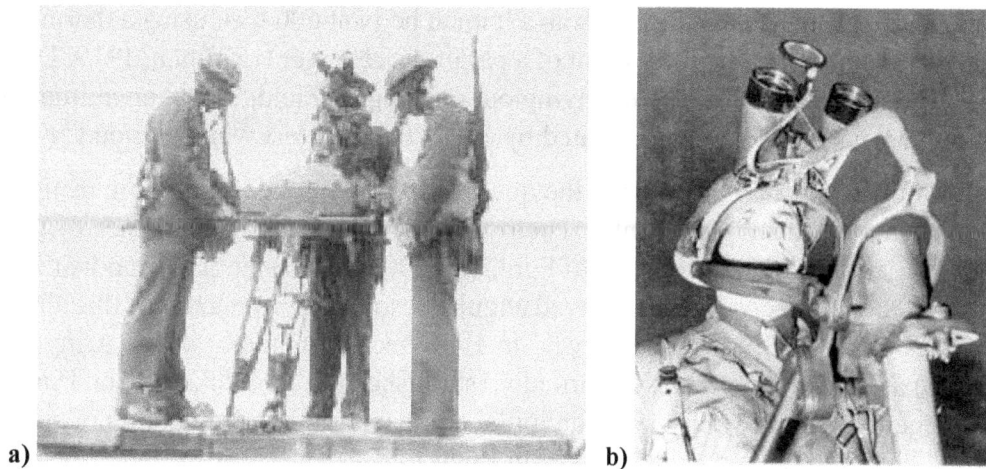

Figure 8-6. a) Three men were typically required to operate the *DEC* station in an antiaircraft role. **b)** Once an aircraft was visually acquired through a pair of binoculars attached to the pan-and-tilt axes of the DEC, the observer took over and tracked the target (US Army photos).

Selsyn signals from the DEC were sent to the slaved searchlight so it would point to where the third operator was aiming the binoculars. In other words, the sound locator enabled initial target acquisition for the observer, who in turn provided higher angular resolution by visually positioning the searchlight (Figure 8-7). From an engineering perspective, one human in this servo loop (the observer) stood in for an as yet unavailable two-dimensional optical detector, while the other two operators served as motor amplifiers and output actuators.

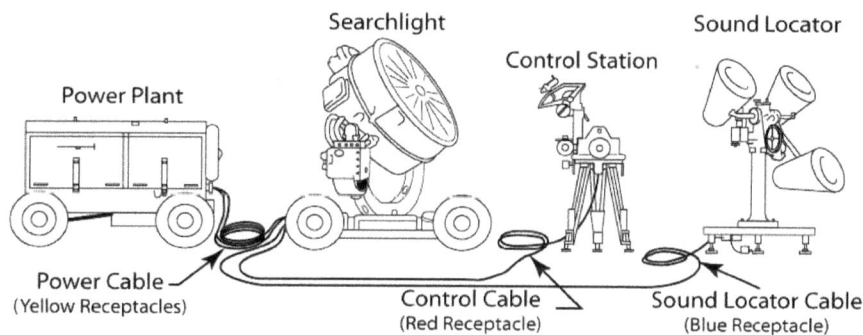

Figure 8-7. A six-cylinder Hercules *JXD* engine (left) provided DC power to the searchlight, where a 78-volt DC dynamotor was used to generate 115-volt 60-Hz AC for the various selsyns on the light, the control station (DEC), and the sound locator (redrawn by Lindsay Seligman from TM-7111, 1942).

While conceptually simple, sound locators were unable to keep up with the ever-increasing speeds of WWII aircraft relative to the rather slow velocity of sound in air (761 mph at sea level and 59° F). To put things into perspective, the top speed of the Mitsubishi *AGM5 Zero* was 346 mph, as compared to less than 140 mph for typical WWI aircraft. The telltale sounds of an approaching *Zero* at 10 miles would not actually arrive at the detector until around 47 seconds later, during which time the aircraft would have flown another 4.5 miles, almost half the intervening distance. The sound locator thus heard where the aircraft had been, as opposed to where it currently was, and the difference could be rather substantial.

As operational speeds continued to increase, the problem would only get worse. Towards the end of the war, both sides had introduced jet aircraft, with Germany's Heinkel *He 162* capable of 562 mph. In recognition of this shortfall, an attempt was made to detect the thermal emissions from aircraft engines, based on earlier work done by Master Signal Electrician Samuel Hoffman. Back in 1918, Hoffman had been able to repeatedly sense the thermal radiation from a human body at 600 feet, using a thermopile (i.e., a series of thermocouples) situated at the focal point of a parabolic reflector (Hoffman, 1919; Davis, 1943). In January 1919, Hoffman had detected a *Jenny* flying out of Langley Field, VA, at an altitude of 3,500 feet, but equally strong thermal signals were generated by even "the slightest whiff of cloud" (Terrett, 1956).

Further work followed during the 1920s in the quest for a truly practical solution, using either bolometers or thermopiles as the sensing elements. There were two perceived advantages driving this strong interest: 1) thermal energy travels at the speed of light, 882,000 times faster than sound; and 2) thermal energy can be more effectively focused for improved angular resolution. In pursuit of this alternative means of aircraft detection, Major William R. Blair, who in 1930 became Director of the Army's Signal Corps Laboratories (SCL) at Fort Monmouth, NJ,[1] formally established "Project 88, Position Finding by Means of Light." Work began in February 1931, with an initial emphasis on the continuation of an active near-infrared scheme that had been reassigned to SCL from Frankford Arsenal (Davis, 1943; Terrett, 1956).

The basic idea was to illuminate the target with a searchlight and detect any reflected energy with a parabolic mirror focused upon a newly invented Thalofide cell.[2] In August, 1932, a Navy blimp was successfully tracked to a distance of 6,300 feet. Subsequent extrapolation out to the desired range of 32,000 feet, however, showed that even an array of four 60-inch lights would be insufficient illumination for reliable operation (Davis, 1943). Additional problems arose due to specular reflection from shiny surfaces, such as the aluminum used in aircraft skins. Consequently, the reflective near-infrared approach was abandoned in 1933, and attention shifted back to the passive far-infrared scheme pioneered by Hoffman in 1918.

About this time, the Eppley Laboratory, Inc., Newport, RI, introduced a dual-element thermopile design for common-mode rejection of heat radiated from clouds, two of which were delivered to SCL in March, 1934 (Davis, 1943):

"In April 1934, an Eppley-compensated thermopile in a 60-inch parabolic metal reflector was set up in a seacoast searchlight mounting at Battery Halleck, Fort Hancock, overlooking the commercial channel to New York. During the ensuing months, tugs, freighters, tankers and small passenger boats were picked up and followed to distances of 5,000 yards in drizzling rain, 10,000 yards in fair weather. Two big liners – the German Bremen and the British Olympic – were tracked on the outbound ocean course to the horizon 20,000 yards away."

In November of that same year, one of two 1934-model Army searchlights acquired by SCL was modified by removing the carbon-arc burner assembly and replacing it with an Eppley-compensated thermopile (Davis, 1943). Further testing of this configuration was done at Navesink Lighthouse in 1935, with the detection system automatically directing the second searchlight to verify accuracy (Zahl, 1972). Meanwhile, unaware of SCL efforts in this area, the Army Corps of Engineers (which held jurisdiction over searchlight development) had in December 1933 let a $100,000 contract with General Electric for "development of a heat-detection device."

When the apparent duplication came to light, it was ultimately decided that the Signal Corps would retain authority to pursue the thermal detection of waterborne targets, while the Corps of Engineers would have responsibility for aircraft detection (Davis, 1943). A comparative test of the prototypes developed by each of these organizations was conducted in 1936 by the Coast Artillery Board at Fort Monroe, VA (Zahl, 1972). Both systems were found to be adequate for detecting ships and subsequently directing a searchlight, but insufficiently advanced for aircraft detection or fire control.

---

1   Now the Communications and Electronics Life Cycle Management Center (C-E LCMC), Fort Monmouth, NJ, previously CECOM.

2   A glass-enclosed photoresistor that experienced a drop in impedance when exposed to heat or light, made of thalium, oxygen, and sulphur fused on a quartz disc (Davis, 1943).

Despite its perceived advantages over sound locators, there were several inherent problems with the thermal approach for detecting and tracking aircraft. For starters, the heat signature of an airplane engine was easily masked by cloud cover or fog, and could also be minimized at the source through proper shielding. Secondly, attaining sufficient detection sensitivity had required a tradeoff (i.e., reduction) in the overall field of view, which aided tracking but seriously complicated initial target acquisition.

Accordingly, in a synergistic combination of competing solutions, the thermal tracker was briefly employed to compensate for the poor angular resolution of early radar equipment. The wider beam of the radar would direct the thermal sensor towards an incoming target, whereupon the thermal tracker took over to steer the tightly focused beam of the searchlight. Figure 8-8 shows a thermal tracker that employed the SCL compensated-thermopile detector installed in a mechanical structure developed by General Electric for the Corps of Engineers, with part of a radar antenna in the background right.

Figure 8-8. Early thermopile-based heat detector developed by SCL installed in a modified searchlight reflector assembly designed by General Electric for the Corps of Engineers, undergoing testing in 1937. Part of the early radar azimuth antenna is visible at far right (US Army photo adapted from Davis, 1943).

Towards the mid-30s, Dr. Harold Zahl, Director of Research at SCL, had meanwhile pioneered a new and innovative detector approach based on a mosaic cluster of pneumatic cells. The principle of operation called for measuring the thermal expansion of an inert gas within each cell in response to incident radiation focused upon the array by the parabolic reflector. A heat-transmitting stibnite (crystalline antimony trisulphide) window about the size and thickness of a dime formed the sensing surface of each cell, which was thus fairly transparent to thermal energy in the 9- to 11-micron region of the spectrum (Zahl, 1972). The other end of each cell was sealed off by a thin flexible membrane that would bulge slightly as the internal gas expanded upon heating.

Since the footprint of incident radiation upon the 61-cell mosaic was tightly focused by the mirror, only a few cells were being heated at any given time, and the lateral position of this thermal concentration was directly related to the target location within the optical field of view. As a consequence, the array not only detected the thermal signature of an aircraft, it also provided X-Y offset information relative to the beam centerline of the receiving optics. This two-dimensional resolution enabled continuous adjustment of mirror elevation and azimuth to keep the heat source centered, thus tracking the target. The principle challenge, of course, was how to measure the infinitesimal thin-film deflections that arose from the thermal expansion associated with each cell.

As Zahl (1972) later explained:

"By placing this thin film into the optical portion of an interferometer, this minute bulge could be observed as Newton's Rings showing up through a viewing telescope. First experimental cells showed a potential of detecting the heat from an airplane at 20-25,000 yards when conditions were good. A mosaic of such cells would have the potential to chart the motion of the aircraft as it moved about within the field of view of the 60-inch mirror."

So in the fledgling days of rather primitive electronics, the required receiver sensitivity for this ambitious thermal tracking application was achieved through use of clever optics, to include a 60-inch aperture and interferometric measurement techniques at the detector. In retrospect, Zahl had essentially produced a simplistic two-dimensional uncooled FLIR (forward-looking infrared) imager, albeit one with extremely limited resolution (an array size less than 8 x 8). Reducing this concept to practice was another classical case of the Army's human-in-the-loop design philosophy, which enabled complex systems to be fielded ahead of their time, well before all elements of the supporting technology were readily available (Mindell, 1995).

Figure 8-9 shows Zahl's improved thermal locator undergoing service testing at Fort Monroe, VA, in 1938, with a third operator observing the cell interference patterns through a telescope-like device (see again Figure 8-9). His job was to continuously adjust the azimuth and elevation angles of the parabolic mirror using hand cranks to keep the heat source centered in the "image." The other two operators provided initial azimuth and elevation control in response to the radar input. (A stand-alone transmit antenna for the *SCR-268-T1* radar prototype is partially visible in the far-right background.) Unfortunately, the late-fall weather at Fort Monroe was consistently cloudy and overcast, which caused the thermal detector performance to suffer considerably (Terrett, 1956).

Figure 8-9. Zahl's improved thermal locator, which employed an array of pneumatic cells situated at the focal point of a 60-inch searchlight mirror, undergoing Service Testing at Fort Monroe, VA, in December, 1938. Note the additional operator required at far right (US Army photo courtesy C-E LCMC).

Meanwhile, however, the capabilities of all-weather radar systems had been steadily improving. The advent of more sophisticated vacuum tubes with increased power output at higher frequencies enabled a shift from 110 MHz to around 240 MHz, resulting in shorter wavelengths with better resolution. More importantly, a new double-array receive-antenna configuration was introduced that enabled a technique known as "lobe switching" for even greater accuracy in azimuth (Terret, 1956):

> "Experiments and dry runs increasingly pointed toward final elimination of the thermal element. Its claim had been accuracy. Project engineers now believed that they could achieve a comparable accuracy without so much trouble. A method known as lobe switching, suggested by the principle of the radio range station, brought about this change."

As a consequence, when the *SCR-268-T3 Searchlight-Control Radar* was introduced into service in early 1939, the thermal locator was no longer needed, as illustrated in Figure 8-10.

Figure 8-10. Artist's concept of the mobile *SCR-268-T3* radar controlling a portable 60-inch searchlight. The DC generator for the searchlight is at far left, while the AC generator for the radar equipment is seen just inside the rear of the truck at lower right (US Army graphic courtesy of C-E LCMC).

The WWII searchlight itself was an amazing piece of equipment, producing a tightly focused 800-million-candlepower beam of light with an effective range of over 5 miles. Like its predecessors, the light source was a high-current electric arc between two carbon rods (Figure 8-11) maintained at the focal point of a highly polished 60-inch parabolic mirror. Made of rhodium-coated brass,[3] the mirror assembly alone weighed 180 pounds, and the combined weight of a portable searchlight and its associated generator was 3 tons.

---

3   The earlier WWI cart and lift-mounted searchlights had used glass mirrors.

Figure 8-11. One of many carbon-rod stubs found scattered around the Fort Rosecrans silo enclosure for WWI-era Searchlight 2, later Searchlight 12 during WWII. This is the larger-diameter positive rod.

The majority of the electromechanical arc-control components were housed in a weatherproof box mounted on the right side of the drum, away from the heat and smoke. Manufactured in large quantities by both General Electric and Sperry Gyroscope, a total of 10,000 portable units were produced between 1932 and 1944 at a cost of $60,000 apiece, in excess of $850,000 each in 2006 prices (Meza, 2006). While most of these mobile searchlights were destined for the European theater, some were employed to augment the fixed-location lights used by coast-defense artillery along US shores.

Since the carbon rods that produced the light-generating arc were gradually consumed by the process, they had to be progressively repositioned to maintain optimal performance. The key issues were arc position (i.e., relative to the focal point of the mirror) and arc length (i.e., the gap between rods). To simplify matters, the designers decoupled and addressed these issues separately in the servo-control loops for the two rods. A continuously running 78-volt DC motor inside the control box was connected via electromagnetic clutches to three output shafts, which actuated rotation and feed of the positive rod as well as feed or retraction of the negative rod (Figure 8-12).

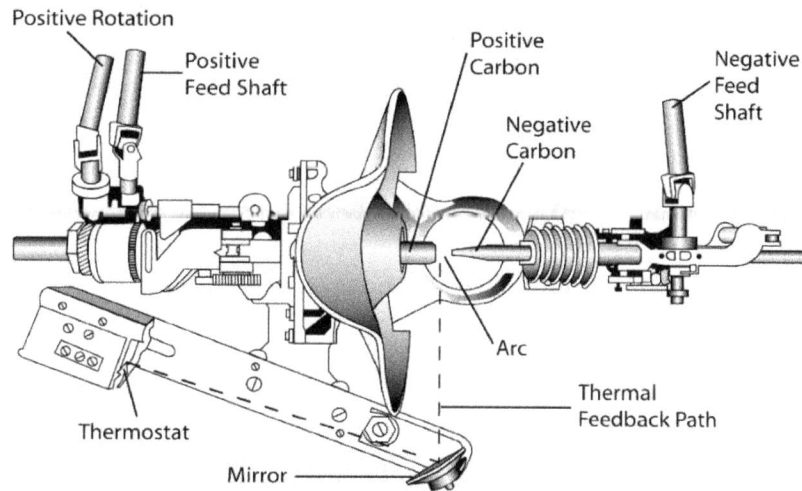

Figure 8-12. Essential elements of the positive (left) and negative (right) rod-feed mechanisms of a GE searchlight (redrawn by Lindsay Seligman from *TM 5-711*, 1943). The small mirror at bottom center focused a portion of the light from the arc on a bimetallic thermostat.

Arc position was the dominant variable, controlled by adjusting the feed rate of the larger-diameter positive carbon, which was inserted from the front of the light through a small hole in the lens. The centroid of the arc had to be rather precisely maintained at the focus of the 60-inch reflector to ensure minimal beam divergence, typically around 0.5 degrees. A rotating cam (driven by the 78-volt motor in the control box) was used to actuate a leaf switch connected to the electromagnetic clutch for the positive-rod-feed mechanism, much like the mechanical points on an older-model automotive distributor switched power to the ignition coil. Each time the leaf-switch contacts were closed by the cam lobe, the clutch coil would be briefly energized to incrementally advance the rod.

An additional closed-loop control strategy, based on the sensed location of the arc, further boosted the feed rate when the above open-loop scheme began to fall behind. The electrical clutch that energized the feed mechanism could also be activated by a bimetallic-strip thermostat mounted on a support bracket on the right-hand side (looking in through the front lens) of the burner assembly as shown in Figure 8-12. A small mirror on the other end of this bracket reflected thermal energy from the arc towards this thermostat, adjusted so the bimetallic strip began to heat up when the positive rod burned back away from the focal point of the 60-inch reflector.[4]

The asymmetric expansion of the two metals that comprised this bimetallic strip caused it to bend, eventually closing a set of electrical contacts that energized the positive feed mechanism. As the tip of the rod advanced, the resulting shift in arc position reduced the amount of thermal energy captured by the mirror feedback path, allowing the thermostat to cool, which in turn de-energized the feed clutch. Meanwhile, arc length was held constant by adjusting the feed position of the negative carbon, accomplished in the GE design by directly sensing the arc current flowing through a magnetic coil. At 78 volts DC, the desired arc length of 1 inch corresponded to a current flow of 150 amps (Meza, 2006).

As the rods were mechanically fed into the arc, the gap would grow smaller, causing the current flow to increase. Similarly, as the carbon burned away, the gap would grow larger with a corresponding drop in arc current. A simplistic (but effective) double-throw center-off relay was directly actuated by the current-sense coil, calibrated to assume the off position when the arc current was 150 amps. If the current decreased below this desired set point, the relay armature would energize an electrical clutch to actuate

---

4   This description applies to the portable GE light only. On the Sperry version, the thermostat was on the inside wall of the drum, just below the control box, and a lens was used instead of a mirror (Meza, 2006).

the negative rod-feed mechanism. If, on the other hand, the arc current grew too large, the relay armature would energize a complimentary clutch to turn the feed mechanism in the opposite direction and thus retract the rod.

The train and elevation motions of the searchlight drum were actuated by a pair of ¼-horsepower DC servomotors, mechanically coupled to their respective axes through spur-gear reduction trains. These motors were each powered by an innovative (and for the most part long since forgotten) electromechanical component known as an amplidyne (short for amplifier dynamotor). An amplidyne was a special-purpose generator that functioned as a DC amplifier, driven by a constant-rpm motor, all mounted in a common housing as shown in Figure 8-13.

Figure 8-13. The amplidyne was an electromechanical DC amplifier that consisted of a constant-rpm motor (right end in photo above) that drove a specialized DC generator (left end) via a common shaft (photo courtesy Bob Meza).

For searchlight applications, three key characteristics were required: 1) very high gain, in order to generate a substantial output voltage from a low-power input signal; 2) quick response, so the output did not lag the input; and, 3) reversible output polarity to support clockwise and counter-clockwise rotation of the load. The rotor output of the generator was directly connected to the load, which in this case were the two searchlight servomotors that positioned the drum in elevation and azimuth.

The generator stator, or field winding, served as the input mechanism that controlled the rotor output, and was thus referred to as the control-field winding. This stator winding was center-tapped to form two separate field coils with a common connecting lead, arranged such that the current in one coil opposed that in the other (TM 5-7111, 1943). If the currents in these two coils were equal, their opposing fields would cancel, resulting in no net magnetic flux being generated by the stator, hence no output voltage would be induced in the spinning rotor. If the control-field currents were not equal, their net difference would produce a resultant flux that would generate a rotor output. As this stator flux increased in magnitude, the rotor output voltage would rise proportionately, causing the searchlight servomotor to turn faster (Figure 8-14).

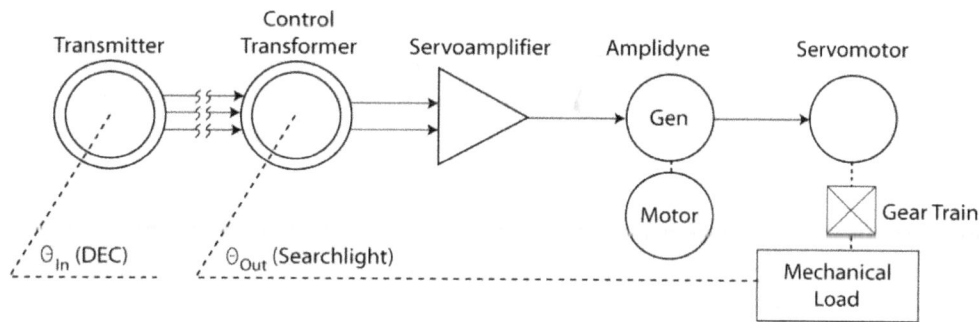

Figure 8-14. In the electromechanical "follow-up" servo-control system employed on WWII searchlights, the amplified control-transformer output was used to drive the control-field input of an amplidyne, which in turn powered the DC servomotor that moved the light. Two such circuits were required for pan and tilt. (drawing by Lindsay Seligman).

The direction of ensuing motion depended upon the polarity of the output voltage, determined by which control-field coil carried the greater current. To better understand how these DC control-field voltages were developed, recall that a *control transformer* compares a pair of selsyn shaft-angle representations and creates a two-wire AC signal proportional to the sine of their difference. The direction of misalignment is indicated by the phase of this output signal with respect to the AC supply that excites the transmitter rotor (i.e., the two are either in phase or 180 degrees out of phase).

As shown in Figure 8-15, the searchlight control-transformer output is connected to input transformer T20, which has two identical secondary windings that feed the control grids of a pair of *6L6* beam-power vacuum tubes. Note that these secondary windings are intentionally wired to be 180 degrees out of phase with each other. Conversely, the dual secondary windings of power transformer T10 are connected to the *6L6* plates in series with the amplidyne control-field windings, but without any phase inversion.

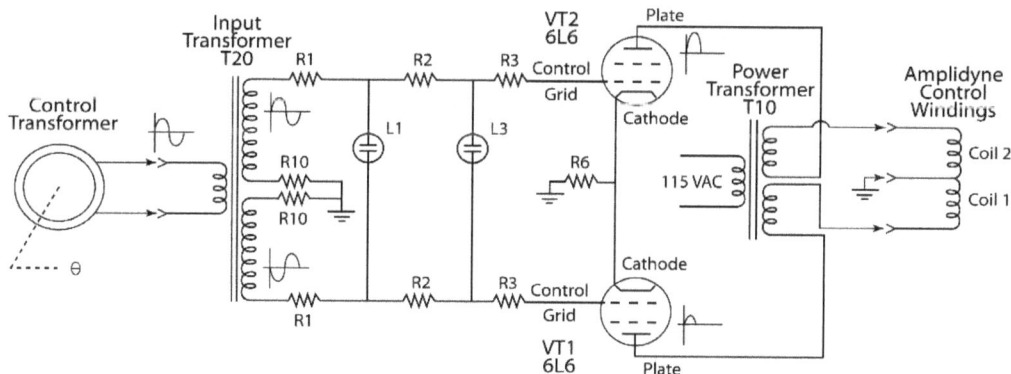

Figure 8-15. Simplified circuit diagram of the electronic servoamplifier that generated the DC amplidyne control-field inputs from the AC control-transformer output (adapted by Lindsay Seligman from *TM 5-7111*, 1943).

With an AC voltage differential applied to the plate and cathode in this fashion, the *6L6* acts like a half-wave rectifier (in addition to a triode amplifier), conducting current only during that half of the AC cycle when the plate is positive. During this time of conduction, the plate-current magnitude is controlled by the level and polarity of the *6L6* control-grid voltage with respect to the cathode. This dual-mode (i.e., rectifier-amplifier) operation produces a modulated half-wave DC output as shown in the schematic of Figure 8-15, which is key to decoding the direction of rotation from the control-transformer output. The plate current for VT1 flows through one half of the amplidyne-generator control-field winding, while that from VT2 flows through the other half.

The *6L6* control grids are initially biased negative with respect to their cathodes. When the control-transformer output is zero, indicating shaft synchronization, the voltage on both grids will be the same, and so the plate currents flowing through the two generator control windings will be equal in magnitude but opposite in direction. The amplidyne rotor output under these conditions is therefore zero, causing the searchlight servomotor to be halted.

As the control-transformer output rises, however, the out-of-phase relationship of the T20 secondary windings causes an increase in grid voltage for one tube and a decrease for the other, with a corresponding increase and decrease in their respective plate currents. The resulting imbalance in the amplidyne control-field windings produces a rotor output voltage that activates the servomotor, repositioning the searchlight drum to restore proper alignment with the locator device. The polarity of the amplidyne output, and hence the servomotor direction of rotation, is determined by the phase of the control-transmitter output with respect to the AC supply.

For example, if the control-transformer displacement from synchronization were in the opposite direction from that producing the T20 sine-wave outputs depicted in Figure 8-15, its output would be shifted 180 degrees in phase. As a result, the sine-wave inputs to the control grids for VT1 and VT2 would also be inverted. The effect on the *6L6* plate currents would therefore be opposite that of the previous case, in terms of which one increased or decreased with a rise in the input signal. The searchlight servomotor, consequently, would be driven in the opposite direction to restore alignment.

For improved resolution (and to overcome dead-band stall), not one but two selsyn transmitters were used on each axis of the later-model locator devices, mechanically geared together with a ratio of 33-to-1, and thus referred to as the high-speed and low-speed transmitters. The outputs of their associated high-speed and low-speed control transformers on the searchlight were mixed in the servo-amplifier input stage, the details of which are omitted in the previously shown simplified schematic of Figure 8-16.

The searchlight control station (i.e., DEC) was equipped with a low-speed data transmission system only, as the human operator could easily compensate for dead-band stall by simply increasing the input command based on visual feedback. A DEC Transfer Switch on the control station actuated four double-pole-double-throw relays on the light to disconnect the DEC selsyns, and instead connect the searchlight directly to the data transmission system of the locator device.

Pre-WWI Searchlights

There is evidence at least four searchlights were in use at Fort Rosecrans towards the turn of the 19th century, as cited in recommendations by the Board on Electric Installations regarding a proposed light and powerhouse project (Roessler et al., 1903):

> "...the searchlight on Ballast Point could be advantageously increased to 30 inches diameter and be located closer to the lighthouse, so as to be further away from the batteries in that vicinity; that the 30-inch searchlight shown in rear of Battery Fetterman and the 24-inch shown in rear of Battery McGrath could be advantageously increased to 36 inches and 30 inches, respectively... The 36-inch searchlight on the heights of Point Lomo (sic) is well designed for general illumination to the west and south and could be advantageously increased to 60-inch diameter."

Roughly 10 years later, still only four searchlights were officially assigned, with three at Fort Rosecrans and the fourth at Fort Pio Pico near Battery Meed on North Island, all strategically positioned to illuminate the harbor entrance.

Flower and Roth (1982) report the first of these was purchased in 1898,[5] but held in storage until 1902. This "truck-mounted" 30-inch light was moved from its original location to a new wooden shelter 200 yards south of Battery McGrath (Figure 8-16). The term "truck-mounted" is a bit misleading, in that this was not a motorized portable searchlight with extended range, since it had to be pushed by hand and trailed a tethered umbilical. Power was furnished by a DC motor-generator in the central powerhouse of Fort Rosecrans, located in a ravine behind Battery Wilkeson (later Calef-Wilkeson). Officially transferred to the artillery on 5 February 1903, the light's primary purpose was illuminating the controlled minefield guarding the harbor entrance.

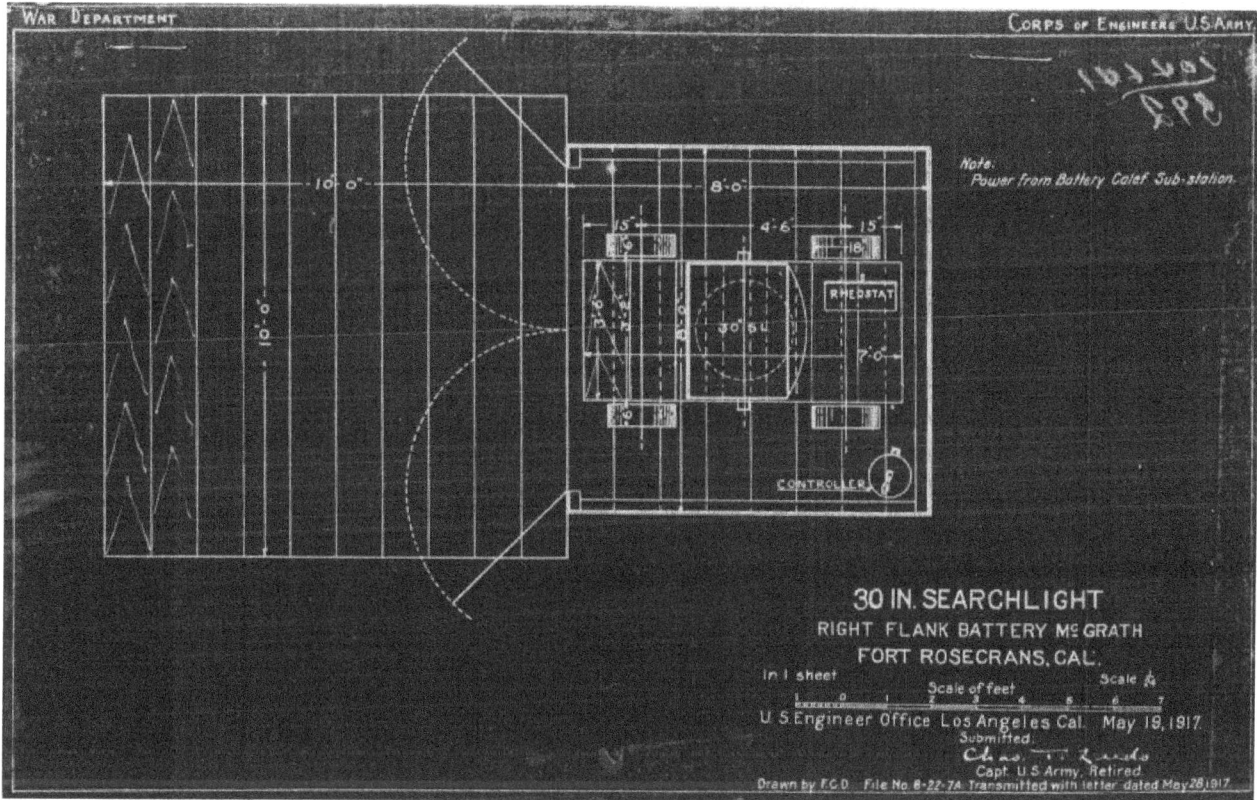

Figure 8-16. Blueprint depicting the shelter for Searchlight 1, located south of Battery McGrath (see again Chapter 2), dated 19 May 1917 (NARA). Note the "Rheostat" on the truck bed and similarity of the controller to that for the 36-inch light shown earlier in Figure 8-1.[6]

A 1919 photo that appears to correlate well with this blueprint is shown in Figure 8-17. For starters, the shelter is clearly just long enough to house the "truck," with no additional floor space to the rear for a motor-generator set, as this light received its DC power from an external source. Secondly, the light itself has the same profile as the turn-of-the century model shown earlier in Figure 8-1, and the purchase date of 1898 is in keeping with the timeframe the smaller 30-inch lights were still available. Thirdly, the rheostat is mounted on the aft end of the truck precisely as shown on the blueprint in Figure 8-16.

---

5   Flower and Roth (1982) cite "General Correspondence 1899:3261/29; 1902: 34966/9, 34966/10."

6   The rheostat was used to adjust the operating voltage at the light to compensate for electrical losses in the power cable, which varied as a function of distance from the power source.

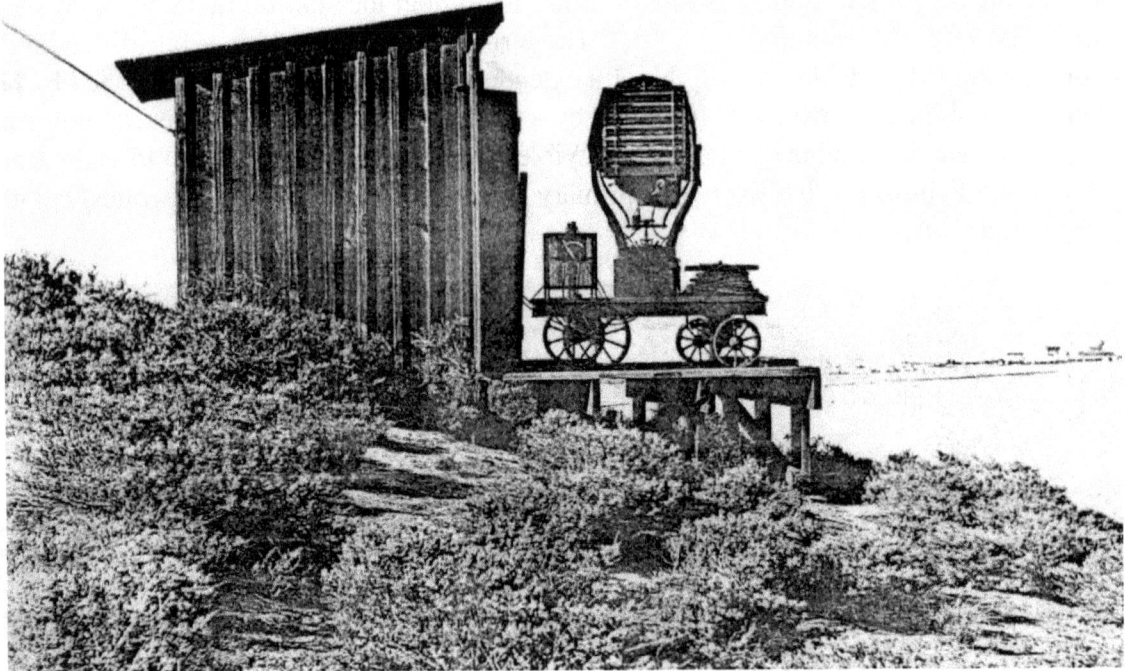

Figure 8-17. This March 1919 photo of "Exterior Shelter Searchlight No. 4" is thought to be the 30-inch light, previously designated Searchlight 1, south of Battery McGrath (US Army photo courtesy Al Grobmeier). Note what appears to be the shoreline of Ballast Point across the water in background.

While on first glance this truck appears to be rail mounted, close inspection of the wheels reveals them to be tangent to the wood deck surface, with what appear to be wooden guide rails added for lateral constraint. An excerpt from the "Project Engineer's Notebook" (EN, 1913) shows the chronological numbering scheme for these first four lights (Figure 8-18a), to which I have appended their respective dates of transfer to the artillery (Figure 8-18b). There is evidence in the same "Searchlight" section of this notebook, however, of a different numbering strategy, as seen in the caption above.

Size	Light No.	Location	Remarks.	**Transferred**
30"	1	Point Loma, south of Batt. McGrath		5 Feb 1903
36"	2	Billy Goat Point, Point Loma.		9 May 1906
60"	3	Southerly end of Point Loma		15 Feb 1911
60"	4	Fort Pio Pico		15 Feb 1911

6

OFFICE OF THE CHIEF OF ENGINEERS, U.S.A. FEB 19 1913 MILITARY SECTION

SAN DIEGO, CAL.

Project calls for 2 60-inch, 1-36-inch, and 1 30-inch lights.

All are on hand, installed, and transferred.

a) b)

Figure 8-18. a) Numerical designation of searchlights as listed in "Engineer Notes, Ledger, Harbor Defense San Diego," 19 February 1913. Fort Pio Pico was across the harbor entrance on North Island. **b)** Appended column showing the date of transfer to the Coast Artillery.

The second installation at Fort Rosecrans during this period was a 36-inch searchlight at Billy Goat Point, just down the slope from Sylvester Road and opposite the present location of WWII Searchlight 19. Officially transferred to the artillery on 9 May 1906, this light was probably the one reportedly situated "on the heights of Point Lomo" (sic) by Roessler et al. in 1903. DC power was provided by a local motor-generator set located inside the wooden shelter (Figure 8-19a), driven by AC power from the central plant substation at Battery Wilkeson. A 1919 photo of what appears to be this 36-inch installation, redesignated Searchlight 3, is shown in Figure 8-19b.

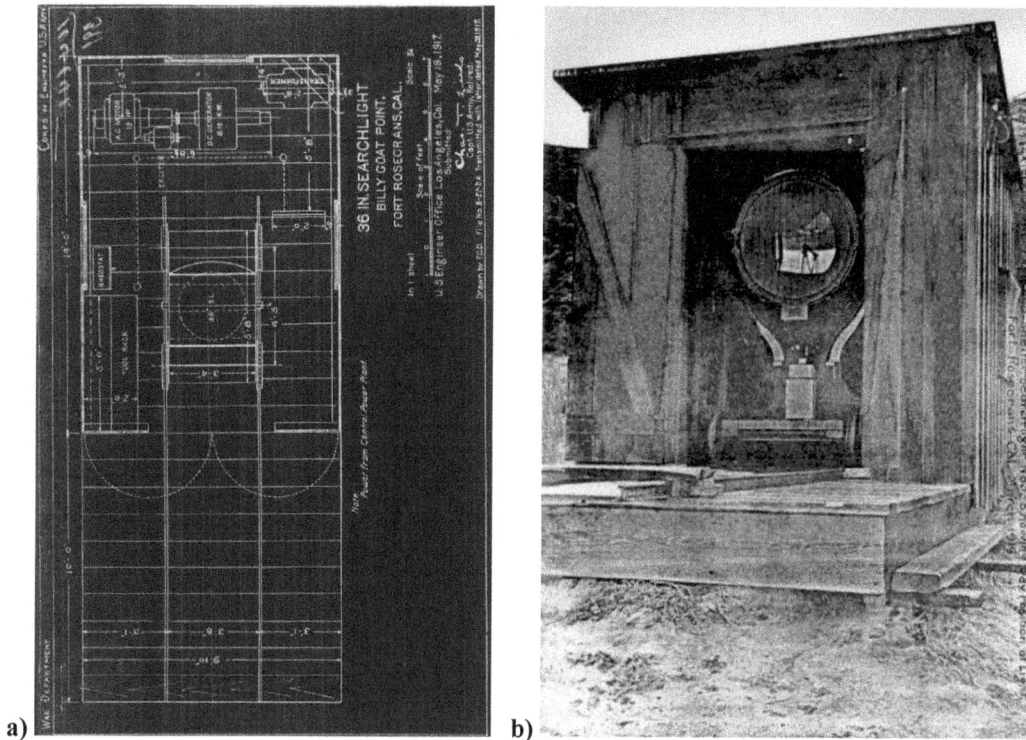

Figure 8-19 a) Mounted on a rail cart, Searchlight 2 was housed in a wooden shelter on Billy Goat Point as shown on this 19 May 1917 blueprint (NARA). Note motor-generator set at rear of shelter, and rheostat relocated from mobile cart to under the left window. **b)** March 1919 photo of "Exterior Power Plant and Shelter Searchlight No. 3" (US Army photo courtesy Al Grobmeier).

The diameter of the projector housing in Figure 8-19b looks to be slightly less than the wheel separation, which mated with a track spacing of 3 feet 8 inches as indicated on the blueprint of Figure 8-19a, and thus is consistent with a 36-inch light. The caption on the right border of the photo indicates the shelter contained some means of power generation, which was not present in the shelter for the 30-inch light. This light was ultimately earmarked for transfer to San Francisco in a "Project Engineer's Notebook" ledger entry that read: "To be considered as assigned to San Francisco, not to be shipped at present." There is no date on this page but it appears from context the entry was made circa 1919.

Another undated "Ledger" entry, this one with handwritten annotations, shows Searchlight 1 as being the 60-inch light at the southern tip of Point Loma, Searchlights 2 and 3 as the 36- and 30-inch lights, respectively, while Searchlight 4 was the 60-inch light at Fort Pio Pico (Table 8-1). This alternative arrangement reflects the artillery convention of assigning numbers from right to left across the observed field of fire, as opposed to chronologically. A "Memoranda" entry dated 12 March 1914 indicated the District Engineer Officer had been specifically requested to provide comments on this "proposed numbering."

Number	Size	Location
1	60 inch	Southern Tip of Point Loma
2	36 inch	Billy Goat Point
3	30 inch	South of Battery McGrath
4	60 inch	Fort Pio Pico

Table 8-1. Later (undated, sometime prior to 1914) numerical designations of Fort Rosecrans searchlights.

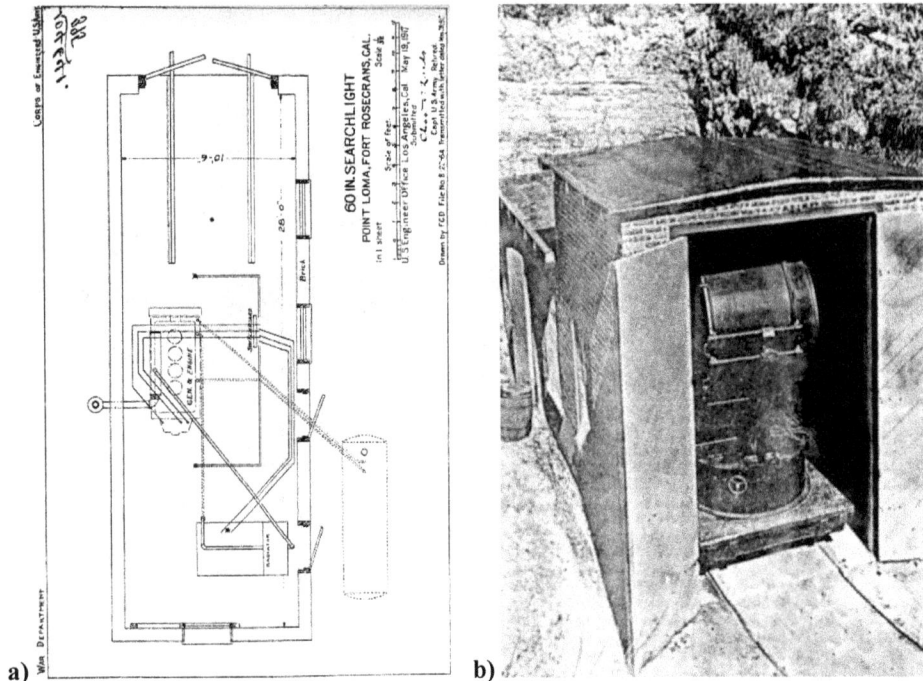

Figure 8-20 a) Blueprint of the original brick powerhouse and shelter for the 60-inch Searchlight 1 located at the southern tip of Point Loma, 19 May 1917 (adapted from Berhow, 2004). **b)** Photo of Searchlight 1 (later Searchlight 3) inside its shelter (US Army photo courtesy Al Grobmeier).

The 60-inch light at the southernmost tip of Point Loma is believed to have been installed sometime between 1909 and 1911, with a combined powerhouse and shelter made of brick (Figure 8-20a). The operating position was a wooden dock situated at the edge of the cliff face, to which the cart-mounted searchlight was pushed along a short section of narrow-gage track extending from the shelter (Figure 8-20b). Information recorded in the Project Engineer's Notebook suggests the light was originally designated Number 1, but was renumbered numerous times around the timeframe of WWI. Photographic evidence also shows the searchlight was replaced at some point with a more modern version, possibly when the powerhouse was rebuilt in 1921.

Searchlight 4 was initially situated at Zuniga Point southeast of Battery Meed at Fort Pio Pico on North Island as shown in Figure 8-21,[7] and transferred to the Coast Artillery on 15 February 1911. The light was powered by the standard 25-kilowatt generator manufactured by General Electric (Miller, 1917), identical to that employed on Searchlight 3. On 12 February 1914, the Coast Defense Commander recommended to the District Engineer Officer that the light be raised from its current level of just 23 feet above mean low water to at least 40 feet above the high tide mark. The word came back on 12 March that same year that funds were not available, but detailed estimates and plans were requested nonetheless (EN, 1914).

---

7   Note the alternate spelling "Zuninga" in Figure 8-21, which may have been the original name, perhaps later changed by the US Geographic Names Board (Grobmeier, 2007).

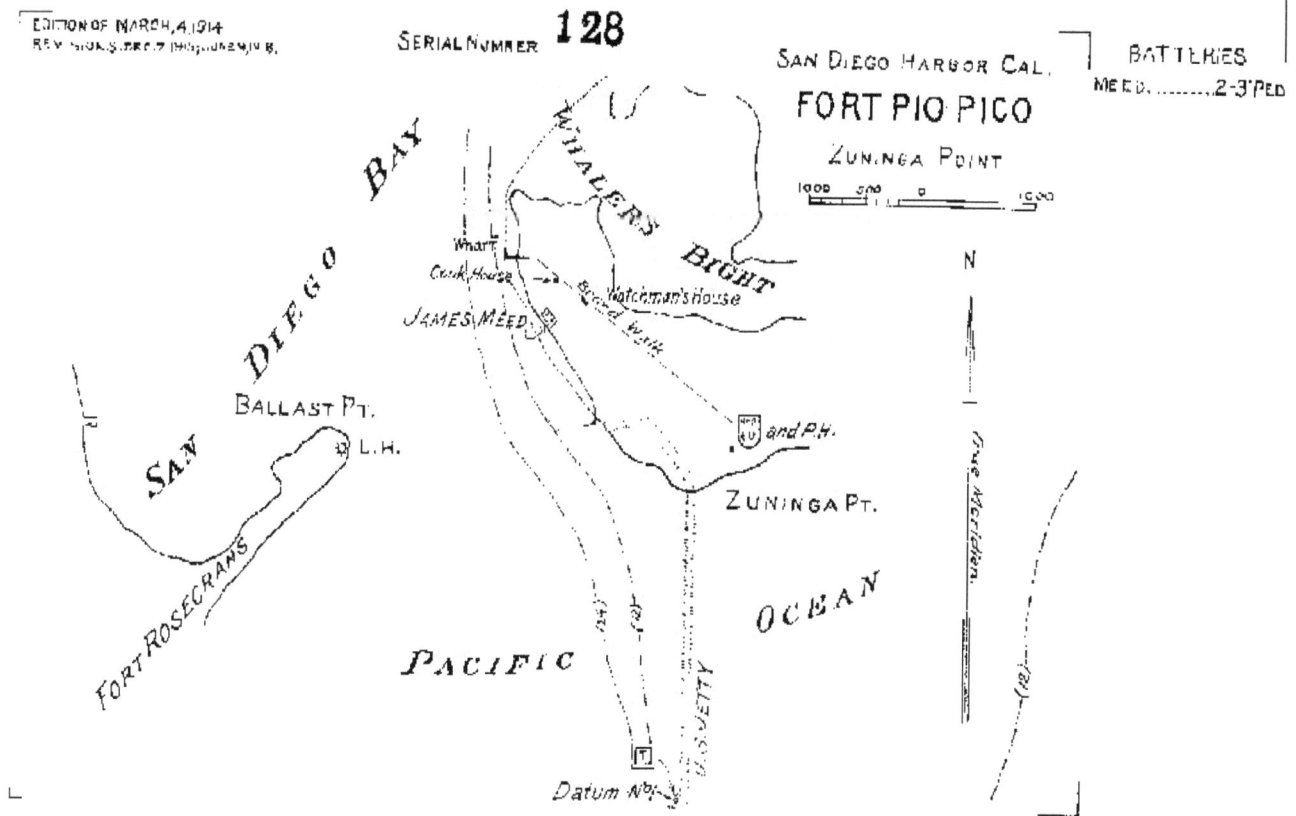

Figure 8-21. Searchlight 4 and its associated powerhouse were located on Zuniga Point (right of center) at the southeast end of a boardwalk extending down from the wharf northwest of Battery James Meed (NARA). This 60-inch light was later moved to the southern tip of Fort Rosecrans in the 1919-1920 timeframe, retaining the same numerical designation.

On 27 May 1914, a Board of Officers convened by Western Department recommended that "the guns now in position at Battery Meed, Fort Pio Pico, be moved to a position at Billy Goat Point. Battery Meed, in its present state, is in great danger of being washed out as the winter storms begin." The Commanding General, Western Department, concurred with this recommendation on 6 July 1914, and although the transfer did not take place until some 4 years later, no further action was apparently taken to raise Searchlight 4. The light and its associated shelter and lift mechanism were authorized for transfer to Fort Rosecrans on 21 April 1917, for eventual reinstallation east of Searchlight 3, as will be discussed in the next section.

An unidentified pre-WWI searchlight is shown in Figure 8-22, reproduced from the Paul T. Mizony Fort Rosecrans Collection of the San Diego Historical Society, Booth Historical Archives, which covers the period 1902-1911. Mizony served at Fort Rosecrans from 1905-1911 as a member of the National Guard of California, Coast Artillery Corps, 8th Company. This particular photo of him standing next to a 60-inch light was supposedly taken in 1908, but no information seems to have been recorded as to the location.

Figure 8-22. Paul Mizony on a wooden platform next to a 60-inch searchlight in 1908 (courtesy Paul T. Mizony Fort Rosecrans Collection, San Diego Historical Society, Booth Historical Photograph Archives).

To further complicate matters, examination of the "Project Engineer's Notebook, 1900-1922" and various other HDSD archival records shows a rather confusing history of conflicting numbering schemes, as seen in Table 8-2.[8] Note, for example, the renumbering associated with the 60-inch light at "South Point Loma." Be that as it may, this searchlight was the only known 60-inch installation reasonably close to Mizony's timeframe and within the boundaries of Fort Rosecrans, presumably installed sometime between 1909 and its transfer to the Coast Artillery on 15 February 1911 (EN, 1911).

Size	Location	1911[9]	1913[10]	1914[11]	1914[12]	1918[13]	1919[14]	1920[15]
30"	South of McGrath	3	1	2	7	1	4	N/A
36"	Billy Goat Point	2	2	3		2	3	N/A
60"	South Point Loma	1	3	1	5	3	1	3
60"	Fort Pio Pico	4	4	4	9	4		N/A

Table 8-2. Chronological representation of various searchlight numbering schemes encountered in the "Project Engineer's Notebook, 1900-1922" for HDSD, which causes some confusion today.

---

8   All lights were typically renumbered whenever new lights were added.

9   *Engineer Notes, Ledger-1, Searchlights*, multiple dates on page up through 15 February 1911.

10 *Engineer Notes, Ledger, Searchlights* dated 19 February 1913, as well as *Journal-1, Searchlights* dated 1 October 1913.

11 *Engineer Notes, Memoranda-1, Searchlights* dated 12 March 1914, which proposed this new numbering scheme.

12 *Engineer Notes, Memoranda-1, Searchlights* dated 20 June 1914, report of proposed upgrade to 11 lights by Board of Officers convened at Fort Rosecrans.

13 *Engineer Notes, Memoranda-2, Searchlights*, 5 March 1918, lights not numbered but listed in this order.

14 As marked on photos from National Archives.

15 As marked on photo from National Archives.

The only other 60-inch possibility in 1908 would have been Searchlight 4 at Fort Pio Pico, which was a sub-post of Fort Rosecrans. As previously discussed, however, that light was not moved from Pio Pico until April 1917, and situated "east of Searchlight 3," which had previously been Searchlight 1. Accordingly, it seems most probable that Mizony is standing next to the original Searchlight 1, which must have been installed in 1908 versus the reported 1909, assuming the photo is correctly dated. On closer examination, there are several clues available in this photograph to help prove or disprove this theory, at least with regard to the actual location.

For starters, what appears to be all sky with no visible water or terrain in the photo background suggests the searchlight is on the edge of a cliff. (In contrast, a similar perspective of Searchlight 4 at an elevation of only 23 feet at Fort Pio Pico would very likely reveal both terrain and water.) The wooden platform upon which Mizony is standing could be a stand-alone dock or the top of a rail cart, but the horizontal safety rail and vertical post seen in the lower right corner of the photo suggest a dock. The dock at the end of the rail track for Searchlight 1 in 1911 had such a safety rail (see later Figure 8-30), given its precarious location on the edge of the 116-foot cliff at the southern tip of Point Loma.

Focusing now on the searchlight, note the transverse vent-motor housing on top, which is orthogonal to the later designs. Most interesting of all, however, is the imagery reflected in the 60-inch parabolic mirror. The semi-circular radial pattern at the very top of the mirror appears to match the rotary selector switch seen on rheostat cabinets later installed inside silo- and tunnel-type type shelters for WWI Searchlights 1, 2, 5, and 6. This reflection is probably from the small device on top of the larger box-like assembly in front of Mizony, which contains a number of vertical cylinders that presumably are wire-wound resistors. The open metal cage surrounding these cylinders seems designed to facilitate convection cooling, which would be required at a typical current flow of 150 amps.

Closer examination of the mirror reflection suggests a rail track disappearing into the background terrain, which is a little easier to see if you turn the page upside down and view the image inverted. The rheostat device, which is on the far-left side of the cart looking landward, appears to be centered over the left track, with the track itself converging towards a dark object that could be the shelter. Due to the nature of a parabolic mirror, the reflection of the sky appears at the bottom of the image while the terrain is at the top. It is interesting to note the slope of the horizon between them, which is also inverted (left to right), suggesting the shoreline is off to Mizony's left, while the terrain rises to his right.

This topography is consistent with the area near Building 15 at the Center today where Searchlight 1 (later redesignated Searchlight 3 in 1913) was located. All things considered, it seems reasonable to conclude that Mizony is indeed posing next to Searchlight 1 at the tip of Point Loma, which now appears to have been installed a year or so earlier than previously thought (i.e., 1908 versus 1909). Some evidence has been found that suggests this light was upgraded to a newer model during WWI, and designated as Searchlight 3.

WWI Searchlights

Eight 60-inch coast-defense searchlights recommended by the Taft Board were installed at Fort Rosecrans between 1918 and 1919 (Thompson, 1991):

- Searchlights 1 and 2 in the northwest corner of the fort.
- Searchlights 3 and 4 on the southern tip of the fort.
- Searchlights 5 and 6 in the southeast corner of the fort.
- Searchlights 7 and 8 at Fort Pio Pico across the harbor entrance on North Island.

For operational versatility, paired searchlights with similar fields of regard were usually installed at different elevations, as explained by Abbot (1902):

> "It is generally conceded that a high position is best for a searching light, but another important condition is that the light be near the shore line. These conditions are often irreconcilable and one or the other must, as a rule, be sacrificed."

These 60-inch lights were housed either in underground silos (also referred to as well- or pit-type shelters) with counter-balanced elevator lifts (Figure 8-23), or above-ground in "tunnel-type shelters," with narrow-gage rail transport to the nearby operating location.

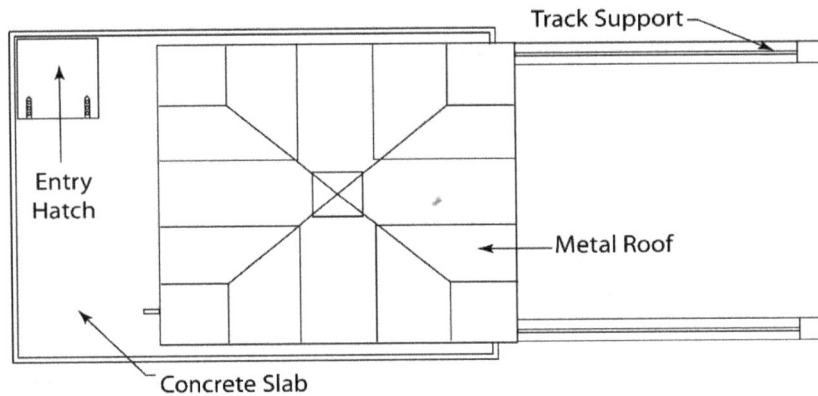

Figure 8-23. Overhead view of an underground silo shelter, showing track supports for the sliding metal roof (adapted from Carey, 2000, courtesy Cabrillo National Monument).

Each of these pairs of fixed lights had an assigned powerhouse containing two gasoline-driven 25-kilowatt DC generators. The maximum separation between a light and its powerhouse was for the most part limited to 600 feet to preclude excessive voltage drop due to the high current draw (150 amps).[16] A series of air-cooled ballast resistors (rheostat) with a rotary selector switch was installed in a small cabinet inside the searchlight shelter to allow final adjustment for variations in the actual distance. The generator (Figure 8-24) produced an output of 110 volts DC in order to provide the required 78 volts at the light itself, taking into account the combined source/return resistance of the cable run.

---

16 In actuality, since 600 feet was the total cable length from generator to light, the physical separation between locations was somewhat less to account for internal wiring within the structures on either end.

Figure 8-24. View of the General Electric 25-KW gas-driven generator as seen from the carburetor side (adapted from Miller, 1917). Note the generator at far left and the large engine flywheel at far right.

The gasoline engine for these units was designed by General Electric in the early 1900s, in response to a competitive solicitation from the Army Corps of Engineers for 60 internal-combustion sets. The 54-horse-power engine produced a continuous DC generator output of 27 KW, with 2 KW required to run the radiator cooling fan, for a net output of 25 KW. An excellent detailed description of this motor-generator set, recently reprinted by the Coast Defense Study Group (CDSG), was provided by Lorimer Miller (1917).

To facilitate rain-water runoff, the underground silos were equipped with a shallow pyramid-shaped sheet-metal roof with a wood frame and steel bracing. A hand-cranked rack-and-pinion drive was used to move the roof out of the way prior to raising the elevator lift. The roof assembly rolled on steel tracks set in concrete for stability as shown in Figure 8-25. A separate entry hatch provided access to steel ladder rungs cast into the concrete wall, which led down to an intermediate platform area approximately 10 feet below. The pit extended down another 5 feet to allow platform-level access to the elevator that supported the light. Several photos of this configuration will be presented later during discussion of the active WWII searchlight installations at Fort Rosecrans.

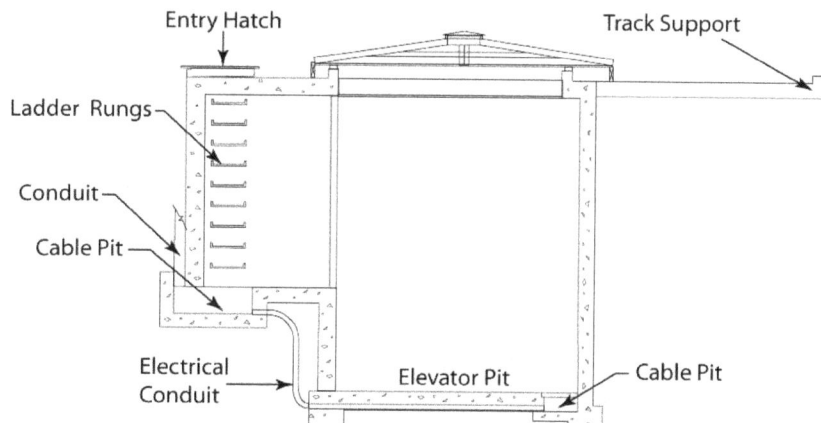

Figure 8-25. Cross-section view (elevator mechanism not shown) of the underground silo shelter for 60-inch Searchlights 1, 2, 4, and 6 (adapted from Carey (2000), courtesy Cabrillo National Monument).

Searchlights 1 and 2

Two new 60-inch lights were installed in underground silos in the northwest corner of Fort Rosecrans, designated Searchlights 1 and 2. As the location of this pair of buried silos has remained on restricted-access Government property, no period photographs have been found. The "Report of Completed Works – Seacoast Fortifications (Searchlights) for Searchlight 1" indicates General Electric projector Serial 21740 was received on 25 May 1920,[17] and the structure was transferred to troops on 14 June 1921. The "Shelter for Searchlights Nos. 1-2-4-6" at Fort Rosecrans blueprint is shown in Figure 8-26.

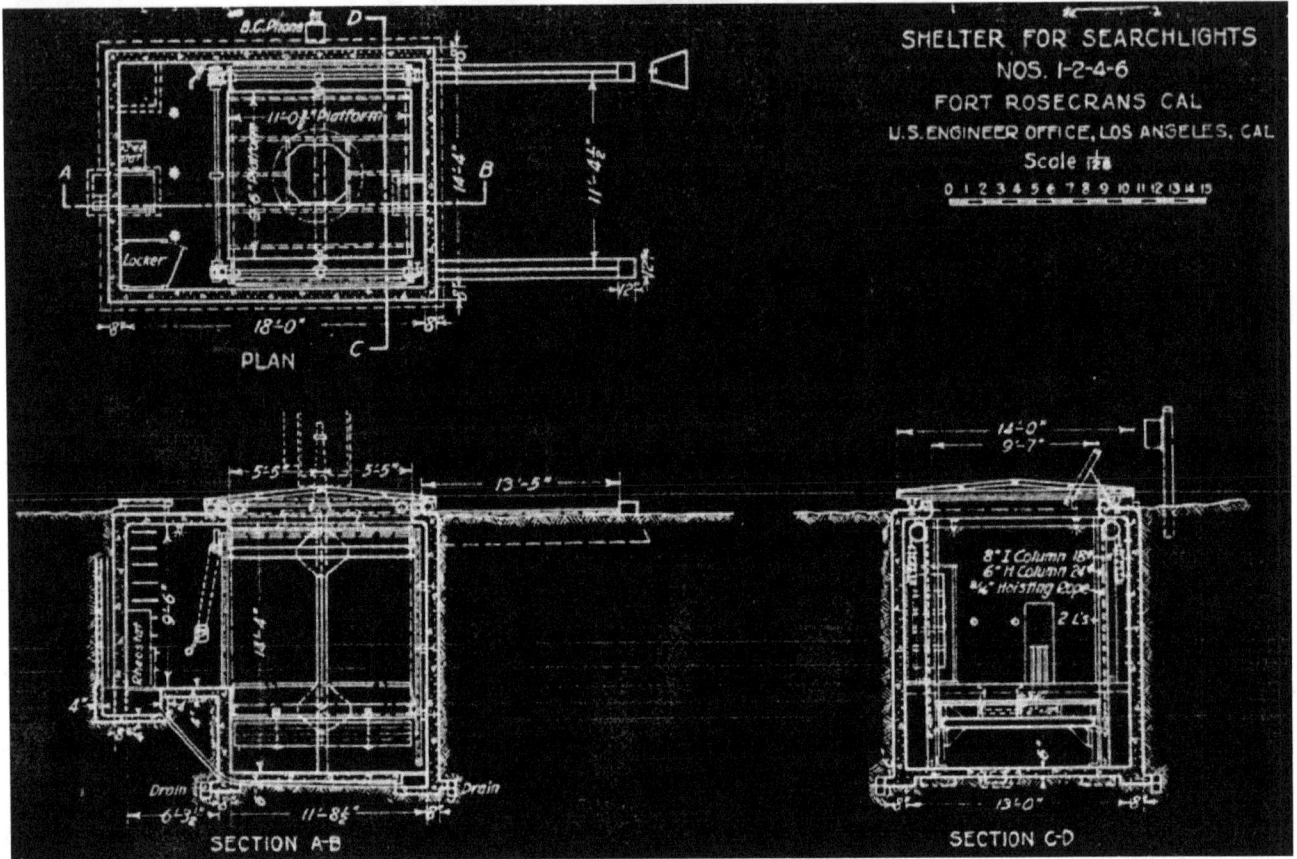

Figure 8-26. The construction details of the underground silo for Searchlight 1, which were the same for lights 2, 4, and 6, are seen in this undated WWI blueprint (NARA).

The previously mentioned Report of Completed Works for this light also indicates the associated controller (Form N-12, serial number 400365) was located "In First Fire Command Sta.", the location of which is unclear. The powerhouse for this light and companion Searchlight 2 was dug into a small ravine upslope from the silo (Figure 8-27). According to the "Project Engineer Notebook (Journal)," the two 25-kilowatt DC generator sets in this powerhouse were turned over to the artillery on 9 November 1920.

---

17 Projector as in light projector, an early term for the searchlight itself.

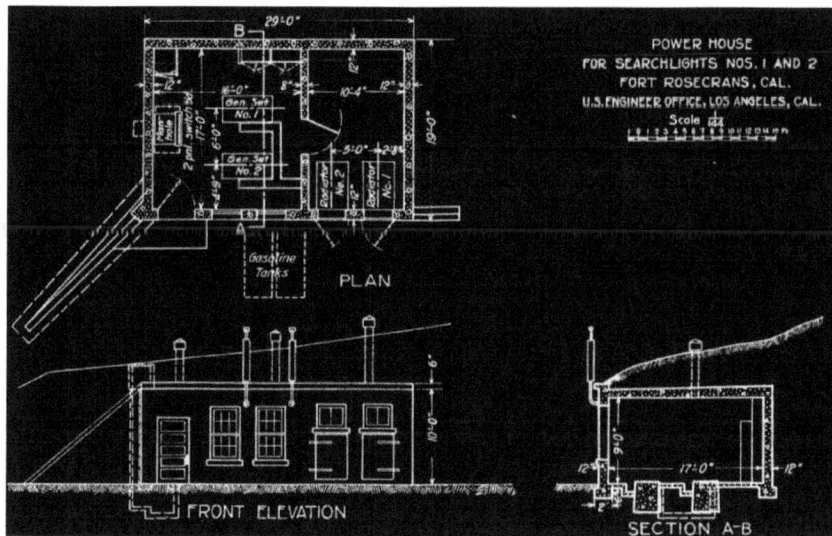

Figure 8-27. Original plan and elevation views of the combined powerhouse for WWI-era Searchlights 1 and 2, redesignated Searchlights 11 and 12 in WWII (NARA). Note the generator room on the left and smaller radiator room on the right in the plan view at upper left, with gasoline tanks in front of the structure.

This still-extant combined powerhouse for these two lights is seen with its dual exhaust stacks in Figure 8-28, dug into the surrounding hillside, looking north circa 1920. This location is south of the northern fence line for Fort Rosecrans and just east of what is now Woodward Road. These two silo installations would be redesignated as Searchlights 11 and 12 during WWII, and are further discussed in a later section.

Figure 8-28. The combined powerhouse for WWI Searchlights 1 and 2 is located just east of what is now Woodward Road in the northwest section of Fort Rosecrans, shown here during construction, circa 1920 (US Army photo courtesy Al Grobmeier). Note the three cone-covered air vents in background center.

Searchlights 3 and 4

During WWI, the original brick powerhouse and shelter for Searchlight 3 at the southern tip of Point Loma (see again Figure 8-20) was rebuilt and enlarged as shown in Figure 8-29 to provide additional power for both Searchlight 3 and a new silo-shelter installation. According to the "Project Engineer's Notebook" (EN, 1917), the existing 60-inch light and lift mechanism for pre-war Searchlight 4 were relocated from their old 1911 silo installation at Fort Pio Pico (see again Figure 8-21), per statement of work dated 21 April 1917.

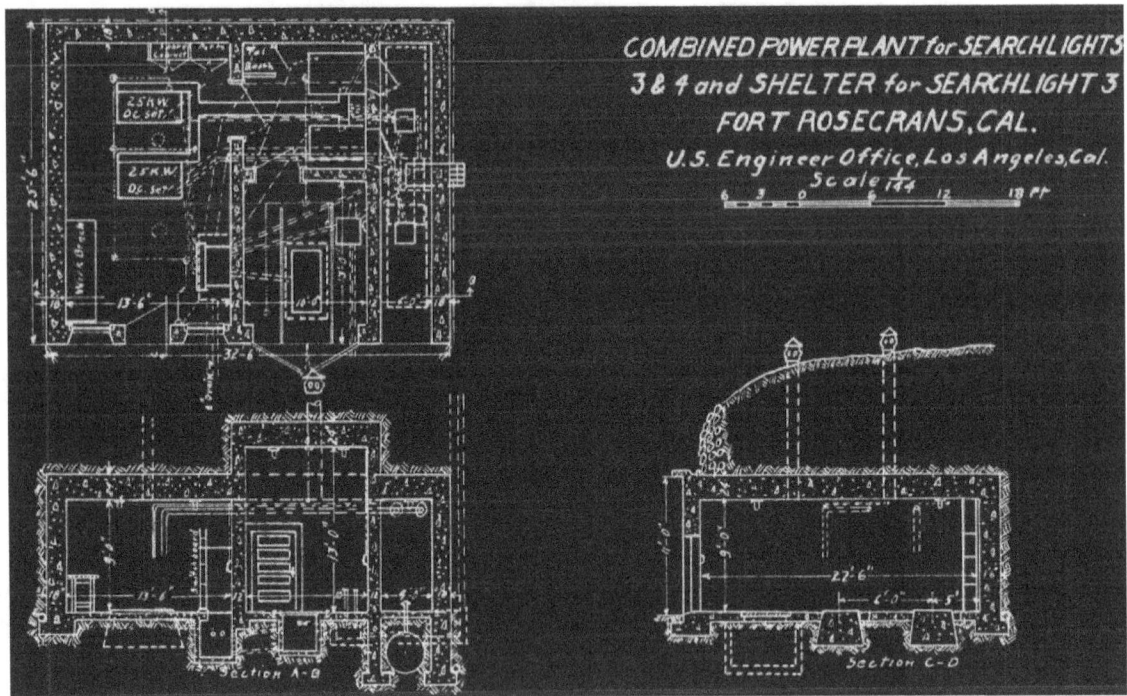

Figure 8-29. The combined powerhouse for Searchlights 3 and 4, which replaced the original brick structure shown in Figure 8-20b, also served as the shelter for Searchlight 3 (NARA).

Figure 8-30 below shows a newer version of Searchlight 3 in its operating position on the cliff edge than that shown earlier with Paul Mizony at this same location in 1908 (Figure 8-22), now with a longitudinal versus transverse vent housing on top. The wooden deck with concrete foundations appears to be new construction, as does the surrounding safety rail. The concrete terminal hut with the power and control connections for this light seen at lower left are original pre-WWI equipment, as is the pole-mounted field-telephone box in the background.

Figure 8-30. Searchlight 3 in operating position, with rail tracks leading back to the combined powerhouse for Searchlights 3 and 4 (undated US Army photo courtesy Al Grobmeier). Note terminal hut for electrical connectors at lower left, with a pole-mounted field-telephone box behind the safety rail in background.

Searchlight 4, relocated from Zuniga Point to the southern tip of Site 8 (Loma) in the 1919-1920 time-frame, retained the same numerical designation in WWII. The underground silo position for this light was along the cliff edge above Searchlight 3 (shown earlier with Mizony in Figure 8-22), just south of the future site for WWII Battery Humphreys, which is seen under construction in Figure 8-31 below, circa 1942. As will be discussed later, the light and its associated lift mechanism were relocated once again during WWII to serve as Searchlight 15 on the west side of Site 8 (Loma).

Figure 8-31. The abandoned remains of the WWI-era Searchlight 4 silo can be seen along the ridgeline at lower left (up arrow), just south of Battery Humphreys construction at lower right, circa 1942 (US Army photo courtesy Cabrillo National Monument). Occluded in this view, Searchlight 3 was located at the cliff edge about halfway down the slope towards the new Coast Guard lighthouse complex (image center).

Flower and Roth (1982) reported finding the remains of this searchlight silo shelter during their 1982 survey of "Navy and Coast Guard Lands on Point Loma":

"This site is the location of searchlight Number 4 of the 1916-1920 generation of harbor defense improvements. The facility consisted of a shelter mounted on an elevating lift.[18] The installation was powered by the powerhouse at searchlight Number 3. During the Second World War, this light was not used."

Their field check results were reported as follows (Flower & Roth, 1982):

"The structure is intact but deteriorating rapidly. Dimensions are 16 by 24 feet; floor to ceiling, 15 feet. This structure varies drastically from other searchlight shelters with elevating lifts... Unlike the other shelters of this type that exhibit a metal roof on tracks, the shelter at this site exhibits a fixed wooden roof. In addition, the light platform is also a fixed wooden structure rather than a platform suspended on cables."[19]

The above description is rather puzzling, in that a "fixed wooden roof" at ground level as seen in Figure 8-31 would not allow the searchlight to be raised to an operating position. Similarly, the statement that "the light platform is also a fixed wooden structure rather than a platform suspended on cables" further confuses the issue of searchlight deployment. It's likely this confusion stems from the fact that this installation was significantly altered when the light was removed and repurposed for use as Searchlight 15 in WWII.

---

18 The searchlight was mounted on the lift, not on the "shelter."
19 This last sentence conflicts with their earlier claim of an "elevating lift."

The survey reported the site location as "172°, 300 feet from NOSC Building 581 at Battery Ashburn," which is in error, as Building 581 is instead adjacent to Battery Humphreys as seen in Figure 8-32. Using the distance scale provided on the source map for this figure (SPAWAR Systems Center Base Map 1999, Point Loma - Area 15), the above location vector points to the southern tip of the lower projection of the Battery Humphreys complex. This was indeed the location of WWI-era Searchlight 4 (see again Figure 8-31), which had a wooden versus concrete silo shelter.[20] The small rectangle at this point in Figure 8-32 was probably intended to represent its location.

Figure 8-32. Building 581 is actually east of and adjacent to Battery Humphreys, per SPAWAR Systems Center "Base Map 1999, Point Loma, Area 15," a portion of which is depicted here. The Searchlight 4 location was at the southern tip of the bottom projection of the Battery Humphreys complex, mislabeled as "Battery Humphrey" above.

Searchlights 5 and 6

Figure 8-33 below shows the early WWI-era plan and elevation drawings of the shelter and terminal hut for the original cart-mounted Searchlight 5. The interior dimensions of this still-extant shelter as drawn were 12 feet wide by 16 feet 2 inches long. The front elevation drawings for both the shelter and its associated terminal hut for some reason show no dimensions. There is no generator indicated in this drawing, as the shelter received underground electrical service from the nearby powerhouse on Sylvester Road, shown later in Figure 8-38.

---

20 The use of wood versus concrete was perhaps a construction expedient rationalized by the lesser likelihood of subterranean water damage due to the elevated location near the cliff edge.

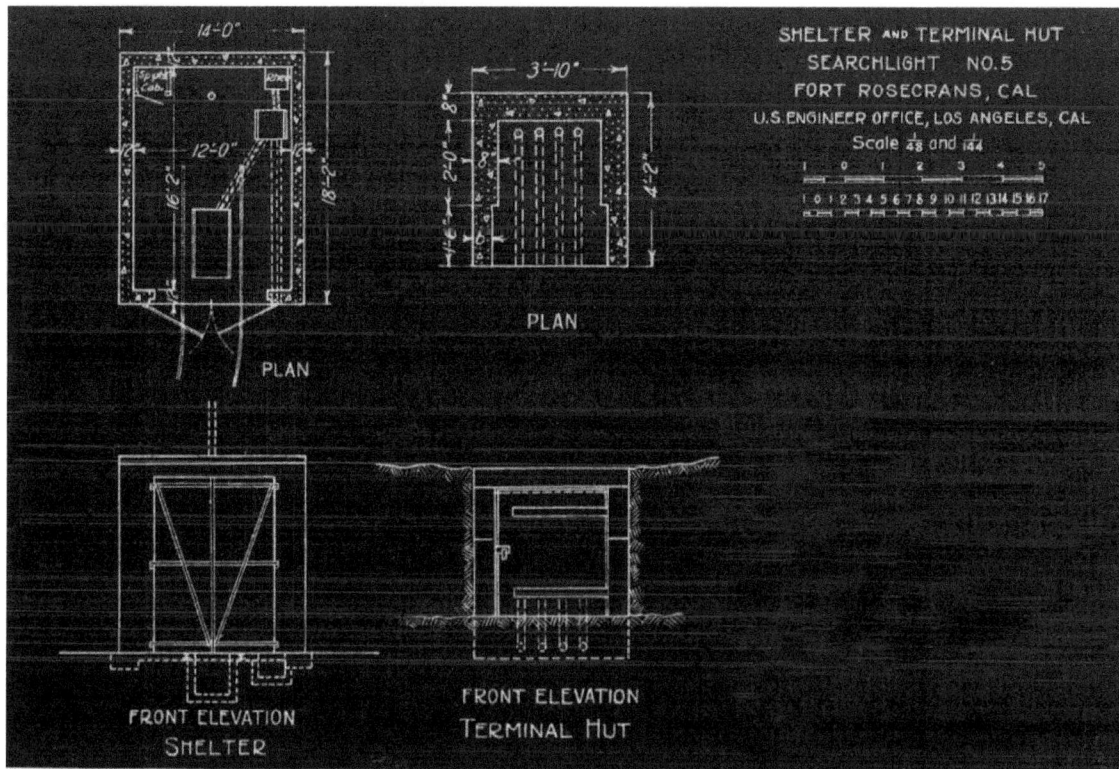

Figure 8-33. Original (undated) blueprint for the shelter and terminal hut for WWI-era Searchlight 5, later upgraded and re-designated as Searchlight 18 in the WWII-era (NARA). The terminal-hut installation at the deployed position appears to have been delayed, however, as further discussed below.

Figure 8-34 below shows WWI-era Searchlight 5 in its deployed position on Sylvester Road, at the end of the short section of narrow-gauge track leading back to the shelter dug into the rising slope at image right, circa March 1920. Note the standard field-telephone communications box attached to the concrete post just right of the light. There is no sign of the terminal hut in this photograph, however, which would have been located outboard of the light at the edge of the downward slope.

Figure 8-34. WWI-era Searchlight 5 (later WWII Searchlight 18) is deployed at the end of its track on Sylvester Road in March 1920 (US Army photo courtesy Cabrillo National Monument). The telephone communications box can be seen just to the right of the light, but the terminal hut is not visible.

The below US Army photos taken on 4 July 1920 show Searchlight 5 in its shelter (Figure 8-35a), as well as deployed, with again no sign of a terminal hut for power and control (Figure 8-35b). This absence would indicate the light trailed its power and control cables along the tracks from the shelter out to its operating position, which was a relatively short distance. It further suggests the terminal hut was installed at a later date, probably sometime just prior to or during WWII, when the light was redesignated as Searchlight 18.

a)　　　　　b)

Figure 8-35. a) Searchlight 5 in its shelter on 4 July 1920. **b)** Searchlight 5 deployed at the end of its tracks, looking northeast on 4 July 1920, with the telephone box on the concrete post at left, but no sign of the terminal hut (US Army photos courtesy Cabrillo National Monument). Note longitudinal vent housing.

Searchlight 6 was installed in a silo shelter at a slightly higher elevation than Searchlight 5 just west of Sylvester Road on the north side of the combined powerhouse, overlooking the harbor entrance as seen in Figure 8-36. The telephone communications box is seen on the concrete post in background left, with North Island just visible through the fog in background right. Note the reinforced horizontal tube at the center of the lens, which facilitated insertion of the positive carbon rod.

Figure 8-36. Searchlight 6 is seen in the raised operating position above its silo shelter on 12 March 1920, above and just west of Sylvester Road (now the Bayside Trail), looking northeast with North Island and San Diego Bay in background.

Figure 8-37 below provides a rare close-up view of the interior of the underground silo for Searchlight 6, circa March 1920. Note the longitudinally-oriented vent housing atop the light, the open storage cabinet at background center, and the partially occluded rheostat panel for adjusting the generator voltage at bottom right. The stacked horizontal counterweight along the wall at background left was connected by cable to the elevator lift via the shaft and pulley mechanism at top center, reducing the burden for the hand-crank operator when raising the light.

Figure 8-37. Interior view of the underground silo shelter for Searchlight 6 in March 1920, later designated Searchlight 19 (US Army photo courtesy Cabrillo National Monument). Note cable-supported elevator counterweight along wall in background left, and part of the electrical rheostat unit in the lower right corner.

Shown in Figure 8-38a below, circa 1920, the combined powerhouse for Searchlights 5 and 6 is more or less identical in construction to that for Searchlights 1 and 2. The entry door and windows for the generator room are seen on the left, with the dual exhaust mufflers extending up above the roofline, and two cone-covered air vents in upper background. The fill pipe and vent for the underground fuel tank arise from the dirt directly beneath the mufflers. The right half of the building with dual air intakes and windows served as the radiator room, as seen in the floorplan presented in Figure 8-38b.

a)

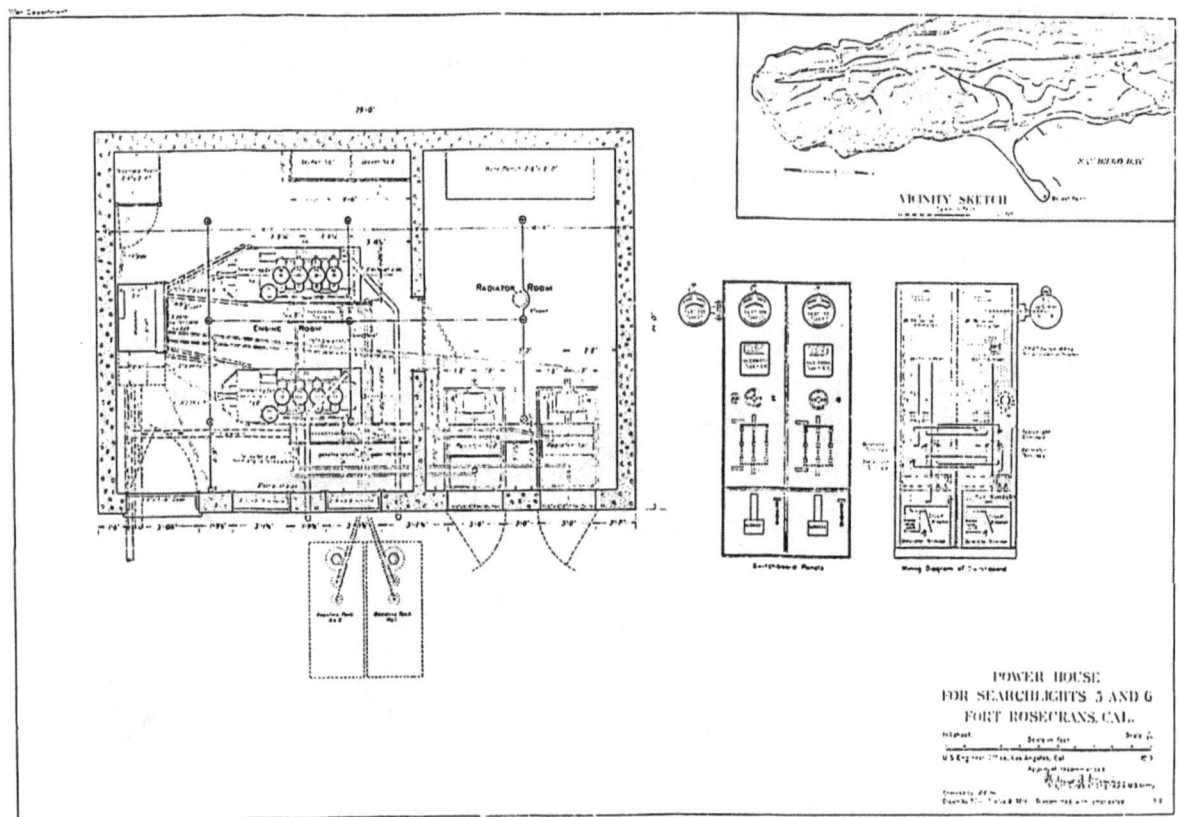

b)

Figure 8-38. a) The dug-in powerhouse for WWI-era Searchlights 5 and 6 (later WWII-era Searchlights 18 and 19) as it appeared on 12 March 1920 (US Army photo courtesy Cabrillo National Monument). **b)** The floorplan of the "Power House for Searchlights 5 and 6, Fort Rosecrans, Cal."

The US Army diagram depicted in Figure 8-39 below, dated 6 May 1911, shows the standard layout and electrical schematics of the WWI-era generator and feeder panels for the powerhouse. The text at far right indicates the generators could be operated individually or in parallel, and that additional panels could be added if more than two generators were installed. These electrical panels were removed when the fixed WWI generators were replaced by standard portable generators in WWII.

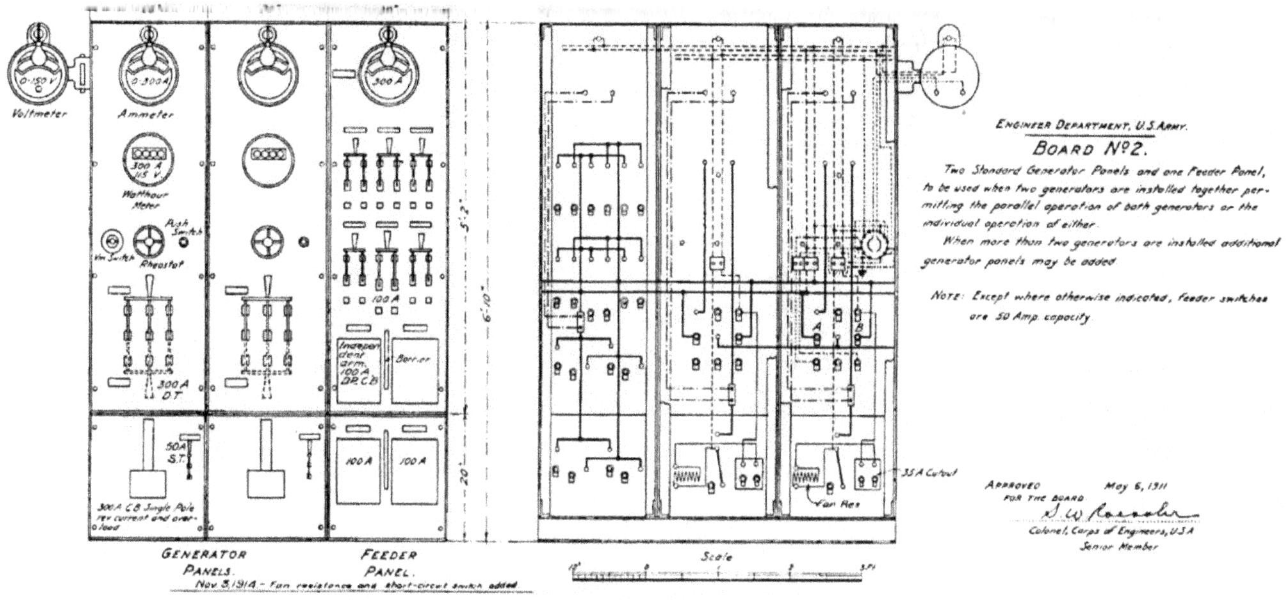

Figure 8-39. Diagram of two standard generator control panels and one feeder panel (left), along with their associated electrical schematics at right (US Army drawing adapted from IB1, 1916).

Searchlights 7 and 8

On 21 April 1917, Searchlights 7 and 8 were authorized for emplacement across the harbor entrance "on a sand spit" at Fort Pio Pico on Zuniga Point.[21] Due to the near sea-level terrain elevation, these lights were to be mounted on elevating towers in order to achieve a higher ground clearance. By the time funds were allocated in July 1919, however, all of North Island had become federal property, Fort Pio Pico had been abandoned, and its two 3-inch guns at Battery Meed were transferred to Battery McGrath at Fort Rosecrans. Accordingly, the proposed locations of Searchlights 7 and 8 were shifted further east on North Island to higher ground as shown in Figure 8-40, where they were installed in underground concrete silos with elevating lifts.

---

21 The alternate spelling "Zuninga" is shown in the map.

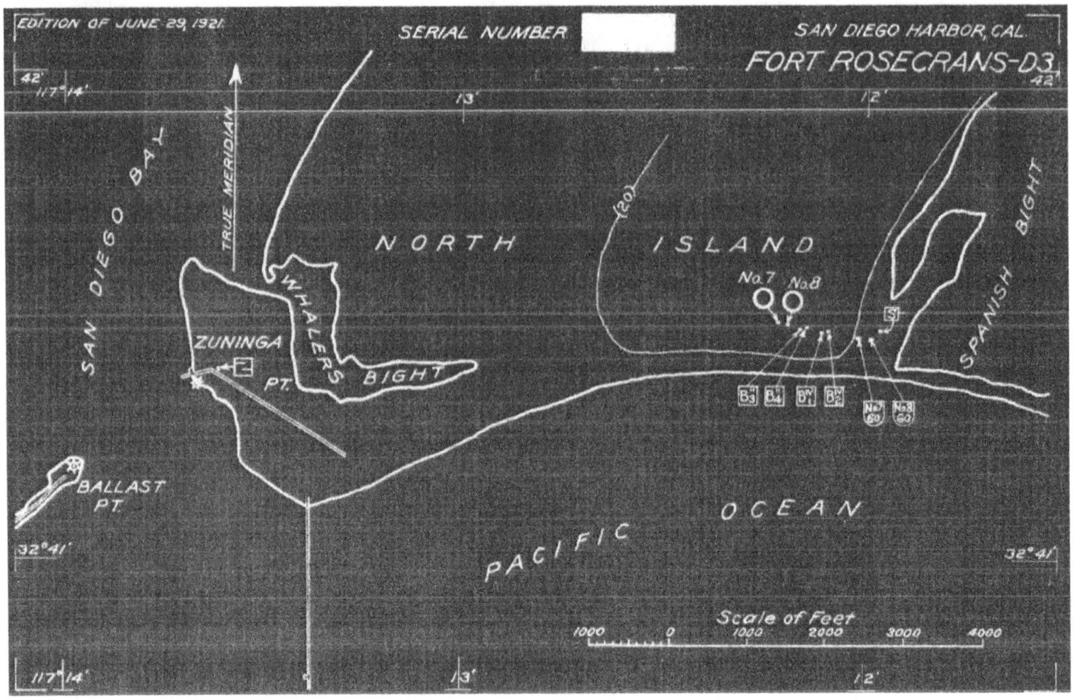

Figure 8-40. Searchlights 7 and 8 (right of center), originally planned for installation at Fort Pio Pico on Zuniga Point (left of center), were moved further east towards Spanish Bight (upper right) when all of North Island became Federal property in 1918 (NARA).

The silo designs for Searchlights 7 and 8 featured a deeper pit to accommodate a more cube-shaped lift design (Figure 8-41). Note these box-like extensions beneath the elevator platform shown in both section views, which increased the interior vertical dimension of the silo structure to 21 feet, 1.5 inches, as opposed to 14 feet, 4 inches for the other silos at Fort Rosecrans. The expectation of flooding at this almost sea-level elevation was the probable reason these lifts received this modification, to keep the light platform somewhat higher relative to the inevitable seepage.

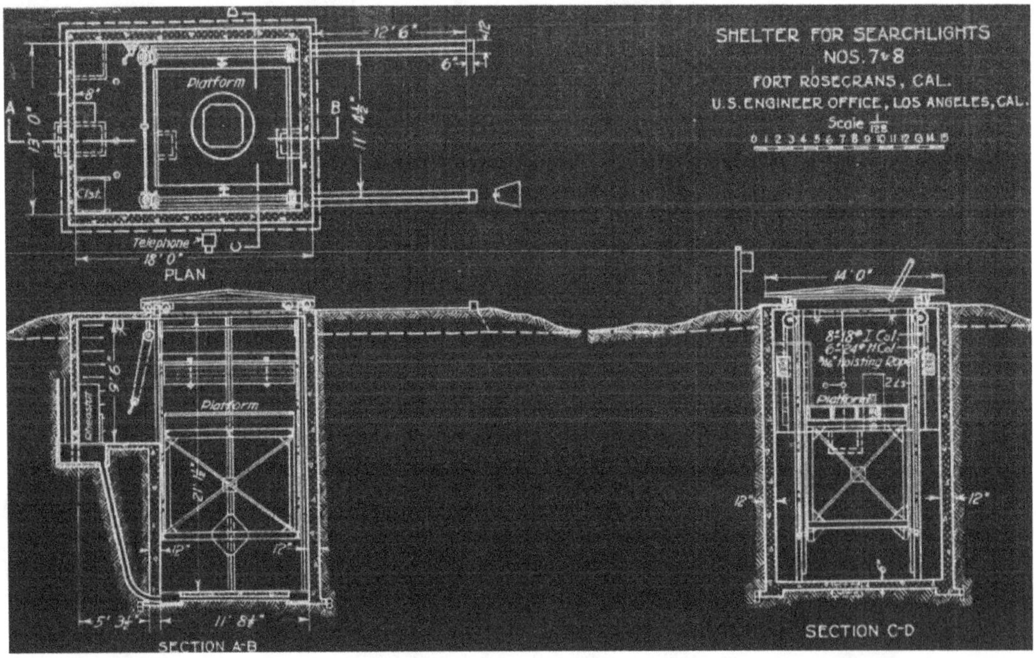

Figure 8-41. Presumably to mitigate water damage, the underground shelters for Searchlights 7 and 8 on North Island featured a box-like elevator design in lieu of the more common configuration previously seen in Figure 8-25. Note what appears to be a drain-sump connection in the lower left corner.

The Searchlight 7 silo was destroyed at some point during post-WWII construction at Naval Air Station North Island, but the pit and lift structure for Searchlight 8 are still in fairly good condition. Coast Defense Study Group members Al Grobmeier and Lee Guidry conducted a site survey of the area in October 1992, reporting their findings as follows (Grobmeier, 1992):

> "Searchlight 7 was located directly in front of the eastern-most admiral's quarters and no longer exists. However, the emplacement for Searchlight 8 still remains approximately 100 foot to the east of the site of number 7. The sliding metal cover is welded shut over the concrete pit and is obviously abandoned, although in good condition."

A photo of Searchlight 8 taken by Al Grobmeier at the time is presented in Figure 8-42.

Figure 8-42. The remains of the Searchlight 8 silo shelter on North Island as seen in 1992, with access hatch in foreground center, and telephone post at upper right (photo courtesy Al Grobmeier).

The powerhouse for lights 7 and 8 on North Island, just southeast of the approach end of Runway 29, is still visible from Rogers Road (Grobmeier, 1992). The interior is typical of the standard layout for such structures, but was situated partly below ground level as shown in Figure 8-43, most likely for purposes of concealment. The three identical powerhouse structures for Searchlights 1 – 6 at Fort Rosecrans saw continued service in WWII and will be further discussed in the next section.

Figure 8-43. The powerhouse for WWI Searchlights 7 and 8 was still intact in 1992 near the end of Runway 29 at NAS North Island, although the air vents and exterior mufflers have been removed (photo courtesy Al Grobmeier).

In summary, the eight 60-inch harbor-defense searchlights in use at the close of the First World War are listed below in Table 8-3. It is interesting to note that the last six of the eight were still primarily concerned with illuminating the harbor entrance versus the Pacific Ocean. The four combined power plants serving these lights, each equipped with two gasoline-powered 25-KW DC generator sets, were transferred to the artillery on 9 November 1920 (EN, 1920).

Number	Location	Elevation	Mount
1	Northwest Fort Rosecrans	190	Elevating Lift
2	Northwest Fort Rosecrans	97	Elevating Lift
3	Southern Tip of Fort Rosecrans	116	Rail Track
4	Southern Tip of Fort Rosecrans	297	Elevating Lift
5	Southeast Fort Rosecrans	218	Rail Track
6	Southeast Fort Rosecrans	144	Elevating Lift
7	North Island	26	Elevating Lift
8	North Island	26	Elevating Lift

Table 8-3. Numeric designations for the eight Taft-era harbor-defense searchlight installations in San Diego. Searchlights 7 and 8 were originally planned for Fort Pio Pico but actually installed near Spanish Bight. Elevations for Searchlights 7 and 8 are assumed to be that for Runway 29, North Island Naval Air Station.

WWII Searchlights

The perceived benefits of mobile versus fixed searchlight positions were well articulated as far back as 1902 (Suter et al., 1902):

- Less predictable location for enemy use as a nighttime navigational reference.
- Temporary off-site deployment without the purchase of land.
- Optimal positioning for both training and operation.
- Substantial cost savings relative to permanent shelters wired to remote generators.

Recall six of the eight WWI-era searchlights had been permanently installed at Fort Rosecrans, with the other two directly across the harbor entrance on North Island.

In order to cover the outer limits of fire of the long-range WWII batteries, it was necessary to extend the searchlight coverage well off the reservation to both the north and south (Annex C, 1946). On 21 June 1940, Battery N, the Regimental Searchlight Battery of the 19th Coast Artillery (Harbor Defense), took charge of ten *M1941* 60-inch Mobile Seacoast Searchlights (Gaines, 1993). These lights had initially been stored in the two Quonset-like storage shelters on the lower cantonment area near Ballast Point (Keniston, 1996), but moved off post immediately after Pearl Harbor "to provide illumination for all the new base-end stations" (HS, 1945).

Most WWII searchlight operators at Fort Rosecrans received their initial training at nearby Camp Callan, which occupied almost 4,000 acres of land just north of La Jolla, CA (Figure 8-44). Established in November 1940 as a Coast Artillery Corp training facility, the site soon became a key Antiaircraft Replacement Training Center for new recruits (Wikipedia, 2019). In addition to coast-defense and antiaircraft searchlight operation and maintenance, training included fire-control procedures and live-fire drills on a variety of weapons, from small arms to larger-caliber 155-millimeter GPF and 90-millimeter guns.

Figure 8-44. Wartime postcard showing the main (south) entrance to Camp Callan off of Torrey Pines Road, at what is now the Torrey Pines Inn (courtesy Cabrillo National Monument). The camp was named for the recently retired Major General Robert E. Callan, US Army.

For the most part forgotten today, the extensive Camp Callan facility was much larger than Fort Rosecrans, with over 297 buildings that included five chapels, three theaters, and five post exchanges (Berhow, 2006). The continuing improvement of radar performance over the course of WWII would eventually make searchlights obsolete, just as post-war missile batteries would later replace the antiaircraft guns themselves. When all further antiaircraft training was transferred to Fort Bliss in June of 1944, Camp Callan shifted its focus to amphibious-assault training.

Figure 8-45. Postcard of a ceremonial searchlight display at Camp Callan, La Jolla, CA (undated US Army photo courtesy Cabrillo National Monument).

Just four searchlight positions were operational at Fort Rosecrans at the beginning of the 20th century, and that number grew to only eight during WWI. By the end of WWII, a total of 27 searchlights were arrayed from Cardiff to the north all the way down to the Mexican border, covering some 50 miles of shoreline as shown in Figure 8-46. In addition, two portable 60-inch searchlights and two 16.2-KW portable generators were kept in reserve (Annex C, 1946), and a 24-inch light was located on the Signal Station at the old Spanish lighthouse on top of Point Loma (see again Chapter 7).

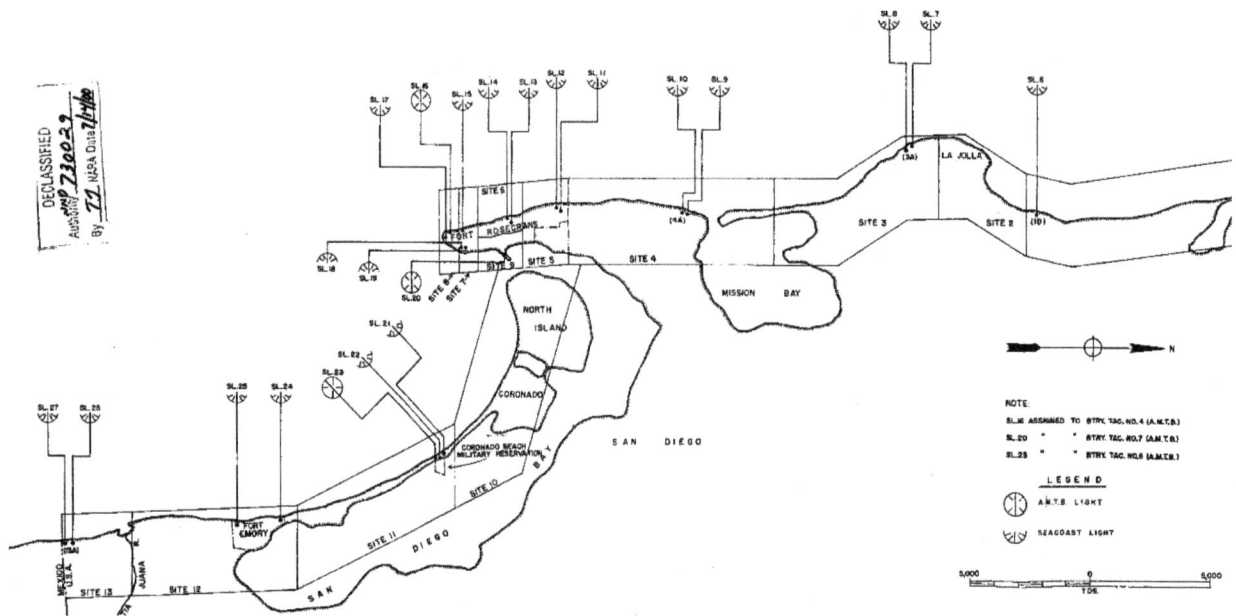

Figure 8-46. A 1946 map of harbor-defense searchlight locations at Sites 1 (upper right) through 13 (lower left), showing three installations specifically designated as Anti-Motor Torpedo Boat (AMTB) lights (see legend bottom right). See expanded detail for Site 1 in Figure 8-48a (NARA).

Searchlights 1 – 10

Per Harbor Defenses of San Diego blueprint "Harbor Defense Elements" dated 1 July 1945, searchlights 1 through 10 were located north of Fort Rosecrans along the Pacific Coast at numerically designated sites as indicated in Table 8-4 below. These ten WWII installations resulted in a renumbering of the Taft-era searchlights presented earlier in Table 8-3. All were locally controlled at their respective sites, directed from the Harbor Defense Command Post.

Number	Site	Location	Mount
1&2	1A	Cardiff	Portable
3&4	1B	Solana	Portable
5	1C	Del Mar	Portable
6	1D	Scripps	Portable
7&8	3A	Neptune	Portable
9&10	4A	Ocean Beach	Portable

Table 8-4. Locations of portable WWII searchlights 1-10 in the civilian communities north of Fort Rosecrans, Sites 1 through 4A, circa July 1945.

The portable searchlights were made by both General Electric and Sperry Gyro at a cost of $60,000 apiece, which included a trailer-mounted 15-KW generator powered by a six-cylinder Hercules *JXD* flathead engine. As previously discussed, a *Distant Electric Control (DEC)* unit allowed for remote operation so that the operator could stand some distance away from the light for better visibility, as backscatter reflection from atmospheric particles made it difficult to see along the beam axis. A General Electric *Model 1942A* portable searchlight, beautifully restored by Bob Meza, is shown in Figure 8-47 (Meza, 2006).

Figure 8-47. This GE *Model 1942A* portable searchlight and generator is the only known restoration to work with a *Distant Electric Control (DEC)* unit, shown at photo center (photo courtesy Bob Meza).

Very little evidence of WWII Searchlights 1 – 10 can be found today, for three reasons: 1) these were portable searchlights and generators, with temporary shelters; 2) the equipment was rendered obsolete by newer technology and removed after the war; and 3), the coastal areas where these lights were deployed have been extensively developed over the past several decades. The photo seen in Figure 8-48b, taken by Al Grobmeier in January 1991, shows the general location of Searchlight 4 in Site 1B (Solana).

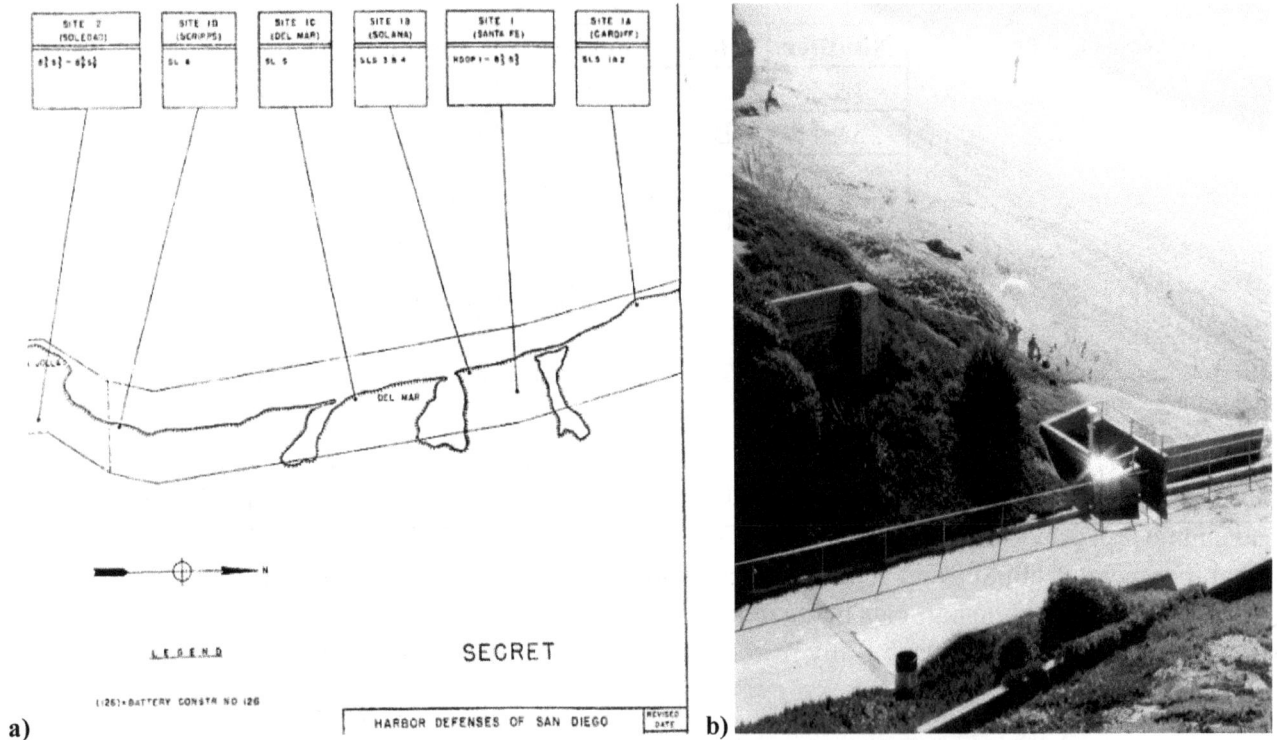

a)

b)

Figure 8-48. a) The location of portable Searchlight 4 is indicated just right of top center in Site 1B (Solana) in this lower-right-corner portion of "Harbor Defense Elements" map dated 1 July 1945. **b)** The general location of portable Searchlight 4 in Solana Beach (January 1991 photo courtesy Al Grobmeier).

Period photos of these ten lights (and/or shelters) have not been located, except for that shown in Figure 8-49, which appears to be the shelter for Searchlight 2 in 1943, situated 300 feet north of Scripps Pier in Site 1-D (Zink and Grobmeier, 1991). This light was later redesignated as Searchlight 6 in 1945 (see again Table 8-4). Note the railroad track running north-south in foreground center, and the overhead power lines up above. Figure 8-50a shows the light's location in Site 1D, with the leased strip of land depicted in Figure 8-50b, extracted from "Land to be Leased in Site 1D, Scripps, City of San Diego" (Annex, 1946).

Figure 8-49. Identified only as "Searchlight 2" in 1943, the railroad track in the foreground and ocean pier at left indicate this to be the above-ground shelter situated at Site 1D (Scripps) for what was later known as portable Searchlight 6 (US Army photo adapted from Wolgast, 1943).

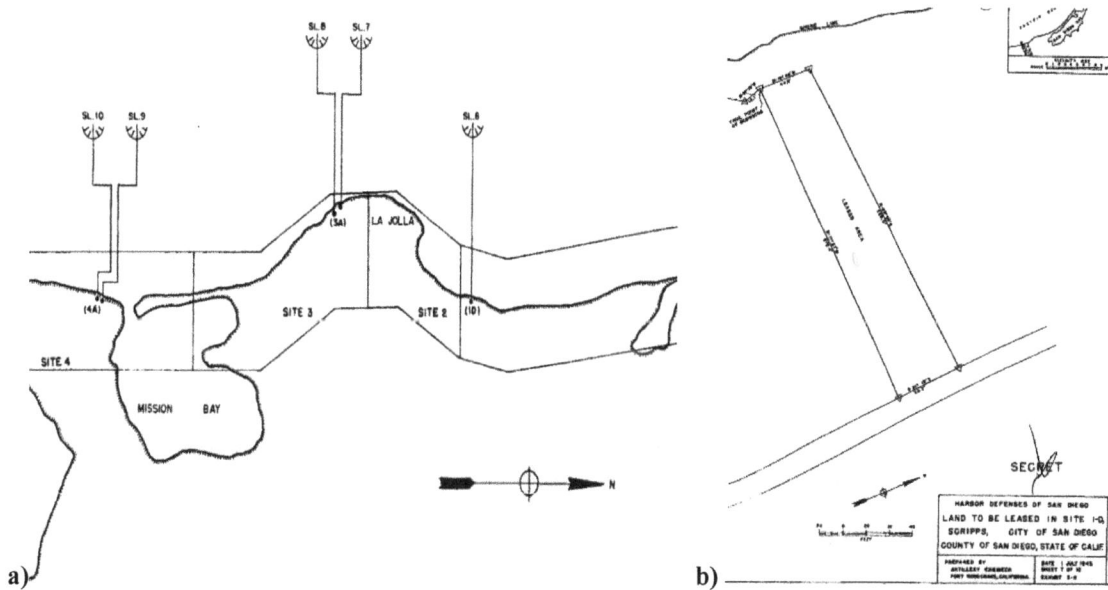

Figure 8-50. a) The operating locations of portable Searchlights 6 through 8 in Sites 1D (Scripps) through 3 (Neptune), circa 1946. **b)** The narrow strip of land leased for Searchlight 6 in Site 1D.

Searchlights 11 and 12

My growing personal interest in the WWI&II infrastructure at what was then the Naval Ocean Systems Center on Point Loma where I worked was triggered by Searchlights 11 and 12 in the northwest corner of what was once Fort Rosecrans. As previously discussed, these two WWI-vintage lights and their combined powerhouse had been constructed at Site 5 during the 1918 to 1919 timeframe, and remained operational until sometime in 1945. The powerhouse was situated right across Woodward Road from our robotics lab, yet known to us only as Building F-17 (Figure 8-51). We had zero awareness of its prior history and no clue regarding the nearby searchlight installations it once serviced.

Figure 8-51. The combined powerhouse for Searchlights 11 and 12 (previously 1 and 2 during WWI) on Woodward Road, circa 1971 (photo courtesy Lee Guidry). Note holes for generator muffler exhausts above window at center, the widened doorway, and the water tank at upper right that resupplied the radiator.

As I was to eventually discover, Searchlight 11 (originally designated Searchlight 1 during WWI) was located high on the slope above and behind Battery Gillespie, just west of its temporary battery commander station. Its companion, Searchlight 12 (formerly Searchlight 2), was located on the western side of Woodward Road, directly across from Building F-17, the combined powerhouse seen in Figure 8-52. Both these 60-inch searchlights were controlled from Battalion Command Post 2, and directed from the Harbor Defense Command Post (HDCP).

Figure 8-52. Front view of the WWI-era powerhouse, circa 2005, repurposed as a storage facility, looking northwest, with Woodward Road just visible in background at upper left. The original earthen backfill (see again Figure 8-28) has been removed from the south (left) side of the building to enlarge the tarmac area.

According to the "Project Engineer Notebook (Journal)," the original pair of 25-kilowatt DC generator sets installed in this powerhouse were turned over to the 19th Coast Artillery on 9 November 1920 (Figure 8-27). The Keniston (1996) survey, however, reports the structure housed "two portable wheel-mounted generators (16.2 KW)." This was also true, as the WWI generators had been later replaced by portable WWII generators, which explains why the door had to be widened. (See also later discussion regarding similar alterations to the identical powerhouse for Searchlights 18 and 19 on the Bayside Trail at Cabrillo National Monument.)

I had worked on Point Loma for 20 years, almost directly across the street from this building, before I ever learned of its legacy. There had been rumors regarding underground searchlight silos in the area, but no mention of a powerhouse. In 2006, my curiosity grew after seeing heavy-gauge armored cable exiting Building F-17 near a ventilation shaft (Figure 8-53a). Another portion of this same type cable had been exposed by erosion near the building-number sign on Woodward Road (Figure 8-53b). In May that same year, I found related evidence in the Flower and Roth (1982) site survey provided by our Public Affairs Officer, Tom LaPuzza, always very supportive of my historical research.

a)   b)

Figure 8-53. a) Remains of a ventilation duct with a conical rain cover and an armored power cable at top of Building F-17. **b)** Exposed section of war-vintage armored cable, which ran west from Building F-17 in the direction of Woodward Road. Building F-31, a small storage facility, has since been demolished.

Intrigued, I began poring over WWII blueprints and photographs, searching westward in the direction of the previously mentioned cable run from F-17. NAVFAC Drawing 1,007,837, a 1967 "General Development Map" of the "Field House Area" where our robotics lab was situated, proved to be especially helpful (Figure 8-54). Running horizontally through the middle of the figure is Woodward Road, which intersects unpaved Vodges Road at bottom right.[22] Searchlight 11 is marked as Building F-23 to the right of Vodges Road, while Searchlight 12 is seen above and just left of the intersection as Building F-16. The "Q.D." circle represents the estimated area of leakage from an earlier sewage spill.[23]

---

22 Vodges Road, no longer extant, was named for Major Anthony W. Vodges, who served as the commanding officer of both the San Diego Barracks and Fort Rosecrans in the early 1900s.
23 The map legend elaborates Q.D. as "Quantity Distance Clearance." A few years after I retired in 2016 from the Center, this contaminated area was the subject of a major soil-remediation project.

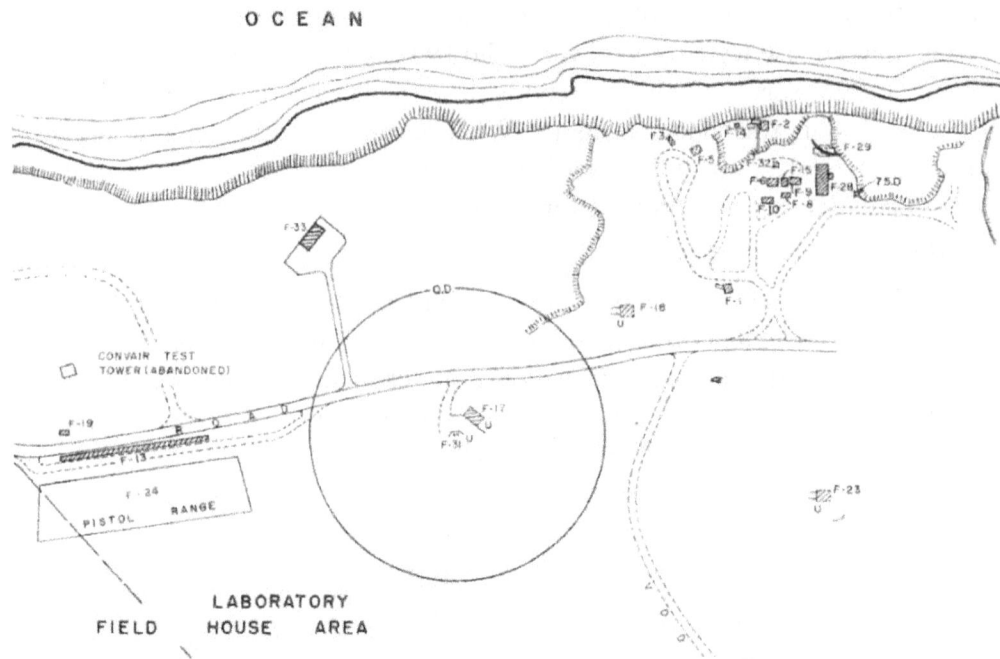

Figure 8-54. This upper right section of NAVFAC Drawing No. 1,007,837, dated 29 September1967, shows the former searchlight powerhouse, now Building F-17, inside the circle at bottom center, with Searchlight 11 (F-23) at bottom right and Searchlight 12 (F-18) above the intersection of Woodward and Vodges roads.

I eventually found photographic evidence of what appeared to have been a path from the F-17 powerhouse to a small rectangular structure of interest as seen in Figure 8-55 below. This intriguing artifact, which I later identified as Searchlight 12 (top right corner), appeared to be situated just north of an imaginary line connecting Buildings F-5 and F-17 at an elevation of about 100 feet. Initial attempts to find it, however, were unsuccessful. It turned out one of my key landmarks, Building F-5, had been relocated sometime after this aerial photo was taken. Assisted by fellow WWII enthusiast Jeff Bowen and his metal detector, we instead referenced off two prominent power poles to continue the search.

Figure 8-55. This portion of a 1954 US Navy photograph taken during site grading for Building F-28 (looking east) shows the location of Searchlight 12 (upper right). The short driveway off Woodward Road at top right leads to the Building F-17 powerhouse. Note the associated footpath trace down to the light.

Back we went into the brush, which over the years had grown shoulder high and quite thick, so progress was slow as I began triangulating off the power poles. Once it appeared we were getting close to our goal, Jeff bent over and spread the branches to get the metal detector down near the surface, and there was the trapdoor for Searchlight 12, less than 6 inches from his hand! We trimmed back some brush to open the unlocked steel hatch (Figure 8-56), and found everything in rather pristine condition, despite almost 90 years of neglect in a corrosive salt-air environment.

Figure 8-56. After carefully cutting away a few branches, the unlocked access hatch was easily raised, revealing the well-preserved interior of the underground silo that once housed Searchlight 12.

Filled with excitement, I carefully descended the steel ladder down into the silo, allowing my eyes to adjust to the dim lighting. Given the structure's age, I was impressed by the well-preserved interior. The steel ladder rungs of thick rebar set into the concrete wall looked a bit rusted but still structurally sound (Figure 8-57). Bolted to the wall on the right side of this ladder was a hand crank for raising and lowering the elevator platform, coupled via a linked chain to a worm-gear mechanism just below the sheet-metal roof.

Figure 8-57. The wall-mounted hand crank next to the access ladder powered the elevator-lift mechanism through a worm-gear drive as seen in Figure 8-58 on the next page.

At the upper end of the chain previously shown in Figure 8-57 is a sprocket on the end of a steel shaft (Figure 8-58) that crosses over to the far wall (Figure 8-59). A worm gear is installed on both ends of this shaft to actuate the thick mating gears directly above. Each turn of the worm gears rotates these larger gears by a one-tooth increment, resulting in a significant mechanical advantage. Such a worm gear is not back drivable, so the lift platform automatically held position when the operator stopped cranking.

Figure 8-58. The chain from the handcrank turns the eastern worm-gear input shaft (upper left), which in turn actuates the elevator-pulley driveshaft to its right to raise or lower the elevator-support cables.

The large gear driven by the worm gear in Figure 8-59 below turns another shaft along the east wall, also with a V-pulley on each end. Looped over these pulleys were two more elevator support cables. One end of each cable is attached to the corner of the elevator below it, while the other is attached to a pair of heavy counterweights hanging in a horizontal position between the elevator and the silo walls. The worm gear shaft can be seen in the upper foreground, crossing over from the east side of the silo.

Figure 8-59. The elevator lift cables pass over V-pulleys on each end of the horizontal shaft at top left. The other ends of these cables are attached to the elevator counterweights (not shown, see Figure 8-61).

A view of the counterweight pulley shafts from above the elevator platform is presented in Figure 8-60. Note also the pinion-gear shaft running across the top of the far wall; the two pinion gears engage a pair of parallel steel racks attached to either side of the peaked sheet-metal roof. Note the left end of this shaft passes through the concrete silo wall in the upper-left corner, where it is chain-driven by an outside handcrank to open or close the roof. The looped control and power cables for the searchlight, which has been removed for salvage, are visible in the background at bottom center.

Figure 8-60. This view of the silo roof shows the two rack-and-pinion mechanisms that were actuated by an operator turning the outside handcrank. This handcrank was chain-coupled to an exterior sprocket attached to the left end of the pinion rod seen running across the top of the south wall in background center.

A small steel ladder on the west wall (opposite the silo entry ladder on the east wall) leads down into the pit area below the lift platform as seen in Figure 8-61. To facilitate maintenance access, the depth of this pit allows the searchlight lift to be lowered flush with the concrete silo floor to the right. One of the horizontal lift counterweights suspended by steel cables can be seen alongside the west wall in background left. There is an identical counterweight suspended on the opposite wall, out of image to the right and behind. Note also the open door of the empty storage cabinet at upper right.

Figure 8-61. Tracy Heath Pastore, then a senior software engineer in the Unmanned Systems Branch (now a Branch Head), descends the pit-access ladder into the elevator pit of Searchlight 12, circa 2006. Note the suspended counterweight running horizontally along the elevator-shaft wall behind the ladder.

The vertical steel I-beam just right of the pit ladder in Figure 8-61 above served as one of four identical guides for the lift, with one of the sliding brackets attached to the bottom of the lift corner just visible at top center. Referring now to Figure 8-62 below, the two cast-aluminum housings at left and center in the photo each contained a single heavy-gauge pin for DC power to the light. The connector housing at far right contained 10 smaller pins for controlling the pan-and-tilt motors that trained and elevated the light. Such original searchlight connectors today are very rare.

Figure 8-62. Searchlight power connectors (left and center) and control connector (right), after extraction from a thin layer of sediment and debris in the southwest corner of the elevator pit, circa 2006.

Close examination of the 1954 aerial photo shown earlier in Figure 8-55 revealed a number of potential artifacts in the vicinity of this newly discovered searchlight silo. Subsequent exploration of the surrounding area uncovered the remains of a machine-gun emplacement, a short wooden footbridge, and a collapsed tunnel south of the silo on the north side of the ravine seen at image right in Figure 8-63. Also visible in this photo but not located during this search was a pair of adjacent rectangular shapes that potentially led to a subterranean space.

Figure 8-63. Artifacts of interest detected in this zoomed-in portion of the 1954 US Navy photograph previously shown in Figure 8-55 included a pair of horizontal rectangles (left up arrow), a wooden footbridge (center up arrow), and a machine-gun emplacement (right up arrow) defending the ravine at lower right.

Now that I knew what to look for, I zoomed in on the upper portion of the original photo and soon found what appeared to be the location of Searchlight 11, situated along a north-south road trace at an elevation of 220 feet, as seen in Figure 8-64. Note the unpaved lower section of WWII-era Vodges Road running up the slope to the east at top center, and the short drive leading to the powerhouse for these lights at upper-right corner (Building F-17). Battery Gillespie is just out of the image to the north (left) and west of Woodward Road.

Figure 8-64. This expanded view of the upper-right portion of the original 1954 US Navy photo reveals the location of Searchlight 11 at upper left, just right of its callout, looking east. Woodward Road runs horizontally through the center of the photo, and unpaved Vodges Road running vertically at top center.

Searchlight 11 upslope in the bushes proved to be much easier to find than its companion Searchlight 12 down below. Vodges Road had long been abandoned after the war, but there were still telltale signs of its former existence. Not even 25 yards up from Woodward Road, I spotted a large washout to my right (Figure 8-65), in which lay an exposed section of the buried power and communication cables that ran from the powerhouse shown earlier in Figure 8-51 up to the higher-level silo for Searchlight 11.

Figure 8-65. This exposed run of four armored power cables and one communications cable from the powerhouse to Searchlight 11 was discovered parallel to the remains of Vodges Road heading east up the slope, circa June 2006. Note Woodward Road downhill running north-south in upper background.

A short walk along a north-south road trace off Vodges Road (see again Figure 8-64) soon revealed a WWI-vintage concrete communications post peeking out from the scrub brush (Figure 8-66a). After pulling back some vegetation, the handcrank and sprocket chain for retracting the silo roof were discovered intact beneath their sheet-metal cover (Figure 8-66b). Sections of rusted chicken wire were found along the roof-retraction track, which had been used to secure painted burlap camouflage across the sheet-metal roof and exposed tracks for aerial concealment during the war.

Figure 8-66. a) The concrete communication post still has the steel brackets attached that supported the field telephone box. **b)** The rusted sheet-metal cover for the handcrank that actuated the rack and pinion for retracting the silo roof lies on its side, with the crank arm buried beneath the leaves at upper left.

Figure 8-67 shows the entry hatch and roof of the silo shelter for Searchlight 11 after cutting away some underbrush, looking southwest towards the Pacific Ocean. Buildings 622 and F-33 of our robotics lab can be seen down near the cliff edge at upper right, circa 2006. The steel tracks just left of the sheet-metal silo roof are covered with brush, as is the concrete slab between this post and the entry hatch at foreground center. Note the small air vent at the peak of the rusted roof near the edge of the brush at top center.

Figure 8-67. View of access hatch (foreground) and steel roof of Searchlight 11, looking southwest towards Buildings 622 and F-33 of the Unmanned Systems Branch along the Pacific coastline, circa June 2006. The remains of what was once burlap-stuffed chicken-wire camouflage are laying on the sheet-metal silo roof.

A small portion of the roof tracks not covered by brush is seen in Figure 8-68, which shows the south-west end of the left track and its concrete foundation fully exposed. The heavy-gauge wire attached to the remnants of a wooden 2x2 suggests the camouflage netting may have been suspended just above the silo tracks. Such a tent configuration would have hidden the otherwise bare area between the tracks when the roof was closed, so as not to give away the light's position from the air. The lack of brush on the downhill side of the shelter probably reflects the underground structure's impediment of subterranean water flow.

Figure 8-68. This southwest corner of the westernmost roof-retraction track was the only portion exposed to view. The rusted chicken wire at foreground center was originally stuffed with painted burlap strips and used for camouflage. Note Pacific Ocean at upper left, and wire attached to remnants of a wooden 2x2 post.

This WWI/WWII searchlight silo was surprisingly well-preserved, given its close proximity to the ocean. The surrounding area was littered with considerable wartime debris, such as chicken-wire-covered wooden frames for camouflage, sections of stone wall to minimize erosion, and a set of wooden stairs leading up to the access path (see also Figure 8-77). The most interesting aspect of the shelter was the interior, and fortunately our Facilities representative had provided a key for the padlocked entry hatch. Figure 8-69 shows the open hatchway after entry. Due to watching too many scary movies as a child, I made sure to lock the padlock above in such a way that the hatch could not be secured with me inside.

Figure 8-69. The paint outline on the north wall to the left of the access ladder is from an equipment cabinet that has been removed. Three holes in the floor approximately 18 inches from the wall indicate it was serviced by electrical cables, also since removed, suggesting this was the rheostat location.

Looking 90 degrees left of the previous view, Figure 8-70 shows a large storage cabinet in the corner, with a wooden rifle rack just to its left against the west wall. The square piece of wood hanging from the light switch on the north wall was a cover for the cable pit below it, where the DC electrical service entered the silo from the down-slope powerhouse off Woodward Road. The northwest corner post of the elevator structure is at far left, with one of four steel cables that supported the lift counterweights just visible along the left edge of the photo.

Figure 8-70. An improvised wooden rifle rack sits against the wall just to the left of the corner storage cabinet in background center. The wooden square suspended from the surface-mounted armored cable at right was a cover for the cable pit immediately below it.

Continuing to the left and looking down into the elevator pit (Figure 8-71), there is a discarded wooden structure at image center, which looked like it served as a supporting base for some unknown item, perhaps the missing rheostat cabinet (see again Figure 8-69). There are also four 2-inch-thick wooden boards on the pit floor, three of which appear to be purposely arranged around the perimeter. This configuration suggests they may have served to keep the lift from sitting on the bottom of the pit, which had a bit of standing water from seepage during the winter rainy season.

Figure 8-71. A rectangular wood-frame base of some sort has been discarded in the middle of the elevator pit. Note searchlight power cables arising from conduits in the floor near the far wall, and the horizontal counterweights for the lift suspended along both side walls behind the vertical I-beams.

The armored cables seen arising from electrical conduits in the elevator pit floor in Figure 8-71 above continue up the wall and loop over to a double-pole-single-throw knife switch that served as a safety disconnect for searchlight power (Figure 8-72a). In the close-up view provided in Figure 8-72b, notice the power cables that ran to the light, which were originally connected to the bottom two terminals of the switch, have been disconnected. This probably happened when the searchlight was removed from the silo for salvage.

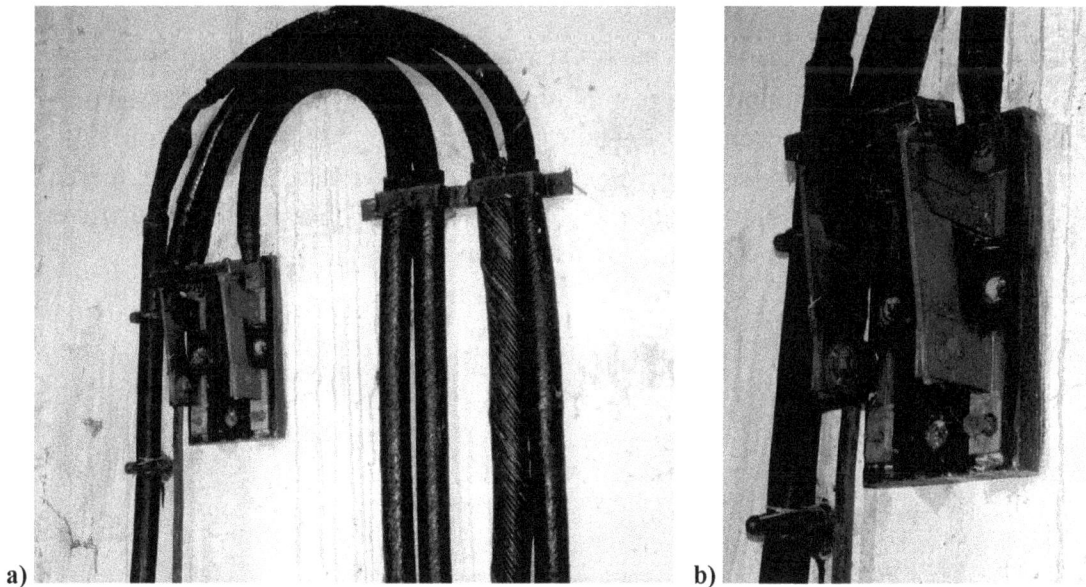

a)   b)

Figure 8-72. a) A double-pole single-throw knife switch served as a safety disconnect for DC power to the light. **b)** This close up of the knife-switch mechanism shows where the power cables leading to the light were once connected to the two copper terminals at the bottom of the switch.

Figure 8-73 below shows the bottom of the searchlight lift platform, with rust apparent on the steel angle-iron structural support for the roof above, but not on the sheet-metal roof itself, which was galvanized. Note the heavy wooden 2x12 boards framing the lift opening beneath the roof, reinforced by three pressure-treated 2x4s below. This unusual reinforcement, which was not found present in the silo for Searchlight 12, appeared to have been added in more recent times, perhaps to stabilize the lift platform when the 60-inch light was salvaged.

Figure 8-73. View looking upwards at the galvanized sheet-metal roof and its rusted angle-iron bracing, as seen through the center opening of the searchlight lift. Note what appears to be a more recently added transverse reinforcement of the lift platform.

Figure 8-74 below, shot from just above the lift platform, provides a good view of the rack-and-pinion configuration that retracted the roof prior to searchlight deployment. The pinion shaft, actuated from outside by an operator turning the handcrank, penetrates the silo through the concrete wall at lower left and runs across the back wall, with a pinion gear on each end. As previously mentioned, these pinion gears engage the teeth of two parallel steel racks attached to the roof frame, thus enabling it to traverse back and forth on its steel rail tracks above the shelter (see also Figure 8-68).

Figure 8-74. The sheet-metal pyramid roof and its steel support frame in the Searchlight 11 silo shelter are seen here from the top of the lift platform. The rack-and-pinion roof-retraction components are visible around the upper perimeter of the concrete silo.

I returned to this site with my son in December 2007 to take a few more photos of the exterior but did not reenter the silo, which was again locked. Surprisingly, the sheet-metal roof and surrounding perimeter had been partially cleared of debris as shown in Figure 8-75. The silo shelter for companion Searchlight 12 is hidden from view by the underbrush just across Woodward Road below, roughly midway between the power pole at background center and the ravine at upper left. The combined powerhouse for these two lights is in the ravine on this side of Woodward Road, obscured from view at left center, directly across from the same power pole.

Figure 8-75. Todd Everett stands on the concrete pad for Searchlight 11 in December 2007. Companion Searchlight 12 was located on the far side of Woodward Road below. The powerhouse was situated in the small ravine this side of Woodward Road, directly across from the power pole at photo center.

A few months later in my research I gained access to a formerly Secret "Harbor Defenses of San Diego" map dated 1 July 1945, which had been recently declassified on 14 July 2000. Prepared by the Artillery Engineer, Fort Rosecrans, CA, this map depicted the "Fire Control Stations at North Fort Rosecrans, Sites 5, 6, 9." The upper right corner of this map showed the locations of Searchlight 11 and 12 (Figure 8-76), and above them in elevation to the east, much to my surprise, a planned *SCR-296-A (RAD 9-127)* radar (see Chapter 9). This latter revelation took the sting out of how much easier my searchlight quest would have been if I'd had this map to begin with.

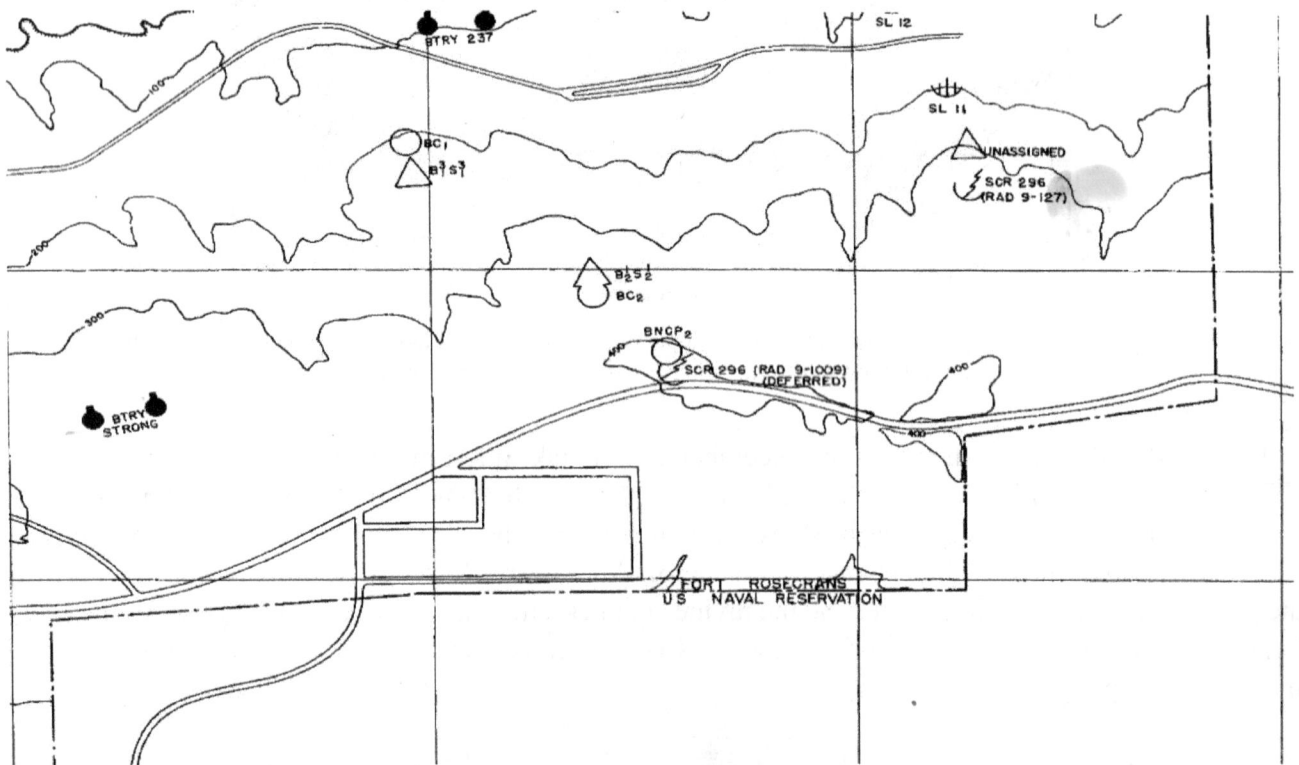

Figure 8-76. Northwest corner of the HDSD map "Fire Control Stations at North Fort Rosecrans, Sites 5, 6, 9," showing the locations of Searchlights 11 and 12 (upper right), along with several other historic points of interest (NARA). North is to the right, with the Pacific Ocean to the west just out of image at the top.

Similarly, I was eventually able to scan a copy of the November 1962 photo seen in Figure 8-77, which was held in the historical collection of our Facilities Department at the Center. The below zoomed-in portion of this aerial image clearly shows both silos for Searchlights 11 and 12 near the northwest corner of Center property, before they had been completely concealed by an overgrowth of dense brush. Even the concrete rail foundations are visible for Searchlight 12, just to the left of Woodward Road. Note the distinctive traces of Vodges Road running southeast from Woodward in the lower-right corner, and the north/south dirt pathway off Vodges Road, with wooden stairs down to Searchlight 11.

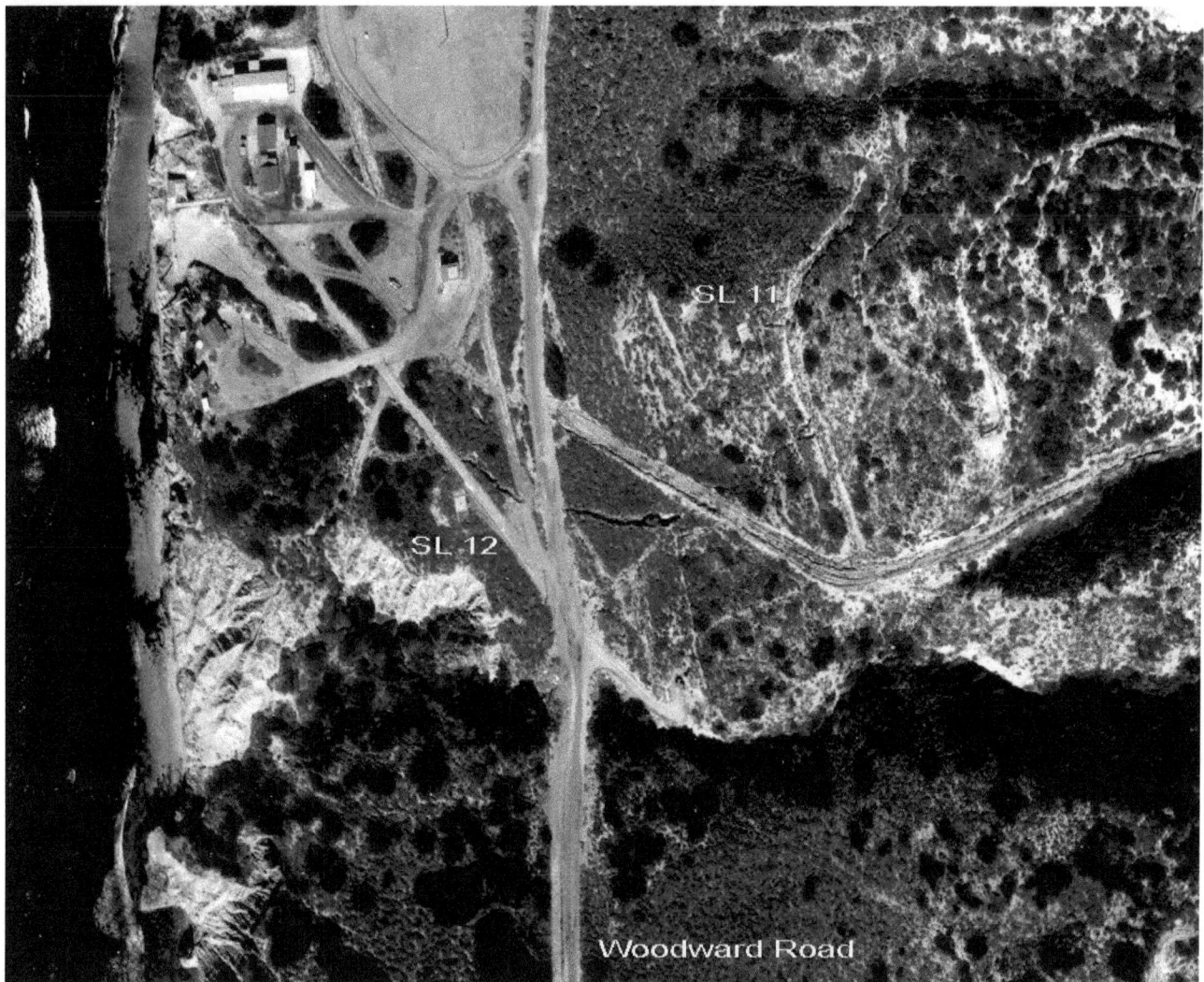

Figure 8-77. The locations of Searchlights 11 and 12 are clearly visible in this 1962 aerial photograph (LSF22 - 11 62), with the Pacific Ocean to the west at left. Note the support rails for retracting the cover of the Searchlight 12 silo, and the wooden stairs descending down to Searchlight 11 from the dirt pathway off Vodges Road (US Navy photo courtesy SSC Pacific).

Searchlights 13 and 14

Portable Searchlights 13 and 14 were situated southwest of the Harbor Defense Command Post/Harbor Entry Command Post (HDCP/HECP) and about 800 yards south of Woodward Road in Site Number 6 (Figure 8-78). Searchlight 14 was located near the cliffs overlooking the Pacific at an elevation of about 115 feet, while Searchlight 13 was 500 feet further inland at an elevation of about 175 feet, east of what is now Gatchell Road. Both these lights were directed and controlled from the HDCP/HECP, with a local controller hut near their operating positions. Their portable generators were housed in two splinter-proof shelters, also east of Gatchell Road but not shown on the map below.

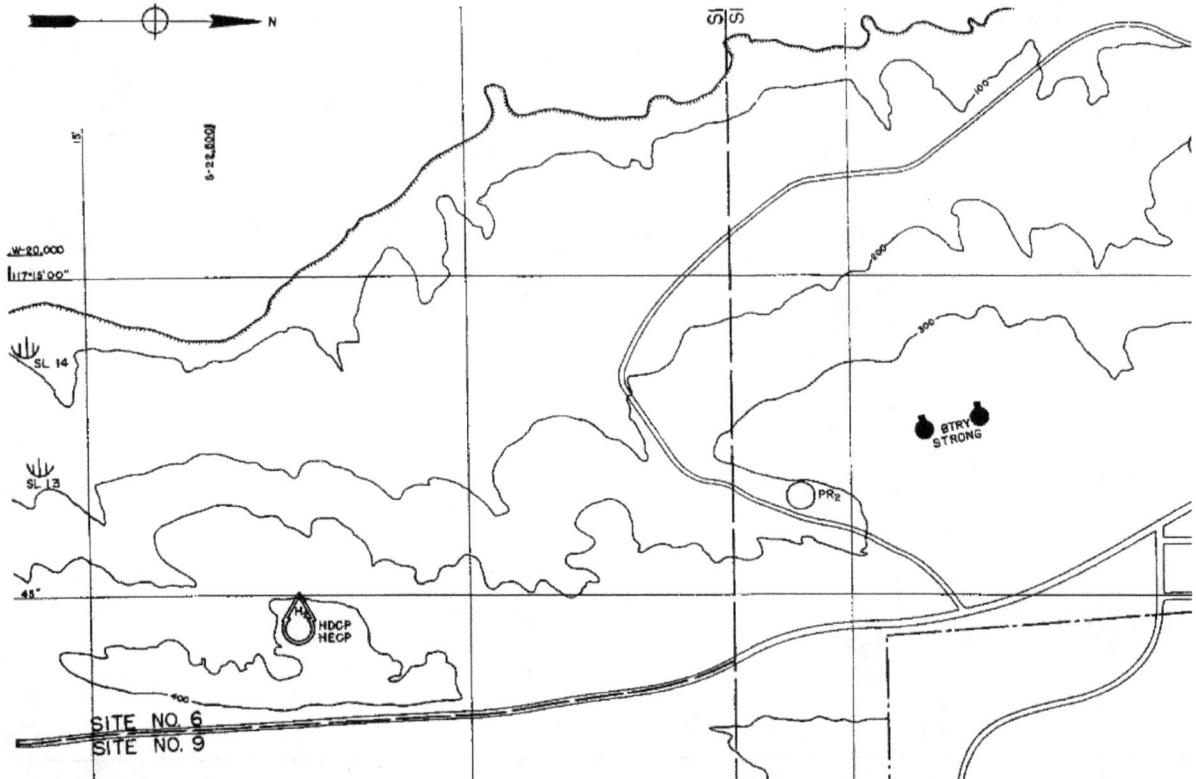

Figure 8-78. Searchlights 13 and 14 (far left) were southwest of the Harbor Defense Command Post in Site Number 6 (adapted from map of "Fire Control Stations at North Fort Rosecrans, Sites 5, 6, 9," 1 July 1946). The Pacific coastline is along the top of the map, with Woodward Road in the upper-right quadrant (NARA).

The shelter for Searchlight 13 was about a third of the way up the ravine between Gatchell Road at far right and Fort Rosecrans National Cemetery at lower left in Figure 8-79, as seen in 1966. About halfway up the dirt access road (upper right quadrant) for this installation can be seen a footpath leading to the local operator's station, a temporary structure made of wood just south of the shelter (see also Figure 8-80). Note the cold-war-era radar installation to the right of the cemetery at bottom center and the HDCP/HDEP observation levels in extreme lower-left corner.

Figure 8-79. This portion of US Navy photo LSF 275-12-66 taken on 9 September 1966 shows the location of the Searchlight 13 shelter on the south side of the ravine running down from Fort Rosecrans National Cemetery (lower left) to Gatchell Road below. Note local operator station directly beyond the shelter.

The photo presented in Figure 8-80 below, looking southwest from the vicinity of the searchlight-shelter doors in 1971, provides a close-up view of the local operator's station for Searchlight 13. Very likely this station also provided local control for the lower-elevation Searchlight 14 location near the cliff edge, depending on which light was in use at the time. Note spillage from atop the corrugated roof in foreground, just downslope from the entrance of the dug-in searchlight shelter, which is out of image to the left.

Figure 8-80. Local controller station for Searchlight 13 and/or 14 as seen in June 1971 from the searchlight shelter doorway (photo courtesy Lee Guidry). The debris field at lower left is eroded rock and clay spillage that once covered the searchlight shelter's corrugated-steel roof, out of image to the left.

The original HDSD blueprints for Searchlight Shelter 13, which was fabricated from 12-guage Multi-Plate Arch and covered in rock and earth, are seen in Figure 8-81. The floor and both rectangular ends were made of reinforced concrete, with large arch-shaped wooden doors on the north end to allow access. "Field wire" for communicating with lights 13 and 14 is depicted running southwest of the Plotting Room for Battery Strong in a 1 July 1945 "Harbor Defenses of San Diego, Main Cable Routings" blueprint (sheet 2 of 4). There were no electrical or communication cables servicing the shelter location, however, which was used for light and generator storage only.

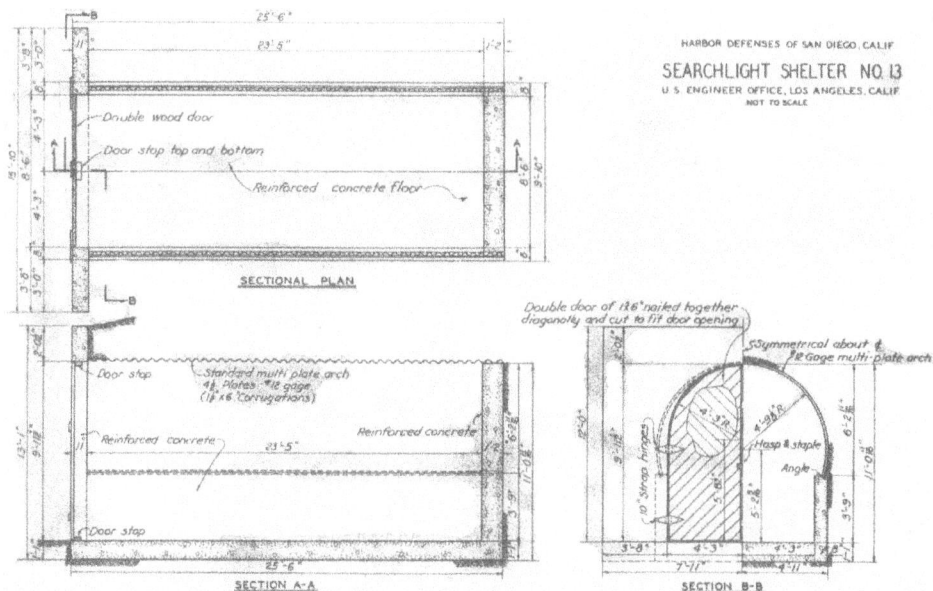

Figure 8-81. Section views for Searchlight Shelter 13, which consisted of a 25.5-foot-long structure of 12-gauge Multi-Plate Arch set atop a short concrete perimeter wall, covered over with dirt (NARA).

A photo of the entrance to Searchlight Shelter 13, taken by CDSG member Lee Guidry in June 1971, is shown below in Figure 8-82. Dug into the south side of an east-west ravine, this structure was fully concealed from the nearby Pacific Ocean, and only a small portion of the entryway can be seen today from Gatchell Road below. Note the double layer of diagonal 1" x 6" wooden planks used in the arched door construction. A much better perspective on the concealment of this shelter can be seen in Figure 8-83, taken 26 years later.

Figure 8-82. Searchlight Shelter 13, looking southwest in June 1971, has been repurposed for use as a Public Works material storage facility (photo courtesy Lee Guidry).

Figure 8-83 provides a later 2006 view of this dug-in searchlight shelter concealed in the ravine. Note the square concrete front wall, which was intended to strengthen the top section weakened by the arched-entry opening, as well as hold back erosion from above, as for example that seen at upper right. The former access road that curved southwest to the right has long since been overgrown with thick vegetation, with only a few traces remaining.

Figure 8-83. The concrete and Multi-Plate-Arch shelter for Searchlight 13 is nestled in a ravine directly below the HDCP/HECP, seen looking southeast in 2006 (photo courtesy Jeff Bowen).

Figure 8-84a shows the north end of this dug-in shelter with one of its large wooden doors partially ajar in 2007, with the interior of the structure seen in Figure 8-84b. Note the corrugated Multi-Plate-Arch roof above the door opening in the concrete end wall, and the stacked pile of steel and concrete pipes stored on the dirt floor. The paint outline on the southern wall (out of image to the rear) indicates the former presence of a double-shelf wooden cabinet that spanned the full width of the interior.

a)   b)

Figure 8-84. a) North end of the shelter as it looked in 2007, showing the westernmost wooden door in the partially opened position. **b)** Interior of the shelter looking north, showing its Multi-Plate-Arch overhead.

Searchlight 14, also a portable 60-inch light, was situated west of Gatchell Road near the cliff edge directly below Searchlight 13, as shown on the map previously depicted in Figure 8-78. Figure 8-85 below shows the HDSD drawing of its associated storage shelter, identical to that constructed for Searchlight 13 in the ravine above. Close examination of a 1971 photo taken by Lee Guidry seems to indicate this buried Multi-Plate-Arch structure was situated on the south side of the ravine just east of Gatchell Road.

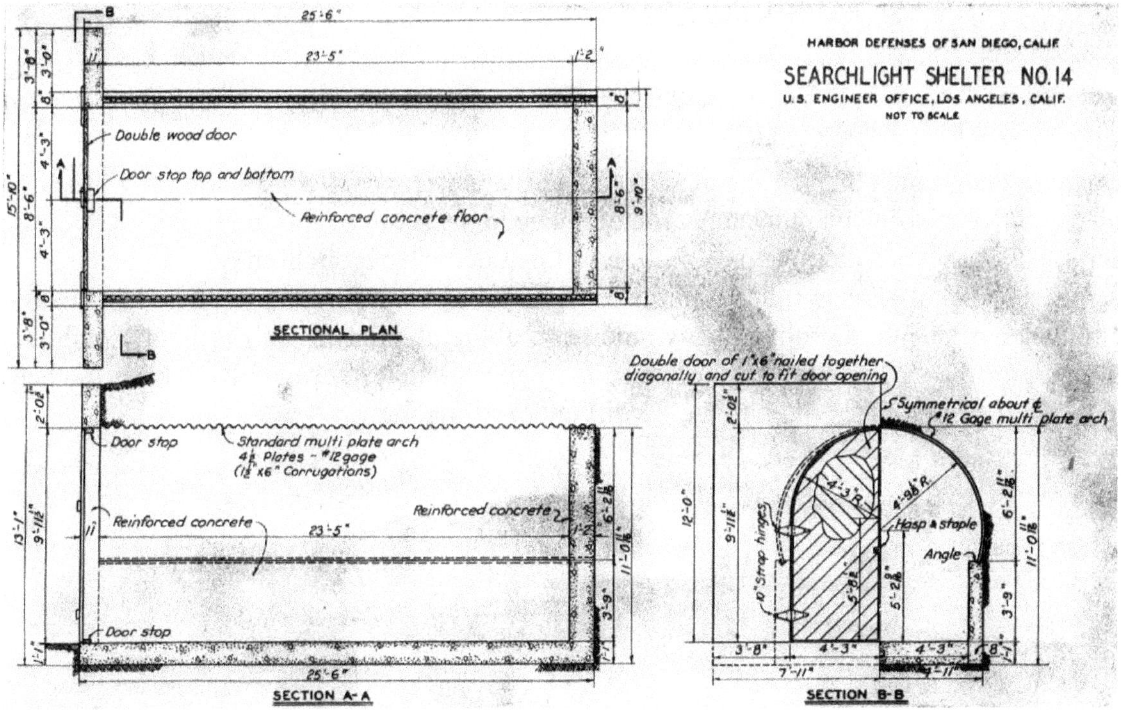

Figure 8-85. The Harbor Defenses San Diego blueprint for Searchlight Shelter 14, which was located at the foot of the ravine below the HDCP/HDEP, just east of Gatchell Road as seen later in Figure 8-86 (NARA).

Presented in Figure 8-86, Guidry's photo suggests the shelter was either removed or filled in and covered over with dirt sometime after the war. This destruction presumably happened during relocation of the San Diego Sewer Plant in 1962, when a large-diameter sewage pipeline was tunneled beneath Point Loma from the east side. The western end of this tunnel emerged just south of Battery Woodward, which was some distance to the north (to the right in the photo below). As the new sewer-plant location is further south along the coast, this area in between was undoubtedly affected by this pipeline installation along Gatchell Road.

Figure 8-86. Location of the former shelter for Searchlight 14 (lower-left quadrant), looking west over Gatchell Road in June 1971 (photo courtesy Lee Guidry). The local controller's shelter was apparently atop the large knoll at background right, as seen in Figure 8-87 below.

Excavation to entrench the sewage pipeline along the western cliff edge started at Battery Woodward and ended at the processing plant location, out of image to the south (right) in Figure 8-87 below. Another 1962 aerial photo shows a massive pipe section of this line being brought in on a flatbed trailer, with the pipe diameter some 50-percent taller than the top of semi-truck pulling the trailer. The destructive path left in the wake of such a dig obviously would have obliterated any WWII artifacts encountered. The sunken nature of the shelter site seen in Figure 8-86 above suggests it may have been impacted by this work and subsequently removed, with the fill dirt later settling.

Figure 8-87. The Searchlight 14 shelter was just east of Gatchell Road on the south (right) side of the ravine below Cabrillo National Cemetery. Note what appears to be the controller's hut on the knoll at bottom right (up arrow). The strip of land along the western cliff edge was extensively excavated and graded in 1962 (US Navy photo courtesy SSC Pacific).

Searchlight 15

The underground silo shelter for Searchlight 15 was situated at an elevation of 103 feet in Site 7 (Cabrillo), west of what is now Gatchell Road on Cabrillo National Monument property. As previously mentioned, the light and lift mechanism were salvaged from the wooden WWI-era silo for Searchlight 4, which was located near the cliff edge south of Battery Humphreys (Figure 8-88), and repurposed for this new installation. During WWII, this light was controlled from battery commander station BC_3 for Battery Ashburn, as directed by the Harbor Defense Command Post.

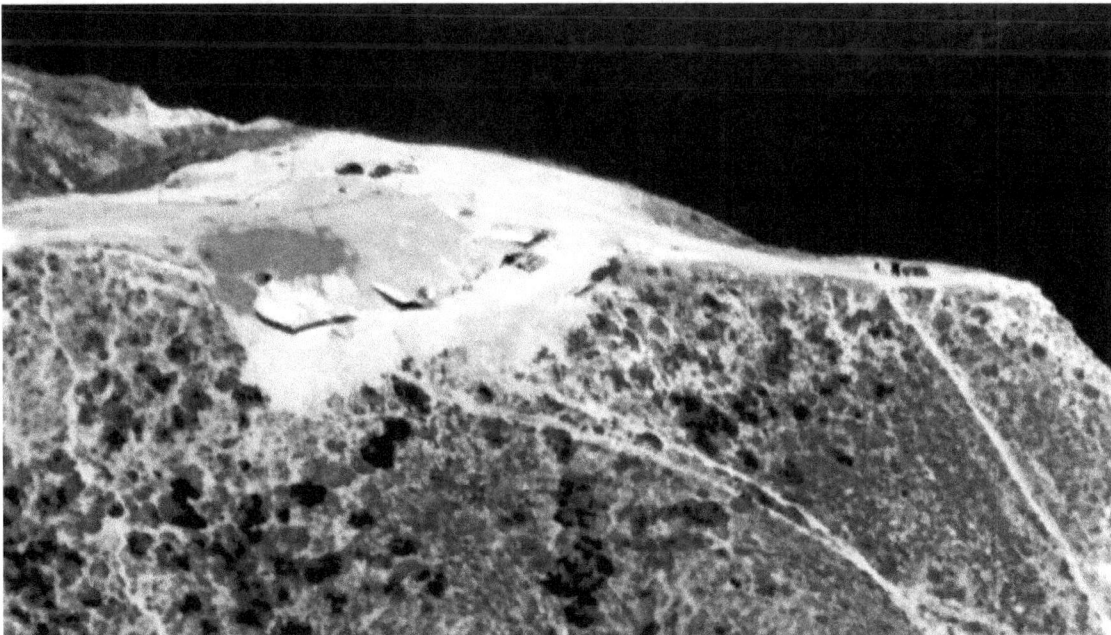

Figure 8-88. The abandoned remains of the WWI silo shelter for Searchlight 4 (upper-right corner) can be seen just south of Battery Humphreys (upper left), seen under construction circa July 1942 (US Navy photo courtesy SSC Pacific). The light and lift mechanisms from this wooden silo were relocated to Site 7 (Cabrillo), as shown later in Figure 8-90.

The HDSD section views prepared by the Los Angeles US Engineer Office for Searchlight 15 (Figure 8-89) are identical to the WWI designs, as the lift mechanism and 60-inch light were from that era. The concrete silo itself was new construction, however, as it was clearly impractical to excavate and relocate the old wooden silo south of Battery Humphreys. That silo instead was repurposed as a temporary support facility for AMTB Battery Cliff, as previously discussed in Chapter 6.

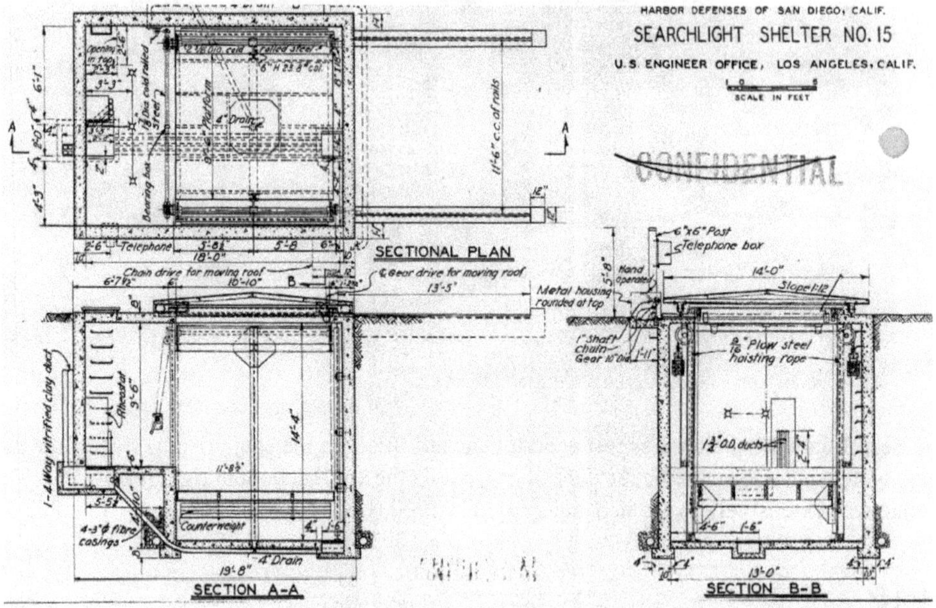

Figure 8-89. The Harbor Defenses San Diego plan and elevation views of WWII Searchlight Shelter 15 off Gatchell Road reflect the standardized silo-shelter design (NARA). The light and lift mechanism were salvaged from the abandoned WWI silo shelter for Searchlight 4, previously shown in Figure 8-88.

Two views of Searchlight 15 looking northeast are shown below, with the camouflaged entry hatch open in Figure 8-90a, and closed in Figure 8-90b. Note the elevated layer of camouflage covering the silo-roof rail tracks at background left in Figure 8-90a. The higher camera perspective is interesting in the latter shot, with the photographer significantly elevated as though standing on something, perhaps a ladder or the bed of a truck. For reference, the small bush directly beneath the light in Figure 8-90a is partially out of the photo in the lower-left corner of Figure 8-90b.

Figure 8-90. a) Rear view of Searchlight 15, looking northeast in January 1944. Note brush camouflage over the retracted roof behind the men at lower left (NARA). **b)** Side view of the light, with a pair of 0.30-caliber water-cooled machine guns sitting on the lift platform, just left of the searchlight base.

Additional HDSD drawings show a dedicated power-plant shelter for this light (Figure 8-91a), which was located just east of the silo. The relative positions are illustrated in Figure 8-91b, which was included in a wartime report on their camouflage scheme prepared by Captain Otto W. Wolgast (1943). Entry to this 17-foot by12-foot subterranean shelter was via an exposed wooden door, shielded by soil-cement wingwalls as shown in section plan C-C below. Chief Ranger Howard Overton (1993) of Cabrillo National Monument reported this underground structure had been filled with sand due to vandalism problems.

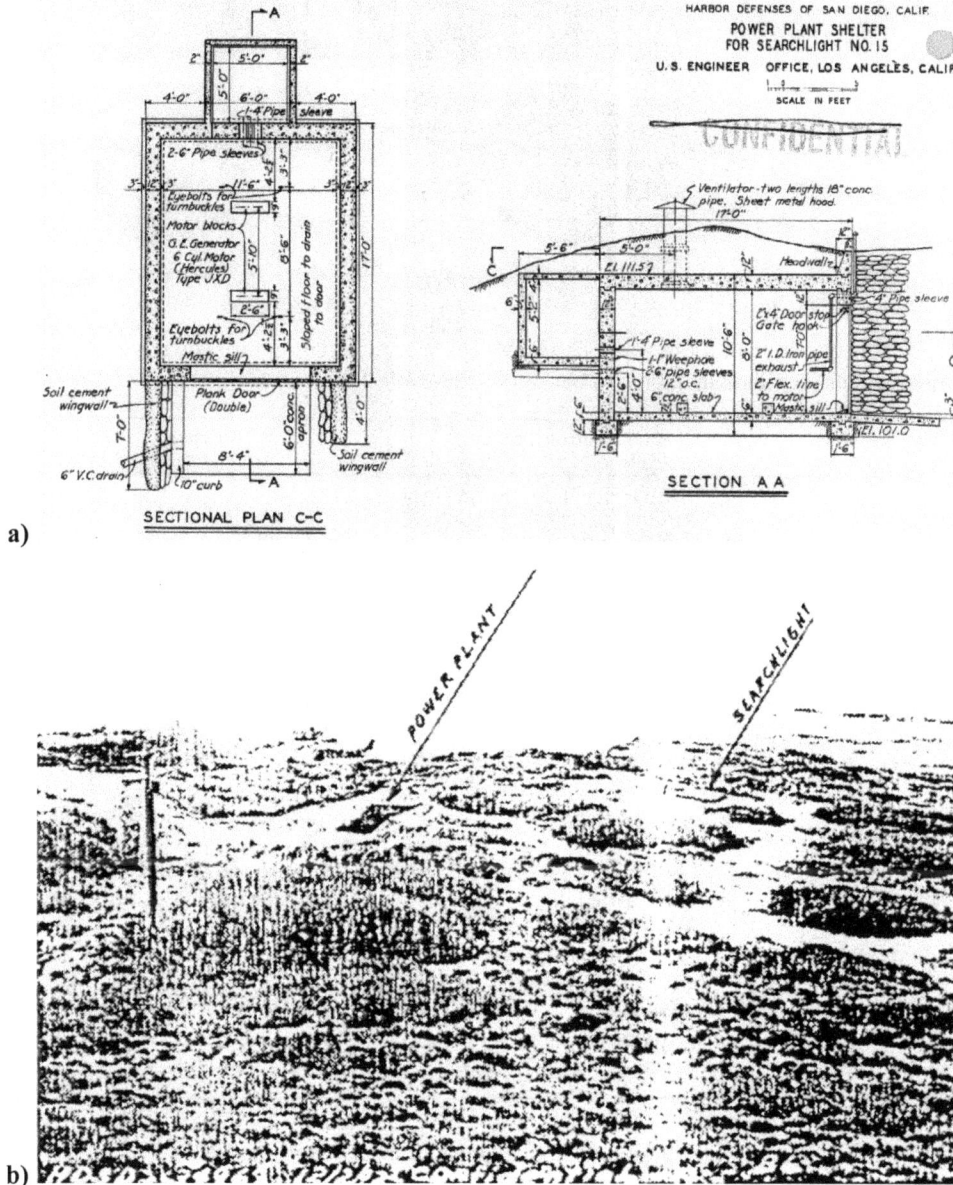

Figure 8-91 a). Section views of the dedicated "Power Plant Shelter" for Searchlight 15, which was some 29 feet long and housed a six-cylinder 110 horsepower Hercules JXD generator. **b)** The powerplant shelter was located just to the east of the silo shelter, with the Pacific Ocean in background (Wolfgast, 1943).

The generator in the powerhouse was a mobile JXD Hercules unit, mounted on concrete blocks and secured by eyebolts. The standard portable General Electric generator that later powered Searchlight 15 was housed when not deployed in an underground (splinter-proof) generator shelter along Cabrillo Road (Thompson, 1991), similar to that shown earlier in Figure 8-81. An annotated image of a GE "Portable Power Plant for Searchlight, 60-inch Model 1942A" that appeared in "Operating Instructions Manual" (TM 5-7115) is reproduced in Figure 8-92 below, with callouts identified in Table 8-5.

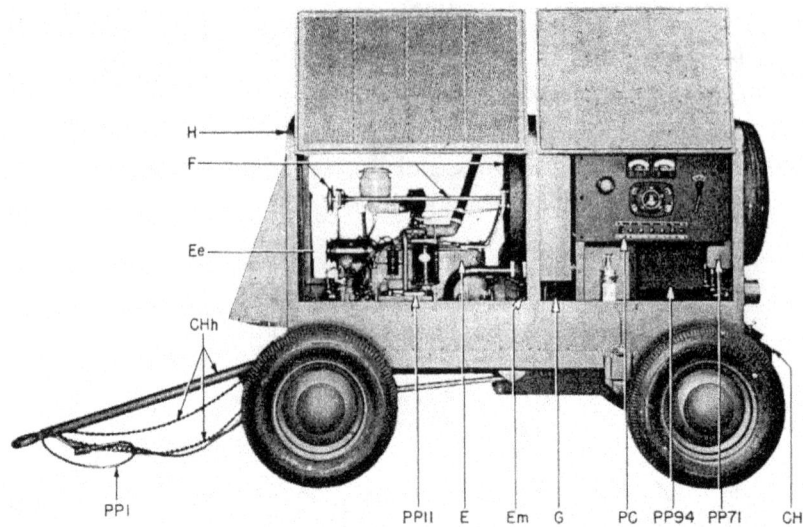

Figure 8-92. The portable power plant for the 60-inch GE Model 1942A searchlight is shown with its access doors open (TM-5-7115, 1-943), with callouts identified in Table 8-5 below.

Ref. No.	Designation	Purpose
E	Engine and accessories	For driving power generator
G	Power generator	For generating electric power for searchlight
CH	Chassis	To support engine-generator assembly, control panel, fuel tank, and running gear
H	Hood with doors	To completely house power plant
PC	Control panel	To support engine and generator controlling devices
F	Fan assembly with fan belt	To draw cooling air through engine radiator
CHh	Drawbar assembly	For towing and steering power plant
PP94	Tool box	To contain tools and spare parts
PP71	Selector switch	To provide for use of either black-out tail lights or service lights
PP1	Tail-and blackout-light jumper cable	To connect tail lights to power source in towing vehicle
PP11	Storage battery	To supply and maintain the six-volt current for starting, lighting, and ignition
Ee	Charging generator	
Em	Regulator	

Table 8-5. Component identification for Figure 8-92 above (adapted from TM-7115, 1943).

The silo shelter for Searchlight 15, overlooking the Pacific Ocean just west of Gatchell Road in Site 7 (Cabrillo), is seen circa 2007 in Figure 8-93 below, well maintained by Cabrillo National Monument. Note the padlocked chains securing the access hatch behind the communication post at left, and one of the roof-retraction tracks extending north to the right. The sheet-metal handcrank cover, partially obscured by the brush, is just visible this side of the northeast (rightmost) corner of the roof.

Figure 8-93. The access hatch (chained shut in background left) for Searchlight 15, looking west, is just beyond the concrete support post (foreground left) for the field telephone, circa 2007.

A close-up view of this handcrank cover is presented in Figure 8-94 below. Note the cover rests on a concrete foundation with a below-ground-level well for the sprocket, which in turn is attached to the end of the pinion shaft that emerges from a small hole in the silo wall. Inside the silo, the pinion shaft engages the two steel racks attached to the bottom of the roof assembly. The current condition of all these components is amazing, given the immediate proximity to corrosive salt-laden air at both locations.

Figure 8-94. Steel cover for the crank handle and sprocket assembly that actuated the rack-and-pinion drive for retracting the steel roof, looking southeast, circa 2007.

Searchlight 16

Assigned to Site 8 (Loma), Searchlight 16 was a portable unit powered by a portable generator. One of three lights not directed from the Harbor Defense Command Post, it was directly assigned to AMTB Battery Cabrillo, Tactical Number 4, with the controller located in BC_4 to the east. Figure 8-95 hypothetically shows Searchlight 16 positioned south of the AMTB Battery, but as this was a portable light, it could have been relocated at will. North is to the right, with the Pacific Ocean (west) across the top of the diagram.

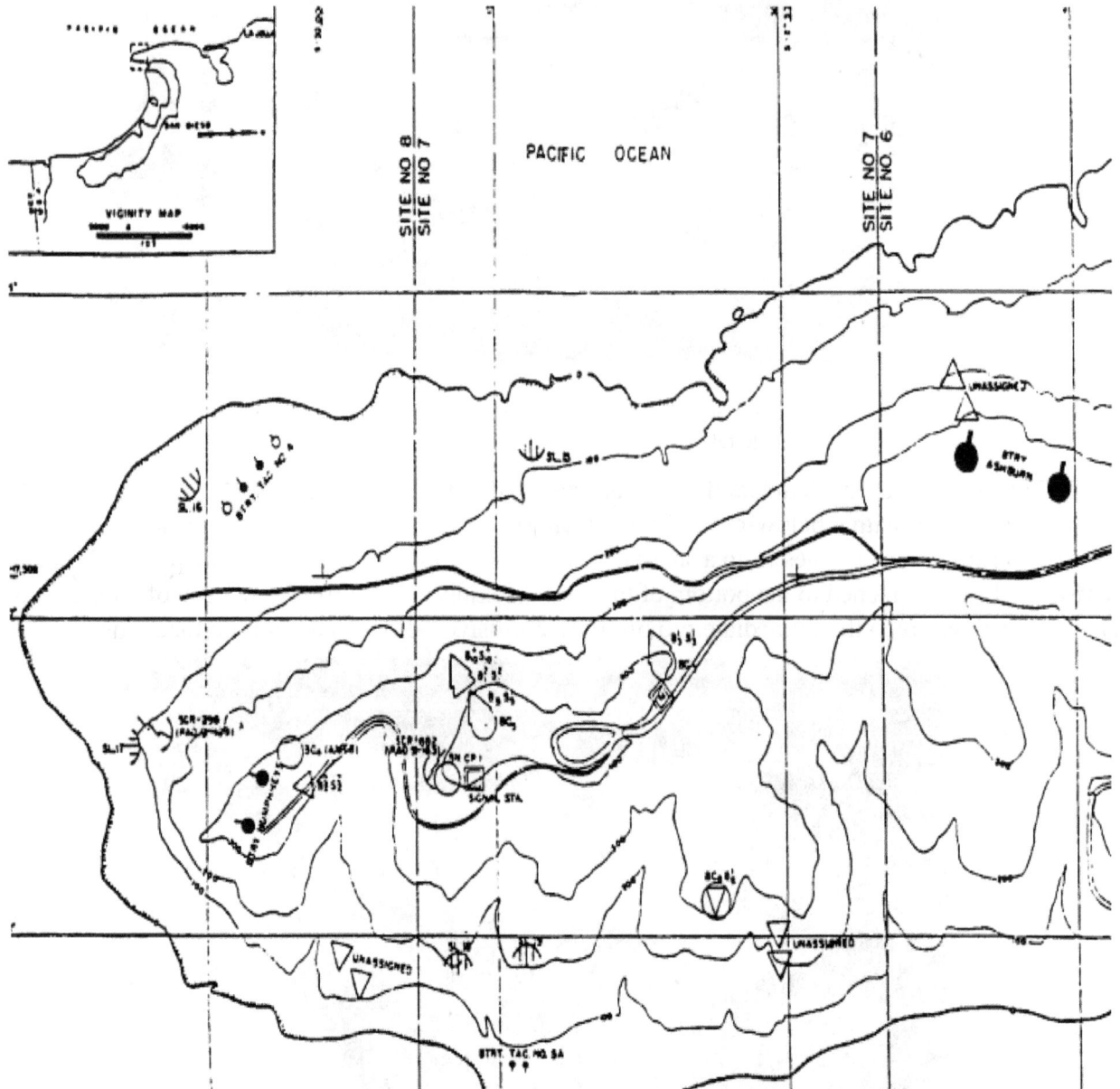

Figure 8-95. This Harbor Defenses of San Diego map notionally shows portable Searchlight 16 just south of Battery Cabrillo in Site 8 at upper left (NARA). Note AMTB battery commander station BC_4 northwest of Battery Humphreys, and the silo shelter for Searchlight 15 near the Pacific Coast at top center in Site 7.

Figure 8-96, a postwar aerial photo of Site 8 (Cabrillo), shows Battery Cabrillo at lower left, circa 1946. The four abandoned Panama gun mounts for Battery Point Loma are directly behind it to the east at center left, between Cabrillo Road above and Gatchell Road running horizontally below. The likely operating position for Searchlight 16 would have been south of AMTB Battery Cabrillo (lower-left corner), so as to provide coverage to both the south and west. The entrance to San Diego harbor is between the Point Loma Peninsula and North Island in the background at upper left.

Figure 8-96. Searchlight 16 would have been near the cliff edge south (right) of AMTB Battery Cabrillo (at lower left) in this aerial photo of 15 July 1946 (US Army photo courtesy Cabrillo National Monument).

Searchlight 17

Searchlight 17 was situated at an elevation of 116 feet within Site 8 (Loma) near what is now Building 15 at the southern tip of Point Loma (see again Figure 8-95). Initially designated as Searchlight 1 (1909-1911), its associated generator was housed in a brick structure that served as a combination powerhouse and shelter for the rail-mounted light. As previously mentioned, this station was renumbered Searchlight 3 in 1911. It retained this same designation during the WWI modernization program (1916-1920), when the brick shelter was replaced with the concrete structure seen in Figure 8-97, which also supported Searchlight 4, approximately 400 feet to the northeast (Keniston, 1996).

Figure 8-97. The "tunnel-type" searchlight shelter and powerhouse is now Building 15, a storage and maintenance facility. The original steel carriage rails for the light are still visible beneath the forklift.

This installation was renumbered Searchlight 17 at the start of WWII, with the presumably upgraded light controlled from $B^2_2 S^2_2$, the secondary base-end station for Battery Strong. Remnants of the searchlight dock along the edge of the cliff still remain, although the rail tracks have been removed and the wooden platform is no longer in place (Figure 8-98). The concrete structure at lower right in the photo served as a foundation for the platform and tracks, while the vertical post in background held the field telephone box. The concrete terminal vault at left provided the remote connection for the searchlight power and control cables.

Figure 8-98. Remains of the searchlight dock foundation (lower right) along the eroding face of the cliff, circa May 2006. Note concrete terminal hut with hinged steel door, partially occluded at photo left, and deteriorating field telephone support post. Both features were previously shown circa WWI in Figure 8-30. Note the effects of years of erosion in comparing the two photos.

Searchlights 18 and 19

As shown in Figure 8-99 below, Searchlights 18 (portable) and 19 (fixed silo installation) were located along a loop in Sylvester Road, on the east side of Site 7 (Cabrillo) in what is now Cabrillo National Monument. Recall Searchlight 18 had earlier been designated as Searchlight 5 during the WWI era. The combined powerhouse (up arrow) for these two lights, originally constructed during WWI, is located between them on the west side of Sylvester Road, now the Bayside Trail.

Figure 8-99. Searchlights 18 and 19 and their combined powerhouse are located on a loop in Sylvester Road (now the Bayside Trail), just below the old Spanish Lighthouse at upper left, circa 1942 (US Army photo courtesy Cabrillo National Monument). Note traces of access paths from the lighthouse area above.

A 5 May 1942 aerial view of Searchlight 18 in its deployed position is seen in Figure 8-100, with the narrow-gauge tracks leading back to its dug-in shelter around the bend in Sylvester Road. What appears to be excavated dirt from the shelter's construction during WWI has been dumped down the side of the ravine at lower right. Note also the upslope path leading to a probable berthing area for the searchlight operators at upper right.

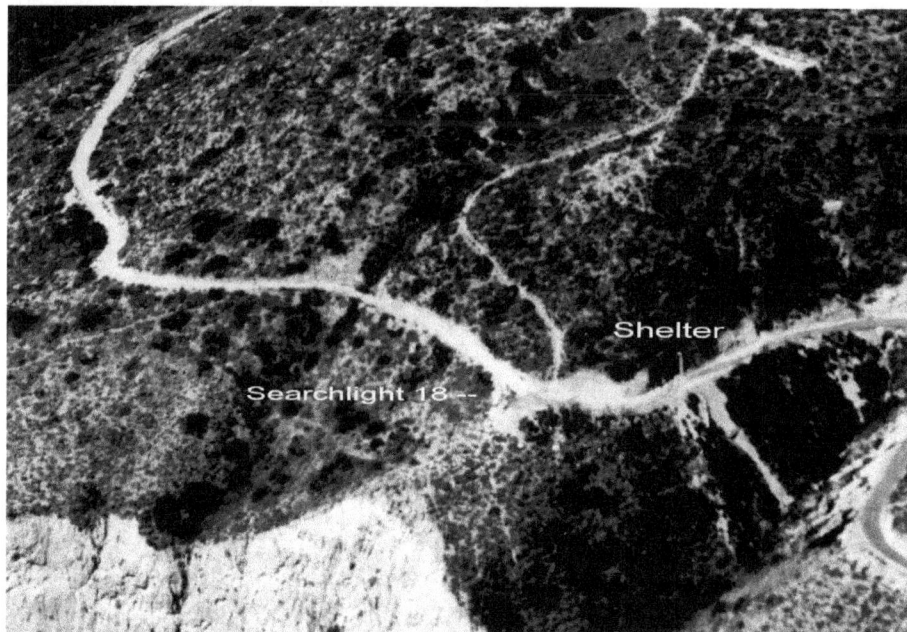

Figure 8-100. The cart-mounted Searchlight 18 is seen in its deployed position at the end on the narrow-gauge tracks leading back to its shelter, circa 5 May 1942. Excavated dirt from the WWI construction appears to run down the side of the ravine at lower right, immediately below the shelter entrance (US Army photo courtesy Cabrillo National Monument).

The still-extant electrical terminal hut for this portable light, the same style as that shown in Figure 8-30 and Figure 8-98, has been fully refurbished by the Cabrillo National Monument staff as seen in Figure 8-101a below. Referring now to Figure 8-101b, the WW II-era terminal hut was serviced by four underground conduits. The 10-pin searchlight-control connector is seen in the upper left corner of the vault interior, with the positive and negative power lugs attached to a Bakelite strip supported by two steel brackets at upper right.

Figure 8-101. **a)** Terminal hut for Searchlight 18 on the east side of the Bayside Trail, looking across the harbor entrance towards North Island on 21 November 2007. **b)** Interior of terminal hut, showing 10-pin searchlight-control connector at upper left, with positive and negative power-connection lugs at upper right.

Figure 8-102. **a)** Shelter for WWI Searchlight 5 (later Searchlight 18) on 12 March 1920, with rheostat cabinet in the right-rear corner (US Army photo courtesy Cabrillo National Monument). **b)** Ranger Charles Schultheis, Chief of Maintenance, stands in the shelter for Searchlight 18 in November 2007.

While viewing the interior of this well-preserved searchlight shelter in November 2017, two electrical conduits were found rising from the floor in front of where the rheostat cabinet is seen in Figure 8-102a above, but the cabinet itself had been removed. The adjustable rheostat may have been replaced by fixed ballast resisters with the introduction of portable generators during WWII. The Cabrillo National Monument had recently acquired the WWII-vintage Sperry *Model 1941A* antiaircraft searchlight seen in Figure 8-102b, similar but not identical to the portable searchlight housed in this shelter during WWII.

US Army Reports of Completed Works (RCWs) indicate all pre-WWII searchlights were to be replaced with the newer portable WWII versions. Figure 8-103 below was taken during the 2006 restoration of the powerhouse to its original WWI configuration along the Bayside Trail, Cabrillo National Monument. Note that security bars had been added to both windows at some point prior to the door being enlarged to accommodate housing the portable WWII generators. The left window was filled with concrete at this time due to its interference with the widened doorway.

Figure 8-103. Powerhouse undergoing restoration to return door and window to original configuration under the supervision of Ranger Charles Schultheis in August 2006 (photo courtesy Jeff Bowen). Note the security bars embedded within the concrete fill being removed from the left window.

A little over a year later, the newly refurbished powerhouse for WWI Searchlights 5 and 6 is seen in Figure 8-104 below, with three ventilator-shaft cones just visible through the brush on the partially buried roof. This impressive restoration effort was assisted by the Center's loan of a WWI-era door found in storage near Battery Woodward, and a surplus window from Building F-17, the former powerhouse on Woodward Road for Searchlights 11 and 12. The single exhaust stack and no exterior muffler, however, are reflective of WWII attributes.

Figure 8-104. Restored powerhouse as seen in November 2007. Note the three ventilator-shaft cones above the roofline, the single exhaust stack, and absence of the exterior mufflers previously shown in Figure 8-38.

The integral mufflers on the new portable generators were connected to a common exhaust line overhead (top center), which penetrated the front of the building through one of the original WWI exhaust ports (Figure 8-105). The unused exhaust port is seen between the restored windows at top right. The plywood sheets on the floor cover discontinuities associated with the original fixed-installation WWI-era generator foundations. Note the WWI color scheme, and the open doorway leading to the adjoining radiator room.

Figure 8-105. The two portable-generator exhausts were connected to a common overhead line that penetrates the east wall just left of the window. Note the small hole in the dark-green stripe at the bottom of the interior wall, just above the plywood, and the missing right-hand door trim, as further discussed later.

Figure 8-106 provides a right-corner view of the generator-room interior, showing the new windows and entry door. The searchlight power connectors are just visible on the floor at bottom right, with the muffler-exhaust discharge pipe in the bottom-left corner. Along the lower wall at far right is a telephone-circuit junction box, with evidence an enclosed phone booth was located in the corner out of image to the right, as seen later Figure 8-107b.

Figure 8-106. Ranger Charles Schultheis, Jeff Bowen, and Rachel TenWolde discuss details of the restoration effort in the generator room on 21 November 2007. Note the searchligh power connectors on the floor in foreground right and the telephone junction box along south wall at photo right.

Figure 8-107 provides a close-up of the porcelain telephone junction box, as well as the outlines of the corner phone booth, which facilitated conversation in a noisy environment whenever one or more generators were online. In the event of cable damage, the telephone junction box in Figure 8-107a allowed the phone-booth circuit to be quickly rerouted through an alternative path in the underground Army communication cable. Note installation point for the telephone inside the booth on the west wall in Figure 8-107b.

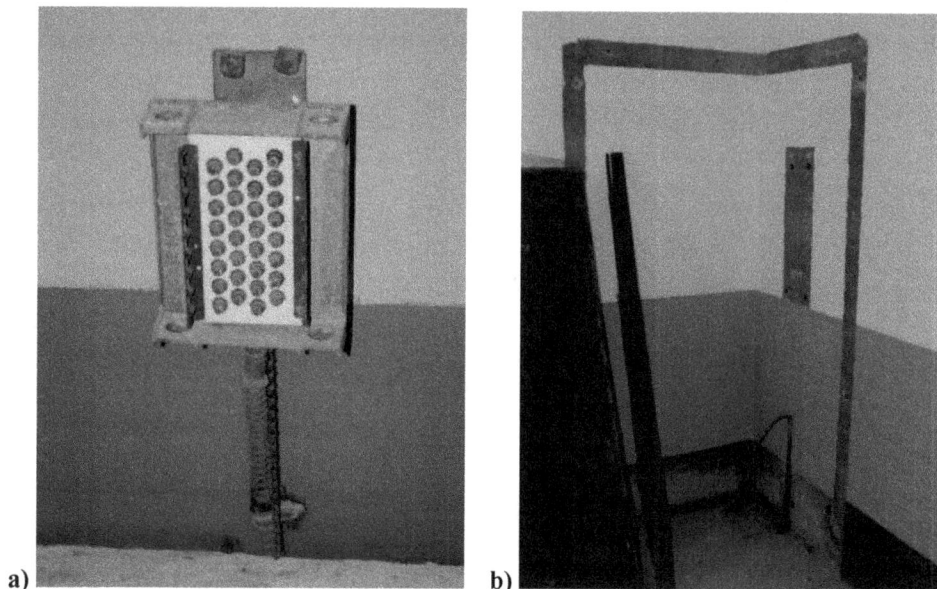

a)   b)

Figure 8-107. a) Close up of porcelain telephone junction box shown earlier in Figure 8-106. **b)** The outline of the original phone booth is preserved in the paint pattern in the southwest corner of the generator room.

Searchlight power connections for the two portable DC generators were via a pair of two-conductor cables that ran in conduits from the generators, emerging through holes in the floor as seen in Figure 8-108a. Note the identical electrical pinout of the two power connectors in Figure 8-108b. The entry door alteration, common-exhaust configuration, and WWII-style connectors confirm the original fixed WWI generator equipment was indeed replaced at some point by portable searchlight power units.

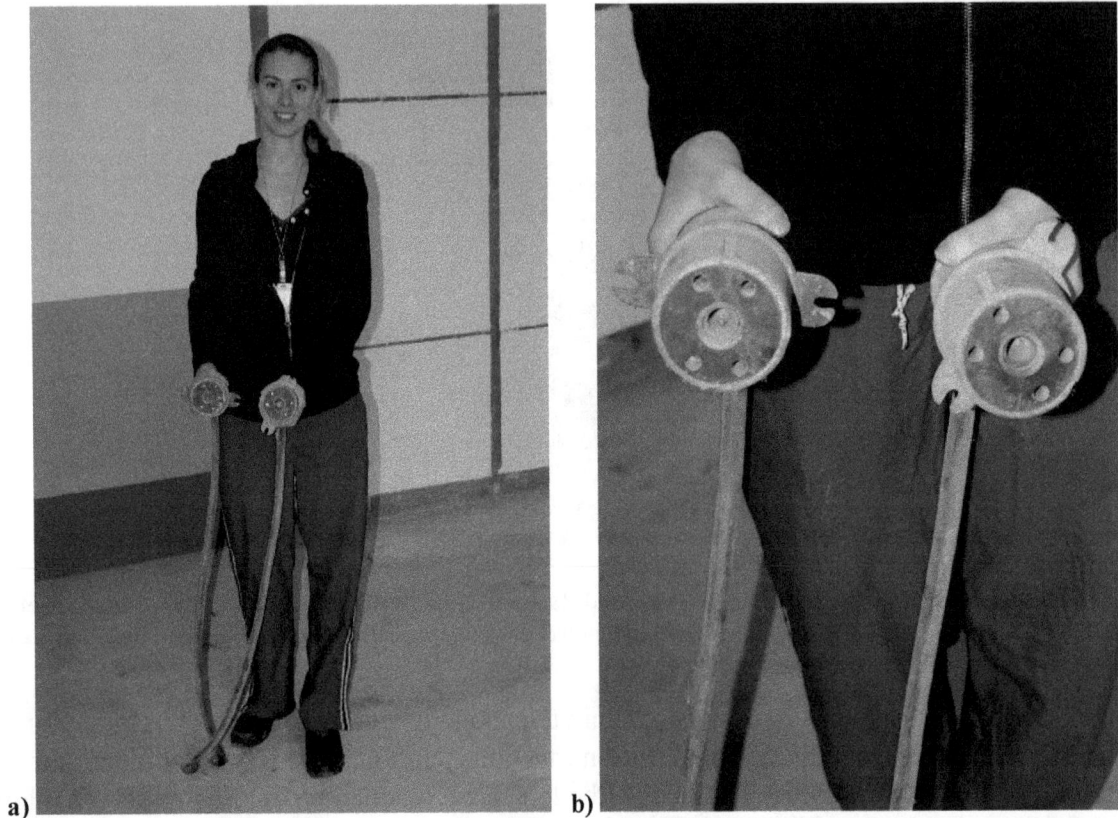

a)   b)

Figure 8-108. a) Rachel TenWolde examines power cables that emerge from floor conduits near the southwest corner of the original foundation footprint. **b)** This close-up view of the extremely well-preserved connectors shows the electrical pinout for the power connections.

Dwight Smith was a Battery K searchlight operator assigned to Searchlight 18 until April or May of 1943. Smith, who bunked in the powerhouse, recalls the poor material condition of the original aging WWI gasoline engines illustrated in Figure 8-109 (Overton, 1993), which were eventually replaced by the newer GE portable units:

"Those old Fairbanks-Morse engines with those flywheels on them. You had a crank that fit on the stub end of the crankshaft. Had a lever to throw the compression off. Old type engines. So you could crank it easy then. One of them was always having trouble with the carburetor. About once an evening, it would catch fire and we'd have to put it out. I guess you couldn't buy anything for them."

Figure 8-109. US Army training graphic indicating the proper way to manually crank the Westinghouse gasoline-driven generator set, which was similar in form and function to the Fairbanks-Morse units initially employed in the powerhouse for WWI Searchlights 1, 2, 5 and 6 (courtesy Cabrillo National Monument).

Smith's recollection of the recurring fire-hazard experience is particularly intriguing in light of an astute observation made by Rachel TenWolde during our site visit on 21 November 2007. Figure 8-110a shows an obvious radial fragmentation pattern in the interior wall surface that originates from the small hole below, previously shown in Figure 8-105, with exit damage on the other side of the wall in the radiator room seen in Figure 8-110b. Additional pockmarks are visible on the ceiling above and on the east wall to the right, extending outward about 3 to 4 feet from the interior wall. The wooden trim is missing from the right side of the interior doorway, but still intact on the left side and top.

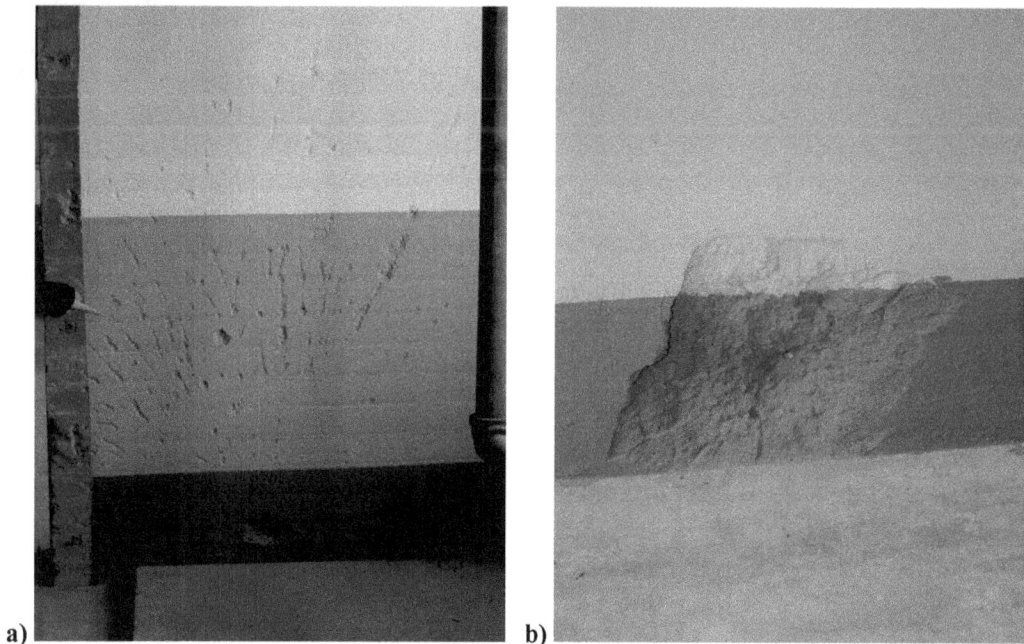

a)   b)

Figure 8-110. **a)** This obvious radial pattern of fragmentation scars in the concrete wall fans out from the larger cylindrical crater below, which measures approximately 2 inches in diameter, since painted over. Note the missing door trim at photo left. **b)** Close up of associated exit damage on rear of wall (in the radiator room).

All the above indicate something exploded near the floor with considerable force, punched a hole almost completely through a 6-inch concrete wall, blew off the nearby door trim, and seriously pock-marked the surrounding concrete walls and ceiling. This explosion drove some type of projectile into the wall to create the crater, which then sent shrapnel radially in all directions. While it's possible the aging WWI engine-generator set was the source of this explosion, it's curious the impact was so concentrated and low to the floor. An alternative and more likely possibility is that someone errantly discharged a weapon.

The nicely restored silo cover for Searchlight 19 is seen in the closed position in Figure 8-111, with its associated steel roof-retraction tracks clearly visible on their concrete foundations in the foreground. Note the well-preserved concrete support post for the no longer extant field-telephone box, and the square entry hatch to its left just behind the silo roof. The companion Searchlight 18 shelter (up arrow) can be seen in the background along the Bayside Trail (formerly Sylvester Road), a portion of which is visible just below the ridgeline to the left of the shelter.

Figure 8-111. Searchlight 19 silo in November 2007, looking south across the ravine towards the shelter for companion Searchlight 18 (up arrow), just visible in background. The powerhouse is northwest of Searchlight 18 along the Bayside Trail, out of image in background right.

Searchlight 20

During WWII, Searchlight 20 was a portable 60-inch unit located in Site 9 (East Rosecrans), assigned to Anti-Motor-Torpedo-Boat (AMTB) Battery Fetterman, Tac. No. 7, on Ballast Point. This light was situated southwest of the battery overlooking the San Diego Harbor entrance as shown in Figure 8-112. No period photos of this searchlight or its shelter have yet been found.

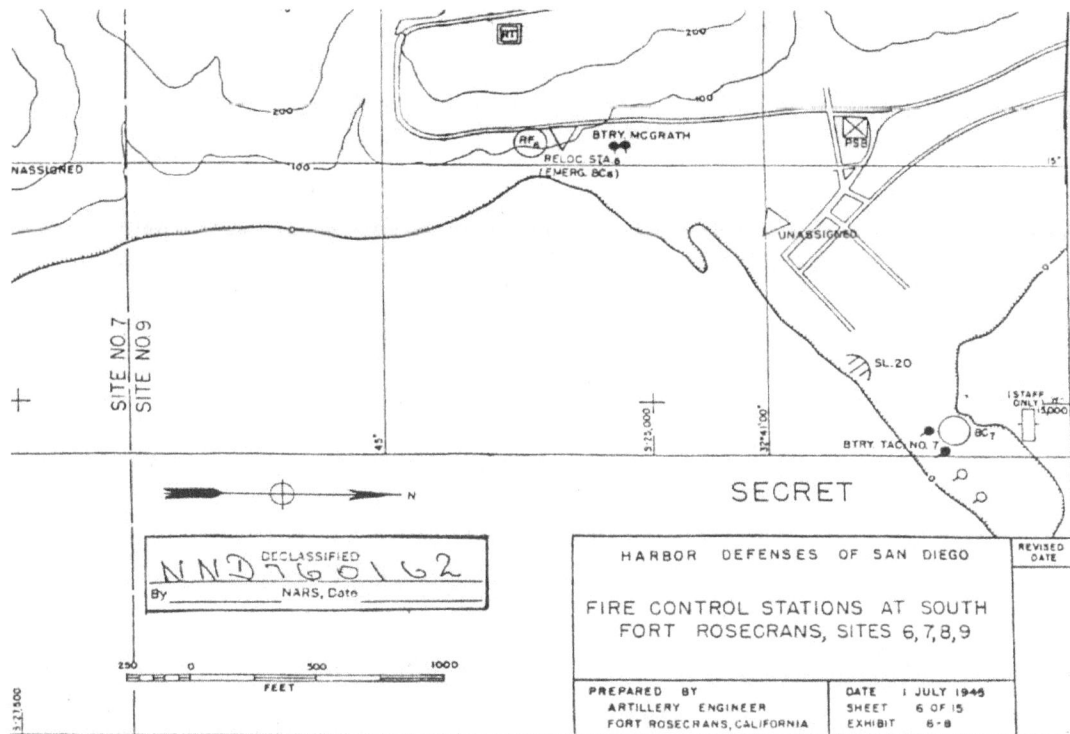

Figure 8-112. Searchlight 20 (center right) was a portable light assigned to nearby AMTB Battery Fetterman, Tac. No. 7, in Site 9 on Ballast Point.

Figure 8-113 is a 1943 photo of the northeast side of AMTB Battery Fetterman, looking southwest towards the Point Loma peninsula. The cottage-type camouflage shelter for the 37-millimeter Gun 2 installation is shown with both the sides and front collapsed, with the roof pushed back onto storage supports to the rear. The identical sloped-roof shelter for Gun 1 in background is seen covered with its camouflage sides erect. Note that Figure 8-112 above notionally shows Searchlight 20 farther to the southwest, where it would be occluded from view in this perspective.

Figure 8-113. In this view of Ballast Point looking southwest, the two 37-millimeter antiaircraft guns at AMTB Battery Fetterman are disguised under collapsible cottage-type shelters. Occluded from view, Searchlight 20 would be further to the southwest behind these structures (adapted from Wolgast, 1943).

Searchlights 21 – 27

Portable Searchlights 21 through 27 were located in Sites 11, 12, and 13a along the east side of the harbor entrance down to the Mexican border (Table 8-6). This stretch of coastline included both the Coronado Beach Military Reservation south of Coronado and the Coronado Heights Military Reservation down closer to the border. Fort Emory was located within the latter (Figure 8-114), home to the 16-inch Battery Gatchell (never armed) and the 6-inch Battery Grant, as previously discussed in Chapter 5. The 90-millimeter guns of AMTB Battery Cortez were closer to the harbor entrance at the former.

Number	Site	Location
21 - 23	11	Silver Strand
24 & 25	12	Emory
26 & 27	13A	Monument

Table 8-6. Portable Searchlights 21 through 27 were located off Fort Rosecrans from Coronado Beach down towards the Mexican Border (Monument).

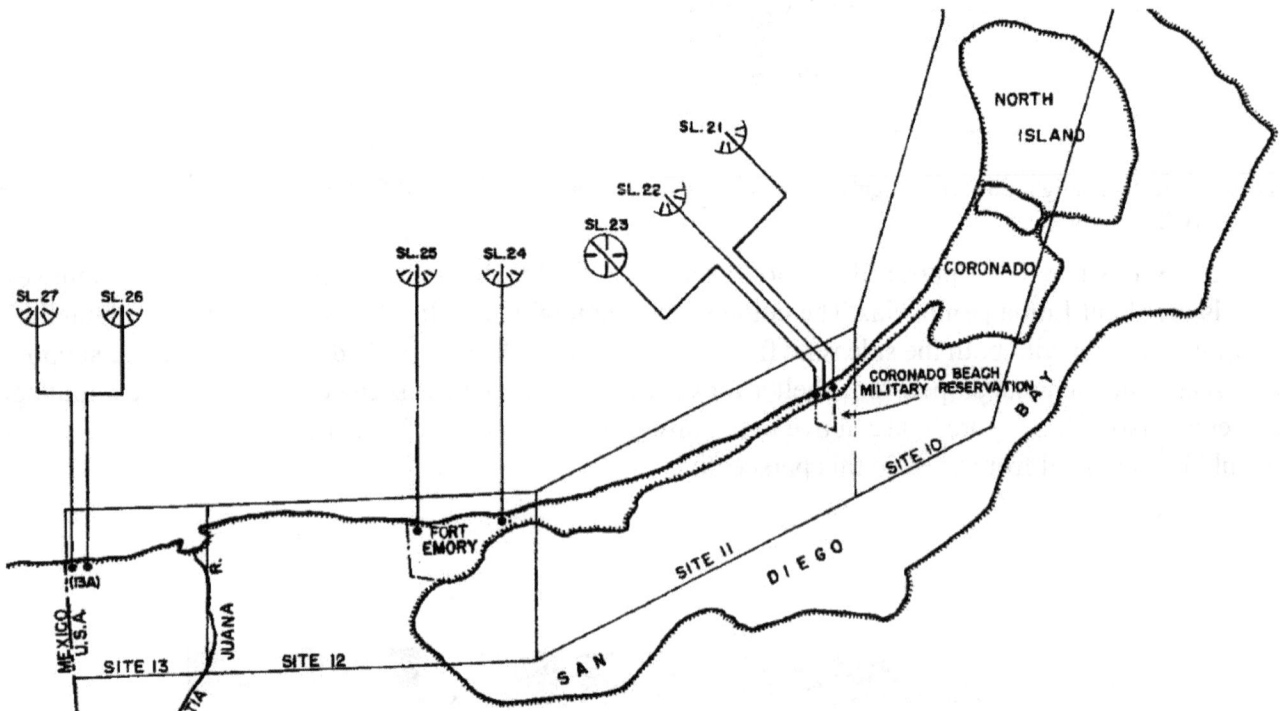

Figure 8-114. Lower left portion of the map depicted earlier in Figure 846, showing the locations of Searchlights 21 through 27 in Sites 11 through 13A. Note the AMTB light symbol for Searchlight 23, and the close spacing of the Searchlight 26/27 pair by the Mexican border at far left, as further discussed later.

Searchlights 21 through 23 in Site 11 were located north to south along the Silver Strand at the Coronado Beach Military reservation below North Island. Searchlight 21 and 22 were locally controlled under the direction of the Harbor Defense Command Post over on Fort Rosecrans. Searchlight 23 was assigned to and controlled from AMTB Battery Cortez, as previously discussed in Chapter 6. Annex C of the HDSD 1946 Annex provides the following brief notation regarding Searchlights 24 and 25 in Site 12 at Fort Emory on the Coronado Heights Military reservation:

"In the vicinity of Fort Emory the terrain is very flat and at an elevation very close to sea level. The two searchlights at this location were placed on the top of sand dunes in order to obtain as much height of sight as possible. The maximum range of these lights is limited to approximately 8,000 yards."

The flat topography and beach-cottage camouflage scheme in Figure 8-115 suggests this unidentified pair of searchlight shelters was situated in the sand dunes along coastal shores, somewhere north of Fort Rosecrans or between Coronado and the Mexican border. The accompanying caption in "Camouflage – Harbor Defenses of San Diego" (Wolgast, 1943) identifies them as Searchlights 19 and 20, which reflects a pre-1945 numbering scheme but effectively eliminates the north-of-fort option. Taking into account their close proximity (see again lower-left corner of Figure 8-114 above) and desolate surroundings, the most likely pair in 1945 would have been Searchlights 26 and 27 by the southern border.

Figure 8-115. This pair of cottage-type shelters was presumably associated with portable Searchlights 26 and 27 at Site 13 on the Mexican border (US Army photo adapted from Wolgast, 1943).

Antiaircraft Searchlights

The three 90-millimeter AMTB batteries (Fetterman, Cabrillo, and Cortez), which had a secondary mission of antiaircraft defense, were each equipped with a dedicated searchlight installation and radar. A HDSD Report of Completed Works dated 6 January 1941 indicates "Shelter No. 1 for Portable AA Searchlights" was constructed approximately 90 feet west of Gun 1 at WWI-era Battery Calef-Wilkeson on Ballast Point (Figure 8-116). This standard above-ground structure was made of corrugated galvanized-steel sheets, equipped with commercial electrical power, and connected to the water mains.

REPORT OF COMPLETED WORKS - SEACOAST FORTIFICATIONS
 (Fire control or Torpedo Structures)

                                                    HARBOR DEFENSES OF SAN DIEGO, CALIF.
                                                    FORT ROSECRANS
                                                     STRUCTURE SHELTER No. 1
                                                         FOR PORTABLE AA SEARCHLIGHTS

Form 2.   Corrected to January 6, 1941.

STRUCTURE:		INSTRUMENTS AND EQUIPMENT:	
Location	Approx. 90 ft. west of Gun No.1, Battery Wilkeson	Type of observing inst. : Type of plotting board	None. None.
Date of transfer	December 30, 1939		
~~t to that date	$3,900.00	DATA TRANSMISSION:	
Type of construction		Type of	None.
(a) Roof	Corr.Galv.Sheets Steel:		
(b) Remainder of bldg.	" " " Frame:		
How concealed	Not concealed.		Note: Single electric circuit. 5 reflector lights.
How protected	Not protected.		Remarks: Transfer of structure to Q.M.Gen. for maintenance approved by Adj.Gen.,1st Ind., 9-11-40, to letter from C.of E. to Adj.Gen., "Transfer of Searchlight Shelters," dated 8-19-40. Historical records submitted to Q.M.,Ft.Rosecrans,Calif.,11-30-40
Height above concealment	15'-4½"		
Height above protection	15'-4½"		
Conspicuous at - yards			
ELECTRIC CURRENT:			
Source of	Commercial		
Kilowatts required	Approx. 1 kw.(See note)		
Type of lighting fixtures	Direct industrial reflectors.		
HEAT:			
How heated	None.		
W~~R & SEWER:			
Connected to water mains	Yes.		
Connected to sewer	No.		
Type of latrine	None.		
Permanent or temporary inst'ln	Permanent		
Present condition	Good.		
REFERENCE:			
Reference of site	Floor el. 13.08		
Reference of instrument axis	-		
Type and capacity of crane	None.		
Max. dim. of reel handled.	-		

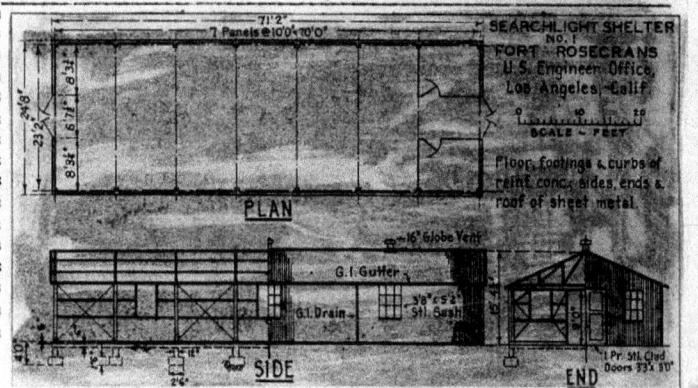

Figure 8-116. Transferred for operational use on 30 December 1939, Shelter No. 1 for Portable AA Searchlights was situated approximately 90 feet west of Gun 1 at Battery Calef-Wilkeson (NARA).

Figure 8-117 shows a 31 December 1940 Report of Completed Works that indicates a similar corrugated-steel structure, "Shelter No. 2 for Portable AA Searchlights," was installed approximately 200 feet northwest of Gun 1 at AMTB Battery Fetterman. This new 90-millimeter battery was constructed on Ballast Point at the site of former WWI-era Battery Fetterman (two 3-inch guns), which had been demolished in 1940. Like Shelter No 1 for Portable AA Searchlights, this above-ground structure was also serviced by commercial power and connected to the water mains. It is unclear how many portable searchlights were assigned to these two shelters for anti-aircraft defense, and where they were operationally deployed.

REPORT OF COMPLETED WORKS - SEACOAST FORTIFICATIONS   HARBOR DEFENSES OF SAN DIEGO, CALIF.
   (Fire control or Torpedo Structures)   FORT ROSECRANS
                                                                        STRUCTURE SHELTER No. 2
                                                                                    FOR PORTABLE AA SEARCHLIGHTS

Form 2.   Corrected to December 31, 1940.

STRUCTURE:		:INSTRUMENTS AND EQUIPMENT:	
Location	:Approx. 200 ft. N.W. :Gun No. 1, Bty. :Fetterman.	: Type of observing inst. : Type of plotting board	:None. :None.
Date of transfer	:November 18, 1940	:DATA TRANSMISSION:	
Cost to that date	:$7516.61	: Type of	:None.
Type of construction	:		
(a) Roof	:Corr.Galv.Sheets - Steel:		
(b) Remainder of bldg.	: " " " Frame :		:Note: Single circuit
How concealed	:Not concealed.		: 7 reflector lights
			: 2 double convenience outlets
How protected	:Not protected.		:Remarks: Transfer of structure to
			:Q.M.Gen. for maintenance approved
Height above concealment	:16'-3 3/8"		:by Adj.Gen.,1st Ind.,9-11-40, to
Height above protection	:16'-3 3/8"		:letter from C.of E. to Adj.Gen.,
Conspicuous at - yards	:		:"Transfer of Searchlight Shelters,"
ELECTRIC CURRENT:	:		:dated 8-19-40. Historical records
Source of	:Commercial.		:submitted to Q.M., Ft. Rosecrans,
Kilowatts required	:Approx.2 kw.(See note.):		:California, 11-30-40.
Type of lighting fixtures	:Direct industrial		
HEAT:	:reflectors.		
How heated	:Not heated.		
WATER & SEWER:	:		
Connected to water mains	:Yes.		
Connected to sewer	:No.		
Type of latrine	:None.		
Permanent or temporary inst'ln:	Permanent.		
Present condition	:Excellent.		
	:		
REFERENCE:	:		
Reference of site	:Floor El. 13.5		
Reference of instrument axis	: -		
	:		
Type and capacity of crane	:None.		
Max. dim. of reel handled.	: -		

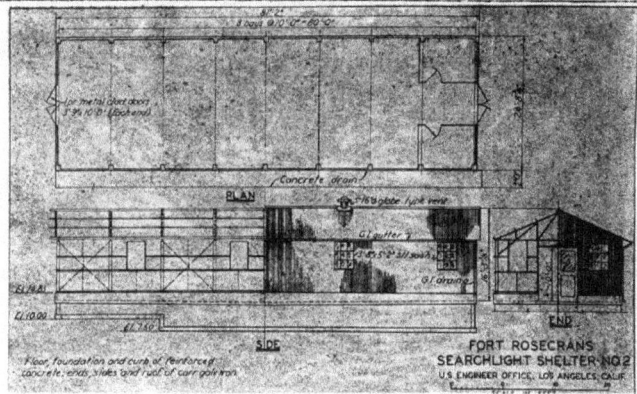

Figure 8-117. Transferred 18 November 1940, Shelter No. 2 for Portable AA Searchlights was situated 200 feet northwest of Gun 2 at AMTB Battery Fetterman (NARA).

9
Coast Defense Radar

> "Give me an army of West Point graduates and I'll win the battle; give me an army of Texas Aggies and I'll win the war!"
>
> *General George S. Patton*

The term "radar," short for radio detection and ranging, was coined by Captain S.M. Tucker of the US Navy, and later adopted by the US Army, Great Britain, and ultimately Canada (Page, 1962; Helgeson, 2006). By the outbreak of World War II, the concept had been around for nearly half a century, having been predicted as early as 1900 by the prolific Serbian-American inventor Nikola Tesla (Fisher, 1989). A German engineer by the name of Christian Hulsmeyer had in fact demonstrated and patented a primitive spark-gap radar-like device in 1904 (Terrett, 1956; Price, 1967).

In 1922, Albert Hoyt Taylor and Leo C. Young of the US Naval Research Laboratory (NRL) produced an improved version based on a continuous-wave oscillator, but did not fully follow up on their idea until some years later, as the necessary high-frequency electronic components did not yet exist (Brown, 1999). A demonstration was eventually performed at NRL on 10 December 1930, attended by Army representatives of the Signal Corps, Coast Artillery, and Air Corps (Terrett, 1956). A Scots physicist by the name of Robert Watson-Watt performed a similar proof-of-concept demonstration near Daventry on 26 February 1935 for the British Air Ministry, detecting an airborne target flying through a fixed beam (Neale, 1985; Fisher, 1989).[1]

By this time, Germany, Italy, France, and the Soviet Union were also pursuing continuous-wave "interference detection" prototypes. The major shortcoming of all these efforts, however, was the lack of any range information; the presence of an approaching aircraft was detected, but distance to the target could not yet be calculated. The invention of pulse-generation and timing circuitry would soon address this shortfall, setting the stage for a feverish international race to further improve performance and maintain technical superiority, which Winston Churchill would later term "The Wizard Wars."

The principle Allied players during WWII were the British and Americans, with Germany and Japan leading the pursuit for the Axis powers. At the outbreak of hostilities, German technology was the most advanced, a fact not fully appreciated at the time and to some extent even today (Jones, 1978):

> "German radar was much better engineered than ours, it was much more like a scientific instrument in stability and precision of performance. The philosophy of using it, however, seemed to have been left to the German Services, and the Luftwaffe in particular made a philosophical mistake by focusing on the wrong objective."

1　Watson-Watt was then superintendent of the Radio Research Station at Slough (Brown, 1999). A.F. Wilkins, a member of Watson-Watt's staff, recalled a 1932 British General Post Office report that mentioned aircraft had interfered with and "re-radiated" radio signals, which led him to suggest the possibility of detecting aircraft through their reflection of transmitted pulses (Fisher, 1989).

The early driving force behind German development was the Kreigsmarine, motivated in part by universally acknowledged British naval superiority. The German start-up company GEMA[2] demonstrated a prototype surface-search radar in Kiel Harbor on 20 March 1934 (Clark, 1997; von Kroge, 2000). Subsequent incorporation of pulsed transmission at higher operational frequencies enabled the detection of aircraft as well as ships, with a maximum range of 50 miles demonstrated in 1936. In 1938, the fledgling company began production deliveries to the German navy of the *Seetakt* early-warning radar, which in the form of *Freya* would also become a key component of Luftwaffe anti-aircraft defense (Clark, 1997).

Great Britain's precarious proximity to the advancing Nazi threat pushed their operational incorporation of radar at a much faster pace than in the US, even though both countries were fairly evenly matched in terms of the technology itself. While Germany initially held the advantage over the Allies in terms of radar performance, they failed to implement an early-warning command-and-control network on par with Great Britain until well into the war.

The British began establishing a structured zone defense incorporating fighter squadrons and antiaircraft artillery as far back as 1923, supported by a civilian "Observer Corps" to report sightings of foreign aircraft (Weitze, 2003). Following experiments at Daventry in 1935, radar was introduced into the concept, and defensive exercises over the next few years produced continuing improvements in both equipment and procedures (Fisher, 1989).

Conventional radio direction finding (RDF), which had seen extensive use in both aerial and maritime navigation since before WWI, was ill suited to the detection of incoming enemy bombers, for three reasons:

    1. It involved tracking an active RF emitter, but incoming enemy aircraft would naturally limit their radio transmissions.

    2. It employed a rotating antenna, which limited its physical size and hence sensitivity, making it harder to detect the weak radar returns.

    3. It typically took up to a minute to get a single fix.

The use of an active radar transmission to generate a reflected-echo signal solved the first problem, while the radio goniometer eliminated the last two, as will be further discussed.

Chain Home Radar

The British "Chain Home" early-warning air-defense radar is widely regarded as the first such system to be brought into effective operational use (Neale, 1985). The fact that an extensive network of *Chain Home* installations was in place protecting the southeast shorelines of the British Isles when war broke out is largely attributable to the pioneering efforts of Sir Robert Watson-Watt, who proposed it to the Air Ministry in January 1935 (Brown, 1999). Given the lack of supporting technology in the late 1930s timeframe, the functional capability of this early radar design was nothing short of remarkable.

Chain Home was a pulsed time-of-flight system employing a static array of horizontal dipoles supported by large towers, yet it could determine target range, bearing, and height out to distances of 120 to 185 miles. The transmitting antenna dipoles were not directly attached to their 360-foot steel towers (Figure 9-1a below), but vertically suspended between them and generally aligned parallel to the coastline. The upper array, consisting of eight end-fed half-wave dipoles spaced half a wavelength apart, was hung at a mean height of 215 feet, resulting in a main elevation lobe of 2.6 degrees and a minimum, or "gap," at 5.2 degrees (Neale, 1985).

---

2   GEMA - Gesellschaft für Elektroakustische und Mechanische Apparate.

Figure 9-1. a) Line of four 360-foot steel *Chain Home* transmitting towers spaced 180 feet apart, equipped with cantilevered maintenance platforms at 50, 200, and 350 feet. **b)** The receiving antennae were supported by 240-foot towers made of wood to avoid electrical interference with the directional dipole stacks.

Additional towers with suspended dipole arrays tuned to alternate wavelengths facilitated rapid switchover in the event of propagation issues, interference, or enemy jamming. Four different operating frequencies between 20 and 55 MHz were originally envisioned for each station, but in the end this strategy was downgraded to a primary and alternate frequency in the 20-30 MHz band (Neale, 1985). To prevent mutual interference, the individual stations were synchronized by the 50-Hz national electric grid, with a rather low pulse-repetition frequency (PRF) of 25 Hz. Alternate PRFs of 12.5 and 50 Hz were also provided (Neale, 1985).

The receiving antenna arrays were mounted on four 240-foot wooden towers, arranged about 250 feet apart in a square pattern (Figure 9-1b above), situated between the transmitting towers and the coastline (Claydon, 2009). Determination of target bearing (azimuth) was made possible through the use of crossed (i.e., orthogonal) dipoles backed by center-switched reflectors, physically aligned for maximum sensitivity along the cardinal headings east and south. The reflected radar signals detected by these two dipole arrays were fed to a Bellini-Tosi radio goniometer, a special-purpose variable transformer that allowed a fixed antenna configuration to be electronically scanned in azimuth.

The radio goniometer (Figure 9-2a) is a simplistic AC signal comparator that employs an adjustable sense coil suspended within two orthogonal stator coils that serve as inputs (Figure 9-2b), electrically similar to the AC selsyn discussed in chapter 8. Maximum coupling between the rotor and a stator coil occurs when the two are parallel, with minimal coupling when they are orthogonal. The signal induced in the sense (rotor) coil is the vector sum of the two dipole-antenna signals applied to the orthogonal stators. To determine target bearing, the operator adjusted the sense-coil knob to achieve a null (i.e., minimum strength) in the returned echo.

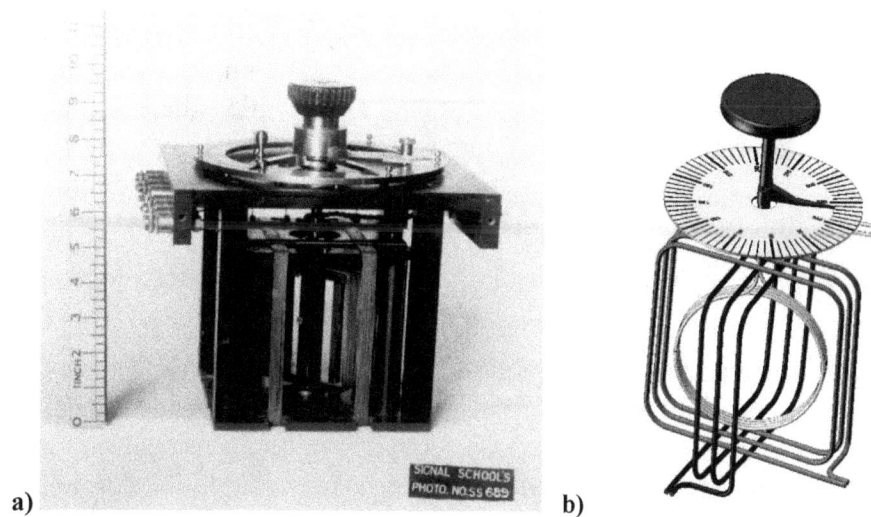

Figure 9-2. a) Invented by E. Belini and A. Tosi in 1907 (Bauer, 2004),[3] the radio goniometer employed a rotating sense coil suspended within two orthogonal stator coils serving as inputs (military.wikia.org). **b)** The rotating sense-coil scale indicated target bearing at the null position (conceptual drawing by Raphael Reyes).

A second goniometer enabled determination of aircraft altitude, as explained by Claydon (2009):

"At the null position, the signal picked up by the rotating sensor coil from one dipole exactly cancels the signal picked up by the other dipole. The goniometer was fitted with a scale indicating the angular position of the sensor coil. This was calibrated to indicate the geographical target bearing under a null condition."

"It was thus possible to establish the angle to the horizontal of a received signal by comparing the output of the upper dipoles with that of the lower dipoles. A radio goniometer was once again used to compare the upper and lower antenna signal strengths. The goniometer reading and the target's slant range were then used to calculate the target's height, using an electromechanical calculator called the "fruit machine.""

While the timely development of *Chain Home* is largely attributed to Sir Watson-Watt, credit for establishing the equally important operational procedures for effective use of the early-warning radar information arguably belongs to Sir Henry Tizard, chairman of the Committee for the Scientific Survey of Air Defense. It was through Tizard's insightful persistence that a series of practice interceptions, later known as the "Biggin Hill Experiment," were conducted over the period of 1936 to 1937 (Clark, 1965):

"The main operational value of an adequate warning and tracking system could be the elimination of standing patrols, with the resulting conservation of men and planes. But this could only be exploited to the full if some completely new technique of interception were devised; and it would, Tizard knew, be unrealistic to hope this could happen amid the confusions of war. A peacetime experiment was therefore necessary."

Using filtered plots from *Chain Home* radar stations along the southeast coast, appropriate operational tactics and procedures were developed to vector Royal Air Force (RAF) fighters towards simulated incoming threats (Neale, 1985). By the outbreak of hostilities in 1939, a sophisticated network of "filter rooms" was already in place to assimilate data from a multitude of *Chain Home* sites, enabling optimal fighter intercept against large numbers of incoming bombers.[4] It was this synergistic combination of *Chain Home* radar and the associated command-and-control network that allowed the seriously outnumbered RAF to hold on and ultimately win the Battle of Britain.

---

3   Robert Watson-Watt later coauthored a paper in 1926 entitled "An Instantaneous Direct-Reading Radiogoniometer" (Watson-Watt & Herd, 1926).
4   Great Britain's first operational "Filter Room" was an experimental setup at Bawdsy Manor in July 1937.

US Equivalent

In contrast, while one of six US Army *SCR-270* air-search radars on the island of Oahu had detected the incoming Japanese assault on Pearl Harbor at a distance of 132 miles, there was no established methodology or infrastructure for processing the data (Berhow, 2004). Some early users reportedly had to run down the road to the nearest gas station to call in their reports, but tactical and administrative phone lines were in place to the new "Information Center" at Fort Shafter by December 1941.

The concept of a coordinated fighter response, however, was another matter altogether in 1941 (Brown, 1999):

> "By December the radars functioned and the plotters moved the markers about the big map board during the few hours a day that they could practice, but in the balcony sat neither fighter control nor liaison officers... There were no tactical lines out of the Information Center to any fighting unit capable of making use of the information. There were no serviced fighters ready for takeoff, no rooms filled with pilots waiting on call."

The two *SCR-270* radar operators at Opana Point dutifully tracked the incoming wave of Japanese bombers for 30 minutes and phoned in their contacts, but it was misinterpreted as a flight of US *B-17s* from the mainland and subsequently ignored (Brown, 1999).[5]

After the war, an *SCR-270* radar (serial number 012) was loaned to the University of Saskatchewan in Canada for imaging aurora in 1949 (Wikipedia, 2006). Returned to the US in 1990, it was restored and put on display at the Historical Electronics Museum in Linthicum, Maryland, near the Baltimore-Washington International airport (Figure 9-3).

Figure 9-3. This trailer-mounted *SCR-270* radar antenna is of the type that detected the incoming Japanese assault on Pearl Harbor, 7 December 1941 (photo courtesy Historical Electronics Museum, Linthicum, MD).

---

5   A fairly accurate portrayal of the *SCR-270* radar setup on Oahu can be seen in the 1970 Twentieth Century Fox movie "Tora Tora Tora."

US Radar Development

The significant period of radar development that produced the functional hardware in place on Oahu that fateful Sunday morning spanned most of the preceding decade. The Naval Research Laboratory (NRL) had pioneered a low-frequency (28.3 MHz) pulsed (versus conventional continuous-wave) radar design (Helgeson, 2006), first demonstrated by Robert Page in December 1934 (Page, 1962; Brown, 1999). By April 1936, Page and Robert Guthrie of NRL had an improved version that could detect aircraft at a distance of 5 miles, and by 6 May they had increased the effective range to 17 miles (Allison, 1981).

NRL's use of short outgoing pulses facilitated subsequent development of a duplexer that eliminated the need for separate and bulky transmit and receive antennae. This ability to operate with a single (monostatic) antenna configuration was a key achievement for shipboard applications, where space and weight were a premium, especially on an elevated mast. In April of 1937, a prototype 200-MHz (1.5-meter) configuration using a directional Yagi antenna temporarily attached to the barrel of a 5-inch deck gun was demonstrated onboard the destroyer *USS Leary*. The results were encouraging, although this early prototype lacked sufficient range to be of much tactical use.

Robert Page recounts the historic radar-development effort as follows (Page, 1962):

"We move ahead now to the summer of 1937. Pulse radar had been operating with phenomenal success for over a year at the Naval Research Laboratory. Sets had been built and operated successfully at 28.6, 50, 80, and 200 megacycles. Many demonstrations had been given to high officials of the Government, and the Navy and the Army. The basic design had been given to the Army's Signal Corps Laboratories, and they were hard at work developing them for their particular needs. The immediate Navy problem was to get higher power pulses at 200 megacycles."

A much-improved version known as the *XAF* was successfully tested on the battleship *USS York* in January 1939, detecting aircraft at a range of 48 miles and surface targets at 10. The Radio Corporation of America (RCA) was brought under contract for 20 *XAF* sets to be put into operational use as *CXAM*, which in turn would evolve into the venerable *SK* shipboard radar (Goebel, 2005).[6] Affectionately known as the "Flying Bedspring," the *SK* would remain the Navy's standard early-warning radar throughout WWII (Figure 9-4).

Figure 9-4. Artist's concept of an *SK* long-wave search radar on a battleship (adapted from FTP, 1943).

---

6   In the US Navy's type-model designation scheme, *Type S* was a *Search* radar, while *K* (eleventh number of the alphabet) indicated the eleventh model (Friedman, 1981).

The other major US player in radar development was the Army Signal Corps Laboratories (SCL) at Fort Monmouth, NJ, headed by Major (later Lieutenant Colonel) William R. Blair (Terrett, 1956). Beginning in 1931, SCL aircraft-detection efforts under "Project 88" took the form of active near-infrared (followed by passive infrared) sensing (see again Chapter 8), which in 1933 gave way to continuous-wave "radio interference" or "beat detection" radar. Towards the end of 1934, however, Blair was beginning to favor the pulsed-ranging technique, hinting at such a future approach in his annual report to the Chief Signal Officer (Terrett, 1956):

> "...projecting an interrupted sequence of trains of oscillations against the target and attempting to detect the echoes during the interstices between the projections."

In late 1935, Blair's newly arrived executive officer, Major Roger B. Colton, persuaded him to send SCL engineer William Hershberger to NRL to check out the Navy's radar program. Hershberger visited NRL in January 1936 and was duly impressed. Upon his return, the aircraft-detection emphasis at SCL shifted from infrared to radar, with the interference-detector approach soon abandoned in favor of pulsed radar. As reported by First Lieutenant H.M. Davis (1943):

> "The main effort of the Signal Corps Laboratories now focused on an attempt to develop a workable detector on the principle proposed by Colonel Blair in 1934 – the system of 'projecting an interrupted sequence of trains of oscillations against the target' – now optimized by the word 'pulse.' Construction of a demonstration pulse detector began at the Signal Corps Laboratories in early 1936 while work was still going on with the interference detector."

SCR-268 Searchlight-Control Radar

The first Army radar application to be addressed was replacing the sound locator for antiaircraft-searchlight control, in response to a requirement generated by the Coast Artillery Corps. A pulsed-system prototype was working by December 1936, with the receiver placed about a mile away from the transmitter and screened by a grove of trees to minimize crosstalk. Slant ranges to commercial aircraft were recorded out to a distance of 7 miles, but no azimuth or elevation information was available, except that the detected targets were somewhere within the radiated beam pattern (Davis, 1943). Without a duplexer, three separate antenna arrays were required for range, azimuth, and elevation, and follow-on work began immediately to optimize and individually test these as stand-alone subsystems mounted on cannibalized sound-locator platforms.

The next step called for integrating the various subcomponents into a complete system consisting of the radar apparatus, a reflective heat detector, and a 60-inch searchlight (Figure 9-5), in preparation for a series of demonstrations at Fort Monmouth, NJ, on 18-19 May 1937. The objective was to automatically direct the searchlight at an incoming aircraft through non-visual means, such that the target was instantly illuminated when the searchlight was energized. In execution, the wider field-of-view radar components made the initial detection, then passed azimuth and elevation readings to the more directional thermal detector via the selsyn indicators of the sound-locator mounts. The thermal detector in turn tracked the aircraft and controlled the searchlight (Davis, 1943).

Figure 9-5. An azimuth antenna mounted on a sound-locator platform is seen undergoing the first service test of the *SCR-268-T1* at Fort Monroe, VA, in October 1938. Note the similarly mounted thermal detector in the background (US Army photo courtesy Communications-Electronics Life Cycle Management Command (C-E LCMC)).

During the first set of VIP demonstrations conducted on 18 and 19 May 1937 for the Chief Signal Officer, the Chief of Coast Artillery, and the Assistant Chief of the Air Corps, inclement weather conditions precluded any use of the thermal detector. The result was only four detected approaches out of seven. With more cooperative weather during the May 26 demonstration for Secretary of War Harry Woodring, the Army Chief of Staff, and the Chief of the Air Corps, the system achieved a perfect score of four detections out of four approaches. Following Woodring's positive response, work began in earnest on what would become the "Searchlight Control Radar" *SCR-268* shown in Figure 9-6 below.[7]

Figure 9-6. A production version of the *SCR-268* searchlight-control radar, with azimuth antenna at far left, IFF equipment *RC-148* on top, transmit antenna at middle right, and elevation antenna at far right (US Army photo courtesy C-E LCMC). Note the three operator seats and oscilloscope displays at image center.

---

7   The acronym "SCR" did not stand for "Searchlight Control Radar," but rather "Set Complete Radio," a model designation adopted by the SCL for radio sets during WWI and applied to radar during WWII to conceal the true nature of the equipment. Components of a set were assigned "RC" (Radio Component) numbers, while individual parts were identified with a "BC" (Basic Component) number.

The *SCR-268* required a minimum of six operators, three of which sat in tractor-like seats in front of their oscilloscope displays at the center of the rotating antenna assembly (Figure 9-7), fully exposed to hostile fire and the elements. Target azimuth and elevation were transmitted to the searchlight via selsyns, while target azimuth, elevation, and altitude could simultaneously be passed to an *M-4*, *M-7*, or *M-9* director for anti-aircraft gun laying (FTP, 1943). Effective range was 24 miles, with 90 miles achieved on occasion (Malone, 1989), but the set was highly susceptible to jamming. Other weaknesses included the large and bulky configuration, poor resolution, insufficient accuracy, and excessive setup time (Orman, 1946).

Given the paucity of available funding and the aggressive schedule to address a critical wartime need, however, the SCL's development of this patriarch of all subsequent Army radar was nothing short of legendary. Although declared obsolete later in the war and replaced by the *AN/TPL-1* for searchlight control, more *SCR-268s* were produced during WWII than any other radar model (Thompson et al., 1957). An upgraded version was introduced as the *SCR-516* (Figure 9-7), which featured a shorter wavelength, mechanically aided tracking, and the new "Plan Position Indicator (PPI)" display (Malone, 2004). The *SCR-516* was intended for ground-controlled intercept (GCI), but only a few saw service before the war ended.

Figure 9-7. Rear view of the *SCR-516* radar without the *RC-148* IFF equipment. The range, azimuth, and elevation operators sat directly behind the three oscilloscope displays on the rotating antenna assembly, which due to its size was extremely difficult to camouflage (US Army photo courtesy C-E LCMC).

SCR-270/271 Early-Warning Radar

Meanwhile, in early May 1937, M.F. Davis, an Army Air Corps lieutenant colonel assigned to a pursuit squadron in the Panama Canal Zone, had sent a letter to the Chief Signal Officer in Washington requesting: "radio equipment… which might be used to detect the presence of aircraft by reflected signals…" (Helgeson, 2006). Subsequent Signal Corps endorsement of this request had prompted Brigadier General Henry H. Arnold (then Assistant Chief of the Air Corps) to attend the previously mentioned *SCR-268-T1* prototype demo at Fort Monmouth on 18-19 May 1937, where he was favorably impressed.

His boss, Major General Oscar Westover, was similarly disposed after a follow-on 26 May demonstration, which led to formal Air Corps endorsement of the early-warning air-search capability. Work on the mobile *SCR-270* radar began in the summer of 1937.[8] An experimental prototype was successfully tested in August 1938 at an offsite SCL facility in Twin Lights, NJ (Figure 9-8), tracking an aircraft out to 78 miles (Vieweger & White, 1959).[9] A single antenna was employed for both transmit and receive, a significant improvement over the earlier *SCR-268* design.

Figure 9-8. Early prototypes of the fixed *SCR-271* (background right) and mobile *SCR-270* (foreground right) radars at Twin Lights, NJ. For their prescribed early-warning role, neither set had provision for determining target elevation (US Army photo courtesy C-E LCMC).

This monostatic configuration was made possible by a duplexer developed by Dr. Harold Zahl, which used appropriately positioned spark gaps to shunt the high-voltage outgoing pulse before it reached the receiver input (Terrett, 1956).[10] A service test model (*SCR-270-T1*) was performing reliably out to 80 miles by June of the following year, demonstrated to the Secretary of War in November 1939, and formally adopted by the Army in May of 1940 (Figure 9-9). A production contract was awarded to the Westinghouse Electronics Division, Baltimore, MD, in August of that same year.

---

8   The *SCR-271* was the fixed-installation version of this radar as shown in Figure 9-8. Early prototypes of the fixed SCR-271 (background right) and mobile SCR-270 (foreground right) radars at Twin Lights, NJ. For their prescribed early-warning role, neither set had provision for determining target elevation (US Army photo courtesy C-E LCMC)..

9   Fearing that German spies were observing their radar work at Fort Monmouth, the SCL had relocated further development to Fort Hancock on Sandy Hook.

10   The much weaker received-echo signal from the antenna would not have sufficient potential to ionize the spark gap, which would simply appear as a high impedance (versus a low-impedance shunt).

Westinghouse had delivered 112 early-warning radar sets prior to the Japanese attack on Pearl Harbor in December 1941 (Vieweger & White, 1959). In accordance with the 1937 request submitted by Lieutenant Colonel Davis, the first two units, both fixed-installation *SCR-271s*, went to guard the east and west entrances to the Panama Canal (SCHP, 1945):

"Summing up the actual disposition of the first long-range radar sets to be produced, two of the *SCR-271* had been sent in 1940 to Panama and two of the *SCR-270* had been delivered to the First Aircraft Warning Company, with another one retained at the Signal Corps Laboratories. In January and February 1941, five more *SCR-270* sets were delivered to the First Aircraft Warning Company. In March two *SCR-271* sets and one *SCR-271A* were earmarked for Hawaii, one *SCR-271A* for Alaska and another for Panama."

Figure 9-9. The mobile *SCR-270* (left) and fixed-installation *SCR-271* (right) early-warning radars at Fort Hancock in Sandy Hook, NJ (undated US Army photo courtesy C-E LCMC).

Range information was read from an "A" scope display (Figure 9-10), but the *SCR-270* operator had to look out the window of the operating trailer to read azimuth information off a calibrated ring on the 8-foot-diameter rotating tower base (Figure 9-11), which was problematic (Thompson & Harris, 1966; Suffield, 1995). The fixed-site *SCR-271*, on the other hand, featured a mechanical coupling between the rotating antenna mast and an azimuth indicator inside the operating shelter.

"A" Scope

Range: Horizontally.
Presence: Vertically.
Bearing: None indicated.
Advantages: Ease in detection, ranging, and determination of target composition.

PPI Scope

Range: Distance from center of scope.
Presence: Intensity.
Bearing: Direction of sweep.
Advantages: Complete picture in a few seconds, shows all objects in true relative positions.

"B" Scope

Range: Vertically.
Presence: Intensity.
Bearing: Horizontally.
Advantages: Good bearings at any range; good bearing resolution or target separation at short ranges as contrasted with PPI; shows all targets at the same time. Similarity to cross-hairs makes it especially good for gunnery.

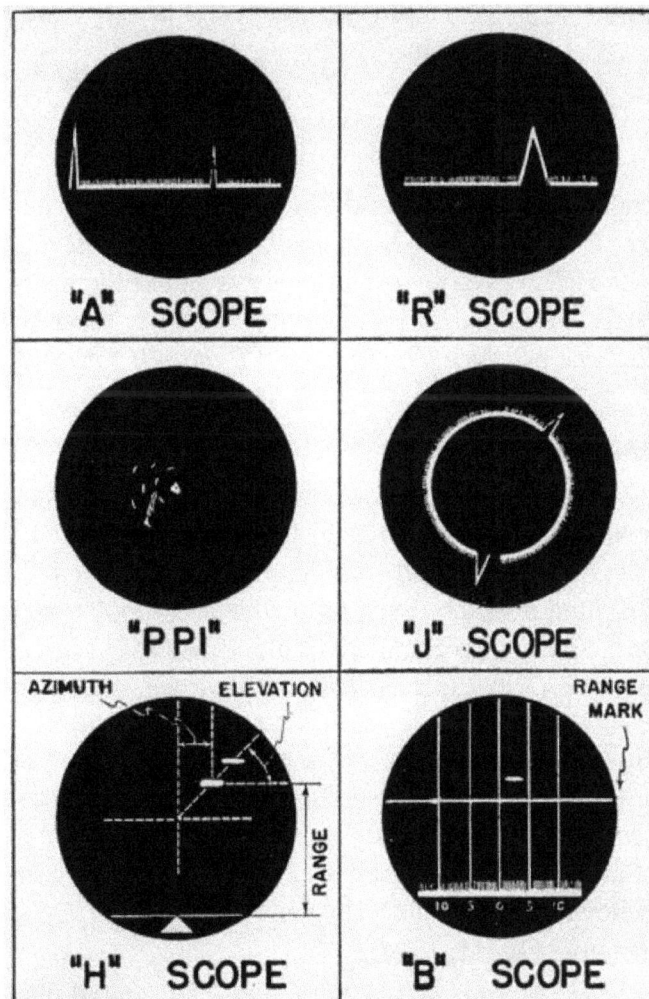

Figure 9-10. Scope presentations (ROM, 1944).

The improved Westinghouse *AP-2* antenna-positioning system, introduced for both sets in 1942, employed a selsyn to electrically transmit azimuth information to a remote indicator. In addition, the operator could now manipulate the antenna orientation in "position-control mode" by turning a hand crank, or in "velocity-control mode" to achieve a desired slew rate (SCHP, 1945). The *AP-2* could also provide an automatic back-and-forth sector sweep at any requested speed.

Figure 9-11. The azimuth scale is seen mounted on the rotating base of a mobile *SCR-270* radar installation in Vanua Levu, Fiji Islands (US Army photo adapted from Thompson & Harris, 1966).

A detailed historical accounting of *SCR-270* and *SCR-271* development and fielding is presented in *The Signal Corps Development of US Army Radar Equipment*, prepared by the Signal Corps Historical Section (SCHP, 1945). In addition, the IEEE History Center presents an excellent summary of the *SCR-270* technical specifics (Suffield, 1995). For a fascinating overview of WWII radar development and employment in general, see also *Technical and Military Imperatives: A Radar History of World War II* (Brown, 1999).

SCR-296-A Fire-Control Radar

The principle gun-laying radar for US coast defense was the *SCR-296-A* (Figure 9-12), developed by the MIT Radiation Lab, more commonly known at the time as the "MIT Rad Lab." The accomplishments of this group in support of the Allied war effort were nothing short of phenomenal, and a number of fascinating accounts have been written over the years, to include a 28-book series funded by the US Government after the war (Lundberg, 2002). Development of the *SCR-296-A* fire-control radar was by necessity delayed due to the much higher priority of the *SCR-270/271* air-search requirement (Malone, 1989).

Radar was heavily classified during the war and not open for discussion, much less photography. To complicate matters, the *SCR-296-A* equipment was declared obsolete on 17 January 1946, almost immediately after war ended, and subsequently removed. As a consequence, there are very few period photos of actual site locations in the San Diego area. Figure 9-12 shows an artist's concept of a typical *SCR-296-A* installation disguised as an elevated water tank. The generator shelter *HO-1-A* in background left housed two *Power Unit PE-84* generators (TM11, 1944).

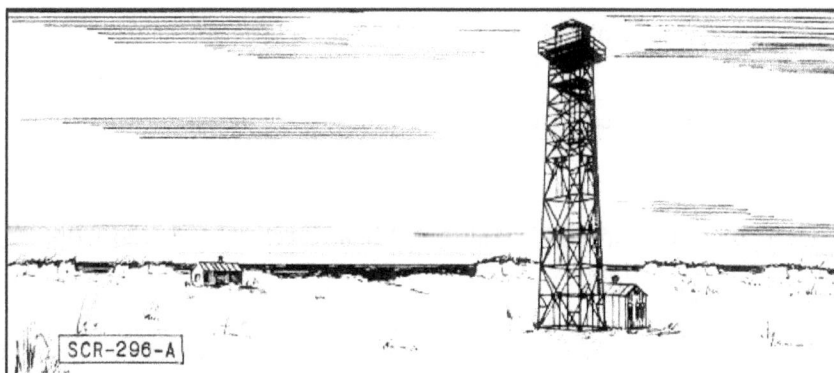

Figure 9-12. a) The *SCR-296-A* fire-control radar was typically mounted on a steel tower with its antenna hidden within a fake wooden water tank at the top (adapted from FTP, 1943). The generator shelter *HO-1-A* is in background left, with the operator shelter *HO-2-A* adjacent to the tower base.

Figure 9-13 shows an early version of an *SCR-296-A* equipment setup inside an operator shelter *HO-2-A*.[11] The two operators sat at the table in front of the five pieces of display equipment identified in Table 9-1, with the azimuth operator on the left and the range operator on the right. The radar equipment was housed in cabinet *BE-82-A* at far left. The *MG-16-A* motor-generator unit on the floor beneath the right end of the table amplified the voltage from the Manual or Rate controls to position the radar antenna (FM, 1943). Seven personnel were required to operate and maintain the radar, to include a chief of section, range operator, range reader, azimuth operator, azimuth reader, power-plant operator, and maintenance man.

Figure 9-13. *SCR-296-A* operating-table layout adapted from *FM 4-95* (FM, 1943), with component identification provided in Table 9-1. The radar equipment was housed in cabinet *BE-82-A* at far left.

11 A detailed technical overview of the *SCR-296-A* radar is presented by Danny Malone (1991).

1	Power Supply *RA-49*	4	Azimuth Indicator *I-110-A*
2	Indicator	5	Range Scope *BC-723*-A
3	Azimuth Scope *BC-718-A*	6	Range Unit *BC-723-A*

Table 9-1. Identification of *SCR-296-A* radar equipment shown in Figure 9-13.

The *SCR-296-A* specifications presented in Table 9-2 below are reproduced from *Field Technical Publication 217* (FTP, 1943).

DESCRIPTION:	Fixed Coast Artillery gun-laying medium-wave radar assigned to modern 6-inch or larger batteries.
USES:	Set is designed to track a surface target in range and azimuth. Data are sent to the plotting room and used in firing. An *SCR-296-A* normally is assigned to one battery, but may furnish data to more. Works with *IFF RC-136A*.
PERFORMANCE AND SITING:	Range is shown on "A" scope. The target is tracked in azimuth with a pip-matching oscilloscope or a zero-center meter. Range accuracy is about ±30 yards while azimuth accuracy is about ± 0.02 degree under the best conditions. The set has a dependable range of 20,000 yards on a destroyer-size target when employed at a height of 145 feet.
TRANSPORTABILITY:	Shipment includes spares and separate generator. When crated the total weight is 91,763 lbs. Largest unit is 5,270 lbs.
INSTALLATION:	*SCR-296-A* includes a tower, an operating building, and two power-plant buildings. The tower is obtainable in heights of 25, 50 75, and 100 feet. Concrete floors must be put in locally.
PERSONNEL:	Operating crew consists of 5 men in addition to a power-plant operator and maintenance man, who should be available at all times.
POWER:	Primary power of 2.3 KW is supplied by *PE-84C* – commercial or auxiliary 110 V, AC single phase. Generator needs high-octane gas.

Table 9-2. *SCR-296-A* Fixed Medium-Wave Coast Artillery Fire Control Set specifications (FTP, 1943).

The full detection range of the *SCR-296-A* is presented as a horizontal line across the range scope *BC-723-A* as seen in Figure 9-14a (FM, 1943):

"The center portion of the sweep may be expanded, so that 4 inches represent about 5,000 yards, this allowing accurate determination of target range and facilitating discrimination between targets differing slightly in range."

"In the center of the expanded portion of the sweep a section of the baseline, having a width which represents about 600 yards, is lowered to form a notch. The range to any target centered in the notch can be read from the range dials."

Detected echoes (pips) appear as vertical deflections in the sweep line (Figure 9-14b).

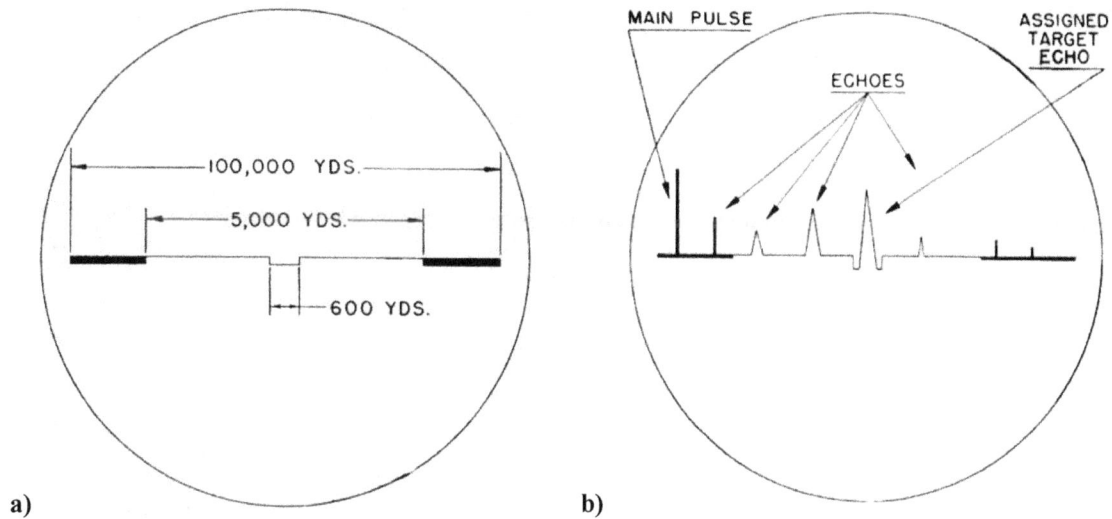

Figure 9-14. **a)** The *BC-723-A* range-scope screen with no signal detected; note 600-yard notch below the baseline. **b)** The range-scope screen with the desired echo signal in the notch (adapted from FM, 1943).

SCR-547 Range-Finding Radar

Introduced in mid-1941 by the National Defense Research Committee (NDRC) and Bell Labs, the *SCR-547* was a trailer-mounted 10-centimeter set intended to measure the slant range to an airborne target, with azimuth and elevation determined by optical tracking. While the maximum theoretical range was 25,000 yards, the effective range was dependent upon visibility, since the set had to be optically directed at the target via a sighting telescope (FTP, 1943). A rotating central column supported two parabolic antennae (transmit and receive) that could be simultaneously adjusted in elevation as illustrated in Figure 9-15.

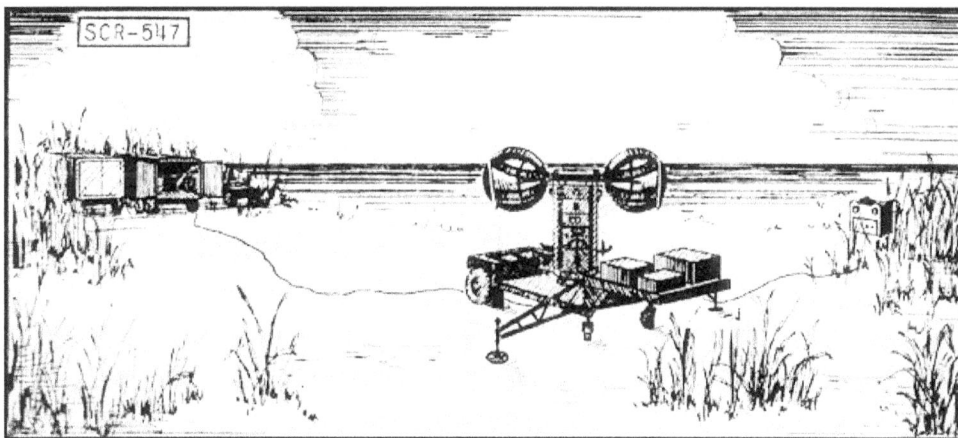

Figure 9-15. The twin parabolic antennae of the *SCR-547* radar gave rise to the nickname "Mickey Mouse" (adapted from FTP, 1943).

Optional power-aided tracking helped maintain a constant rate of traverse to keep the target within the optical field-of-view of the telescopic sight (Malone, 1990). The *SCR-547* could also be used to determine horizontal range to surface targets; this dual-mode capability made it a potential candidate for controlling the 90-millimeter anti-motor torpedo boat (AMTB) batteries discussed in Chapter 6, which had a secondary mission of antiaircraft defense (Malone, 1990).[12]

---

12 Slant range was previously provided to the *M-9 Gun Director* by the stereoscopic *M-1 Height Finder*, which proved to be less than adequate except under ideal conditions (Malone, 1990).

SCR-584 Early-Warning / Anti-Aircraft Radar

The *SCR-584* early-warning and antiaircraft gun-laying radar was one of the most successful radar projects developed by the MIT Rad Lab during the war (Bragg, 2005). A postwar excerpt from a tribute to the *SCR-584* radar in the former McGraw-Hill magazine *Electronics* read as follows (EM, 1945):

"High on the list of electronic achievements of the war is the *SCR-584* radar. A microwave set developed primarily for accurate fire control of 90-millimeter anti-aircraft batteries, the *584* served this basic purpose from Anzio to the end of the war. It also served as an early-warning radar against approaching enemy aircraft, as a ground control for low-flying fighter aircraft in the advance across France, and detected the motion of transportation along roads and the flight of enemy shells and mortars in the Italian campaign."

Performance specifications for this 10-centimeter radar are listed in Table 9-3.

Wavelength	10 centimeters
Magnetron	*2J32*
Peak Output	250 kilowatts
Pulse Width	0.8 microseconds
Pulse Repetition Frequency	1707 pulses/second
Antenna Diameter	6 feet
Beamwidth	4 degrees (-3 dB)
Maximum Range:	
PPI Search	70,000 yards
Auto-Tracking	32,000 yards
Potentiometer Data (gun control)	28,000 yards
Minimum Range	500-1000 yards
Range Error	25 yards
Elevation coverage	- 9.8 to + 88.9 degrees
Azimuth Coverage	360 degrees continuous
Angular Error	0.06 degrees

Table 9-3. *SCR-584* performance specifications from *TM 11-1324* and *TM 11-1524* (Bragg, 2005).

As recalled by former *SCR-584* project manager Lee L. Davenport (1991) in a detailed oral-history interview conducted by John Bryant, IEEE History Center:

"The *M-4* and *M-7* gun directors to which we initially fed data were mechanical units that existed at the start of the war. They were normally fed by optically tracking the target with slant-range information coming from an optical rangefinder. The Signal Corps hoped to introduce night firing capability with the forthcoming *SCR-584* while still keeping the optical inputs in reserve. Also planned was a shift to a high-precision electronic gun director, the *M-9*. It was not known when the *M-9* would be available so *584* units had to be capable of working with both types of directors."

A typical *SCR-584* setup with a 90-millimeter gun battery is presented in Figure 9-16a, with a cutaway view of the *K-78-A* trailer interior in Figure 9-16b.

SCR-584 OPERATING WITH GUN BATTERY

A. POWER GENERATOR M7
B. RADIO SET SCR-584
C. DIRECTOR M9
D. TRACKER FOR DIRECTOR M9
E. BATTERY 90mm.

a)

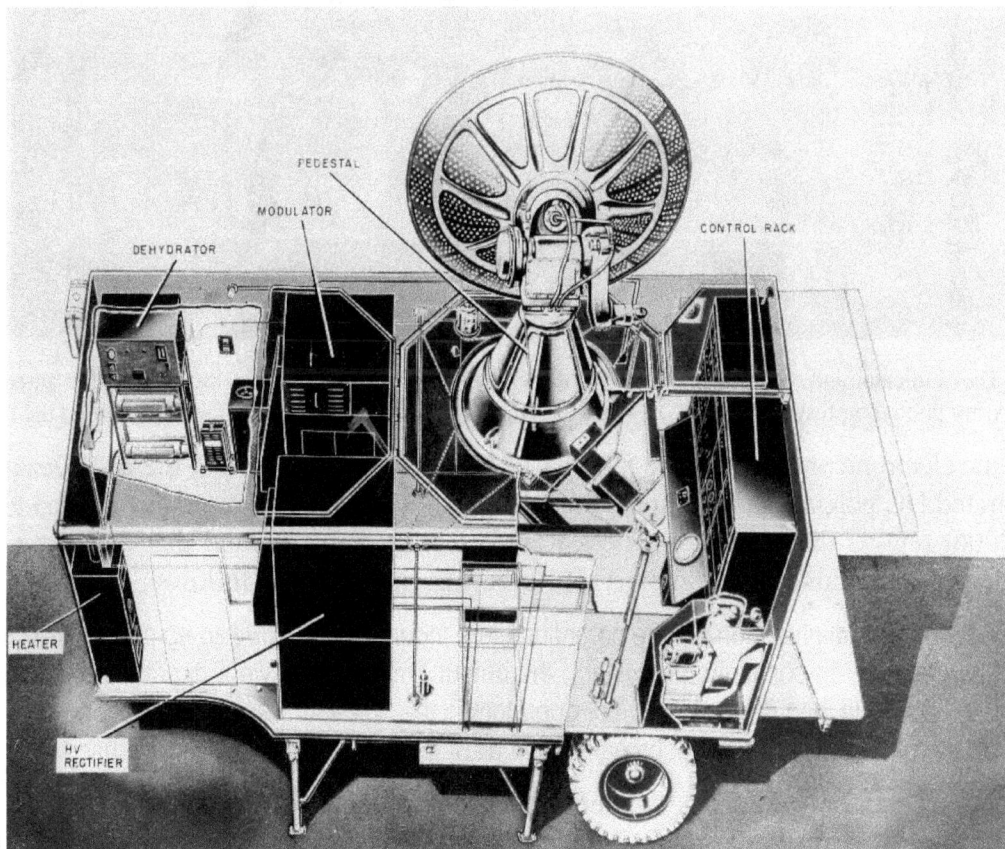

b)

Figure 9-16. **a)** Typical *SCR-584* setup for supporting a 90-millimeter antiaircraft battery. **b)** The principle components of the *SCR-584* were mounted inside a *K-78-A* trailer with a lift mechanism that allowed for antenna stowage during transit (US Army images courtesy C-E LCMC). See also Figure 9-17.

The operator console for the *SCR-584* radar is shown in Figure 9-17, with the PPI scope seen on the sloped table in front of the right chair. The azimuth crank is directly above the PPI scope on the equipment rack, with the elevation crank just to its left. Immediately above these two cranks are the "A" and "B" scopes. The range crank and aided-range-tracking control are at the bottom of the large panel in the equipment rack above the left chair. The automatic/manual tracking switch is midway between the two operator positions, on the table just left of the PPI scope.

Figure 9-17. The "indicator and control panel" *PN-24* of the *SCR-584* inside the *K-78-A* trailer, with the antenna in the raised position (US Army photo courtesy C-E LCMC). The PPI scope is on the sloped table surface just behind the right chair.

To facilitate electronic interface to the tube-based operational amplifiers of the *M-9 Director*, the *SCR-584* incorporated DC potentiometer outputs for range, azimuth, and elevation, as opposed to AC selsyns. Spalding (2006) reports these potentiometers were 3 feet in diameter, and that the op amps had to be hand-zeroed every few minutes. Davenport (1991) recalls the design as follows:

"The *M-9* was fed data from three large potentiometers mounted in the *584* so no pointer-matching human link would be needed. To complete the system, the output from the *M-9* went directly to servo-driven guns and fuze (sic) cutters; in the final form, only the operators in the *584* and the gun crew to load shells would be required. To the best of my knowledge, by early 1944 all of these final elements were in use."

The combination of those synergistic elements made for a lethal performance indeed. The following excerpt from "American Ack-Ack," written by the famed war correspondent Ernie Pyle, provides some insight on automatic target acquisition (Pyle, 1944):

"A gunner turned a switch on the side of the gun, and it went into remote control. From then on a mystic machine at the far end of the field handled the pointing of the gun, through electrical cables. It was all automatic. The long snout of the barrel began weaving in the air and the mechanism that directed it made a buzzing noise. The barrel went up and down, to the right and back to the left, grinding and whining and jerking. It was like a giant cobra, maddened and with its head raised, weaving back and forth before striking. Finally the gun settled rigidly in one spot and the gun commander called out, "On target! Three rounds! Commence firing!"

The US Navy expressed an early interest in an *SCR-584* derivative to replace their *CXAM* early-warning radar (Goebel, 2009):

> "The prototype of the ocean-going *SCR-584*, the "CXBL", was mounted on the new carrier USS *Lexington* in March 1943,[13] while the production version, the "SM," built by General Electric, was operational on the carriers USS *Bunker Hill* and USS *Enterprise* by October 1943. The fast schedule was possible because the *M-9* director wasn't required for the application, though the shipboard installation was complicated by the need to gyrostabilize the antenna."

A light-weight version of the *SCR-584* would later be introduced as the *SCR-784* (MMR, 2007), mounted on the *K-84* searchlight trailer shown in Figure 9-18 (TM 9, 1947*)*.

Figure 9-18. The light-weight mobile *SCR-784* radar (not shown) was mounted on the *K-84* searchlight trailer shown above (US Army photo).

Figure 9-19 shows an *SCR-584* and its generator positioned south of Building F-5 at the U.S. Navy Electronics Lab in 1949, presumably in support of the Lab's post-war harbor-defense role (Fisher, 1951). An equipment trailer for another type of radar has been placed upon the roof of the wooden shed at right, Building F-5. The cubical enclosure on top of the steel tower at left appears to be capable of rotation in azimuth. This area overlooking the Pacific Ocean at the north end of the laboratory, which supported extensive test and evaluation of various radar systems following WWII, was designated as the "Electromagnetic Radiation Equipment RDT&E Range."

---

13 This would be the Essex-class USS *Lexington* (CV-16), which replaced the earlier USS *Lexington* (CV-2) lost in the Battle of the Coral Sea in May 1942.

Figure 9-19. An *SCR-584* is seen at photo center south (left) of Building F-5 at the US Navy Electronics Lab in 1949. For reference, Building F-5 and the tower-mounted antenna structure left of the *SCR-584* are seen again in the 1954 aerial photo presented in Figure 9-20 (US Navy photo Courtesy SSC Pacific).

The aerial photo presented in Figure 9-20 shows this same area 5 years later in 1954, looking east, with a black up arrow denoting the earlier 1949 *SCR-584* test site at bottom right. Other radar antennae can be seen on the new elevated platforms by Building F-5, on the ground near Building F-28, and on the rooftop of Building F-1, with Woodward Road in upper background. The lower-level site overlooking the Pacific Ocean has been improved with stair access down to several additional wooden structures above the beach. The underground silo for WWII Searchlight 12 is seen in the upper-right corner.

Figure 9-20. Aerial view looking east in 1954 during site grading near Building F-28 (left center). The *SCR-584* and its generator seen in the 1949 photo presented in Figure 9-19 had been situated in the lower-right corner (black up arrow), just south (right) of Building F-5 (US Navy photo Courtesy SSC Pacific).

An article in the 12 September 2008 *San Diego Union-Tribune* regarding the convicted spy Morton Sobell, who as an electrical engineer passed military secrets to the Soviets during WWII, includes the following interesting statement (Roberts, 2008):

> "One device mentioned by Sobell, the *SCR-584* radar, is believed by military experts to have been used against U.S. aircraft in Korea and Vietnam."

After his former classmate Julius Rosenburg was arrested for similar treason, Sobell fled to Mexico under a false name, but was captured in 1950 and found guilty of treason in 1951. He was eventually released from prison in 1969 (Roberts, 2008).

SCR-582/682 Surface-Search Radar

The previously discussed *SCR-296-A* was a dedicated fire-control radar for a specific gun battery, as opposed to a surface-search radar. The *SCR-582* (Figure 9-21), a fixed harbor-surveillance radar with 360-degree coverage, could be used to detect incoming surface threats, but its accuracy and update rate were insufficient for automated fire-control. The *SCR-582* (or the mobile version, *SCR-682*) would provide approximate range and bearing to a target vessel, whereupon the *SCR-296-A* took over, scanning back and forth within an operator-controlled sector of interest. If surface-search radar was not available, initial range and bearing to the target had to be optically determined through triangulation, then passed to the *SCR-296-A*.

Figure 9-21. Artist's concept of a fixed *SCR-582/682* radar installation (adapted from FTP, 1943). The *XT-3* prototype had been introduced in December 1941 and tested at Fort Dawes in Boston Harbor (Malone, 2004).

According to the *SCR Components* website of the Army Signal Center, Ft Gordon, GA,[14] the *SCR-582* was a 10-centimeter surveillance radar for use in locating targets on the surface of the water, effective to 90,000 yards.[15] The transportable *SCR-682* version (Table 9-4) reportedly achieved a maximum range of 240,000 yards. Although designed for surface-search, this more powerful model also turned out to be fairly useful for long-range over-water detection of low-flying aircraft (Clement, 1948).[16] Both sets were compatible with the Navy's Precision Plan Position Indicator (PPI), which could portray multiple targets in relationship to natural harbor features (Malone, 2004).

---

14 This website is apparently no longer available.

15 Malone (2004) reports the maximum effective *SCR-582* range was listed as 35,000 yards postwar, with a range accuracy of plus or minus 25 yards and an angular accuracy of plus or minus 2 degrees.

16 The *SCR-582* was modified with a tilting antenna dish for aircraft detection, which became the *SCR-682* (Thompson, et al., 1957).

DESCRIPTION:	Transportable Seacoast Artillery Microwave Radar.
USES:	Used by 155-mm and 8-inch railway battalions. PPI gives indication of surface craft and low flying planes. Uses *IFF RC-282*.
PERFORMANCE AND SITING:	Set covers 360°. For a height of 100 feet, range will average 50,000 yards or greater. At times, under favorable conditions, very much greater maximum ranges will be achieved from sufficient heights of site. Minimum range is about 500 yards. Azimuth accuracy is of the order of ±2 degrees depending somewhat on the range scale in use. The range accuracy thus may be considered ± 3 percent of the range scale in use. The range scales are: 10,000; 40,000; 160,000; and 320,000 yards. Set should be sited not less than 30 feet above the surrounding terrain, and preferably at least 100 feet above sea level. The optimum height of site is between 150 and 500 feet.
TRANSPORTABILITY:	A transportable version of the *SCR-582*, *SCR-682* can be carried in two standard 2 1/2 -ton 6x6 cargo trucks.
INSTALLATION:	Set normally is operated from a portable tower furnished with the set. The operating components are located near the tower in a *Shelter HO-17* or housing constructed by the using troops from materials obtained locally. No transportation is furnished with the set.
PERSONNEL:	One man operates the set and another plots the position of detected targets. These men relieve one another every 30 minutes to reduce fatigue. For 24-hour operation a chief of section, five operators, two power plant operators, and one maintenance man are required.
POWER:	Requires about 1.28 KW, 120 VAC, 60 cycles, single phase.

Table 9-4. The transportable *SCR-682* specifications as listed in *Field Technical Publication 217*, which indicated the set would be available for issue the latter part of 1943 (FTP, 1943).

As recalled by WWII veteran and *SCR-582* operator Dan Vesper in an e-mail to webmaster Joe Stevens of the Kodiak Military History Museum (Vesper, 1999a):

"An amber rotatable disc was attached to the front of the PPI. An engraved radial line was etched or inscribed on the face of the amber disc. When a target was observed, the operator could rotate the amber disc until the etched line centered on the target. The actual azimuth or direction to the target could then be read from an engraved or painted scale, in degrees located around the outer edge of the PPI. The CRT (Cathode-Ray Tube) used in the PPI had a long persistent phosphorous coating on the inside face."

This persistence caused target-echo pips to remain visible on the PPI display for some 5 to 10 seconds as the antenna rotated.

The *SCR-582* "A" Scope was located on a panel in the equipment rack immediately above the PPI Scope on the Operator's Console as shown in Figure 9-22 (Vesper, 1999b):

"A target (ship) would appear as a vertical spike on the scope at a relative distance from the left side of the scope as the distance indicated on the PPI Scope. The "A" Scope gave a much sharper indication of the target. If two ships were traveling close together, they could appear as a single target on the PPI, but could be distinguished as two targets on the "A" Scope."

Figure 9-22. An *SCR-582* operator console (foreground right) at Alexai Point, Attu, courtesy Joe Stevens (US Army photo, NARA). The "A" scope is in the equipment rack directly above the PPI display on the sloped panel. The *RC-182* IFF equipment is seen against the far wall in background.

In operation, the *SCR-582* surveillance-radar crew would pass the range and bearing of any perceived surface targets to the *SCR-296-A* operators by telephone (Malone, 1991):

> "This coaching was necessary due to the poor target discrimination of the radar and its narrow beam, which meant that the antenna had to be pointed almost exactly at the target."

Figure 9-23 shows the interface between the *SCR-582* surface-search radar and the *SCR-296-A* fire-control radar for the designated gun battery. Note annotation on horizontal dotted line in the upper-left corner: "This direct liaison line may be desirable."

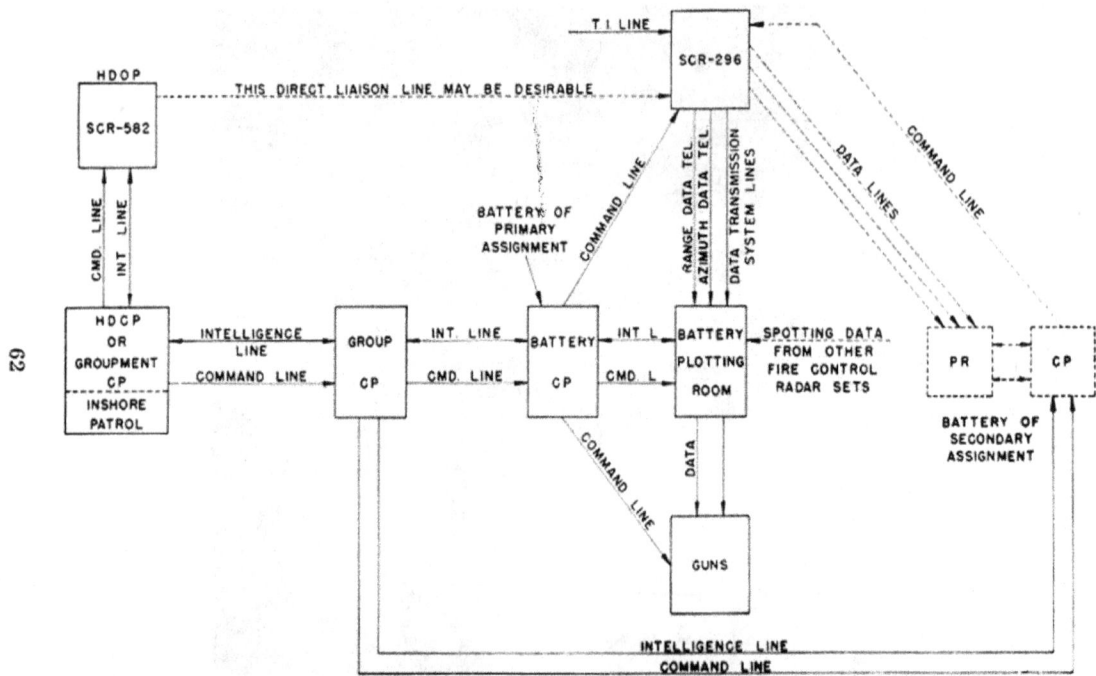

FIGURE 28.—Communications system.

Figure 9-23. Schematic diagram illustrating the telephone connectivity between the *SCR-582* surface-search radar and an *SCR-296A* gun-laying radar, adapted from page 62 of *FM 4-95* (FM, 1943).

Coast Defense Radar Installations

Thompson (1991) reported a total of six *SCR-296-A* fire-control radars were ultimately installed in the San Diego area:

"Coastal radar, SCR296A, was added to the harbor defenses early in 1943. The first three sets were erected near Battery Strong in north Fort Rosecrans, near the new lighthouse at Point Loma, and near the Mexican border. Before the year was out, three additional sets were authorized: La Jolla, Fort Rosecrans, and Fort Emory."

Cross correlating Thompson's chart of fire-control stations with wartime "Harbor Defenses of San Diego" blueprints showing radar symbols near known fire-control structures gave a general idea of location. Reproductions of the Reports of Completed Works for these installations, downloaded from the *Coast Defense Study Group* website (CDSG, 2006), later provided additional information of a very specific nature. Piecing together all the evidence yielded the following summary (Table 9-5):

Project	Type	Location	Assignment	Transferred
9-126	*SCR-296-A*	Mexican Border	Strong	12 January 1945
9-127	*SCR-296-A*	North Fort Rosecrans	Woodward	15 March 1944
9-129	*SCR-296-A*	By New Lighthouse	Humphreys	4 September 1944
9-163	*SCR-682*	Signal Station	HDCP	19 December 1944
9-1007	*SCR-296-A*	Fort Emory	Grant	4 October 1944
9-1008	*SCR-296-A*	La Jolla	Ashburn	14 September 1944
9-1009	*SCR-296-A*	Battalion CP 2	Gatchell	(Deferred 1944)

Table 9-5. Summary of radar installations and their assigned gun batteries in the San Diego area towards the end of WWII.

Regarding Thompson's reported locations for the first three radars assigned to HDSD, radars 9-127, 9-129, and 9-126 correspond to: "The first three sets were erected near Battery Strong in north Fort Rosecrans, near the new lighthouse at Point Loma, and near the Mexican border." Similarly, radars 9-1008, 9-163, and 9-1007 would seem to agree with: "Before the year was out, three additional sets were authorized: La Jolla, Fort Rosecrans, and Fort Emory," except that set 9-163 was an *SCR-682* as opposed to an *SCR-296-A* (Figure 9-24). Installation of the *SCR-296-A* (9-1009) for Battery Gatchell was never begun as the battery construction was halted in February 1944.

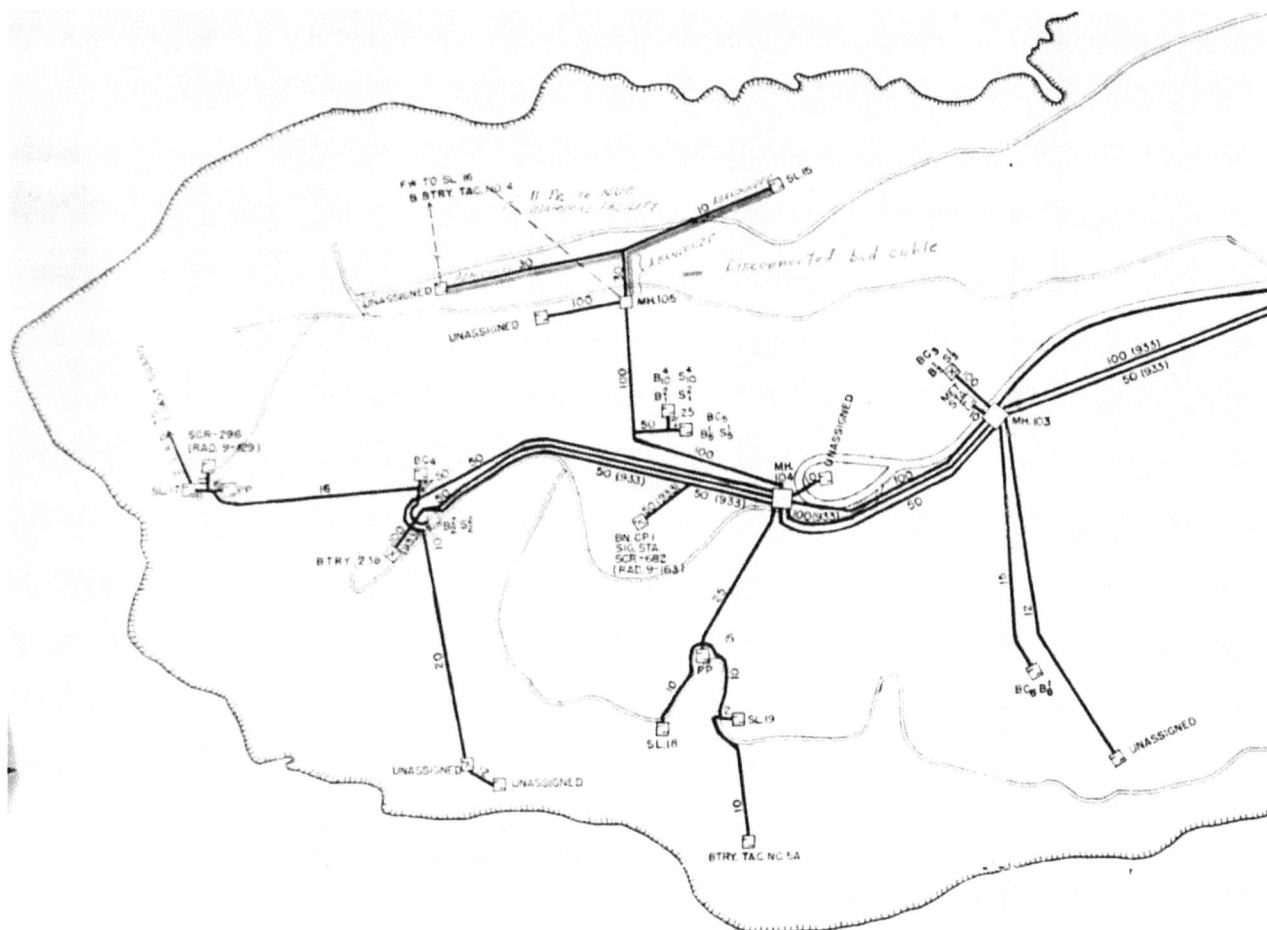

Figure 9-24. Portion of "Harbor Defenses of San Diego, Main Cable Routings" (sheet 3 of 4), dated 1 July 1945 (with north to the right), showing an *SCR-296-A* radar east of the new lighthouse (far left), and an *SCR-682* located near the Signal Station atop the Battalion One Command Post south of the old lighthouse (center).

Annex B (Fire Control) of the 1946 *Supplement to the Harbor Defense Project, Harbor Defenses of San Diego*, provided the following post-war assessment of *SCR-296-A* performance (Annex B, 1946):

"Their accuracy in range and azimuth readings is close to that obtained from horizontal base tracking except that in areas close to the coastline land reflections distort and blank out the reflected echoes. The examination of many observed courses shows that the average error in azimuth is less than .15 degrees and in range less than 30 yards for ranges up to 15,000 yards. The sets constantly give good reflections from destroyer targets at ranges to 30,000 yards. The sets with more height of site usually give greater ranges with a given target. Stationary land targets and moving targets over 100,000 yards away have been picked up on all the sets..."

Further discussion of the HDSD installations will be provided in the following sections.

SCR-682 Radar 9-163 at Signal Station

The surface-search radar that provided early-warning target information to the various San Diego fire-control radars was an *SCR-682,* installed in the wooden Signal Station Tower just south of the old lighthouse. The multi-purpose tower had several levels as shown in Figure 9-25, accommodating the radar antenna, a signal house platform, a portable-searchlight deck, and a "radio room" that actually housed the Army's underwater hydrophone-monitoring equipment. This tower was situated directly above the northern escape hatch of the underground Battalion One Command Post described in chapter 7.

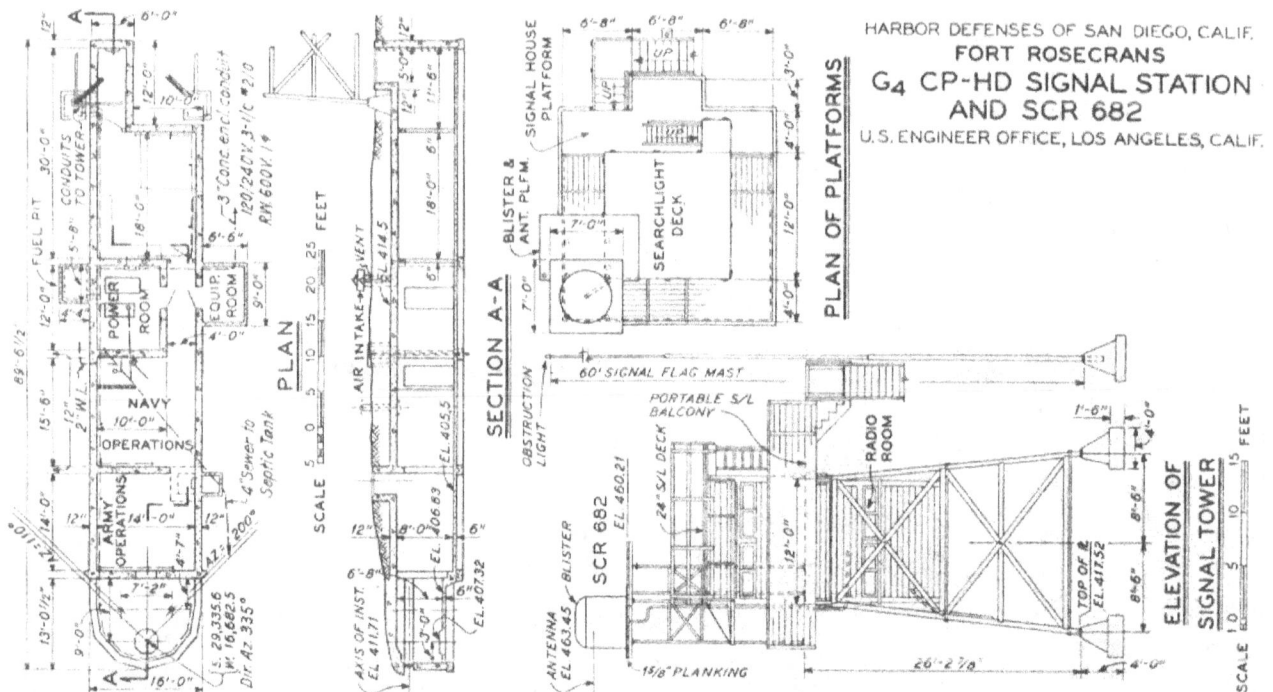

Figure 9-25. The *SCR-682* surface-search radar antenna was mounted on the top level of the wooden signal tower (lower right), which was situated directly above the rear hatch of the underground Battalion One (formerly Group 4) Command Post depicted in the plan view at far left (NARA).

Radar set *SCR-682* was originally fielded as a mobile version of the fixed harbor-surveillance *SCR-582* radar, but when the latter equipment was discontinued, the *SCR-682* was used for both mobile and fixed installations. The original authorization for the Signal Station installation in Site 7 at Fort Rosecrans, in fact, called for an *SCR-582*, but 3 months later referred to the set as an *SCR-682* (Annex B, 1946):

"SCR-682 project RAD 9-163 is authorized by 5th Ind, OCE, file 655 (San Diego) CM 8108 SPEOF, dated 21 September 1942, to letter, HQ HDSD, file 660.2 SECRET, dated 31 July 1942, subject: "Transmittal of Local Board Proceedings, HDSD, dated 18 July 1942, covering Radio Set SCR-582."

The primary source of power was commercial 60-cycle 110-volt AC, with a backup gasoline-driven 6.3-KVA generator. An additional 3-KVA generator supplied emergency power to the Battalion One Command Post, with both generators located in the "Power Room." According to the post-war "Historical Sketch" of the HDSD project (HSD, 1945), the surveillance radar was completed and in operation by August 1944, providing the Harbor Entry Control Post a very important intelligence means (Figure 9-26). The facility was officially transferred to the artillery on 19 December 1944, some 3 years after the Japanese attack on Pearl Harbor (RCW, 1945c).

Figure 9-26. The small elevated platform at upper left on top of the Signal Tower (see again chapter 7) supported the radome for the *SCR-682* radar antenna, not yet installed (US Army photo courtesy Cabrillo National Monument).

The following description of procedures associated with the new harbor-surveillance radar was provided by Major General George S. Prugh (1988), who as a young Army officer had been assigned to Fort Rosecrans during the war:

"We were no longer dependent solely upon visual observation. We also had an IFF system – Identification Friend or Foe – whereby we could challenge a vessel electronically and obtain a coded reply. And we had encoding and decoding apparatus with which we could communicate with US Naval vessels that wished to maintain a security cover. The normal routine was established with our new radar equipment for the Harbor Defense to scan the harbor approaches out as far as we could go, to plot all vessels passing or entering within our range, to check against a list of expected traffic (sent to us from the Naval District), to interrogate an unknown with IFF, and then to track unknowns that came within the range of our guns."

The performance of the San Diego *SCR-682* surface-search radar was summarized after the war as follows (Annex B, 1946):

"SCR-682 (RAD 9-163) located at the tip of Point Loma, Site 7, operates satisfactorily. The set constantly gives good reflections from destroyer targets at ranges up to 50,000 yards. Land reflections have frequently been observed at 240,000 yards and moving targets at 140,000 yards. Increased ranges usually accompany rises in air temperature. All water areas around Point Loma can be observed except that targets inside San Diego Bay cannot usually be separated from the many returns from nearby structures. The Coronado Islands create a dead area from azimuth 351° to 10°, range 28,500 to 36,000 yards, and from azimuth 10° to 20°, range 26,000 to 30,000..."

The wooden signal tower and *SCR-682* radar installation were declared surplus and removed shortly after the war. A steel radio tower (502) was later erected west of and adjacent to what had been the Navy Operations room as shown in Figure 9-27. The shadow of Tower 502 falls upon the northern escape hatch of the underground Battalion One Command Post on the left side of the photo. A post-war radar installation is seen just beyond the Observation Room at photo center. In the close-up image of this radar presented in Figure 9-28, it is apparent that the antenna rotates in azimuth with a fixed elevation, indicating its function was harbor surveillance as opposed to air search.

Figure 9-27. Looking northeast in April 1954, Tower 502 is seen on its new concrete foundation pad just left of center, just west of the Battalion One Command Post bunker (US Navy photo courtesy SSC Pacific).

Figure 9-28. Undated (presumably also 1954) close-up view of the postwar surface-search radar installation at the former Battalion One Command Post (US Navy photo courtesy SSC Pacific).

Given the obvious strategic value of the site, the former Battalion One Command Post was later taken over by the US Coast Guard for continued harbor surveillance. The floorplan of this underground bunker (as of 7 June 1983) is presented in Figure 9-29. Note the observation room at far left, and the radar room just left of image center.

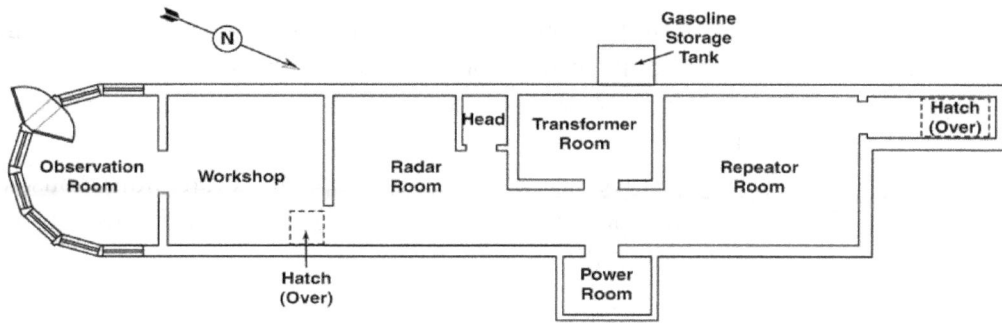

Figure 9-29. Bunker floorplan per 7 June 1983 (redrawn by Lindsay Seligman from US Coast Guard "VHF-FM High Site Point Loma" blueprint 7111-87).

SCR-296-A Radar 9-126 for Battery Strong

One of the first *SCR-296-A* fire-control radars authorized for the Harbor Defenses of San Diego was located within Site 13 at the Mexican border, assigned to the 8-inch Battery Strong, with secondary assignment to Battery Grant. It was brought on line in October of 1943 (HSD, 1945), but not formally transferred to the artillery until 12 January 1945, just 7 months before the formal Japanese surrender on 14 August. Due to this delayed availability, Radar 9-127 (Woodward) was initially used for Battery Strong (Keniston, 1996), until such time as Battery Woodard was transferred.

At least part of the reason for the completion delay of Radar 9-126 (Strong) was the fairly rugged terrain at this remote Site 13 location, which necessitated construction of an access road and installation of a 370-foot incline hoist (Figure 9-30). There was also no commercial power available in this area, which required installation of a 12,000-volt transmission line and a pole-mounted transformer as depicted in the "Location Plan" detail at upper left in the diagram below.

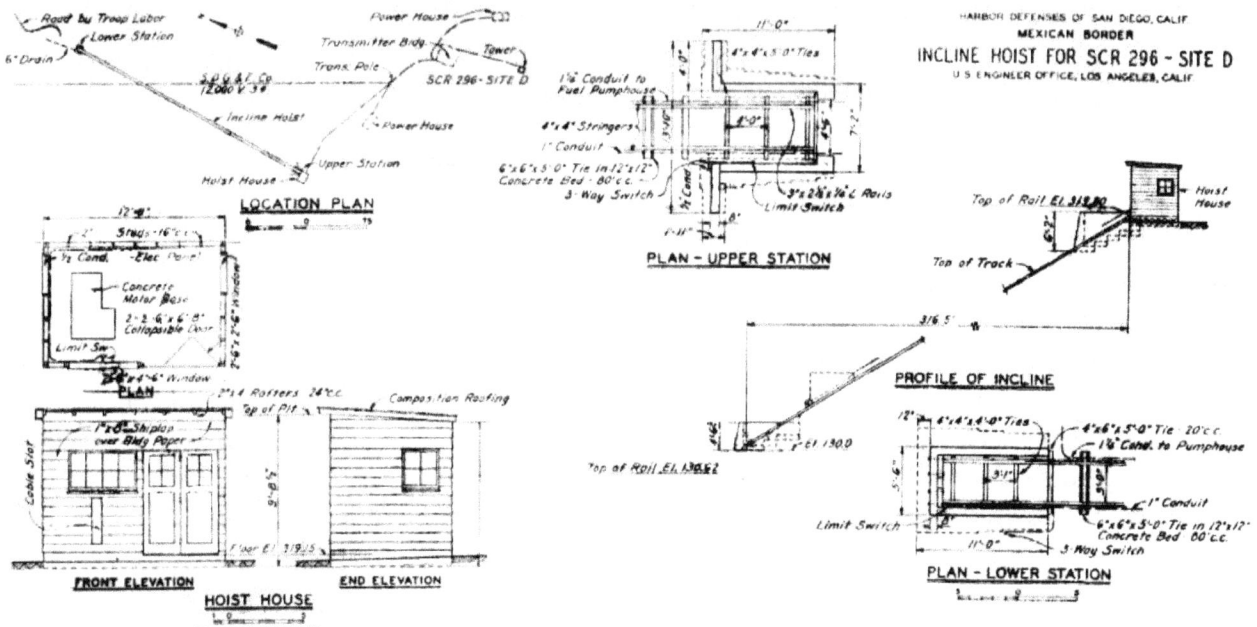

Figure 9-30. An electric incline hoist was installed to facilitate access to the elevated *SCR-296-A* radar installation by the Mexican border in Site 13 (NARA).

Built by Otis Elevator Company, the incline hoist was powered by a 10-HP three-phase electric motor (RCW, 1945a). The inside and top faces of the wooden rails were covered by lengths of 3-inch by 2.5-inch angle iron to provide a wear-resistant mating surface for the flanged wheels of the lift cart. These steel-clad rails were in turn supported by wooden 4x4 ties spaced 4 feet apart, with a 6x6 tie set in 12 inches of concrete every 80 feet for added stability. The track ascended the 30-degree rise along an approximate true bearing of 185 degrees, climbing 190 feet in altitude to the "upper station" hoist-house elevation of 319 feet (Figure 9-31). Electrical wiring for the up/down control and limit switches was enclosed in a 1-inch conduit adjacent to the westernmost rail.

Figure 9-31 a) View of the inclined rail leading to the upper station, looking southeast in 2008. **b)** Todd Everett inspects the remains of the Otis Elevator winch unit and concrete foundation of the hoist house at the upper station. The 370-foot track descended almost due north along a 30-degree grade to the lower station.

The upper-station hoist house was situated just 15 feet northeast of base-end and spotting stations $B^2_{10}S^2_{10}$ and $B^4_3S^4_3$ for Batteries Ashburn and Grant, respectively (Figure 9-32), approximately 225 feet northwest of the *SCR-296-A* antenna tower. This custom-built structure featured 1x8 wooden shiplap sheeting over a 2x4 frame, with a south-sloping composition roof. It could be accessed for equipment inspection and maintenance through a folding wooden door on the north side. Two small windows faced north and west, with a vertical slot for the winch cable centered over the tracks. Remains of the winch spool and worm gear were found by Lee Guidry in the late 1990s at the foot of the incline rail.

Figure 9-32. Remains of counter-balanced entry hatch to $B^4_3 S^4_3$, with the concrete foundations of the upper-station hoist house in background, looking northeast across Goat Canyon towards Spooners Mesa in 2008.

The *SCR-296-A* radar tower was located approximately 115 feet north of Harbor Defense Observation Post 3 (HDOP-3), only 240 feet from the international border as shown in Figure 9-33. The antenna was disguised as a water tank atop a standard 25-foot steel Signal Corps tower at a site elevation of 340 feet, for a total antenna height of 373.8 feet (RCW, 1945b). From this vantage point the radar had a sweeping view of San Diego Bay and approaches from the tip of Point Loma well down the Mexican coast to the south. One area of occlusion was behind the Coronado Islands to the southwest, but most of this region was beyond the range of Battery Strong's 8-inch guns anyway (Annex B, 1946).

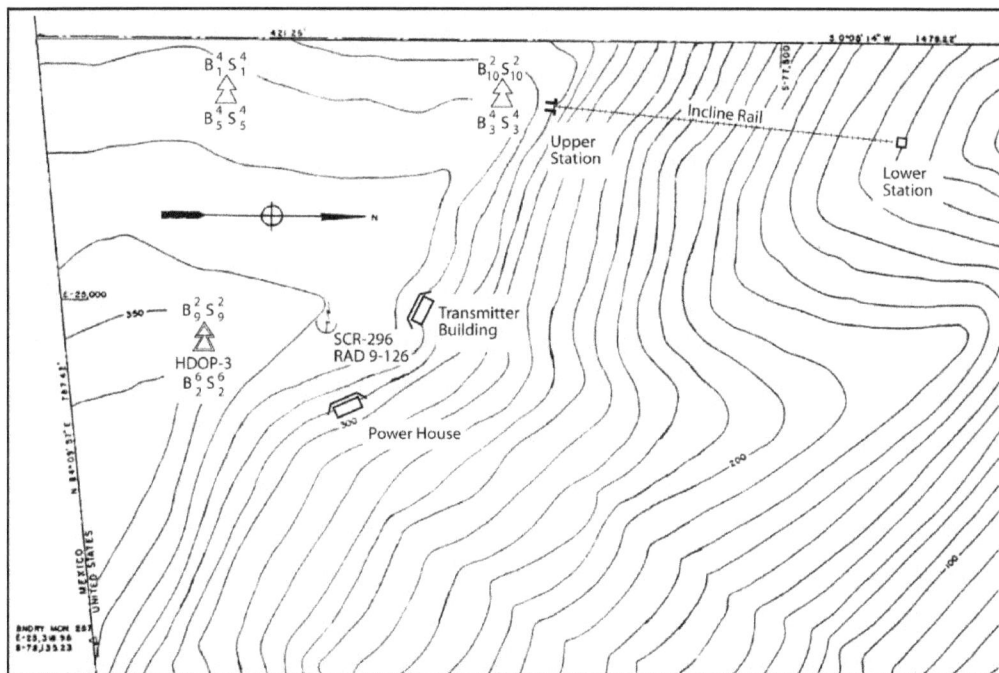

Figure 9-33. Portion of 1 July 1945 map from "Report of Completed Works, Fire Control Stations at Border, Site 13," augmented to show approximate locations of the radar buildings and incline rail (NARA). The *SCR-296-A* radar is due north of *HDOP-3* at an elevation of 340 feet, with the international border at far left.

The transmitter/receiver building was situated approximately 65 feet north of the antenna tower on a small excavation cut into the sheltered northeast side of the ridge (Figure 9-34). A concrete drain gutter was constructed around the excavation perimeter to redirect runoff in the event of rain. A set of wooden stairs on the west end facilitated access to the lower level at an elevation of 320 feet. Remains of the 30-foot utility pole for commercial power were found lying just north of the concrete foundation slab. The power house for the 25-KVA auxiliary generator was similarly emplaced down slope about 50 feet to the east of the tower, serviced by a concrete stairway (see again Figure 9-33).

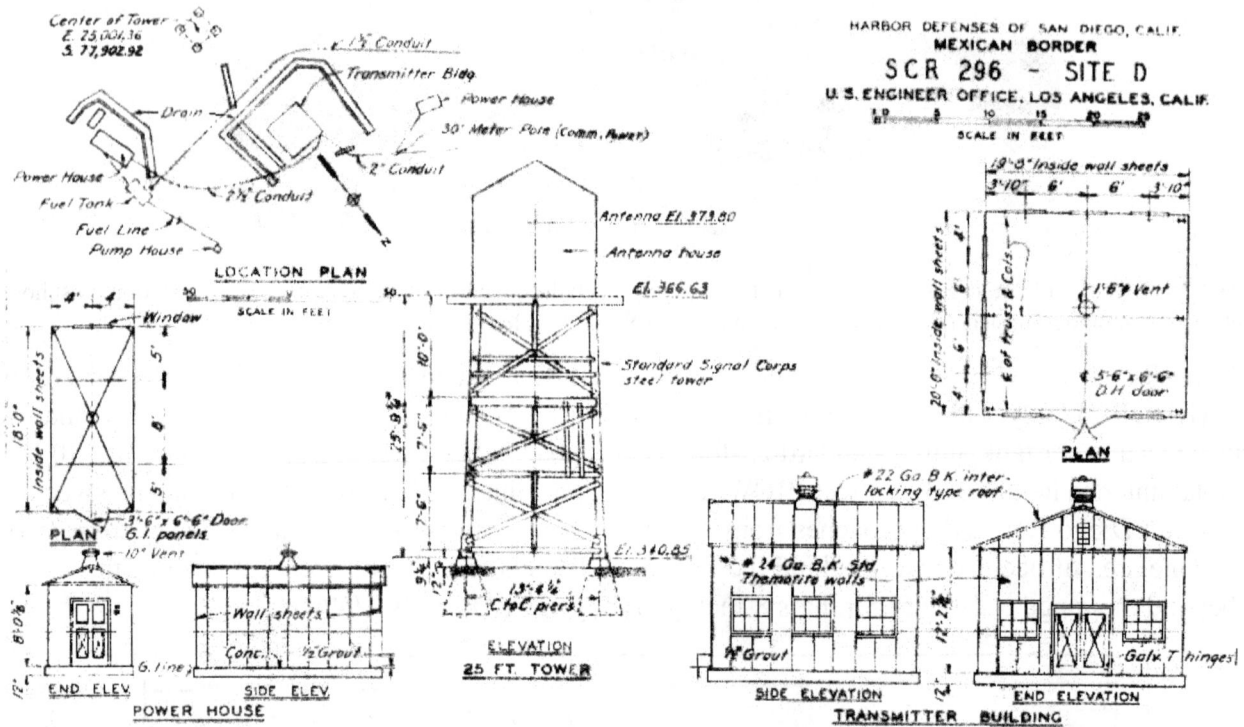

Figure 9-34. Disguised as a water tank, the *SCR-296-A* fire-control radar at Site 13 near the Mexican border was assigned to Battery Strong at Fort Rosecrans (NARA).

The *SCR-296-A* tower-foundation pillars are seen in Figure 9-35. The transmitter/receiver site is shown later in Figure 9-36, just down the northeast slope in the background, with the power house out of photo to the right. The entire compound, including the three base-end stations, was protected by a chain-link fence, the remains of which are still in evidence. The cost of security fencing ($2,538) and getting commercial power extended to this remote location ($2,331) was over 20 percent of the total expenditure ($20,870), not including the additional charge for the incline hoist ($5,198).

Figure 9-35. The four *SCR-296-A* radar-tower foundation pillars are seen looking north across the Tijuana Estuary towards Imperial Beach in March 2008, with the Point Loma peninsula on the far horizon at upper left. The transmitter/receiver building was once located just down the slope at top center (Figure 9-36).

Figure 9-36. Former location of the *SCR-296-A* transmitter/receiver building looking northwest, downslope due north from the tower foundation seen in Figure 9-35. Note exposed pipes at photo center and lower-right corner. No evidence was found of the power-house foundation (see again Location Plan in Figure 9-34).

SCR-296-A Radar 9-127 for Battery Woodward

This *SCR-296-A* installation at Site No. 5 was the one Thompson (1991) referred to as "near Battery Strong in north Fort Rosecrans." The "Harbor Defense of San Diego, Main Cable Routings" blueprint (sheet 2 of 4) dated 1 July 1945 identifies the radar as "*SCR-296, #9-127.*" The U.S. Navy Electronics Laboratory blueprint "Outside Telephone Plant Record Drawing" dated 5 April 1954 further identifies it as the "Woodward Radar." The location is due east and just south of a line defined by utility poles 80, 81, and 82 on the northwest corner of Fort Rosecrans. The tower was located approximately 15 feet northwest of the radar equipment building (Figure 9-37), while the power house for the auxiliary 25-KVA generator was approximately 22 feet to the east.

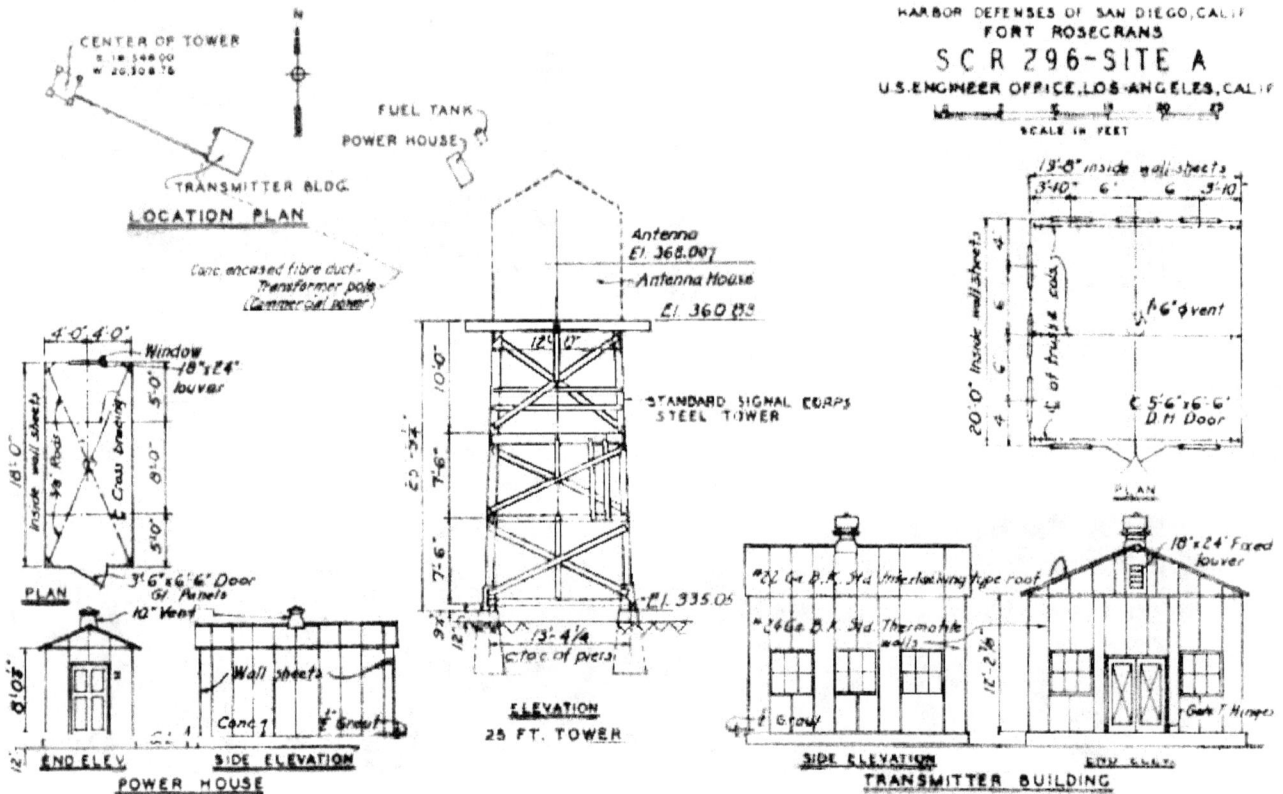

Figure 9-37. The antenna for the *SCR-296-A* fire-control radar assigned to Battery Woodward was situated on a 25-foot steel tower at a terrain elevation of 335 feet (NARA).

This radar was the first to come on line at Fort Rosecrans (Figure 9-38), going "on the air" on 17 February 1943 (HSD, 1945). It initially supported the 7-inch Battery Strong (its secondary assignment), since the 6-inch Battery Woodward was not proof fired until November of that same year (HSD, 1945). It was also the first HDSD radar used for fire control during live target practice, this time with the 5-inch Battery Gillespie on 18 May 1943. "Results compared favorably with those which had previously been obtained by visual methods" (HSD, 1945).

Figure 9-38. The *SCR-296-A* radar tower for Battery Woodward, disguised as a water tower, is seen west of Catalina Boulevard in the upper-right corner (1946 US Army photo courtesy SSC Pacific). See also Figure 9-39. Note the San Pedro radio tower at upper left.

An intriguing statement concerning this radar installation appears in the post-war *Historical Sketch of the Harbor Defenses of San Diego* (HSD, 1945):

"This radar was on the air in time to track the battleship task group which simulated a bombardment of Point Loma as a part of their training for the Aleutian counter-invasion. Together with a similar Navy set at the radar laboratory, complete tracks of the ships' movements was (sic) obtained, although because of fog they were visible for only a short time. It was the first use of radar in such a joint Army-Navy problem."

Keniston (1996), however, suggests this second radar was the Army's *SCR-296-A* fire-control radar for Battery Humphreys, discussed in the next section.

Figure 9-39. Aerial view looking southeast in 1946, showing the transmitter/receiver building immediately behind the radar tower (lower right quadrant), with the much larger San Pedro radio tower in background left. Catalina Boulevard is at extreme upper left (undated US Army photo courtesy SSC Pacific).

This radar was transferred to the Coast Artillery on 15 March 1944, and declared obsolete on 17 January 1946. The Flower and Roth (1982) survey reported the concrete remains of a radar foundation were found 15 feet west (down slope) of the temporary wooden base-end station for Battery Gillespie. In actuality, the radar site is at the crest of the hill just east of this wooden structure, the temporary battery commander station for Batteries North and Gillespie. The rightmost foundation pillar seen in Figure 9-40a is visible from Woodward Road down below. After trimming away some brush, four smaller concrete pillars were found in the center of the foundation layout as seen in Figure 9-40b. This central cluster of pylons was not present at the border-site radar for Battery Strong (see again Figure 9-35).

Figure 9-40. a) The concrete foundation pillars of the Woodward radar, circa 2007, looking southeast. The middle pillar in background center is one of four smaller pillars not shown in the RCW blueprints of Figure 9-37. **b)** Clearing away some brush revealed a cluster of four smaller pillars at the center of the foundation.

Following the war, the northwest corner of Fort Rosecrans became a hotbed of radar research at the U.S. Navy Electronics Lab. This entire area overlooking the Pacific was soon designated as the "Electromagnetic Radiation Equipment RDT&E Range," as seen in Figure 9-41. Battery Woodward (Building F-12) is located just east of the shoreline in map column 27, with Battery Strong in column 24 at extreme lower left. Buildings F-33 and F-36 seen in column 29 would eventually become part of our expanding unmanned systems compound at SSC Pacific.

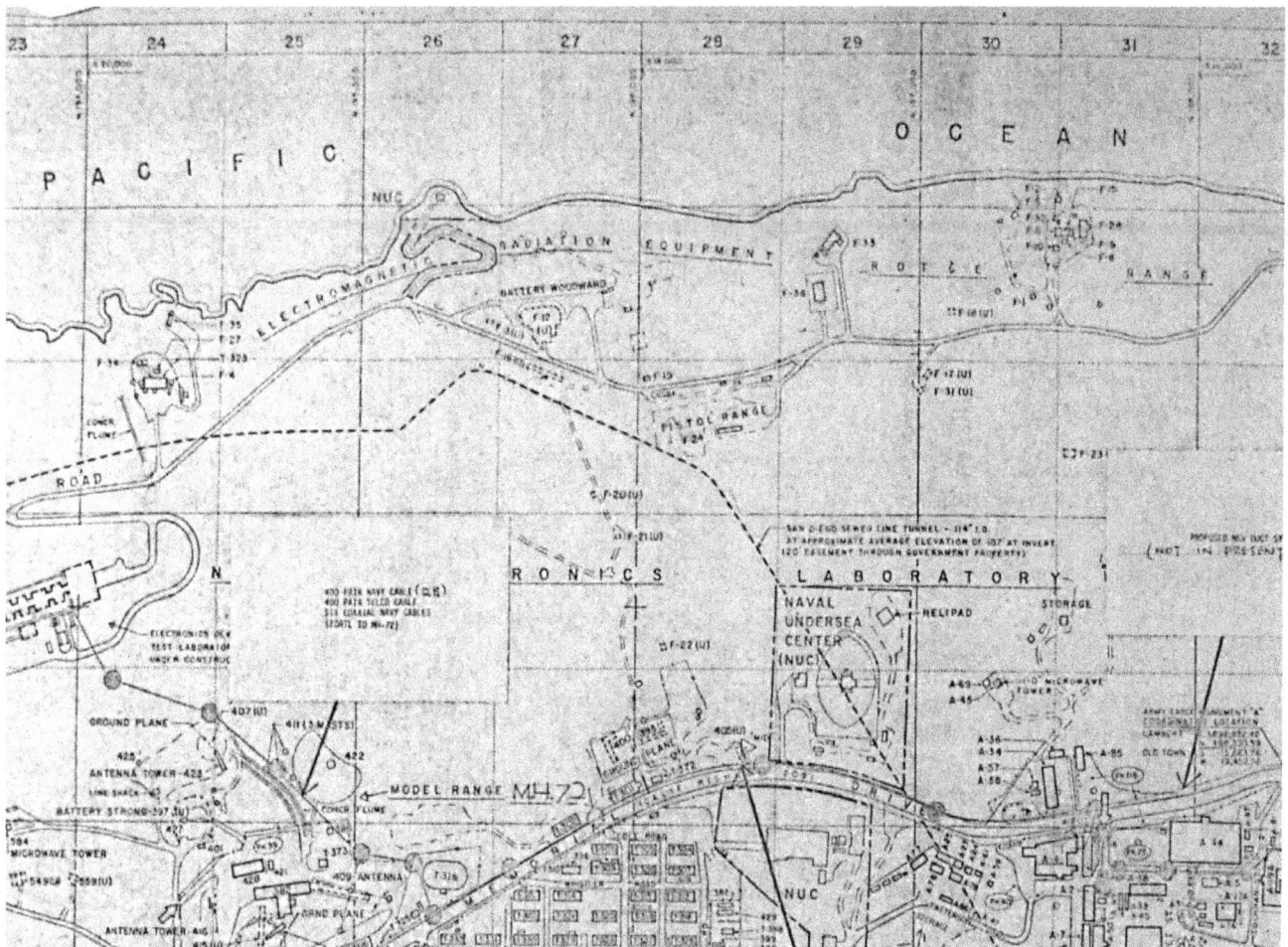

Figure 9-41. Battery Woodward (Building F-12) is just east of the shoreline in map column 27, with Battery Strong in column 24 at extreme lower left. The Unmanned Systems Branch, which initially occupied Building F-36 (upper portion of column 29), expanded to three branches and a total of 12 buildings in what previously had been the "Electromagnetic Radiation Equipment RDT&E Range" at the Navy Electronics Laboratory.

SCR-296-A Radar 9-129 for Battery Humphreys

Another *SCR-296-A* fire-control radar was located within Site 8 near Building 15 at the southern tip of Point Loma, just east of the new lighthouse and down-slope from its assigned Battery Humphreys. Its secondary assignment was the 6-inch Battery Grant at Fort Emory. The "Harbor Defenses of San Diego, Main Cable Routings" blueprint (sheet 3 of 4) dated 1 July 1945 identifies this radar as an "SCR-296, #9-129" as seen in Figure 9-42. It was officially handed over to the Coast Artillery on 4 September 1944 (RCW, 1945b), and declared obsolete just 16 months later on 17 January 1946.

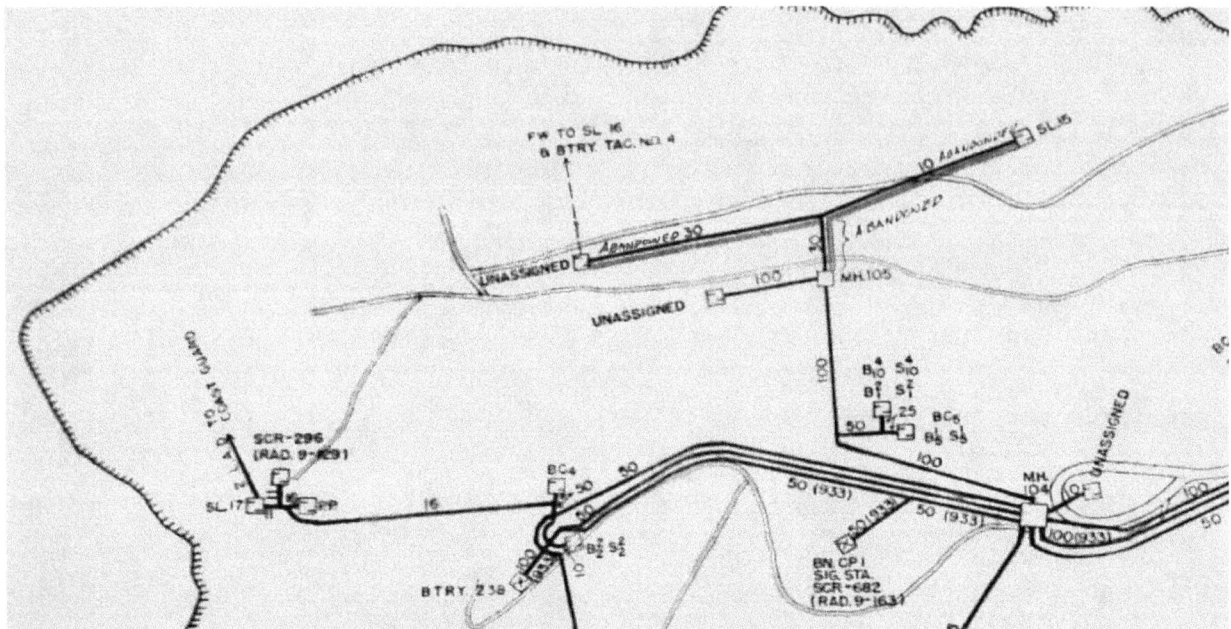

Figure 9-42. The above portion of "HDSD Main Cable Routings" blueprint (sheet 3 of 4), dated 1 July 1945, shows the *SCR-296-A* radar location near Searchlight 17 in the Building 15 area (lower-left corner), northeast of the new Coast Guard light-house (left center) on the southern tip of Point Loma.

Disguised as a water tank, the radar antenna was situated atop a 50-foot Signal Corps tower at a site el-evation of 129 feet, for a total antenna height of 187 feet (RCW, 1945b). The radar equipment was housed in a 24-gauge *Thermotite*™ structure just south of the antenna tower, and the 25-KVA auxiliary generator was located in a similar shelter to the west (Figure 9-43). Commercial power (single-phase 120/240-volt AC) was provided via a subterranean 2.5" duct from an underground transformer vault north of the power house.

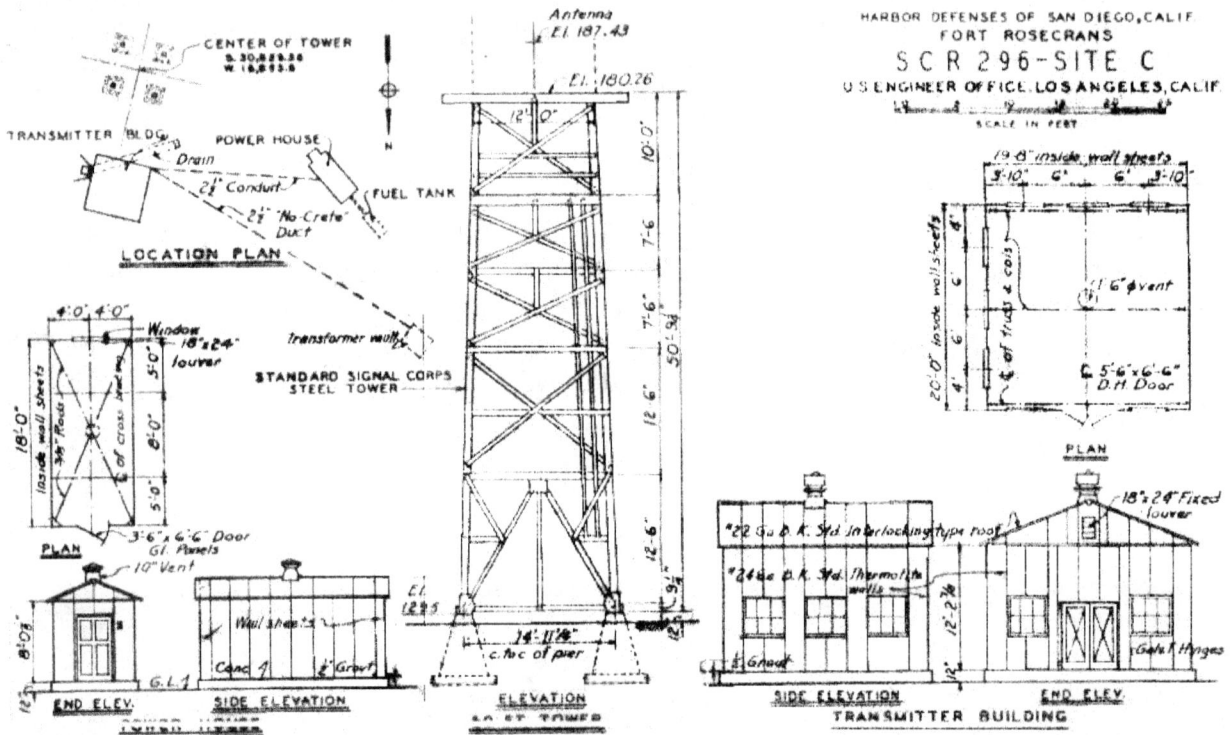

Figure 9-43. The *SCR-296-A* antenna for Battery Humphreys was mounted on a 50-foot tower near Searchlight 17 at the southern tip of Point Loma (NARA).

The transmitter building can be seen in the shadow of the *SCR-296-A* tower at image center in Figure 9-44 below, with a set of wooden stairs extending down the hillside to the smaller power house (down arrow) to the west. Note the small structure on the steep cliff edge at the top of the long stairs in the upper right corner, below which is the operating dock for Searchlight 17 (up arrow). Situated at the very edge of the cliff, this structure may have been replaced after the war by that shown in the foreground of Figure 9-45a.

Figure 9-44. The *SCR-296-A* antenna is at photo center in this 1946 photo, its shadow falling on the transmitter building just to the north. The power house (down arrow) is left of center, with the Searchlight 17 dock (up arrow) on the steep cliff edge at right (US Army photo courtesy SSC Pacific).

Figure 9-45a below shows this former *SCR-296-A* site, looking northwest in October 1985, with the radar building seen in the upper-right corner and one of the tower-foundation pillars at image center. The small wooden structure seen in foreground center (not present in wartime photos of this location or Figure 9-44 above) was apparently used by the Navy after the war to plot the approaches of ships entering the harbor. Figure 9-45b shows a Navy maneuvering board that was found hanging on the back wall of this structure in 2006.

Figure 9-45. a) The *Thermotite*™ radar building (upper right) is still standing in this October 1985 photo, north of the concrete tower-foundation pillar at image center. **b)** In 2006, a Navy maneuvering board was found hanging inside the wooden structure seen in foreground of Figure 9-45a (photo courtesy Jeff Bowen).

There is post-war photographic evidence that the *SCR-296-A* radar at this location had been upgraded to an *AN/MPG-1* set.[17]　Relative to the *SCR-296-A* radar it replaced, the *AN/MPG-1* provided a significant increase in performance, especially when used in conjunction with the *Director M-8* (Orman, 1946). Maximum search range on large targets (i.e., cruisers or above) was 50,000 yards, while tracking could be accomplished out to 28,000 yards, with an accuracy in the latter case of ±10 yards in range and ±0.1 degree in azimuth.　A later upgrade of the set's range unit yielded the *AN/FPG-2*, which could track out to 50,000 yards to accommodate the maximum range of the 16-inch gun batteries (Clement, 1948).

An undated print for "Building 15 Area Communication Lines" (Figure 9-46) shows a termination box for the following stations:　1) Active Sonar, 2) Searchlight, 3) Ploting (sic) Table, 4) MPG-1 Radar, and 5) Bldg. 15 East End.　The plotting table may have been inside the small wooden structure seen in the foreground of Figure 9-45a.　The *1946 Annex* summarizing assets in place at the conclusion of the war lists the original *SCR-296-A* at this site, with no mention of an *MPG-1* radar.　Clement (1948) reported: "When the *AN/MPG-1* was under development, it was contemplated that most of the sets would be transferred to the harbor defenses after the war."

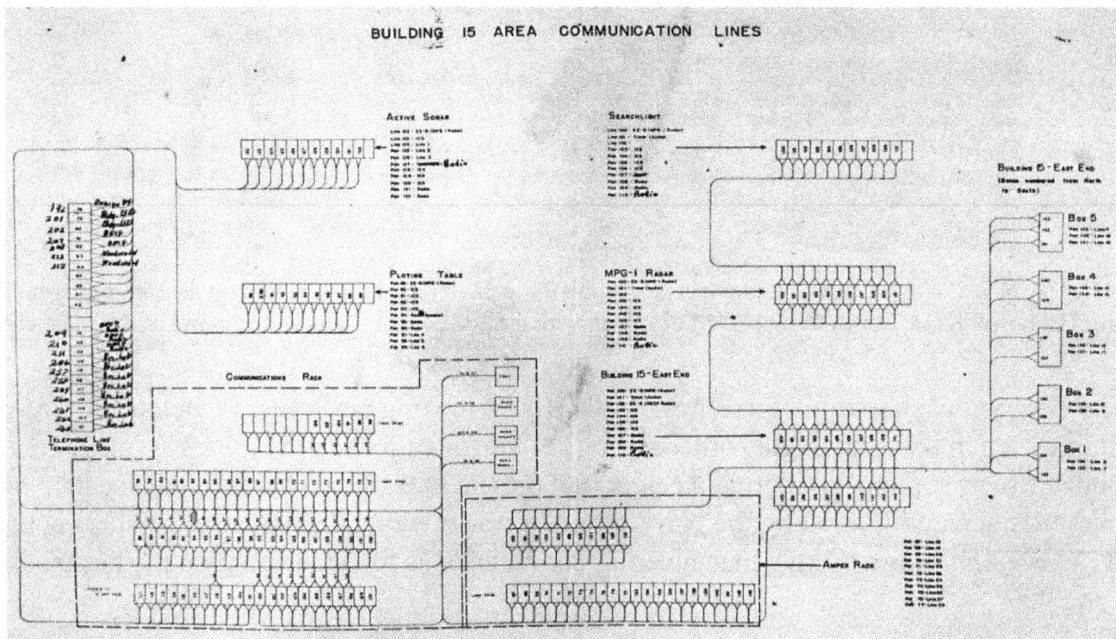

Figure 9-46. This undated print of the "Building 15 Area Communication Lines" shows a termination strip for an "MPG-1 Radar" just right of center.　Note also the termination for "Active Sonar" at upper left.

Additional insight is provided in paragraph 5 of a U.S. Navy Electronics Laboratory memorandum requesting additional telephone lines for their Harbor Defense Section (Fisher, 1951):

"These telephone lines are needed for the prosecution of BuShips assigned problem 6M1. The individual sections of this communication network are primarily needed for the following reasons:

a. The 100 pr. from Bldg. 15 to MH-205 to be used principally for transmission of data from the underwater detection center to the HECP which will be located in Battery 235.

b. 75 pr. of this group will tie Bldg. 15 to the HECP.

c. 30 pr. from the radar center, BNCP-1, will tie into the lines to the HECP for transmittal of radar data, infrared data, etc.

d. The remaining 25 pr. of the original 100 pr. will connect Bldg. 15 to the main frame Bldg. 4. Part of these lines will transmit evaluation data for the Human Factors Division personnel in Bldg. 321.

17　This Army-Navy (AN) equipment nomenclature replaced the earlier Set Complete Radio (SCR) series.

e. The 20 pr. terminating in unassigned sighting stations just to the east of BNCP-1 will serve to bring information from a sonar ship's noise monitor to be installed near the ship traffic channel.

f. The lines terminating in the North Island sighting stations are to be used for photo-theodolite station synchronization and communications lines.

g. The 12 pr. terminating in the recording laboratory, Bldg. 17, will be used for recording data for special tests of units now installed at sea, their terminating equipment being installed at Bldg. 15."

Note the statement in subparagraph "a" above indicating the Harbor Entrance Command Post (HECP) "will be located in Battery 235." This was clearly an error, in that Battery 235 was at Fort San Jacinto in Galveston, TX, and not at Fort Rosecrans. In addition, paragraph 4 of this same memorandum makes reference instead to "Battery 237" versus "Battery 235" regarding the use of existing Army telephone cables. Battery 237 is also cited in additional memorandums, notes, and the schematic diagram seen in Figure 9-47, all of which suggest the Navy's post-war HECP was at least initially in Battery Woodward (BCN 237).

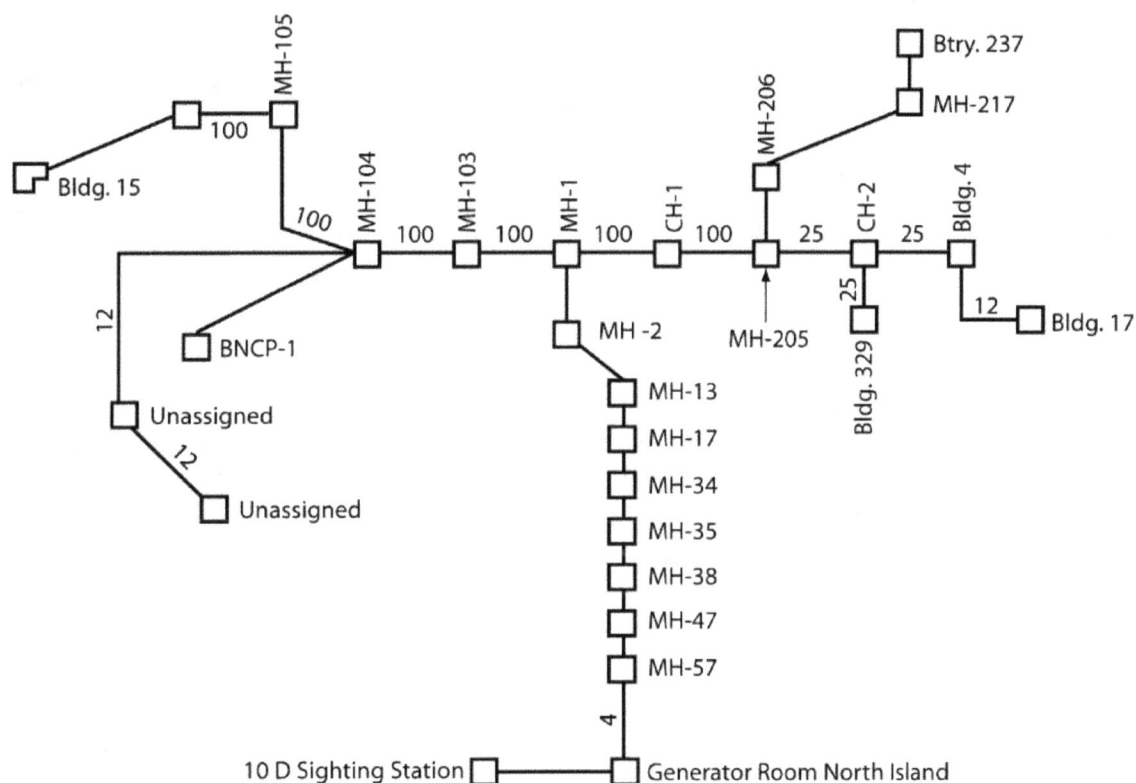

Figure 9-47. Undated schematic (circa 1951) of the communications network requested by the Harbor Defense Section of the US Navy Electronics Laboratory (Fisher, 1951), redrawn for improved legibility. Battery Woodward (237) is depicted at upper right, with Building 15 at upper left (see also Figure 9-48a).

The "underwater detection center" mentioned in subparagraph "a" of the Fisher memorandum above is probably Building 15 of the U.S. Navy Electronics Lab, located on the edge of the cliff at the very southern tip of Point Loma (Figure 9-48a). Note the numerous cables running down the cliff face, presumably to an underwater acoustic-detection array. The earlier WWII radar building is directly behind the tower in the upper-right corner. Figure 9-48b provides a better view of this post-war AN/MPG-1 radar, with the shelter/powerhouse for Searchlight 17 at center right. Note what appears to be operational radar antennae on the flatbed trailer in the bottom-right corner.

Figure 9-48. a) Numerous cables are seen descending down the cliff edge from Building 15 (left of center), looking north in April 1954. **b)** This closer view better shows the *AN/MPG-1* radar on its grid-like tower support in front of the original *SCR-296-A* radar building (US Navy photos courtesy SSC Pacific). Note the shelter and powerhouse for Searchlight 17 in the upper-right corner.

The standard grid-like *AN/MPG-1* tower base seen in Figure 9-48b above allowed it to be installed upon the existing concrete pillars of the earlier *SCR-296-A* radar foundation. Figure 9-49, a rather hazy photo of the southern tip of Point Loma shot from some distance to the south in 1969, shows this tower-mounted *AN/MPG-1* radar antenna was still present in the Building 15 area. The large white structure seen behind and to the right of this tower is the original WWII radar building, while the small white rectangle just to its right is the Navy's post-war plotting shack shown earlier in the foreground of Figure 9-45a, which is also visible in the upper-right corner of Figure 9-48a above.

Figure 9-49. The post-war *AN/MPG-1* radar installation is up slope next to the original *SCR-296-A* radar building, right of Building 15 (large white structure just right of center) in this May 1969 photograph (US Navy photo courtesy SSC Pacific). The new Coast Guard lighthouse compound is at far left.

Only two of the four original *SCR-296-A* radar pylons are still in existence (Flower & Roth, 1982): "Two large concrete pillars located within the center of the present compound were radar mounts." According to personnel working at this Building 15 compound in 2006, the original radar-foundation pillars had been repurposed for a large three-legged Navy radio tower built well after the war, but one of these pillars was later removed to facilitate further construction (Figure 9-50). Note the more-recent harbor-surveillance radar tower near the former Gun 1 position of Battery Humphreys in background at upper left.

Figure 9-50. According to site personnel, the two remaining concrete tower foundations shown here near Building 15 in April 2006 were used for a three-legged radio tower built after the *AN/MPG-1* radar was removed. Note the newer harbor-surveillance-radar tower by Battery Humphreys in the upper-left corner.

Figure 9-51 below shows this same area from a more easterly vantage point, looking north, with the harbor-surveillance radar in the Battery Humphreys compound in the upper-left corner. The open doorway in the background at lower left leads to the former shelter and powerhouse for Searchlight 17. The small wooden building atop the concrete-covered hillside in foreground right is the previously mentioned plotting shack used years ago by the U.S. Navy Electronics Lab in their post-war harbor-defense role. The Navy Maneuvering Board found inside this structure, shown earlier in Figure 7-45b, was used to plot the position of surface contacts in support of this mission.

Figure 9-51. As seen in April 2006, the hillside at right has been covered with concrete to control erosion, with the U.S. Navy Electronics Lab's post-war plotting shack still standing. The entrance to the former WWII shelter and powerhouse for Searchlight 17 is at lower left.

SCR-296-A Radar 9-1007 for Battery Grant

The combined base-end station and *SCR-296-A* radar tower for Battery Grant (Tac. No. 10) was situated at an elevation of 29 feet in Site 12 at Fort Emory. The multi-purpose tower also supported a stacked pair of cylindrical steel enclosures, with the upper level serving as the battery commander station and the lower level as the base-end and spotting station. These fire-control structures were originally designated BC_4 and $B^1_4 S^1_4$ (Figure 9-52), later changed to BC_7 and $B^1_7 S^1_7$ (RCW, 1944a), and ultimately finalized as BC_{10} and $B^1_{10} S^1_{10}$. The radar antenna was located at the very top of the tower at an elevation of 146.8 feet, with the entire affair meant to appear as an elevated water tank.

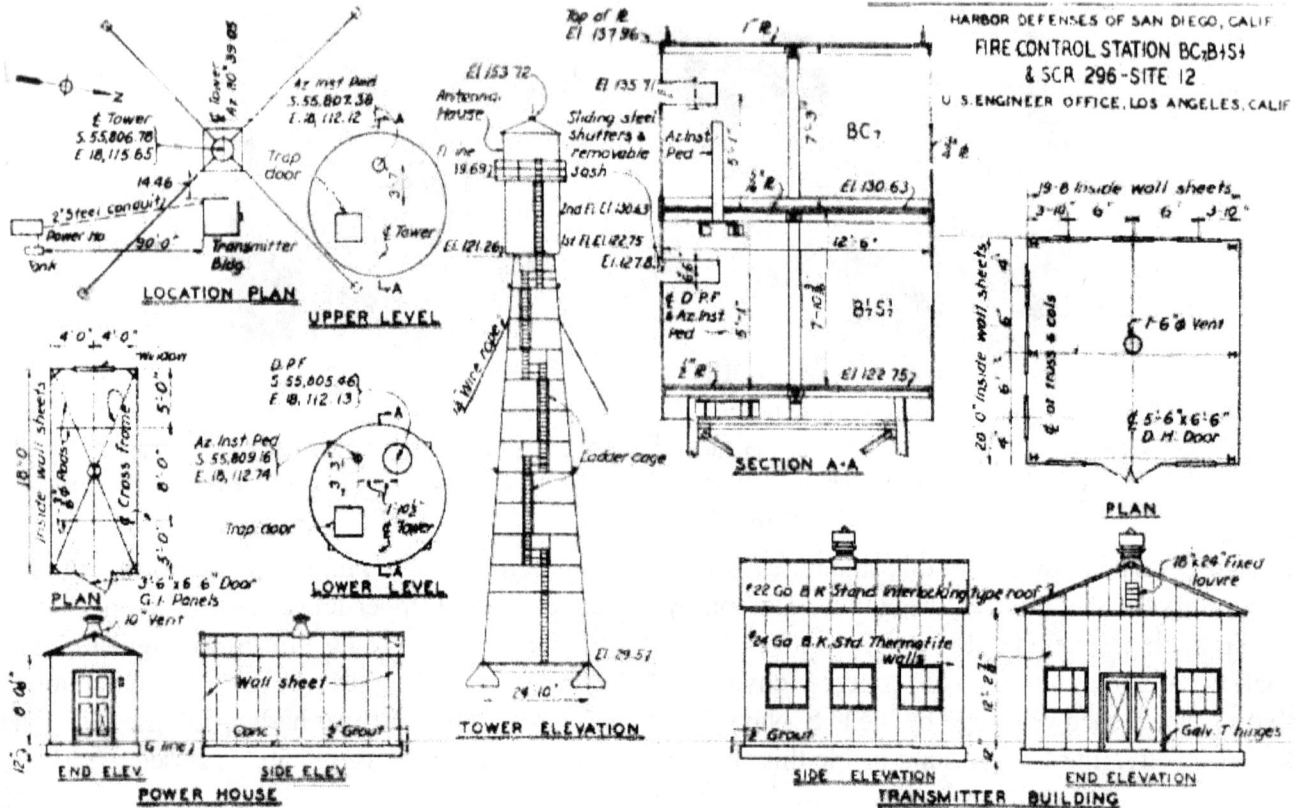

Figure 9-52. The *SCR-296-A* antenna for Battery Grant was mounted at the top of a two-level steel tower that also supported BC_4 and $B^1_4 S^1_4$, later BC_7 and $B^1_7 S^1_7$, and ultimately BC_{10} and $B^1_{10} S^1_{10}$ (NARA).

Battery Grant's *SCR-296-A* radar tower in Site 12 is seen in the foreground of Figure 3-53 below, circa 1943. The two *Thermotite*™ shelters for the radar equipment and 25-KVA auxiliary generator have apparently not yet been constructed. This facility went on the air in June of 1944 (HSD, 1945), and was officially transferred to the Coast Artillery on 4 October 1944, 6 months after Battery Grant came on line in late April of that same year. The secondary assignment for this radar was Battery Humphreys (Tac. No. 5).

Figure 9-53. Scanned photocopy of a microfiche image showing the three 100-foot fire-control towers at Site 12 just east of the rail tracks, looking north, with the *SCR- 296-A* radar antenna housing at the very top of the nearest structure at photo center (US Army photo adapted from Wolfgast, 1943).

The 1 July 1945 HDSD map depicting "Fire Control Stations at Fort Emory, Site 12," presented in Figure 9-54 below shows the 100-foot tower location for this *SCR-296-A (9-1007)* radar in the lower-left corner. The 6-inch Battery Grant is seen directly above this tower to the west. This multi-level tower installation also included battery commander station BC_{10} and base end and spotting station B^1_{10} S^1_{10} for Battery Grant (see again Figure 9-53). The 16-inch Battery Gatchell (134) is at image center to the north of Battery Grant.

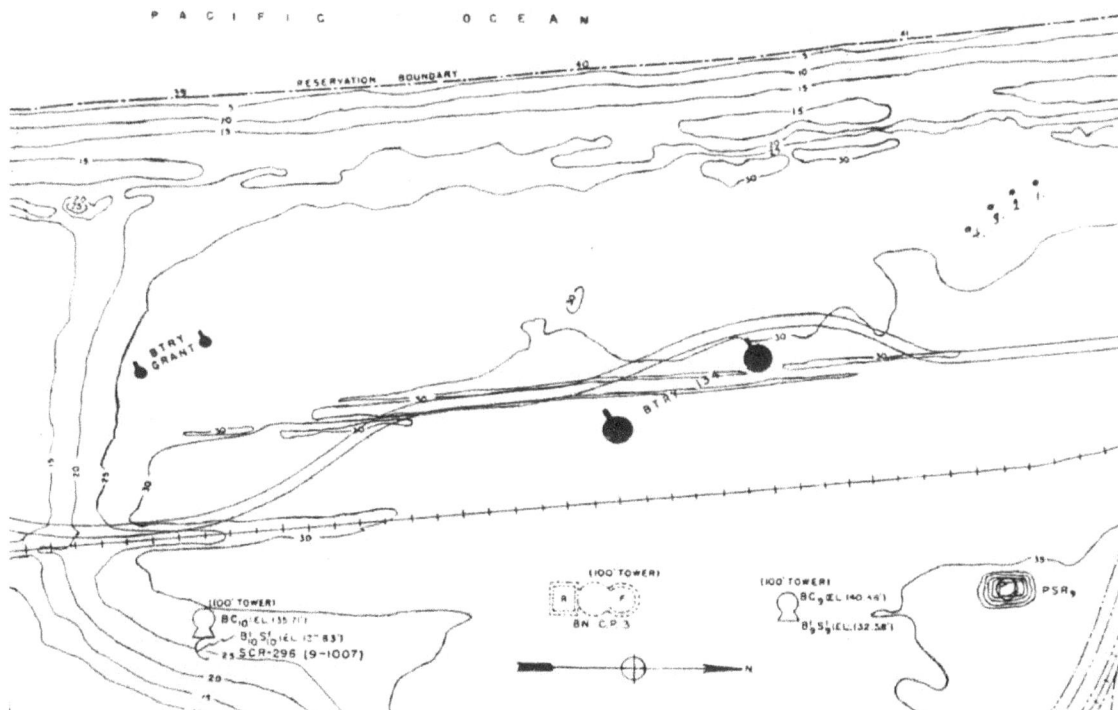

Figure 9-54. The three towers east of Batteries Grant and Gatchell (134) are seen along the lower portion of a 1 July 1945 map of "Fire Control Stations at Fort Emory, Site 12," with the *SCR-296-A* radar in the lower-left corner (NARA). Note railway shown earlier in Figure 9-53 running north-south just west of the towers.

Ernie C. Johnson, who served as a staff sergeant in the range-section plotting room for Battery Grant, describes an interesting early test of this radar's ability to effectively provide fire-control data, held before an audience of "all the big shots from all up and down the coast" (Overton, 1993). As the demonstration began to unfold, the radar range data proved fairly stable, but unbeknownst to the assembled VIPs, "azimuth was jumping all over the place." The radar azimuth theoretically should have been identical to that visually obtained from the Warner-Swayze telescope operator in base end and spotting station B^1_{10} S^1_{10}, since the latter was directly beneath the radar in the same tower.

As further recalled by Johnson (Overton, 1993):

"Well, I was a'sweatin cause the captain was on the phone and I was on the phone in the process and he was all over me because we couldn't get a satisfactory course. I reached up to the PBX switchboard, I switched two switches, which sent visual azimuth down to the arm setter and the radar azimuth down to the guy that was recording. Well, visual azimuth is accurate, absolutely. The course straightened right out, see. I told the captain the course was satisfactory. We went to fire and we had a dandy target practice. They decided, well, radar is the way to go. Had they known that, they'd court martialed me."

SCR-296-A Radar 9-1008 for Battery Ashburn

The *SCR-296-A* fire-control radar for Battery Ashburn was located on an almost 5-acre tract of undeveloped land (Figure 9-55), assessed at the time at a mere $6,000, in "an exclusive residential district" known as Hermosa at Site No. 3 in La Jolla (Annex H, 1946). The antenna was mounted inside a wooden shelter on a 100-foot Signal Corps tower (*Type TR-18*), with the entire configuration disguised as a water tank.

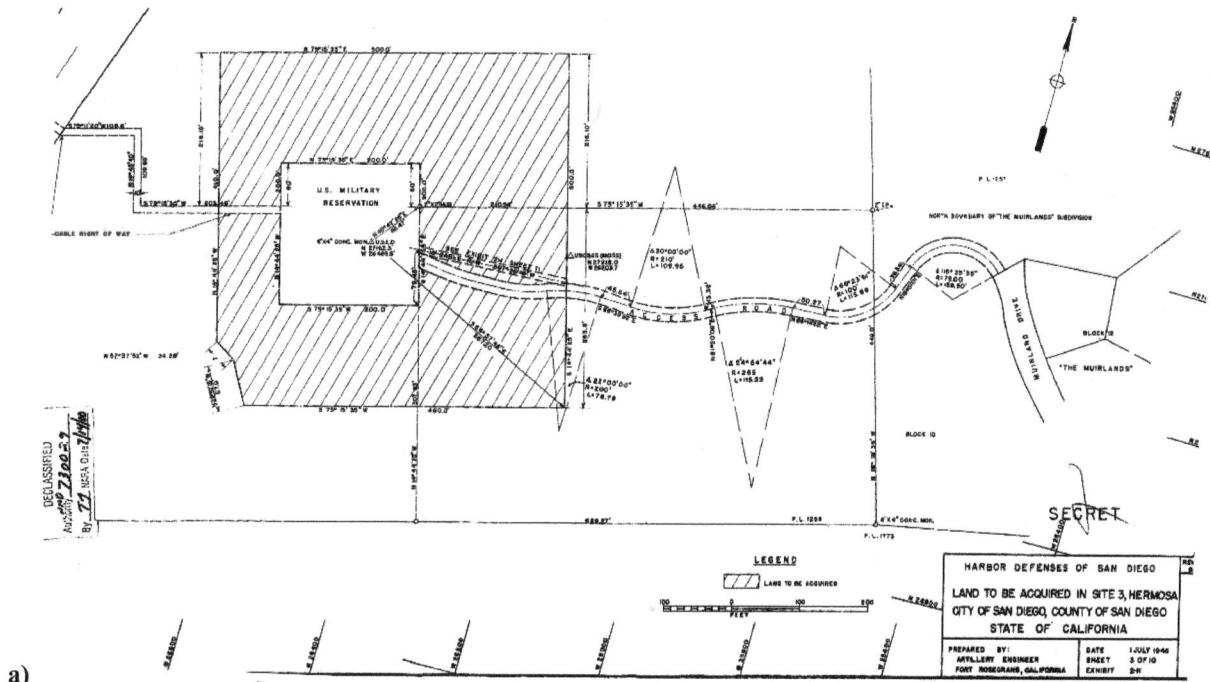

a)

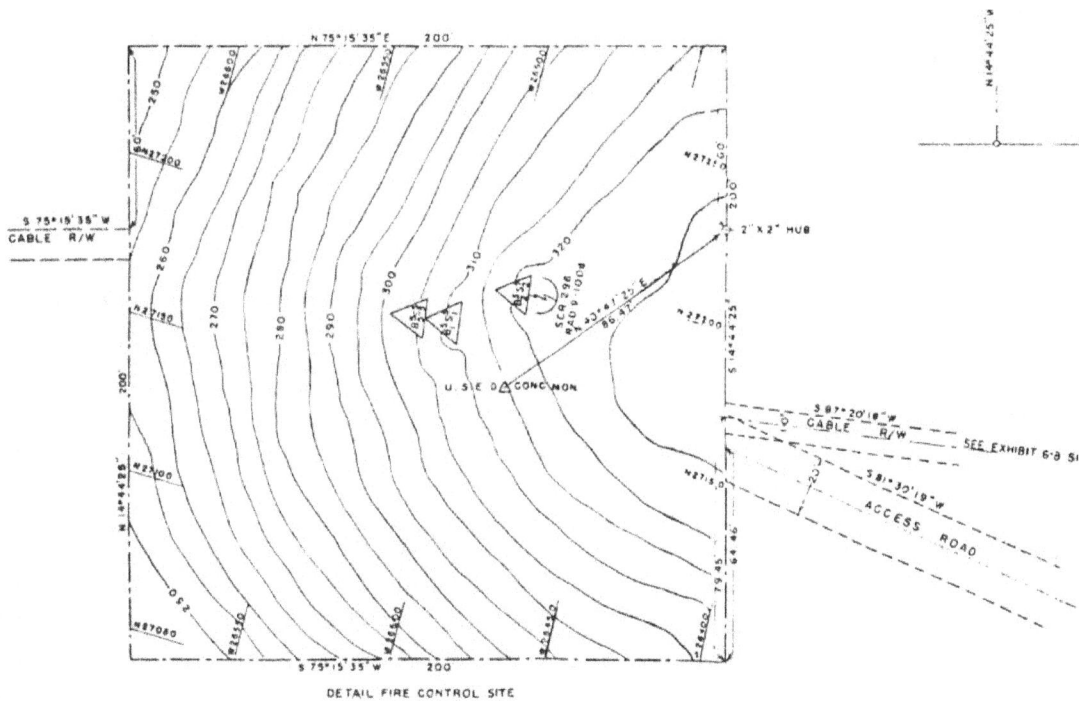

b)

Figure 9-55. a) The *SCR-296-A* fire-control radar installation at Site 3 (Hermosa) in La Jolla was accessed from Muirlands Drive at far right (NARA). **b)** Detail map of the *SCR-296-A* radar installation (right of center) at the Hermosa compound (NARA).

The two 24-gauge *Thermotite*™ shelters for the transmitter/receiver equipment and the 25-KVA emergency generator were designed to look like nondescript beach cottages (Figure 9-56). The facility came on line in March of 1944 (HSD, 1945), and was officially transferred on 14 September of that same year, with its secondary assignment being Battery Woodward (Tac. No. 1). The steel tower and radar equipment were declared obsolete on 17 January 1946 and subsequently removed, although the buildings themselves were temporarily retained (RCW, 1944b).

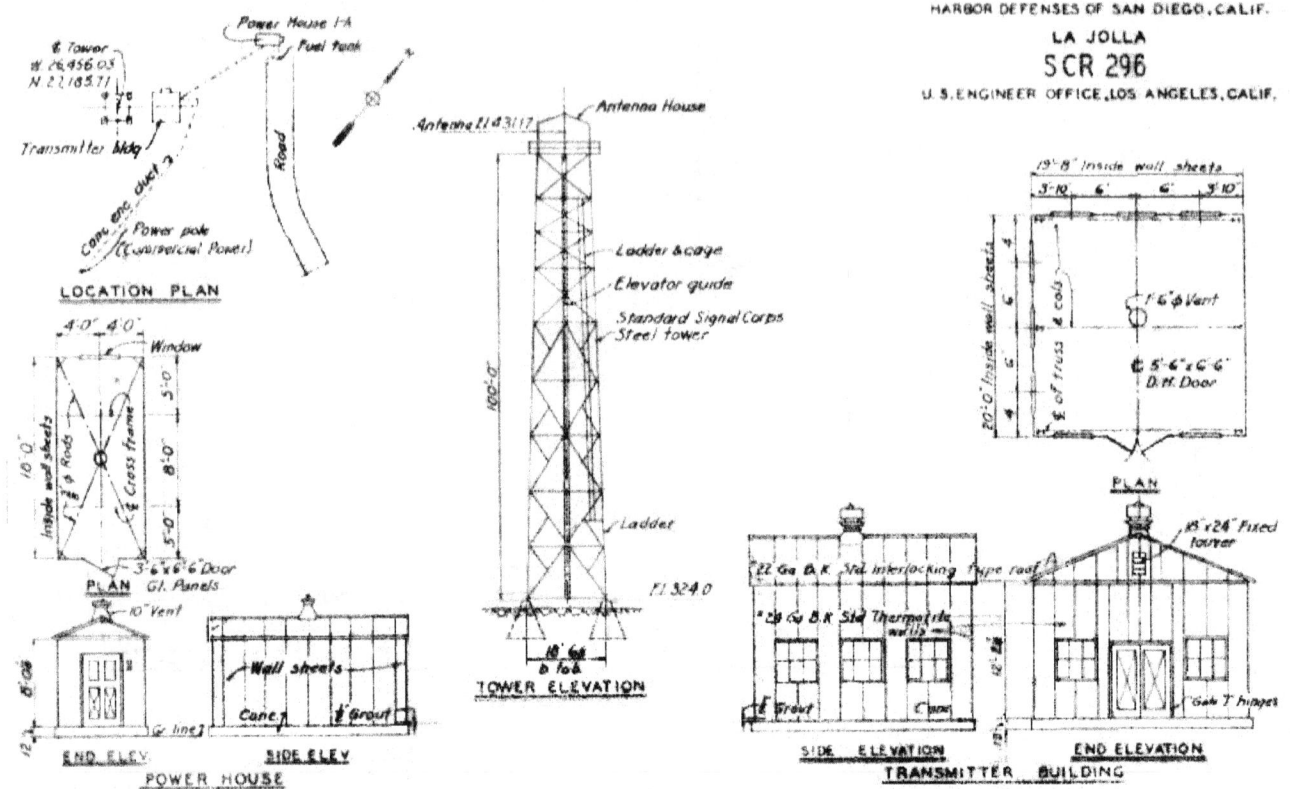

Figure 9-56. The antenna seen at the center of this drawing for Battery Ashburn's *SCR-296-A* installation in La Jolla (Hermosa) was disguised as yet another water tower (NARA).

SCR-296-A Radar 9-1009 for Battery Construction 134 (Gatchell)

Following severe Japanese losses at Midway in June of 1941, the threat of a west-coast attack on the United States was greatly diminished, and continued to decline as the Pacific war progressed favorably for the Allies. In February 1944, the War Department ordered work on some parts of the Harbor Defense San Diego modernization project deferred, to include mounting of the guns and carriages as well as installation of the director and power plant on the otherwise completed 16-inch Battery Construction 134 (HSD, 1945).

As a consequence, Battery Gatchell's *SCR-296-A* fire-control radar (9-1009) was never installed at its planned location adjacent to Battalion 2 Command Post in Site 5 at Fort Rosecrans (Figure 9-57). This same position on the west side of Catalina Boulevard was apparently later used for some type of radar equipment after the war, however, circa 1951. The 22 March 1954 U.S. Navy Electronics Laboratory "Outside Telephone Plant Record Drawing" (sheet 2 of 5) shows "Radar Mount A28 1-51" just south of "Cable Hut No. 2" which was adjacent to the Battalion 2 Command Post.

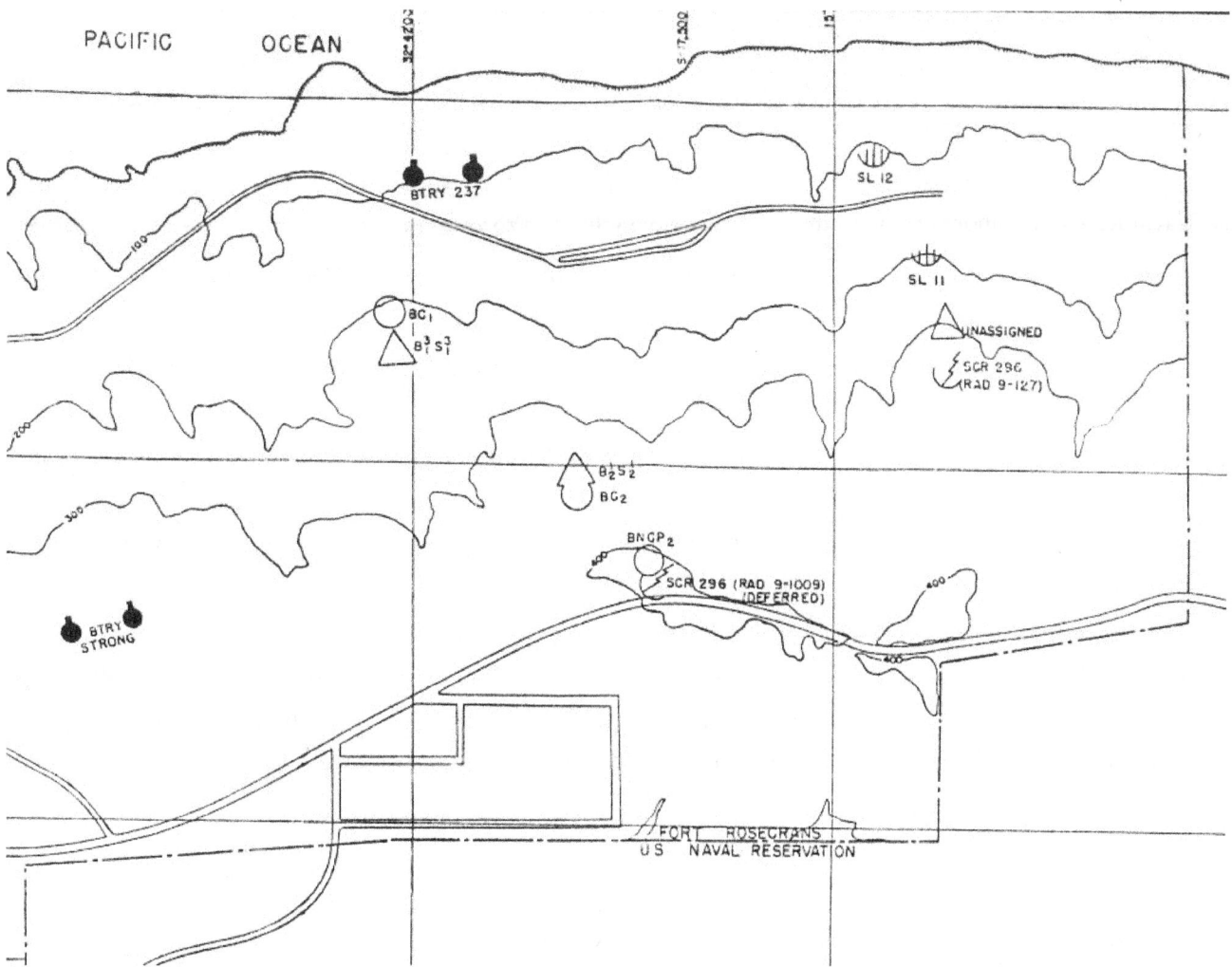

Figure 9-57. The 1 July 1945 "Fire Control Stations at North Fort Rosecrans" map (sheet 5 of 15) identifies the radar just south of the Battalion 2 Command Post (bottom center) in Site 5 as "*SCR-296* RAD 9-1009 (DEFERRED)". A post-war radar was reportedly situated in this same location in 1951.

References

Abbot, F.V., "Searchlights," *Supplement to Mimeograph No. 39*, Office of the Chief of Engineers, United States Army, 17 December, 1902.

Allison, D.K., *New Eye for the Navy: The Origin of Radar at the Naval Research Laboratory*, NRL Report 8466, Naval Research Laboratory, Washington, DC, 29 September, 1981.

Annex B, "Annex B, Fire Control," in *Supplement to the Harbor Defense Project, Harbor Defenses of San Diego*, HDSC-AN-45, 30 April 1946.

Annex C, "Annex C, Searchlights," in *Supplement to the Harbor Defense Project, Harbor Defenses of San Diego*, HDSC-AN-45, 30 April, 1946.

Annex H, "Annex H, Real Estate Required," in *Supplement to the Harbor Defense Project, Harbor Defenses of San Diego*, HDSC-AN-45, 30 April 1946.

Annex, "1946 Annex," *Supplement to the Harbor Defense Project, Harbor Defenses of San Diego*, HDSD-AN-45, prepared 1 July 1945, distributed 30 April, 1946.

AP, "A Photo-History of Atlas Precursors (PDF)," Vol. 0, p. 23, https://forum.nasaspaceflight.com/index.php?action=dlattach;topic=26915.0;attach=829469, retrieved 12 April, 2017.

Apple, R.M, Van Wormer, S. and Cleland, J.H., *Historic and Archeological Resources Protection Plan for the Point Loma Naval Complex*, KEA Environmental, Inc., San Diego, CA, December, 1995.

ASL, *Artic Submarine Lab*, "Timeline," https://www.public.navy.mil/subfor/uwdc/asl/Pages/default.aspx, retrieved 19 October, 2019.

ASM, "Long Range Communications," US Army Signal Museum, Fort Gordon, GA, http://www.gordon.army.mil/ocos/museum/AMC/radio2.asp, retrieved 25 August, 2007.

Axelson, D., "World War II Era Bunker to Be Demolished on New Navy Coastal Campus," *Eagle & Times*, http://www.imperialbeachnewsca.com/news/article_65b8f6da-c15d-11e5-a77c-63455fe9a363.html, 22 January, 2016.

Bauer, A.O., "HF/DF: An Allied Weapon Against German U-boats, 1939 - 1945," http://jproc.ca/rrp/hfdf1998.pdf, Diemen, Netherlands, 27 December, 2004.

Berhow, M., "The Battery Humphreys Accident of January 29, 1944 at Fort Rosecrans, San Diego, California, as Remembered by 1st Lt. Hazen White," Coast Defense Study Group Journal, Vol. 8., No. 4, pp. 9-10, November, 1994.

Berhow, M., *American Seacoast Defenses, A Reference Guide*, Coast Defense Study Group Press, 2004.

Berhow, M., e-mail correspondence, 3 January, 2008.

Berhow, M., e-mail correspondence, 4 April, 2020.

Berhow, M.A, "Book Manuscript," e-mail correspondence, 20 April, 2020.

Bishop, E. C., *Prints in the Sand: The U.S. Coast Guard Beach Patrol During World War II*, ISBN 0-929521-22-6, Pictorial Histories Publishing Co., Missoula, MT, 1989.

BNP, *Naval Sonar*, "SOFAR, Harbor Defense, and Other Sonar Systems," Ch. 16, pp. 288-289, NAVPERS 10884, Bureau of Naval Personnel, Washington, DC, 1953

Bogart, C.H., *Controlled Mines: A History of Their Use by the United States*, Merriam Press, Bennington, VT, 2008.

Boyd, C., and Yoshida, A., *The Japanese Submarine Force and World War II*, ISBN 1-55750-15-, Naval Institute Press, Annapolis, MD, 1995.

Bragg, S., "The SCR-584 Radar Tribute Page," https://web.archive.org/web/20051125124544/http://www.ham-hud.net/darts/scr584.html, 30 June, 2005.

Breur, W.B., *Secret Weapons of World War II*, ISBN 0-471-20212-6, John Wiley, New York, NY, 2000.

Brown, L., *Technical and Military Imperatives: A Radar History of World War II*, Taylor and Francis, New York, NY, 1999

BTDF, "Bellini-Tosi Medium Frequency Direction Finder," http://www.airwaysmuseum.com/B-T%2MFDF%201.htm, retrieved 13 November, 2019.

BWU, "Japanese Unit 731: Biological Warfare Unit," *World War II in the Pacific*, http://www.ww2Pacific.com/unit731.html, 21 January, 2001.

CAJ, "A Design for 155-mm Gun Emplacements," *Coast Artillery Journal*, Vol. 71, No. 4, 1929.

Calderwood, T.D., "Highlights of NRL's First 75 Years," Naval Research Laboratory, Washington, DC, http://www.nrl.navy.mil/content.php?P=HIGHLIGHTSOF75TH, pp. 6-7, October, 1998.

Carey & Co., *Historic Structures Report for Harbor Defense Structures*, 460 Bush Street, San Francisco, CA, www.cr.nps.gov/history/online_books/cabr3/index.htm, updated 6 April, 2005.

Carey, "Cabrillo Historic Structures Report for Harbor Defense Structures," available online at: http://www.cr.nps.gov/history/online_books/cabr3/index.htm, Prepared for National Parks Service by Carey and Co. Inc., San Francisco, CA, April 15, 2000.

CDSG, "Coast Artillery: Fire Control," *Coast Defense Study Group*, https://cdsg.org/coast-artillery-fire-control/, retrieved 12 July, 2019.

CHA, *The Military of Point Loma*, Cabrillo Festival Historic Seminar, Cabrillo Historical Association, San Diego, CA, 1985.

Chatelain, G.P., "Former Army Tactical Cable Mount Soledad to Mexican Border; comments on," letter from Officer in Charge, US Naval Harbor Defense Unit San Diego, US Naval Base, San Diego 52, CA, to Commandant, Eleventh Naval District, Ser: 169, 11 September, 1956.

Cheney, M., *Tesla: Man Out of Time*, ISBN 0-88029-419-1, Dorset Press, New York, 1981.

Chesson, F.W., "Signal Corps Radio Sets," http//pages.cthome.net/fwc/SCR.HTM, Rev: 9 February, 2000.

CHL, "Chain Home Low Radar System," http://www.radarpages.co.uk/mob/chl/chl.htm, retrieved 6 October, 2007.

Clark, G.C., "Deflating British Radar Myths of World War II," DTIC Accession Number: ADA397960, Research Paper, Air Command and Staff College, Maxwell Air Force Base, Montgomery, AL, 36112, 1 March, 1997.

Clark, R.W., *Tizard*, Library of Congress CCN 65-12911, MIT Press, Massachusetts Institute of technology, Cambridge, MA, 1965.

Claydon, M., "Britain's Early Warning Radar System of WWII," http://s110605900.websitehome.co.uk/radar/radarmain.htm, retrieved 17 January, 2009.

Clement, A.W., "Seacoast Artillery Radar," *The Coast Artillery Journal*, No. 91, pp. 8-12, May-June, 1946.

CNM, "Photographic Tour of the Base End Station," Cabrillo National Monument, San Diego, CA, https://www.nps.gov/cabr/learn/historyculture/photographic-tour-of-the-base-end-station1.htm, retrieved 17 August, 2019.

Conn, S. and Fairchild, B., "The United States and Mexico: Solidarity and Security," Chapter 13, *The US Army in World War II: The Framework of Hemisphere Defense*, http://www.ibiblio.org/hyperwar/USA/USA-WH-Frame/USA-WH-Frame-13.html, Office of the Chief of Military History, Washington, DC, 1960.

CSMM, "The Shelling of Ellwood," *California and the Second World War*, California State Military Museum, http://www.militarymuseum.org/Ellwood.html, retrieved 25 November, 2006.

Davenport, L.L., Electrical Engineer, an oral history conducted by John Bryant, IEEE History Center, Rutgers University, New Brunswick, NJ, 12 June, 1991.

Davis, E.G.B., "Increase of Height of Wing-Jetty, Fort Pio Pico," memorandum 444-5 A.E.O., Headquarters, Coast Defenses of San Diego, Fort Rosecrans, CA, National Archives, (DC), Record Group 77, Entry 103, 2 September, 1913.

Davis, E.G.B., Corps of Engineers, San Francisco, CA, status report to Brigadier General John M. Wilson, Chief of Engineers, US Army, Washington DC, National Archives, (DC), Record Group 77, Entry 103, 29 October, 1896.

Davis, E.G.B., Corps of Engineers, San Francisco, CA, status report to Brigadier General John M. Wilson, Chief of Engineers, US Army, Washington DC, National Archives, (DC), Record Group 77, Entry 103, 17 June, 1897.

Davis, H.M., "Early Research and Development: 1918-1937," *The Signal Corps Development of US Army Radar Equipment*, http://infoage.org/Dav-1-ch1.html, 1943.

Deirmengian, C., "Synchros and Resolvers: Part I," *Sensors*, pp. 31-38, April, 1990.

DoA, "The USS Navy Submarine Nautilus Crosses the North Pole," *Defenders of America,* card series, National Biscuit Company, New York, NY, 1959.

Duchesneau, R.E., "A Brief History of American Seacoast Fortifications: 1830-1945," http://www.geocities.com/Pentagon/Base/3495/FortHistory.html?20064, retrieved 4 May, 2006.

Dunning, C., "Point Loma: Two Centuries of Sentry Duty," *Traditions: San Diego's Military Heritage*, Vol. 1, No. 2, pp. 5-11, July, 1994.

Dunning, C., letter to A.H. Grobmeier, 2 February, 1995.

Edwards, H., "Fort Stevens Target of Jap Submarine's Guns," *The Oregonian*, city edition, Portland, OR, p. 1, 23 June, 1942.

Ellis, M., "The War off California," *Teemings*, Issue 16, http://pursam.org/teemings.issue16/michaelellis.html, retrieved 25 November, 2006.

EM, "The SCR-584 Radar, Part I," Electronics War Report, *Electronics*, McGraw-Hill, pp.104-109, November, 1945.

EN, "Electrical Installations," *Engineer Notes: 1900 -1922*, Project Engineer's Notebook, Journal, Harbor Defense San Diego, reproduced by Coast Defense Study Group, 1920.

EN, "Searchlights," *Engineer Notes: 1900 -1922*, Project Engineer's Notebook, Harbor Defense San Diego, reproduced by Coast Defense Study Group, 1913.

EN, "Searchlights," *Engineer Notes: 1900 -1922*, Project Engineer's Notebook, Memoranda, Harbor Defense San Diego, reproduced by Coast Defense Study Group, 1914.

ENG, "Electrical Installations," *Engineer Notes: 1900 -1922*, Project Engineer's Notebook, Harbor Defense San Diego, reproduced by Coast Defense Study Group, 1911.

Engel, N., "The Journal of San Diego History," *San Diego Historical Society Quarterly*, J. MacMullen, Ed., Vol. 6, No. 4, https://sandiegohistory.org/journal/1960/october/ballast/, October, 1960.

Everett, H.R., "Postwar Transition," Chapter 6, *Unmanned Systems of World Wars I and II*, MIT Press, Cambridge, MA, pp. 598-599, 2015.

Everett, H.R., *A Pictorial History of the Code 717 Unmanned Systems Group: Air, Land, and Sea*, Volume 1: 1970–1999, Technical Document 3289, SSC Pacific, https://apps.dtic.mil/dtic/tr/fulltext/u2/1008314.pdf, April, 2016.

Everett, H.R., *Sensors for Mobile Robots: Theory and Application*, ISBN 1-56881-048-2, A.K. Peters, Ltd., Wellesley, MA, June, 1995.

Everett, H.R., *Unmanned Systems of World Wars I and II*, MIT Press, Cambridge, MA, 2015.

Fisher, D.E., *A Race on the Edge of Time*, , Paragon House, New York, NY, 1989.

Fisher, L.C., "Procurement of telephone cable pairs for use of the Harbor Defense Section at NEL," Code 318 Memorandum to LT (JG) J.E. Bryant, Code 170, Naval Electronics Lab, San Diego, CA, 10 December, 1951.

Flower, D., and Roth, L., *Cultural Resource Inventory, Archeology/History/Architecture, Navy and Coast Guard Lands Point Loma*, Officer in Charge of Construction, Naval Ocean Systems Center, San Diego, CA, October, 1982.

FM 4-15, *Seacoast Artillery: Fire Control and Position Finding*, Coast Artillery Field Manual, War Department, Washington, DC, https://ia802600.us.archive.org/21/items/Fm4-15/Fm4-15.pdf, 2 November 1943.

FM 4-91, *90-mm Gun, Fixed Mount: Service of the Piece*, Coast Artillery Training Bulletin, Vol. 2, No. 16, Coast Artillery School, Fort Monroe, VA, July, 1943.

FM 44-4, *Employment of Antiaircraft Artillery Guns*, Field Manual, War Department, Washington, DC, June, 1945.

FM, *Service of the Radio Set SCR-296A*, FM 4-95, Coast Artillery Field Manual, War Department, Washington, DC, available at: http://www.kadiak.org/radios/scr-296-a.pdf, 15 September, 1943.

Fogler, R.H., "Army Tactical Communication Cable SCD-936-Fort Rosecrans to Mount Soledad and the Hermosa Leg; proposed transfer from the Department of the Army to the Department of the Navy," letter from Assistant Secretary of the Navy (Material) to Assistant Secretary of Defense (P&I), 2 May, 1955.

FortWiki, "Battery AMTB –Fetterman," www.fortwiki.com/Battery_AMTB_Fetterman, retrieved 12 April, 2019.

Freeman, P., "NALF San Clemente Island," San Clemente Island Range Complex, retrieved from: http://www.militarymuseum.org/NAASSanClemente.html, 3 August, 2007.

Friedman, N., *Naval Radar*, Naval Institute Press, Annapolis, MD, 1981.

FTP-217, *U.S. Radar, Operational Characteristics of Radar Classified by Tactical Application*, Radar Research and Development Subcommittee of the Joint Committee on New Weapons and Equipment, Washington, DC, available online at http://www.ibiblio.org/hyperwar/USN/ref/Radar/index.html#contents, 1 August, 1943.

Fulmer, R., personal communication cited by Flower and Roth above, 1982.

Gaines, W.C., "19th Coast Artillery (Harbor Defense) Regiment," in *The 19th Coast Artillery and Fort Rosecrans: Remembrances*, Overton, H.B., Ed., National Park Service, Cabrillo National Monument, January, 1993.

Gaines, W.C., "Notes on the Harbor Defenses of San Diego," in *The 19th Coast Artillery and Fort Rosecrans: Remembrances*, Overton, H.B., Ed., National Park Service, Cabrillo National Monument, January, 1993.

GCI, "Type 7 Final GCI Radar," http://www.radarpages.co.uk/mob/gci/type7.htm, retrieved 6 October, 2007.

Gillespie, H.T., letter to Ranger Brett Jones, Cabrillo National Monument, San Diego, CA, (in Overton, 1993), 4 July, 1979.

Gilmore, J.C., 3rd Indorsement to Davis, 1913, "Increase of Height of Wing-Jetty, Fort Pio Pico," National Archives, (DC), Record Group 77, Entry 103, 29 September, 1913.

Glaze, K., *The Illustrated Fort Rosecrans: A Reference Guide to the Army's Coast Artillery Corps in San Diego*, 2nd edition, self-published, San Diego, CA, 2017.

Goebel, G., "The Fire Balloons," http://www.faqs.org/docs/air/avfusen.html, 1 June, 2002.

Goebel, G.V., "SCR-584 Revisited," *In the Public Domain*, Vol. 2.0.3, Ch. 6, http://www.vectorsite.net/ttwiz_06.html, 1 May, 2009.

Goebel, G.V., "The Wizard Wars: WW2 & the Origins of Radar," *In the Public Domain*, Vol. 2.0.8, http://www.vectorsite.net/ttwiz.html, 1 February, 2005.

Green, J.A., "Periscope Reunited with WWII German U-Boat," *Navy Newsstand*, www.news.navy.mil, story number NNS030207-14, 7 February, 2003.

Grobmeier, A.H., "Additional Info Chapt. 5," e-mail correspondence, 22 July, 2007c.

Grobmeier, A.H., "Additional Info Chapter 6," e-mail correspondence, 22 July, 2007c.

Grobmeier, A.H., "Chapter 4, Searchlight Stations," e-mail correspondence, 21 July, 2007.

Grobmeier, A.H., "Chapter 5 Gun Batteries," e-mail correspondence, 22 July, 2007b.

Grobmeier, A.H., "Chapter 6 Gun Batteries," e-mail correspondence, 22 July, 2007b.

Grobmeier, A.H., "HDSD Remnants Discovered at NAS North Island," *Coast Defense Study Group News*, Vol. 6, No. 4, p. 15, November, 1992.

Grobmeier, A.H., "Postwar Use of Battery Construction Number 134," Coast Defense Study Group News, Vol. 4, No. 2, p. 27, May, 1990.

Grobmeier, A.H., "Upcoming Harbor Defense San Diego Book," e-mail correspondence, 19 July, 2007a.

Grobmeier, A.H., "Update of HDSD Remnants Discovered at NAS North Island," *Coast Defense Study Group News*, Vol. 7, No. 1, pp. 6-7, February, 1993.

Grobmeier, A.H., annotated notes from library search of White Pages, San Diego phonebook, 5 February, 1995.

Grobmeier, A.H., letter to Jack R. Skaggs, Edmond, OK, USMC practice batteries on Fort Rosecrans, 25 September, 1995.

Grobmeier, A.H., research notes on San Diego newspaper article dated 28 July 1948, received 3 July, 2008.

Groundspeak, "Guth, Edwin F., Co. Complex - St. Louis, Missouri - U.S. National Register of Historic Places on Waymarking.com," http://www.waymarking.com/waymarks/WMMW8_Guth_Edwin_F_Co_Complex_St_Louis_Missouri, 20 August, 2006.

Guerlac, H.E., ed., *Radar in World War II: The History of Modern Physics*, American Institute of Physics, 1987.

Guhan, C., "This and That: Quirky Facts about Santa Catalina Island," retrieved from: http://www.ecatalina.com/article_quirky.cfm, 2007

Guidry, L., Coast Defense Study Group, oral interview, 14 July, 2007.

Guidry, L., field conversation during Fort Rosecrans site survey, April, 2008.

Guidry, L., oral interview, San Diego, CA, 14 July, 2007.

H505, *History of the 505th Aircraft Control and Warning Group, 1 July – 31 December 1948*, December, 1948b.

H505, *History of the 505th Aircraft Control and Warning Group, 28 March – 30 June 1948*, June, 1948a.

HABS, "Fort Rosecrans, Mining Casemate, Point Loma, San Diego, San Diego County, CA," *Historic American Buildings Survey*, HABS CAL, 37-SANDI,29-A-, https://www.loc.gov/item/ca1607/, retrieved 28 October, 2019.

Hammond, J.H., and Purington, E.S., "A History of Some Foundations of Modern Radio-Electronic Technology," Proceedings, Institute of Radio Engineers, Vol. 45, pp. 1191-1194, September, 1957.

Harris, S.H., *Factories of Death: Japanese Biological Warfare, 1932-1945, and the American Cover-Up*, Routledge, London, 1994.

Hartwig, T., "Light demolition carrier "Goliath" (Sd.Kfz.302/303a/303b)," retrieved from: http://www.geocities.com/CapeCanaveral/Lab/1167/egoliath.html, 2006.

Hase, 1918a: Lt. Col. W.F. Hase, Memorandum for the Board, February 4, 1918, F32, Box 31, OCE, RG 77, NA, Pacific Southwest Region.

Hase, 1918b: Memorandum for the Board, February 23, 1918, File 662. B (San Diego) I, February 1918-December 1938, Box 13311, OCE, RG 77, WNRC.

Hathaway, R., Van Wormer, S., Agee, B., Schilz, A., "Fort Rosecrans, Mining Casemate, Point Loma," *Historic American Buildings Survey*, HABS CAL, 37-SANDI,29-A-, https://www.loc.gov/item/ca1607/, 10 March, 1987.

HCP, "Fort Pio Pico," *Historic California Posts*, http://www.militarymuseum.org/FtPioPico.html, retrieved 2 July, 2006.

HCP, "Historic California Posts," http://www.militarymuseum.org/HistoryPosts.html, 2007.

HDP, "Gun Defenses," Part IV, Paragraph 2, Harbor Defense Projects (Seacoast Artillery Installations) for the *Harbor Defenses Included in the San Diego – San Pedro Area*, February, 1933.

Helgeson, D., "SCR-270 Historical Notes," http://www.radomes.org/museum/equip/SCR-270.html, 2006.

HG, "Flush Deck Destroyers: All-Purpose Ships," Haze Gray Photo Feature, retrieved from http://www.hazegray.org/features/flushdeck/, 13 November, 2006.

Hill, N.F., *Expose! A History of Searchlights in WWII*, ISBN 0-86439-160-9, Boolarong Publications, Bowen Hills, Brisbane, 1993.

Hinds, J.W., "San Diego's Military Sites," typescript, 1815 Lilac Lane, Alpine, CA, 1991.

Hinds, J.W., "United States Army/Army Air Forces: San Diego, California, World War II," typescript, 1815 Lilac Lane, Alpine, CA, 2004.

Hine, F.T., Ward, F.W., *The Service of Coast Artillery*, Goodenough & Woglom, New York, NY, 1910.

Hofbauer, M., "Panzerfaust: WWII German Infantry Anti-Tank Weapons," p. 3, Panzerschreck, http://www.geocities.com/Augusta/8172/panzerfaust3.htm?20061, August, 1998.

Hoffman, S.O., "The Detection of Invisible Objects by Heat Radiation," *Physical Review*, XIV, p. 154, April, 1919.

Hogg, I.V., Allied Artillery of World War Two," Crowood Press, Wiltshire, Great Britain, 1998.

Houston, J., "Report on Proposed Transfer of Cable Line and Rights of Way at Fort Rosecrans, La Jolla-Hermosa Fire Control Station, and La Jolla AWS Station," Management and Disposal Branch, Real Estate Division, Los Angeles District, 2 December, 1955.

Howeth, L.S., *History of Communications-Electronics in the United States Navy*, prepared for the Bureau of Ships and Office of Naval History, Government Printing Office, Washington, DC, 1963.

HS, "History of Harbor Defenses of San Diego," Annex B to Appendix VIII, WDC, HS, Vol. 6, Pt. 6, 15 September, 1945.

HSC, *History of the Southern California Sector*, Western Defense Command, (HIST-SCS-WDC), 12 December, 1945.

HSD, "History of Harbor Defenses of San Diego," Annex B, Appendix 2B, Chapter 2, *History of the Southern California Sector, Western Defense Command*, 12 December, 1945.

ILC, *Synchro Conversion Handbook,* Third Printing, ILC Data Device Corporation, Bohemia, NY, April, 1982.

Johnson, C.R., "U.S. Navy Electronics Laboratory, San Diego, California; telephone distribution cable," Memorandum from District Public Works Officer to Commandant, Eleventh Naval District, San Diego, CA, 17 January, 1951.

Jones, D.L., Letter, available online at www.tijuanaestuary.com/High%20School%20Curriculum/Activities/HIS%20lesson%204.pdf, 8 September, 2000.

Jones, R.V., *Most Secret War*, Hamish Hamilton, London, England, 1978.

Keen, R., *Direction and Position Finding by Wireless*, http://www.jumpjet.info/Pioneering-Wireless/eBooks/maturity/1922c.pdf, Wireless Press, Ltd., London, 1922

Keniston, "Fort Rosecrans: Point Loma Coastal Defenses," National Register of Historic Places Nomination, prepared by Keniston Architects, San Diego, CA, December, 1995.

Keniston, *Fort Rosecrans: Point Loma Coastal Defense*, National Register of Historic Places Nomination, Prepared for Naval Command Control and Ocean Surveillance Center, RDT&E Division, San Diego, CA, May, 1996.

Klein, C., "Attack of Japan's Killer WWII Balloons, 70 Years Ago," *History Stories*, https://www.history.com/news/attack-of-japans-killer-wwii-balloons-70-years-ago, 29 August, 2018.

Kopp, C., "The Dawn of the Smart Bomb," *The Air Power Australia Website*, http://www.ausairpower.net/WW2-PGMs.html, July, 2006.

Kurtz, T.R., "Army surplus cable from the North boundary of CIC Team Training Center, Point Loma, South to the Mexican border; use of," letter from Commandant, Eleventh Naval District, to District Public Works Officer (DE-200), Ser 5150/34E, 30 July, 1952.

L'Ecuyer, L., "SCR-270/271," http://www.pinetreeline.org/rds/rds270.html, 14 June, 2004.

Lafrenz, W.F., *Coast Artillery Journal*, Vol. 87, No. 4, pp. 62-63, July-August, 1944.

Laird, R.T., Everett, H.R., "Early User Appraisal of the MDARS Interior Robot," American Nuclear Society, *8th Topical International Meeting on Robotics and Remote Systems*, Pittsburgh, PA, 25-29 April, 1999.

LaPuzza, T.J., "Some Suggested Corrections," e-mail correspondence, 11 May, 2020.

Lequin, M., "The Gun Emplacements of Point Loma," *San Diego Reader*," https://www.sandiegoreader.com/news/1977/apr/28/phantoms-fortress/, 28 April, 1977.

Levy, R.S., *History of the 204th Antiaircraft Artillery Group: 1870 – 31 July 1944*, Spanish Village, San Diego, CA, 29 December, 1944.

Linder, B., "Coronado's Big Gun Defenses," *Field Guide to Coronado History*, https://coronadohistory.org/static/media/uploads/historical%20coronado/Field%20Guide/fg_%E2%80%93_coronado%E2%80%99s_big_gun_defenses.pdf, retrieved 21 October, 2019.

Lockwood, H.W., *San Diego's Hysterical History*, Carrol, W., ed., ISBN 0-910390-67-3, Coda Publications, Raton, NM, 2003.

Lougher, H., in "History," *Ballast Point Lighthouse*, https://en.wikipedia.org/wiki/Ballast_Point_Lighthouse, originally retrieved from "Historic Light Station Information and Photography: California" Coast Guard website, 12 April, 2011.[1]

Lundberg, K.H., "The MIT Rad Lab Series," http://web.mit.edu/klund/www/books/radlab.html, November, 2002.

MacLeod, R.M., *Science and the Pacific War: Science and Survival in the Pacific, 1935 - 1945*, Springer, p. 222, December, 1999.

Malone, D., "Seacoast Artillery Radar," Berhow, M.A, ed., *American Seacoast Defenses: A Reference Guide*, Second Edition, Coast Defense Study Group Press, McLean, VA, p. 398, 2004.

Malone, D.R., "Addendum to Seacoast Artillery Radar: 1938-1946," *Coast Defense Study Group News*, Vol. 4, No. 2, pp. 37-39, May 1990.

Malone, D.R., "Seacoast Artillery Radar," *American Seacoast Defenses, A Reference Guide*, M. Berhow, ed., Coast Defense Study Group Press, McLean, VA, 2004.

Malone, D.R., "Seacoast Artillery Radar: 1938-46," *Coast Defense Study Group News*, Vol. 3, No. 5, pp. 1-11, November, 1989.

Malone, D.R., "The SCR-296A Radar," *Coast Defense Study Group News*, Vol. 5, No. 2, http://www.militarymuseum.org/SCR296A.pdf, May, 1991.

MC, "Major Commands," Research Division, Organizational History Branch, available at: http://afhra.maxwell.af.mil/rso/major_commands.html, retrieved 23 September, 2007.

McNaught, A., "Unit 731: Japan's Biological Force," *BBC News World Edition*, http://news.bbc.co.uk/2/hi/programmes/correspondent/1796044.stm, 1 February, 2002.

---

1   The original Coast Guard website is no longer available.

McPhee, J., "Balloons of War," *New Yorker*, vol. 71, No. 46, pp. 52-60, 29 January, 1996.

Meyler, J.J., Corps of Engineers, Los Angeles, CA, status report to Brigadier General John M. Wilson, Chief of Engineers, US Army, Washington DC, National Archives, (DC), Record Group 77, Entry 103, 27 March, 1899a.

Meyler, J.J., Corps of Engineers, Los Angeles, CA, status report to Brigadier General John M. Wilson, Chief of Engineers, US Army, Washington DC, National Archives, (DC), Record Group 77, Entry 103, 21 June, 1899b.

Meyler, J.J., Corps of Engineers, Los Angeles, CA, status report to Brigadier General John M. Wilson, Chief of Engineers, US Army, Washington DC, National Archives, (DC), Record Group 77, Entry 103, 24 August, 1900.

Meyler, J.J., Corps of Engineers, memorandum to Brigadier General John M. Wilson, Chief of Engineers, US Army, Washington DC, National Archives, Record Group 392, 27 April, 1901.

Meza, Bob, "Bob's Searchlight," http://www.geocities.com/mepurina/gotit2.html?20063, 2006.

Mikesh, R.C., *Japan's World War II Balloon Bomb Attacks on North America*, Smithsonian Institution Press, Washington, DC, 1973.

Miller, L.D., "The 25 Kw. Gasoline Driven Generating Set Used in the Coast Defenses of the United States," *Journal of the United States Artillery*, Vol. 48, pp. 54-79, 1917.

Mindell, D.A., "Anti-Aircraft Fire Control and the Development of Integrated Systems at Sperry, 1925-1940," *IEEE Control Systems*, pp. 108-113, April, 1995.

Mitchell, M., addendum to Moriarity and Wisdom (1989), Bureau of Land Management, Palm Springs South Coast Field Office, North Palm Springs, CA, November, 1997.

MM, "California Military History: California and the Second World War," California State Military Museum, http://www.militarymuseum.org/HistoryWWII.html, retrieved 3 August, 2007.

MMR, "Mobile Military Radar," http://www.mobileradar.org/radar_descptn_2.html, retrieved 30 September, 2007.

Moeller, S.P., "Activation and Deployment – the 1950s," *Vigilante and Invincible*, http://www.redstone.army.mil/history/vigilant/chap2.html, Chapter 2, p. 5, 2006.

Moore, J., "McGrath, Hugh J., Maj.," *Together We Served*, https://army.togetherweserved.com/army/servlet/tws.webapp.WebApp?cmd=ShadowBoxProfile&type=Person&ID=306039, retrieved 23 October, 2019.

Moriarty, J. and Wisdom, W., "Otay Mountain Bunkers," prepared for US Department of the Interior, Bureau of Land Management, Palm Springs South Coast Field Office, North Palm Springs, CA, 30 August, 1989.

Murdock, S., "AAF WWII site in San Diego," e-mail correspondence with A.H. Grobmeier, 11 May, 2004.

NAF, "Fort Pio Pico," *Harbor Defenses of San Diego*, https://www.northamericanforts.com/West/ca-south2.html, retrieved 21 October, 2019.

NavWeaps, "7"/45 (17.8 cm) Mark 2," *Naval Weapons, Naval Technology and Naval Reunions*, http://www.navweaps.com/Weapons/WNUS_7-45_mk2.php, retrieved 9 June, 2019.

NBC, "NBC – Who We Are," Naval Base Coronado, San Diego, CA, http://www.nbc.navy.mil/index.asp?fuseaction=information.infoWhoWeAre, retrieved 7 August, 2007.

NCSSD, "Naval Radio Communications in the Eleventh Naval District," *Naval Communications Station San Diego (NPL)*, http://www.navy-radio.com/commsta/ todd-sandiego-01.pdf, retrieved 30 March, 2020.

Neale, B.T., "CH – The First Operational Radar," *The GEC Journal of Research*, Vol. 3, No. 2, pp. 73-83,1985.

NNS, "Museum of Science and Industry 'Launches' U-505 Submarine on Final Tour," *Navy Newsstand*, story number NNS040409-2, 9 April, 2004.

NPS, "Radar Station B-71," *Aviation: From Sand Dunes to Sonic Booms*, National Park Service, http://www.nps. gov/nr/travel/aviation/rad.htm, retrieved 23 September, 2007.

NSG, "NAVSECGRU Stations Past and Present," https://www.navycthistory.com/CI_Stations_past_and_present_alphabetical_4.html, retrieved 30 March, 2020

NSGA, "NSGA Imperial Beach AN/FRD-10 HF/DF Array," http://wikimapia.org/80760/NSGA-Imperial-Beach-AN-FRD-10-HF-DF-Array, updated 2017.

NTS, "New York," *US Naval Activities World War II by State*, https://ibiblio.org/hyperwar/USN/ref/USN-Act/ NY.html, retrieved 20 June, 2019.

Offord, B.W., Russel, S.D., Weiner, K.H., *Laser Processing of Silicon on Sapphire (SOS) for Fabrication of Bipolar Transistors*, Technical Report 1376, Naval Ocean Systems Center (NOSC), San Diego, California, November, 1990.

Ordway, F.I., Wakeford, R.C., *International Missile and Spacecraft Guide*, McGraw-Hill, New York, NY, 1960.

Orman, L.M., "Radar: A Survey," *The Coast Artillery Journal*, Vol. LXXXIX, No. 2, pp. 2-7, March-April, 1946.

Overton, H.B., "No Dead Areas in Battery Bluff's Sights," *Traditions: San Diego's Military Heritage*, Vol. 1, No. 2, pp. 6-7, July, 1994.

Overton, H.B., *Fort Rosecrans Revisited*, National Park Service, Cabrillo National Monument, unpublished draft, October, 1995.

Overton, H.B., *The 19ᵗʰ Coast Artillery and Fort Rosecrans: Remembrances*, National Park Service, Cabrillo National Monument, January, 1993.

Overton. H.B., "Battery Point Loma," *Fort Guijarros Quarterly*, Vol 2, No. 2, Summer, 1988.

Page, R.M., *The Origin of Radar*, Anchor Books, Doubleday, Garden City, NY, 1962.

Payette, P., *American Forts Network* website, http://www.geocities.com/naforts/, 2006.

Peregrine, "History in the Making," https://www.psemi.com/pdf/PSemi_History_Book.pdf, 3 April, 2020.

Perry. T., "Periscope of German WWII Sub Surfaces," *Los Angeles Times*, https://www.latimes.com/archives/la-xpm-2002-sep-13-me-nazisub13-story.html, 13 September, 2002.

Pickard, G., "The Japanese WWII Balloon Attack on America No One Remembers," https://www.topsecretwriters.com/2017/03/japanese-wwii-balloon-attack-america-no-one-remembers/, 8 March, 2017.

Pierson, L.J., "Mount Soledad Base End Station," 39ᵗʰ Annual Conference, Council on America's Military Past (CAMP), Los Angeles, CA, 4-8 May, 2005.

Price, A., *Instruments of Darkness: The History of Electronic Warfare*, Charles Scribner's Sons, New York, NY, 1967.

Prugh, G.S., "Denims, Pinks, and Greens," in *The 19ᵗʰ Coast Artillery and Fort Rosecrans: Remembrances*, Overton, H.B., ed., National Park Service, Cabrillo National Monument, January, 1993.

Pugh, G.S., "Denims, Pinks, and Greens," pp. 161-175 in Overton (1993), October, 1988.

Pyle, E., "American Ack-Ack," *Brave Men*, Grosset & Dunlap, New York, NY, 1944.

Radomes, "LASHUP: 1948-1952," LASHUP Radar System, http://www.radomes.org/cgi-bin/museum/acwlashup.cgi, retrieved 23 August, 2007a.

Radomes, "SCI Chronological Military History," http://www.radomes.org/museum/documents/SanClementeIslandAFSCAhistroy.html, retrieved 23 August, 2007b.

Raines, R.C., "Information on AWS Radar Sites," e-mail correspondence, 10 October, 2007.

RCW, "Battery Calef-Wilkeson," *Report of Completed Works*, Coast Defenses of San Diego, Calif., updated to 15 September, 1925.

RCW, "BC$_7$ and B$^1_7$ S$^1_7$ - SCR 296 (Combined)," Report of Completed Works, Harbor Defenses of San Diego, 8 November, 1944a.

RCW, "Combined G4CP, HD Signal Station and SCR-682," Report of Completed Works, Harbor Defenses of San Diego, 16 November, 1945c.

RCW, "Harbor Defense Command Post, Harbor Entrance Control Command Post," *Report of Completed Works*, Harbor Defenses of San Diego, California, Fort Rosecrans, corrected to 20 January, 1943.

RCW, "Incline Hoist – SCR 296, Site D," Report of Completed Works, Harbor Defenses of San Diego, 7 March, 1945a.

RCW, "SCR-296 (Site "C")," Report of Completed Works, Harbor Defenses of San Diego, 15 March, 1945b.

RCW, "SCR-296 La Jolla," Report of Completed Works, Harbor Defenses of San Diego, 22 November, 1944b.

Rea, J.W., "Transfer of Harbor Defense Facilities, San Diego Area, Army Tactical Communications Cable SCD-936-Fort Rosecrans to Mount Soledad and the Hermosa Leg, from the Department of the Army to the Department of the Navy," memorandum from Chief, Bureau of Yards and Docks to Commandant, Eleventh Naval District, 6 March, 1956.

Rizzo, J., "Japan's Secret WWII Weapon: Balloon Bombs," *National Geographic*, https://www.nationalgeographic.com/news/2013/5/130527-map-video-balloon-bomb-wwii-japanese-air-current-jet-stream/#close, 27 May, 2013.

Robert, H.M., Barlow, J.W., Gillespie, G.L., "Mimeograph No. 39, Subject: Searchlights Required in Defense of Coast of U.S.," Board of Engineers, Army Building, New York, NY, 8 March, 1901.

Roberts, R.B., "Fort Pio Pico," *Historic California Posts, Camps, Stations, and Airfields*, http://www.militarymuseum.org/FtPioPico.html, updated 6 February, 2016.

Roberts, R.B., *Encyclopedia of Historic Forts: The Military, Pioneer, and Trading Posts of the United States*, Macmillan, New York, 1988.

Roberts, S., "Rosenberg Trial Figure Admits Spying for Soviets," *San Diego Union-Tribune*, p. A-3, 12 September, 2008.

Roessler, S.W., "Third Supplement to Mimeograph No. 39, Subject: Test of Searchlights at Portland, Me.," War Department, Office of the Chief of Engineers, 14 September, 1904.

Roessler, S.W., Goethals, G.W., Taylor, H., "Second Supplement to Mimeograph No. 39, Subject: Project for Electric lighting at Fort Rosecrans, Cal.," Office of the Chief of Engineers, 2 January, 1903.

Rogers, D.J., "How Geologists Unraveled the Mystery of Japanese Ballon Bombs in World War II," http://ww-web.umr.edu/~rogersda/forensic_geology/, retrieved 23 November, 2006.

ROM, "General Radar Principles," *Radar Operator's Manual*, Radar Bulletin No. 3, Part 1, p. 1-43, http://www.ibiblio.org/hyperwar/USN/ref/RADTHREE/RADTHREE-1.html, 5 August, 1944.

Roper, J.W., "Transfer of Harbor Defense Facilities, San Diego area, from the Department of the Army to the Department of the Navy," letter from Commandant, Eleventh Naval District to Chief of Naval Operations, Ser 817/49, 21 August, 1952a.

Roper, J.W., "Transfer of Harbor Defense Facilities, San Diego area, from the Department of the Army to the Department of the Navy," letter from Commandant, Eleventh Naval District to Chief of Naval Operations, Ser 1071/49, 6 November, 1952b.

RP, "Detailed Plan for the Retrenchment of Fourth Air Force Control Group Installations," Enclosure 1, *History of the Fourth Air Force of the Continental Forces, 2 September 1945 to 20 March 1946*, 20 March, 1946.

Ruhge, J., "Fort Emory," *Historic California Posts, Camps, and Stations*, http://www.militarymuseum.org/FtEmory.html, updated 8 February, 2016.

Ruhge, J.M., "Battery Cliff, Batter Bluff, and Battery Channel," *Historic California Posts, Camps, Stations, and Airfields*, https://www.militarymuseum.org/ATMBBCC.html, 23 June, 2017a.

Ruhge, J.M., "Fort Pio Pico: Battery James Meed," *Historic California Posts, Camps, Stations, and Airfields*, https://www.militarymuseum.org/BtyMeed.html, 23 June, 2017b.

Ruhge, J.M., "Fort Rosecrans: Batteries Calef and Wilkeson," Historic California Posts, Camps, Stations and Airfields, http://www.militarymuseum.org/BtyCalef-Wilkeson.html, retrieved 24 October, 2019d.

Ruhge, J.M., "Fort Rosecrans: Battery Fetterman," *Historic California Posts, Camps, Stations and Airfields*, http://www.militarymuseum.org/BtyFetterman.html, retrieved 21 October, 2019a.

Ruhge, J.M., "Fort Rosecrans: Battery James Meed," *Historic California Posts, Camps, Stations and Airfields*, https://www.militarymuseum.org/BtyMeed.html, retrieved 22 October, 2019b.

Ruhge, J.M., "Fort Rosecrans: Battery John White," *Historic California Posts, Camps, Stations and Airfields*, http://www.militarymuseum.org/BtyWhite.html, updated 23 June 2017.

Ruhge, J.M., "Fort Rosecrans: Battery McGrath," Historic California Posts, Camps, Stations and Airfields, http://www.militarymuseum.org/BtyMcGrath.html, retrieved 23 October, 2019c.

Ruhge, J.M., "Fort Rosecrans: Battery Fetterman," *Historic California Posts, Camps, Stations, and Airfields*, https://www.militarymuseum.org/BtyFetterman.html, 23 June, 2017c.

Ruhlen, G., "Fort Rosecrans, California," *Journal of San Diego History*, Vol. 5, No. 4, https://sandiegohistory.org/journal/1959/october/fortrosecrans/, October, 1959.

Ruhlen, G., "Fort Rosecrans," *Historic California Post, Camps, Stations and Airfields*, www.militarymuseums.org/FtRosecrans.html, updated 8 February, 2016.

Ruhlen, G., *Historic California Posts, Fort Rosecrans*, www.militarymuseum.org, 2006.

SCHP, "Long Range Radar – SCR-270 and SCR-271," Part III, *The Signal Corps Development of US Army Radar Equipment*, Signal Corps Historical Project A-3, Signal Corps Historical Section, Washington, DC, November, 1945.

Schwartz, O.B., Grafstein, P., *Pictorial Handbook of Technical Devices*, Chemical Publishing Co, Inc., New York, NY, pp. 272-275, 1971.

Scott, J.M., *Rampage: MacArthur, Yamashita, and the Battle of Manila*, Norton, 2018.

Sears, A.B., *History of the 4th Antiaircraft Command: Headquarters and Headquarters Battery, 9 January 1943 to 1 July 1945*, Aquatic Park, San Francisco, CA, 20 July, 1945.

Sears, A.B., *History of the 4th Antiaircraft Command: Headquarters and Headquarters Battery, 9 January 1943 to 1 July 1945*, Aquatic Park, San Francisco, CA, 20 July, 1945.

Sebby, D.M., "Half Moon Bay Radar Site," Historic California Posts, Stations and Airfields, http://www.military museum.org/HMBAWSS.html, retrieved 14 August, 2007.

Shettle, M.L., "Naval Auxiliary Air Station, San Clemente Island," *Historic California Posts*, http://www.military-museum.org/NAASSanClemente.html, retrieved 3 August, 2007.

Showley, R.M., "Hindsight: Fortifications Celebrate 200 Years," *San Diego Union-Tribune*, p. D-10, February, 1996.

Skaggs, J.R., annotations to Grobmeier's Fort Rosecrans map, Edmond, OK, received 10 October, 1995.

Skylighters, "How Were World War II Searchlights Used: A Story of Locators, Lights, & Ack-Ack," http://www.skylighters.org/howalightworks/index.html, 2006.

Smith, B.F., *The Evaluation of Building 258 (AMTB Battery Fetterman) at the Naval Submarine Base San Diego, for Eligibility for Nomination to the National Register of Historic Places*, Martinez-Wong Associates, Inc., (tDAR ID: 339691), 1986.

Smith, B.W., "Fire Control and Position Finding," *American Seacoast Defenses: A Reference Guide*, Berhow, M.A., ed., pp. 257-264, Coast Defense Study Group, McLean, VA, 2004.

Smith, B.W., "Introduction: Searchlights, an Early Perspective," Editorial Introduction in Mimeograph No 39, Robert et al. (1901), Reprinted by the Coast Defense Study Group, August, 1999.

Springer, R.M., "Fort Rosecrans: Big Gun Hiding," *San Diego Magazine*, Vol. 6, No. 2, pp. 19-23, 1930.

Springer, R.M., "Fort Rosecrans: Big Gun Hiding," *San Diego Magazine*, Vol. 6, No. 2, pp. 19-23, 1930.

Stanton, J., "Battery AMTB - Cabrillo," *FortWiki Historic U.S. and Canadian Forts*, www.fortwiki.com/ Battery_AMTB_-_Cabrillo, 7 January, 2019d.

Stanton, J., "Fort Emory," *FortWiki Historic U.S. and Canadian Forts*, www.fortwiki.com/Fort_Emory, 7 January, 2019c.

Stanton, J., "Fort Pio Pico," *FortWiki Historic U.S. and Canadian Forts*, www.fortwiki.com/Fort_Pio_Pico, 7 January, 2019b.

Stanton, J., "Battery AMTB – Fetterman," *FortWiki Historic U.S. and Canadian Forts*, www.fortwiki.com /Battery_AMTB_-_Fetterman, 7 January, 2019a.

Stanton, J., "Battery Calef-Wilkeson," http://fortwiki.com/Battery_Calef-Wilkeson, updated 7 January, 2019.

Stanton, J., and Thayer, B., "Battery Strong," *Fort Wiki History*, http://fortwiki.com/Battery_Strong, 7 January, 2019.

Stanton, J., Beck, J., "Battery Russell," http://www.fortwiki.com/Battery_Russell, updated 7 January, 2019.

Stanton, J., Thayer, B., "Battery Fetterman," http://fortwiki.com/Battery_Fetterman, updated 7 January, 2019b.

Stanton, J., Thayer, B., "Battery Meed," http://fortwiki.com/indexphp?title=Battery_Meed&oldid=125288, updated 7 January, 2019d.

Stanton, J., Thayer, B., "Battery White," http://fortwiki.com/Battery_White_(2), updated 7 January, 2019a.

Stanton, J., Thayer, B., "Fort Pio Pico," http://www.fortwiki.com/Fort_Pio_Pico, updated 7 January, 2019c.

Sturgeon, W.J., "San Clemente Island, Chronological Military History (1934-2000), retrieved from http://www.scisland.org/aboutsci/history/history.php, 7 August, 2007a.

Sturgeon, W.J., personal interview, Crown Point, San Diego, CA, 8 November, 2007.

Sturgeon, W.J., telephone conversation, 30 August, 2007b.

Suffield, F.G., "The SCR-270 Radar," available from IEEE History Center website, http://www.ieee.org/organizations/history_center/milestones_photos/opana.html, 1995.

Suter, C.R., Mansfield, S.M., Raymond, C.W., Fremont, J.C., Birnie, R., Pratt, S., "Supplement to Mimeograph No. 39, Subject: Searchlights," Office of the Chief of Engineers, United States Army, 17 December, 1902.

Tays, G., "Fort Rosecrans, Registered Landmark #62," California Historical Landmark Series, for State of California, Department of Natural Resources, Division of Parks, Berkeley, CA, 1937.

Terret, D., *The Signal Corps: The Emergency*, United States Army in World War II, The Technical Services, Office of the Chief of Military History, Department of the Army, Washington, DC, 1956.

Terrett, D., *The Signal Corps: The Emergency, United States Army in World War II, The Technical Services*, Office of the Chief of Military History, Department of the Army, Washington, DC, 1956.

Tesla, N., "The Problem of Increasing Human Energy," *Century Magazine*, June, 1900.

Thompson, E.N., *The Guns of San Diego*, Historic Resource Study, Cabrillo National Monument, http://www.cr.nps.gov/history/online_books/cabr/index.htm, San Diego, CA, 1991.

Thompson, G.R., Harris, D.R., *The Signal Corps: The Outcome*, Office of the Chief of Military History, Department of the Army, Washington, DC, 1966.

Thompson, G.R., Harris, D.R., Oakes, P.M., Terret, D., *The Signal Corps: The Test, United States Army in World War II, The Technical Services*, Office of the Chief of Military History, Department of the Army, Washington, DC, 1957.

TM 5-7111, *Operating Instructions Manual: Searchlight, 60-inch Model 1942A and Control Station*, General Electric Co., Schenectady, NY, November, 1943.

TM 9-1570, "Ordnance Maintenance," *Plotting Boards for Seacoast Artillery*, Technical Manual, War Department, Washington, DC, https://archive.org/stream/TM9-1570/TM9-1570_djvu.txt, 8 May, 1942.

TM 9-1623, *Height Finders, 13 1/2 - Ft., M1 and M1A1*, Ordnance Maintenance, War Department Technical Manual, War Department, Washington, DC, pp. 6-15, 27 November, 1943.

TM 9-1675, "Ordnance Maintenance," *Azimuth Instruments M1910 and M1910A1 (Degrees)*, Technical Manual, War Department, Washington, DC, https://ia801307.us.archive.org/18/items/TM9-1675/TM9-1675.pdf, 15 September, 1941.

TM 9-1685, "Ordnance Maintenance," *Depression Position Finder M1907*, Technical Manual, War Department, Washington, DC, https://ia800304.us.archive.org/6/items/TM9-1685/TM9-1685.pdf, 14 October, 1941.

TM 9-2300, *Artillery Materiel and Associated Equipment*, Department of the Army Technical Manual, Washington, DC, 1949.

TM 9-235, *37-mm AA Gun Materiel*, War Department Technical Manual, War Department, Washington, DC, 24 January, 1944.

TM 9-345, *155-mm Gun Materiel, M1917, M1918 and Modifications*, Technical Manual, War Department, Washington, DC, 15 July, 1942.

TM 9-428, *6-Inch Seacoast Materiel: Guns M1903A2 and M1905A2; Barbette Carriage M1,* Technical Manual, War Department, Washington, DC, 12 June, 1943.

TM 9-618, *Generator Unit M7*, War Department Technical Manual, War Department, Washington, DC, p.7, 30 July, 1943.

TM-1695, "Ordnance Maintenance," *Depression Position Finder M1*, Technical Manual, War Department, Washington, DC, https://ia601204.us.archive.org/20/items/TM9-1695/TM9-1695.pdf, 21 November, 1941.

TM11, "Power Units PE-84, PE-84-A, PE-84C, PE-84D," TM 11-915, War Department Technical Manual, United States Government Printing Office, Washington, DC, 15 June, 1944.

TM9, "Military Vehicles," TM 9-2800, Department of the Army Technical Manual, pp. 175-217, United States Government Printing Office, Washington, DC, 27 October, 1947a.

Tribune, "Arsenal-Like 'Old Blockhouse' Overlooking La Jolla Being Turned into Communications Hub for Colleges," San Diego Tribune, p. A-34, 21 November, 1965.

Vesper, D., "SCR-582 Info #4," e-mail correspondence to Joe Stevens, http://www.kadiak.org/long_is/scr582.html, 4 May, 1999a.

Vieweger, A.L. and White, A.S., "Development of Radar SCR-270," *C&E Digest*, http://www.monmouth.army.mil/historian/, November, 1959.

Von Kroge, H., *GEMA: Birthplace of German Radar and Sonar*, Taylor and Francis, 1 January 1, 2000.

Walding, R., "United States Harbor Defenses," *Indicator Loops*, www.indicatorloops.com/usn_fishers.htm, retrieved 19 June, 2019.

Watson-Watt, R., "British Radar in WW2," *Discovery*, (reproduced online by Patrick Lockerbie), https://www.science20.com/chatter_box/british_radar_ww2_sir_robert_watsonwatt-111296), September, 1945.

Watson-Watt, R., Herd, J.F., "An Instantaneous Direct-Reading Radiogoniometer," https://digital.nls.uk/scientists/archive/75132037, read before Wireless Section, Radio Research Board, 3 March, 1926.

WD, "Combined Commanding Officer / Officer in Charge War Diaries, US Fleet Training Base / Naval Aviation Facility, San Clemente Island, CA, 7 December 1941 – 31 January 1943," National Archives and Records Administration, College Park, MD, 1943.

Webber, B., *Silent Siege – III: Japanese Attacks on North America in World War II*, ISBN 0-936738-73-1, Webb Research Group, Medford, OR, October, 1997.

Weitz, K.J., "Command Lineage, Scientific Achievement, and Major tenant Missions," Volume I, *Keeping the Edge: Air Force Materiel Command Cold War Context (1945-1991)*, Headquarters Air Force Materiel Command, Wright-Patterson Air Force Base, Ohio, August, 2003.

White, W.S., "US Army Signal Corps at Camp Cactus," Chapter 4, *Santa Catalina Goes to War: World War II, 1941-1945*, ASIN: B0006S38EQ, White Limited Editions, 2002.

Wiki, "Jerry Astl," Wikipedia, https://en.wikipedia.org/wiki/Jerry_Astl, updated 28 September, 2019.

Wikipedia, "90 mm Gun M1/M2/M3," https://en.wikipedia.org/wiki/90_mm_Gun_M1/M2/M3, updated 6 April, 2019.

Wikipedia, "Camp Callan," https://en.wikipedia.org/wiki/Camp_Callan, retrieved 13 May, 2019.

Wikipedia, "Depression Range Finder," https://en.wikipedia.org/wiki/Depression_range_finder, retrieved 4 July, 2019.

Wikipedia, "Lightwell," https://en.wikipedia.org/wiki/lightwell, retrieved 26 October 2019.

Wikipedia, "SCR-270 radar," http://en.wikipedia.org/wiki/SCR-270_radar, 2006.

Wikipedia, "Searchlight," http://en.wikipedia.org/wiki/Searchlight, 2006.

Wikipedia, *Kerrison Predictor*, https://en.wikipedia.org/wiki/Kerrison_Predictor, updated 2 April, 2018.

Willard, J.H., engineers journal compiled by Major J.H. Willard, US Army Corps of Engineers, Record Group 77, Entry 220, National Archives, undated.

Willard, J.H., Major, US Army Corps of Engineers, "Fire Control," *Journal*, Record Group 77, Entry 220, National Archives, Washington, DC, 1920.

Williams, P., and Wallace, D., *Unit 731: Japan's Secret Biological Warfare in World War II*, ISBN: 0029353017, Free Press, New York, NY, March, 1989.

Wolfgast, O.W., "Camouflage: Harbor Defenses of San Diego," Fort Rosecrans, San Diego, CA, 1943.

WSM, "Our Boats," http://www.steamboat.co.uk/index.php?page=boats, Windermere Steamboats Museum, Cumbria, England, retrieved 5 November, 2006.

Wynne, L.P., "Ramblings of an Old Vector Vendor," *The COL Wynne Story – Part 1*, JOURNAL –Stories, 610th AC&W Squadron, Radar Sites – Southern Japan, 6 April, 2005.

Zahl, H.A., *Radar Spelled Backwards*, Vantage Press, New York, NY, 1972.

Zink, R.D., Grobmeier, A.H., "Forts of Wherever," *Coast Defense Study Group News*, Vol. 5, No. 1, p. 91, February, 1991.

APPENDIX A
Chronological listing of the San Diego Naval Laboratory names

1940 - Navy Radio and Sound Laboratory (NRSL)

1945 - U.S. Navy Electronics Laboratory (NEL)

1967 - Naval Command Control and Communications Laboratory Center

1967 - Naval Electronics Laboratory Center (NELC)

1977 - Naval Ocean Systems Center (NOSC)

1992 - Naval Command, Control and Ocean Surveillance Center Research, Development, Test and Evaluation Division (NRaD)

1997 - Space and Naval Warfare Systems Center San Diego (SSC San Diego)

2008 - Space and Naval Warfare Systems Center Pacific (SSC Pacific)

2019 - Naval Information Warfare Center Pacific (NIWC Pacific).

Appendix B

Gun Batteries Built for the Harbor Defenses of San Diego

FORT EMORY Coronado Heights Former Navy radio facility, currently used for training

Battery name	# of guns	caliber	carriage type	service years	status/notes
#134	2	16″	C-BC	1944 NC	unofficially named Gatchell
Grant (#239)	2	6″	S-BC	1943-1946	
Imperial	4	155 mm	PM	1942-1943?	
AMTB Cortez	2	90 mm	Fixed	1943-1946	Coronodo Beach MR, Silver Strand, destroyed

FORT PIO PICO North Island 1906-1919 (to Navy 1935) Naval Air Station

Battery name	# of guns	caliber	carriage type	service years	status/notes
Meed	2	3″	Ped	1902-1919	guns to Rosecrans, destroyed
AMTB Pio Pico	2	90 mm	Fixed	1942-1943	guns later moved to AMTB Fetterman

FORT ROSECRANS Point Loma 1899-1958 Navy Base, Natl. Cem. Cabrillo N.M. /MD, MC

Battery name	# of guns	caliber	carriage type	service years	status/notes
Ashburn (#126)	2	16″	C-BC	1943-1948	modified
Whistler	4	12″	Mortar	1916-1942/	guns from DeSoto, modified
White	4	12″	Mortar	1916-1942	
Wilkeson	2	10″	DC	1900-1943	
Calef	2	10″	DC	1900-1943	
Strong	2	8″	BCLRN	1941-1946	
#237	2	6″	S-BC	1943-1946	unofficially named Woodward
Humphreys (#238)	2	6″	S-BC	1943-1946	
Zeilin	2	7″	NP	1937-1943	buried
Gillespie	3	5″	NP	1937-1943	empl. 1 destroyed, 2 & 3 overgrown
McGrath	2	5″	BP	1900-1943	repl. by 2-3″ guns from Pio Pico
Fetterman	2	3″	MP	1900-1917	destroyed July 1940
AMTB Fetterman	2	90 mm	F	1943-1946	destroyed
AMTB Cabrillo	2	90 mm	F	1943-1946	destroyed
North	4	155 mm	PM	? -1941	
Point Loma	4	155 mm	PM	1941-1943?	

Distinctive Unit Insignia, 19th Coast Artillery Regiment (Harbor Defense)

Appendix C

Military Maps of the Harbor Defenses of San Diego

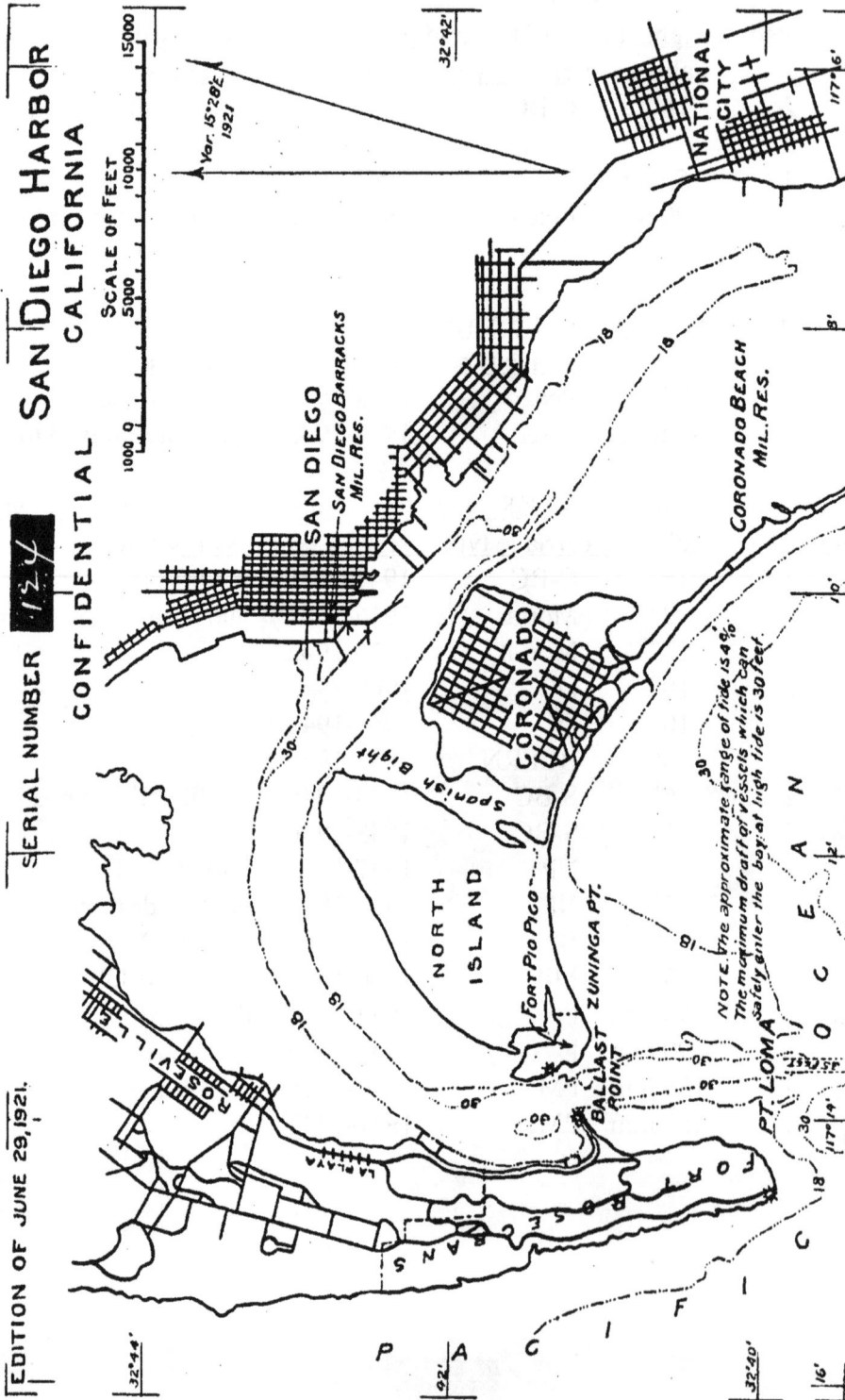

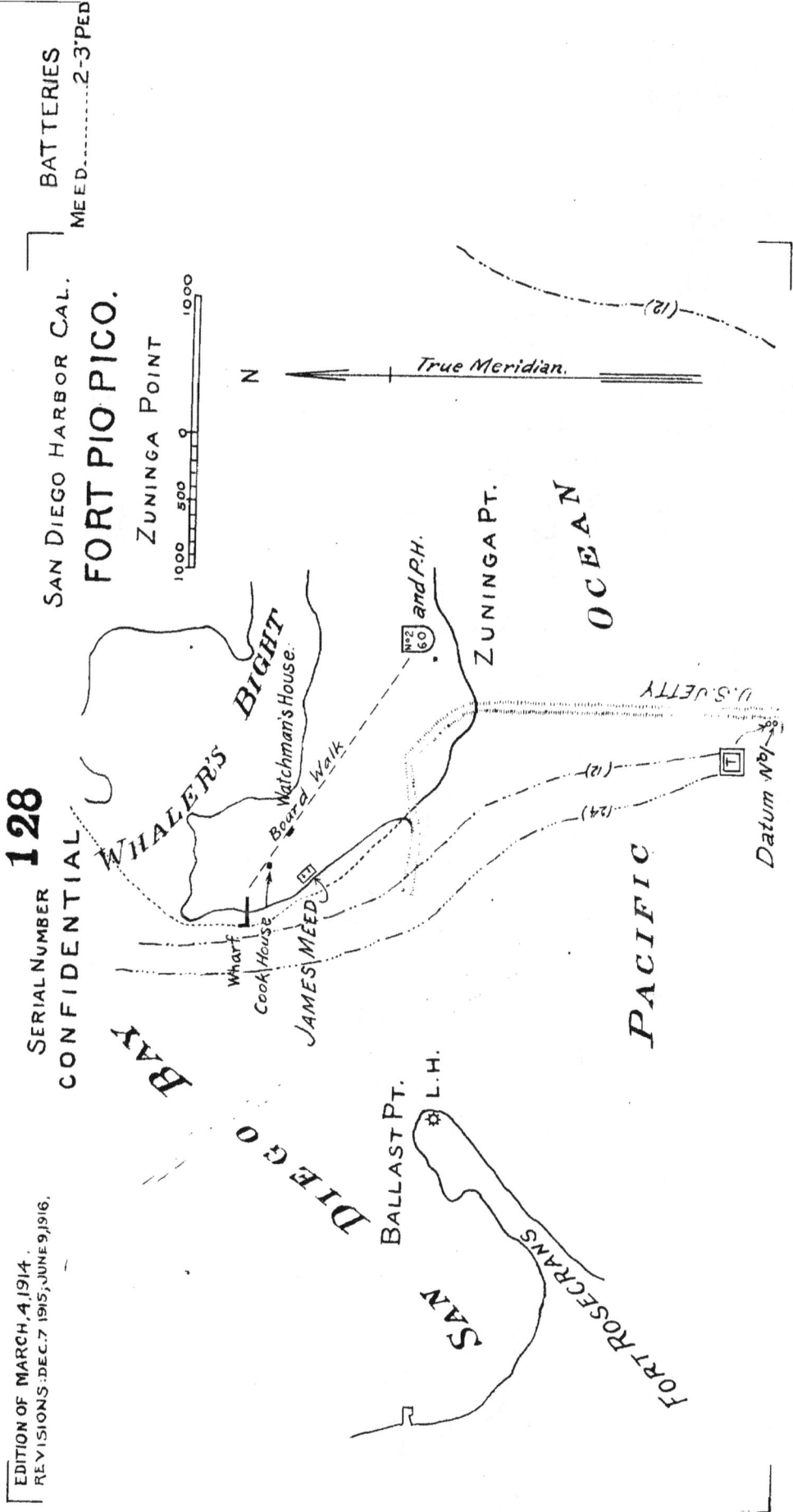

SERIAL NUMBER **128**
CONFIDENTIAL

San Diego Harbor Cal.
FORT PIO PICO.
Zuninga Point

BATTERIES
MEED............2-3"PED.

EDITION OF MARCH, 4,1914.
REVISIONS: DEC.7 1915; JUNE 9,1916.

SAN DIEGO BAY

WHALER'S BIGHT

Watchman's House.
Board Walk
Wharf
Cook House
JAMES MEED

N⁰2
60 and P.H.

ZUNINGA PT.

OCEAN

PACIFIC

U.S.JETTY
Datum N⁰.

BALLAST PT.
L.H.

SAN DIEGO

FORT ROSECRANS

True Meridian.
N

1000 500 1000

FORT ROSECRANS, CAL.

SERIAL NUMBER 124

GENERAL MAP. CONFIDENTIAL

EDITION OF JUNE 29, 1921.

True Meridian

BATTERIES

WHITE	4-12"M.
WHISTLER	4-12"M.
WILKESON	2-10"Dis.
CALEF	2-10" "
FETTERMAN	2-3"B.P.
McGRATH	2-3"PED.

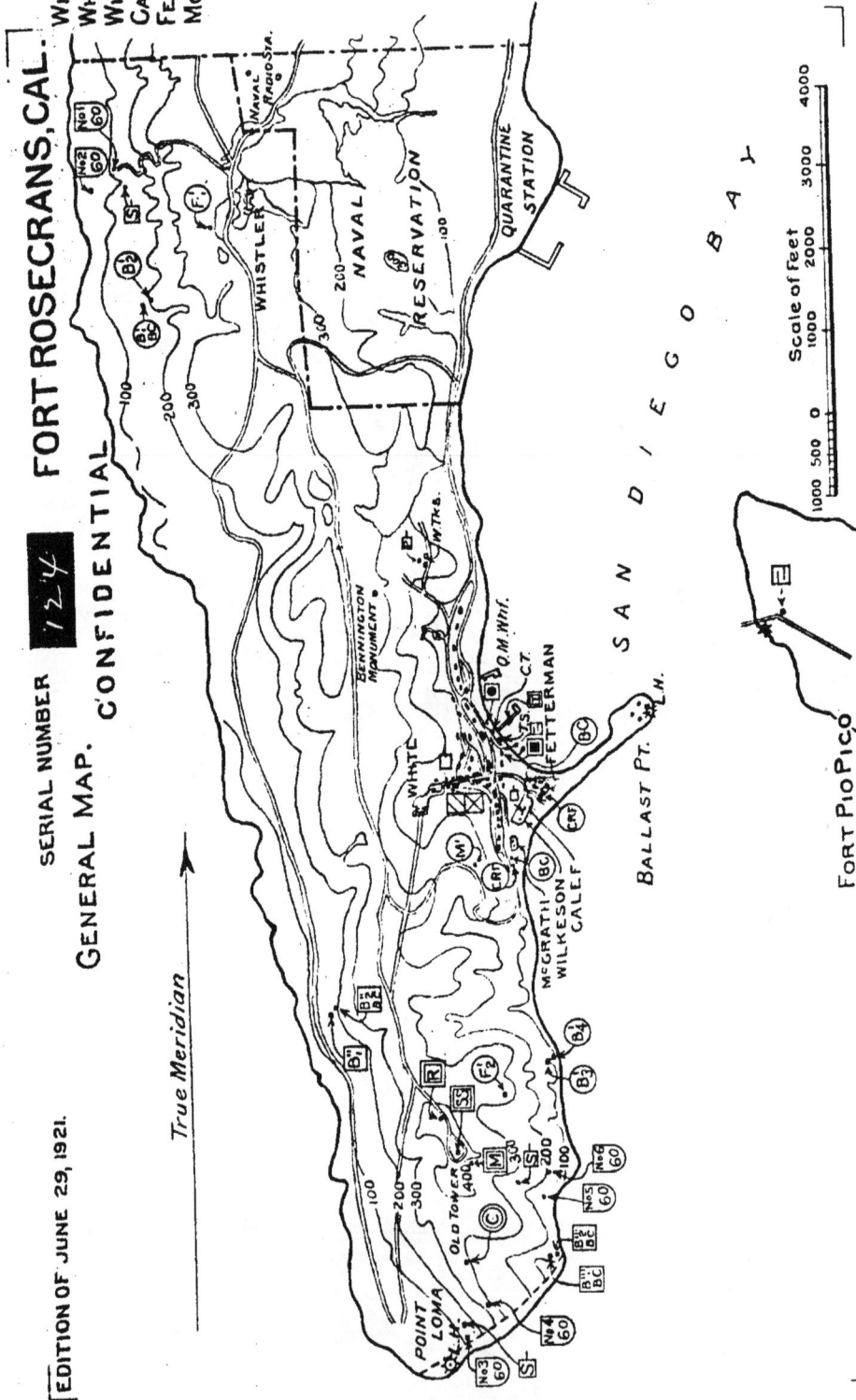

NAVAL RADIO STA.

WHISTLER

NAVAL RESERVATION

QUARANTINE STATION

BENNINGTON MONUMENT

O.M.Whf.

FETTERMAN

WHITE

BALLAST PT.

McGRATH
WILKESON
CALEF

SAN DIEGO BAY

POINT LOMA

OLD TOWER

Scale of Feet

1000 500 0 1000 2000 3000 4000

FORT PIO PICO

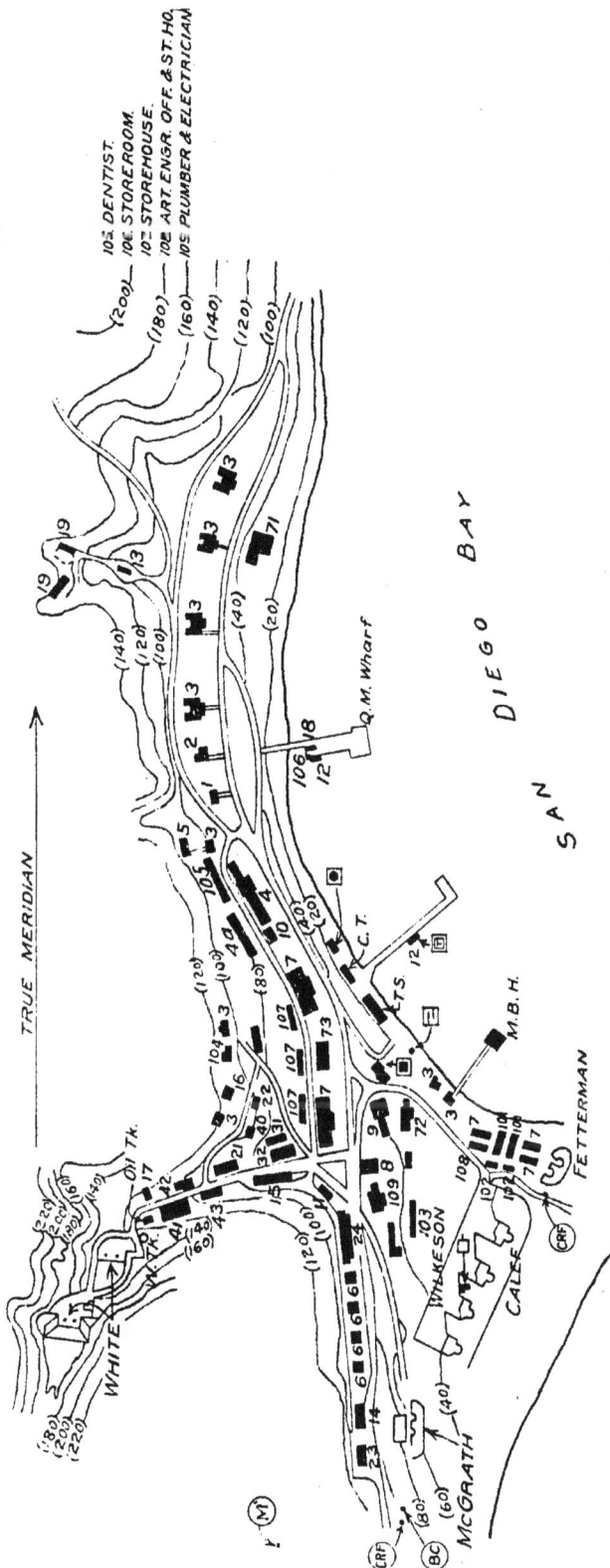

SAN DIEGO HARBOR, CAL.

FORT ROSECRANS — DI

CONFIDENTIAL

SERIAL NUMBER 474

EDITION OF JUNE 29, 1921.

Scale of Feet

BATTERIES

WHITE	4-12"M.
WILKESON	2-10"Dis.
CALEF	2-10"
McGRATH	2-3"Ped.
FETTERMAN	2-3"B.P.

105 DENTIST.
106 STOREROOM.
107 STOREHOUSE.
108 ART. ENGR. OFF. & ST. HO.
109 PLUMBER & ELECTRICIAN

LEGEND

1. ADMINISTRATION BLDG.
2. COMMANDING OFF. QRS.
3. OFFICERS' QUARTERS.
4. HOSPITAL.
4a. HOSPITAL WARD.
5. HOSPITAL SERGEANT.
6. N.C. OFFICERS QRS.
7. BARRACKS.
8. GUARD HOUSE.
9. POST EXCHANGE.
10. FIRE STATION.
11. BAKERY.
12. BOAT HOUSE.
13. OIL HOUSE.
14. WAGON SHED.
15. BLACKSMITH SHOP AND COAL SHED.
16. CIV. EMPLS. QRS.
17. TRANSFORMER AND SWITCHBOARD BLDG.
18. BATH HOUSE.
19. CHICKEN HOUSE.
21. Q.M. WORKSHOP.
22. Q.M. & COM. BLDG.
23. Q.M. STABLES.
24. Q.M. & COM. STO. HO.
31. ORD. REPAIR SHOP.
32. ORD. STOREHOUSE.
40. E.D. OFFICE.
41. E.D. SHOP.
42. E.D. STOREHOUSE.
71. E.D. REPAIR SHOP.
72. SCHOOL, E.&R.
73. SERVICE CLUB.
101. OFFICERS' CLUB.
102. MESS HALL.
103. LAVATORY.
104. M.T. GARAGE.
104. ART. ENGR. QRS.

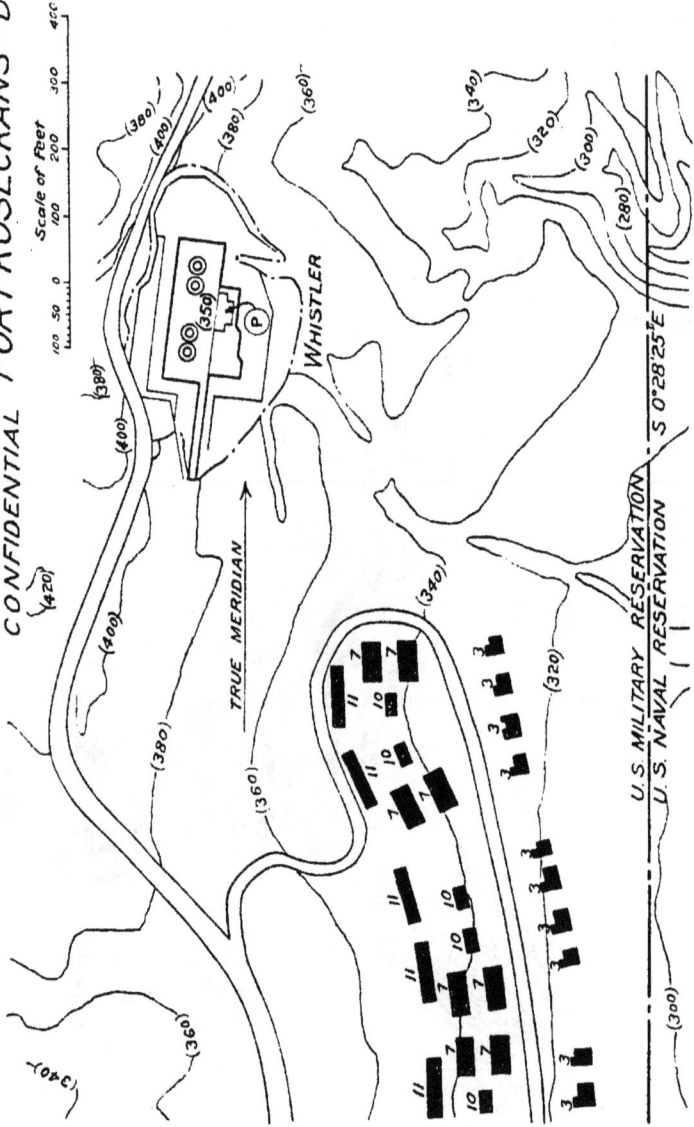

LEGEND.

1.
2.
3 OFFICERS' QUARTERS
4.
5.
6.
7. BARRACKS.
8.
9.
10 LAVATORY.
11. MESS HALL & KCHN.

EDITION OF JUNE 29, 1921.

SERIAL NUMBER

CONFIDENTIAL   FORT ROSECRANS - D2

SAN DIEGO HARBOR, CAL.

BATTERIES.

WHISTLER......4-12"M.

Scale of Feet

TRUE MERIDIAN

WHISTLER

U.S. MILITARY RESERVATION

U.S. NAVAL RESERVATION

S 0°28'25" E

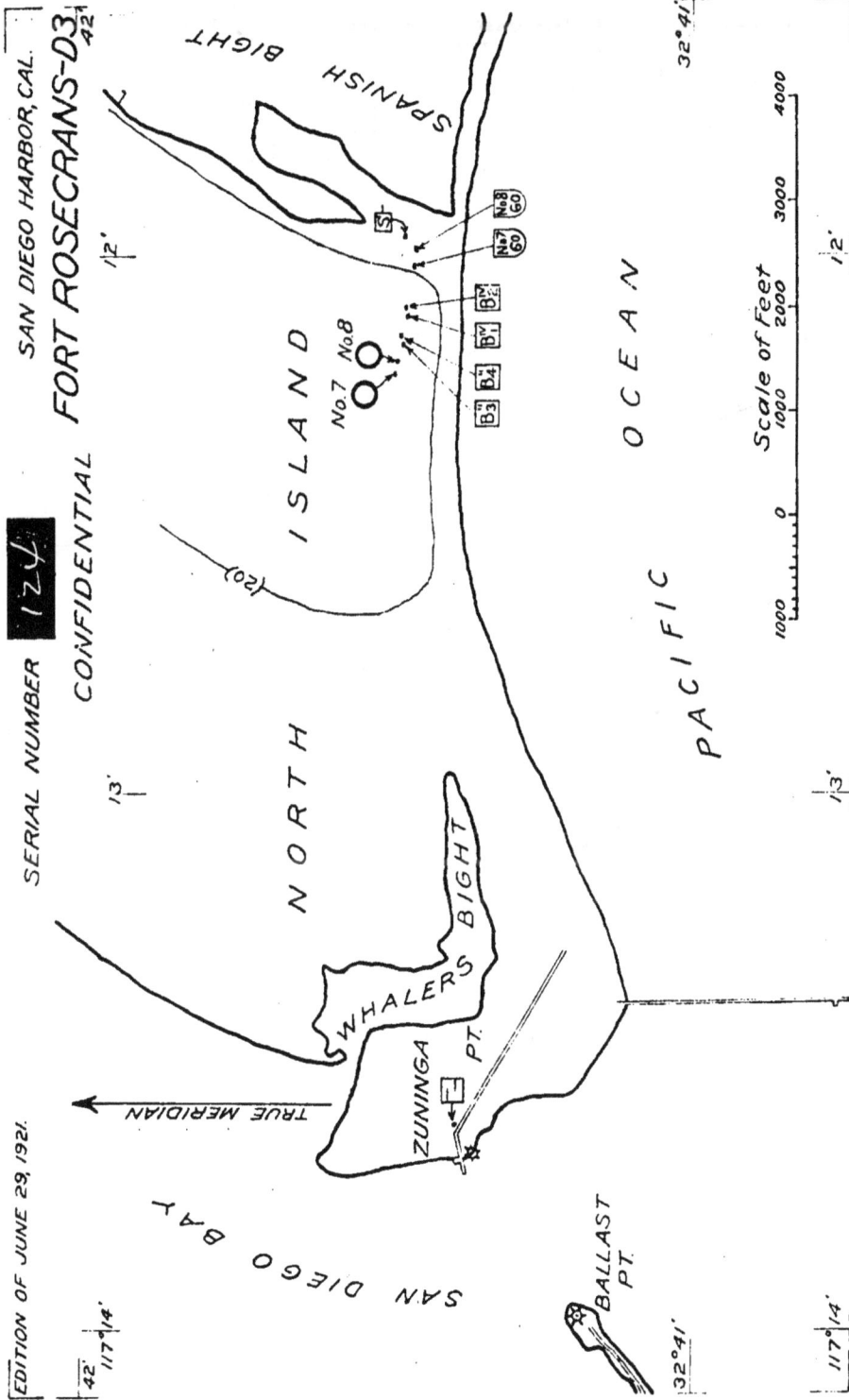

SAN DIEGO HARBOR, CAL.

FORT ROSECRANS—D3

CONFIDENTIAL

SERIAL NUMBER

EDITION OF JUNE 29, 1921.

TRUE MERIDIAN

NORTH   ISLAND

SPANISH   BIGHT

WHALERS   BIGHT

ZUNINGA   PT.

SAN DIEGO BAY

PACIFIC   OCEAN

BALLAST PT.

Scale of Feet

No.7   No.8

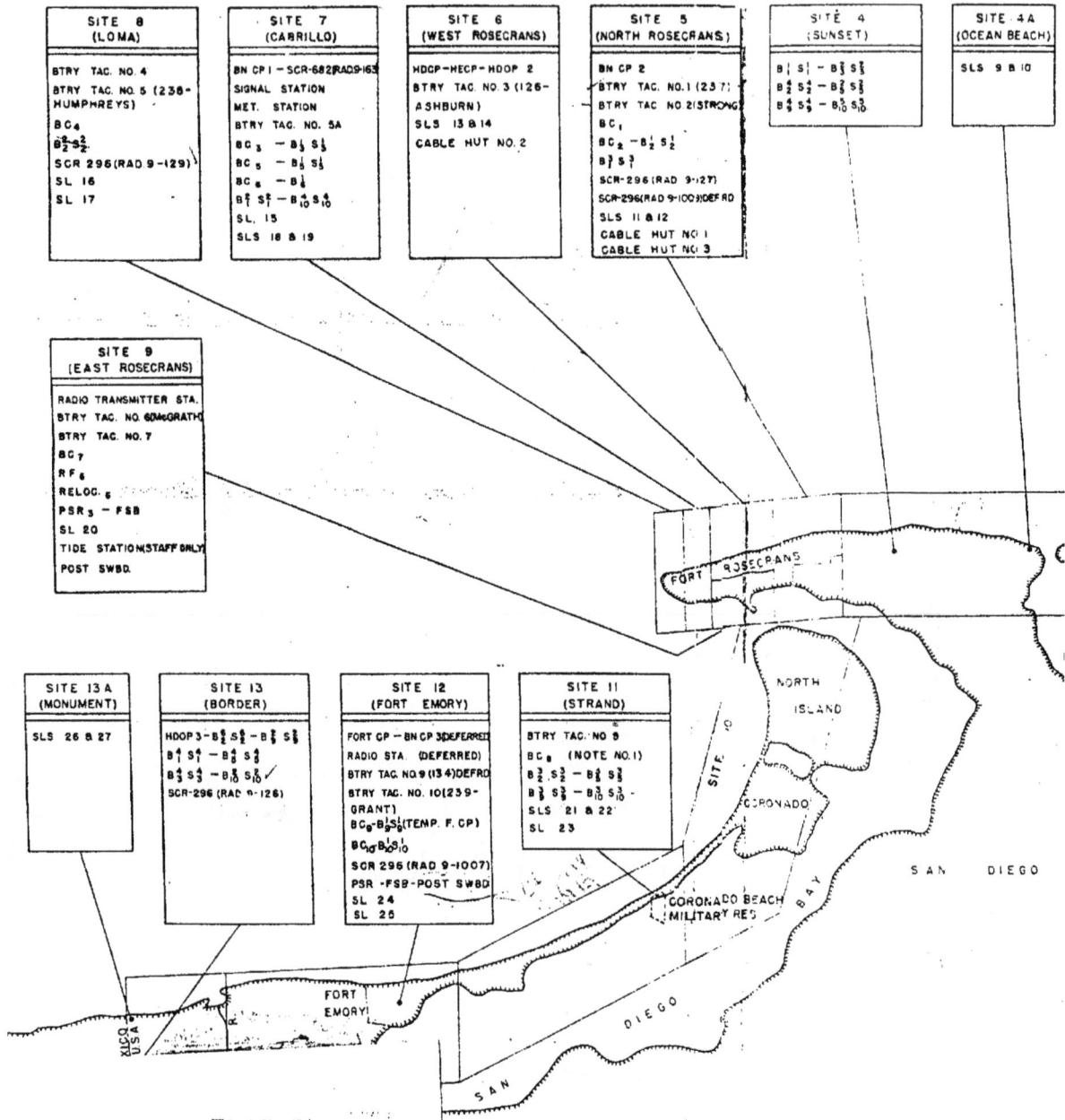

SITE 8 (LOMA)
BTRY TAC. NO. 4
BTRY TAC. NO. 5 (238-HUMPHREYS)
BC 4
$B_2^4 S_2^4$
SCR 296 (RAD 9-129)
SL 16
SL 17

SITE 7 (CABRILLO)
BN CP I — SCR-682(RAD9-163)
SIGNAL STATION
MET. STATION
BTRY TAC. NO. 5A
BC 3 — $B_3^1 S_3^1$
BC 5 — $B_5^1 S_5^1$
BC 6 — B_6^1
$B_7^4 S_7^4 - B_{10}^4 S_{10}^4$
SL 15
SLS 18 & 19

SITE 6 (WEST ROSECRANS)
HDCP - HECP - HDOP 2
BTRY TAC. NO. 3 (126-ASHBURN)
SLS 13 & 14
CABLE HUT NO. 2

SITE 5 (NORTH ROSECRANS)
BN CP 2
BTRY TAC. NO.1 (237)
BTRY TAC. NO.2 (STRONG)
BC 1
BC 2 — $B_2^1 S_2^1$
$B_7^1 S_7^1$
SCR-296 (RAD 9-27)
SCR-296(RAD 9-100)DEFRD
SLS 11 & 12
CABLE HUT NO. 1
CABLE HUT NO. 3

SITE 4 (SUNSET)
$B_1^1 S_1^1 - B_3^2 S_3^2$
$B_2^4 S_2^4 - B_6^2 S_6^2$
$B_4^4 S_4^4 - B_{10}^5 S_{10}^5$

SITE 4A (OCEAN BEACH)
SLS 9 & 10

SITE 9 (EAST ROSECRANS)
RADIO TRANSMITTER STA.
BTRY TAC. NO. 6(McGRATH)
BTRY TAC. NO. 7
BC 7
RF 6
RELOC. 6
PSR 3 — FSB
SL 20
TIDE STATION(STAFF ONLY)
POST SWBD.

SITE 13 A (MONUMENT)
SLS 26 & 27

SITE 13 (BORDER)
HDOP 3 - $B_3^4 S_3^4 - B_3^2 S_3^2$
$B_4^4 S_4^4 - B_2^4 S_2^4$
$B_3^3 S_3^3 - B_{10}^3 S_{10}^3$
SCR-296 (RAD 9-126)

SITE 12 (FORT EMORY)
FORT CP — BN CP 3(DEFERRED)
RADIO STA. (DEFERRED)
BTRY TAC. NO.9 (134)DEFRD
BTRY TAC. NO. 10(239-GRANT)
BC 9-$B_9^1 S_9^1$(TEMP. F. CP)
BC 10-$B_{10}^1 B_9^1$
SCR 296 (RAD 9-1007)
PSR - FSB-POST SWBD
SL 24
SL 25

SITE 11 (STRAND)
BTRY TAC. NO 9
BC 8 (NOTE NO.1)
$B_2^3 S_2^3 - B_4^1 S_4^1$
$B_3^1 S_3^1 - B_{10}^2 S_{10}^2$
SLS 21 & 22
SL 23

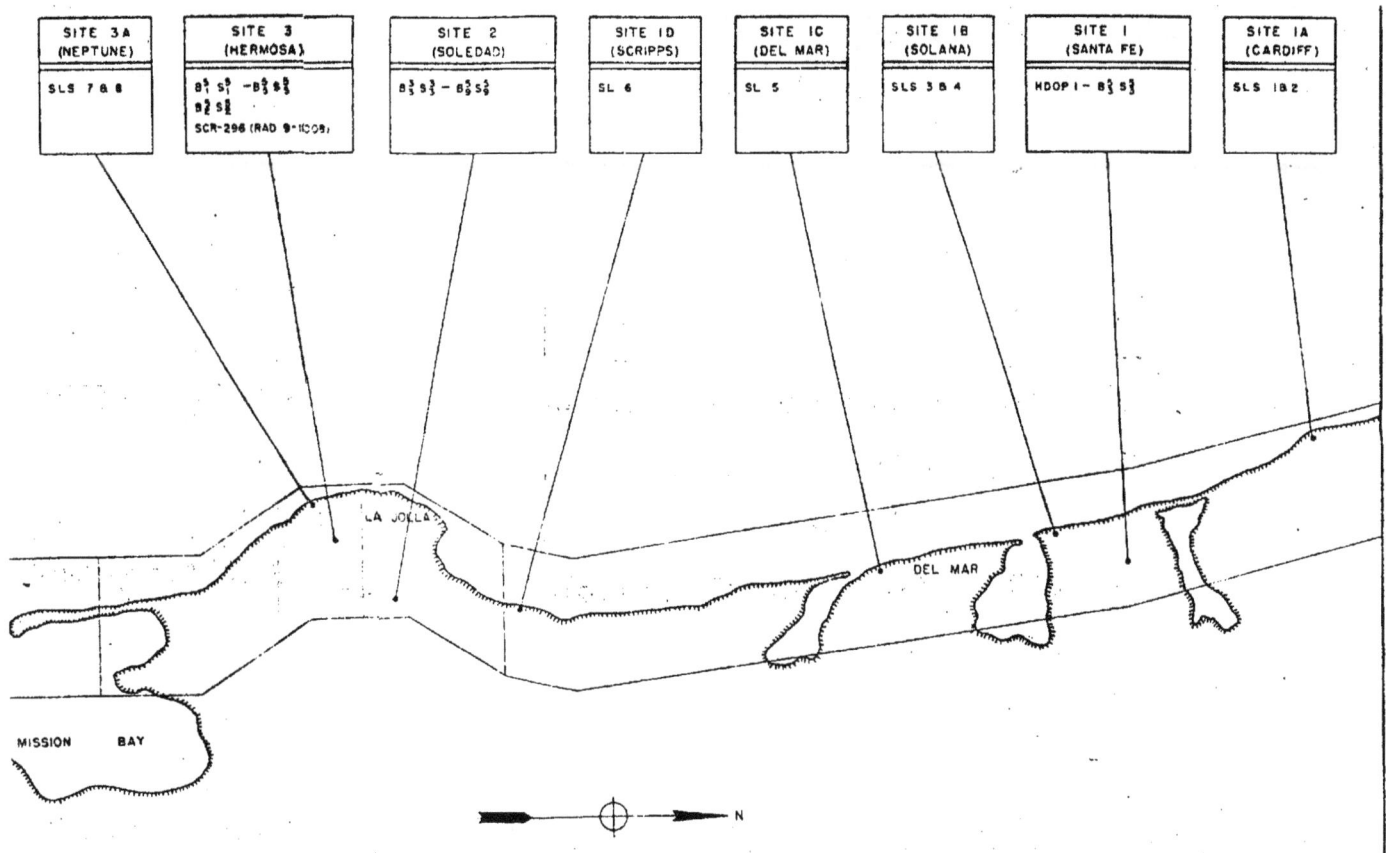

SECRET

LEGEND

(126) = BATTERY CONSTR. NO. 126

NOTE-1- BC TO BE CONSTRUCTED WHEN REQUIRED

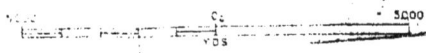

HARBOR DEFENSES OF SAN DIEGO	REVISED DATE
HARBOR DEFENSE ELEMENTS	

ER	DATE 1 JULY 1945
ALIFORNIA	EXHIBIT 1-A

HARBOR DEFENSES OF SAN DIEGO

FIRE CONTROL STATIONS AT SOUTH
FORT ROSECRANS, SITES 6, 7, 8, 9

Map Incomplete

DECLASSIFIED
NND790162

SECRET

HARBOR DEFENSES OF SAN DIEGO

FIRE CONTROL STATIONS AT NORTH
FORT ROSECRANS, SITES 5, 6, 9

PREPARED BY   DATE 1 JULY 1945

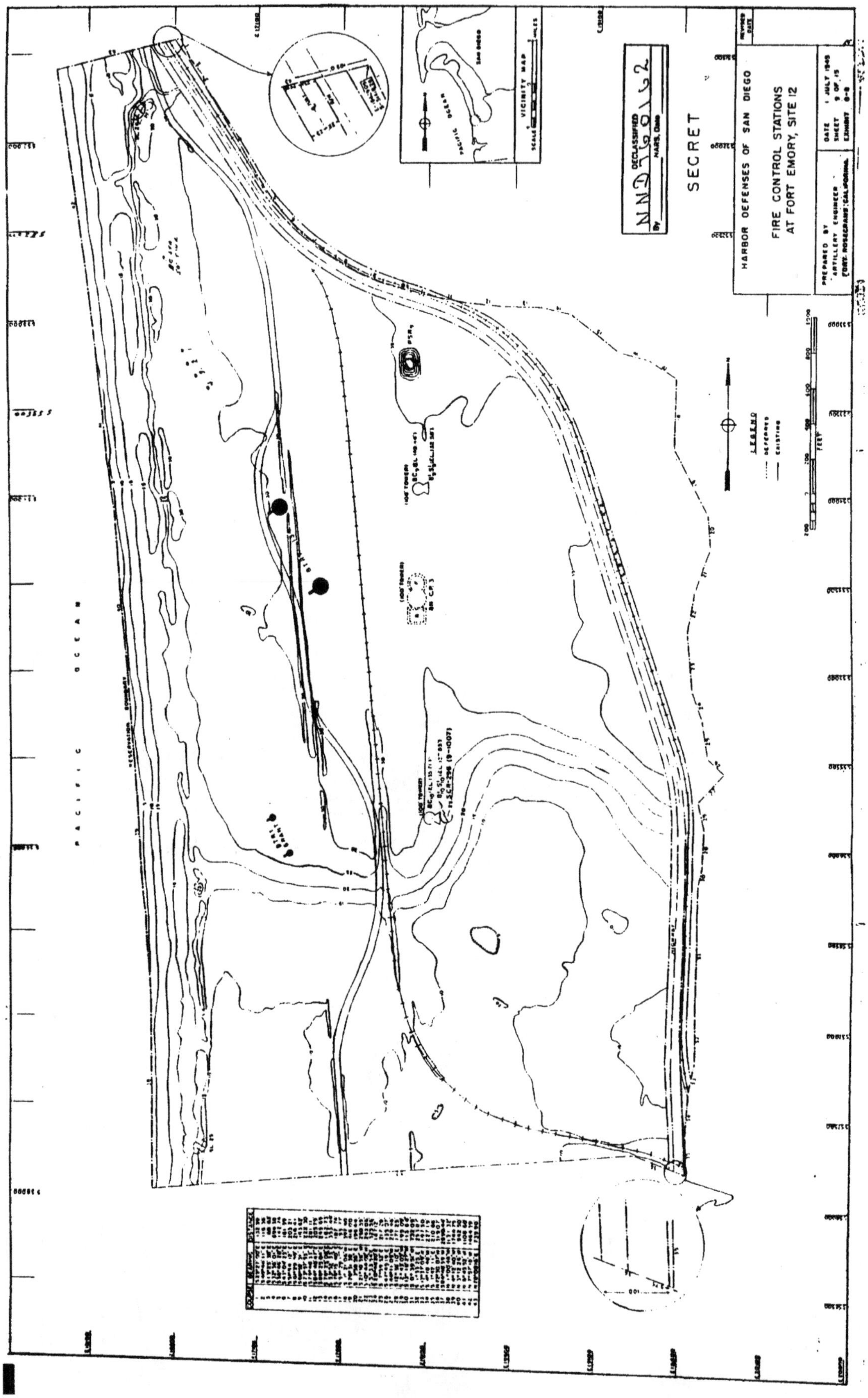
HARBOR DEFENSES OF SAN DIEGO
FIRE CONTROL STATIONS
AT FORT EMORY, SITE 12
SECRET
DECLASSIFIED
VICINITY MAP
PACIFIC OCEAN
LEGEND
FEET

Commander (Ret.) H.R. (Bart) Everett, USN, is the former Technical Director for Robotics at the Space and Naval Warfare Systems Center Pacific in San Diego, CA. In this capacity, he also served as technical director for the Idaho National Laboratory's Advanced Unmanned Systems Development program, technical director for the U.S. Army's Mobile Detection Assessment Response System program, and chief engineer for the USMC Ground Air TeleRobotic System program.

Other books by H.R. Everett:

Sensors for Mobile Robots: Theory and Application, A.K. Peters, Ltd., Wellesley, MA, 1995.

Navigating Mobile Robots: Systems and Techniques (with Borenstein and Feng), A.K. Peters, Ltd., Wellesley, MA, 1996.

Unmanned Systems of World Wars I and II, MIT Press, Cambridge, MA, 2015.

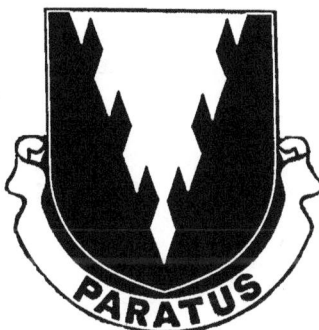

Coat of Arms for the Harbor Defenses of San Diego

Coast Artillery Journal Volume 70 (January 1929) page 72.

Shield: *Azure,* a pile raguly *or.*
Crest: On a wreath of the colors *or* and *azure*
an anchor proper (grayish) behind an eight-pointed mullet of rays *or.*
Motto: *Paratus* (Prepared).

The blue shield and the yellow pile are symbolic of the blue ocean and the yellow land of Point Loma. The place was first visited by the Spaniards, Cabrillo in 1542, and the edges of the pile are made raguly (ragged) as the Spanish flag at that time bore a cross.

The crest (not shown) symbolizes the hardest fought battle of the Mexican War in California, near San Diego, at San Pasquale, December 6, 1846. General Stephen W. Kearny commanded the Americans, consisting of one company of the First Dragoons, a few sailors sent by Commodore Stockton from San Diego, and a volunteer company from San Diego. The anchor commemorates Stockton's sailors, and Kearny's Dragoons wore on their helmets the eight pointed gold star of rays.

Index

The Coast Defense Study Group, Inc. (CDSG) is a tax-exempt corporation dedicated to study of seacoast fortifications. CDSG's purpose is to promote and encourage the study of coastal defenses, primarily but not exclusively those of the United States of America. The study of coast defenses and fortifications in- cludes their history, architecture, technology, strategic and tactical employment and evolution. The primary goals of the CDSG are the following:

- Educational study of coast defenses
- Technical research and documentation of coast defenses
- Preservation of coast defense sites, equipment and records for current and future generations
- Accurate coast defense site interpretations
- Assistance to groups interested in preservation and interpretation of coast defense sites
- Charitable activities which promote the goals of the CDSG

Membership is open to any person or organization interested in the study or history of the coast defenses and fortifications. Membership in the CDSG will allow you to attend the annual conference, special tours and receive the CDSG's quarterly journal and newsletter. For more information on the CDSG, please visit the CDSG website at cdsg.org or contact us at 24624 W. 96th Street, Lenexa, KS 66227-7285 USA, Attn: Quentin Schillare, Membership.

The CDSG Fund supports the efforts of the Coast Defense Study Group by raising funds for preservation and interpretation of American seacoast defenses. The CDSG Fund is seeking donations for projects sup- porting its goals. Donations are tax-deductible for federal tax purposes as the CDSG is a 501(c)(3) orga- nization, and 100% of your gift will go to project grants. Major contributions are acknowledged annually. The Fund is always seeking proposals for the monetary support of preservation and interpretation projects at former coast defense sites and museums. A one-page proposal briefly describing the site, the organization doing the work, and the proposed work or outcome should be sent to the address below. Successful proposals are usually distinct projects rather than general requests for donations. Upon conclusion of a project a short report suitable for publication in the CDSG Newsletter is requested. The trustees shall re- view such requests and pass their recommendation onto the CDSG Board of Directors for approval. Send donations and grant requests to: CDSG Fund c/o Terry McGovern 1700 Oak Lane McLean, VA 22101- 3326 USA or use your credit card via PayPal on the cdsg.org website.

The CDSG ePress
The CDSG Press

CDSG Books and CDSD Gear ($ domestic / $ International) prices include domestic/international postage $US currency only (cash, check, money order or credit card via PayPal at the cdsg.org store), allow 6-8 weeks for delivery.

- *Notes on Seacoast Fortification Construction* by Col. Eben E. Winslow, 1920, 428 pp. 1994 reprint HC with drawings $45/$60
- *Seacoast Artillery Weapons Technical Manual* (TM) 9-210 by U.S. War Dept. 1944, 202 pp. 1995 reprint PB $25/$35
- *The Service of Coast Artillery* by F. Hines & F. Ward, 1910, 736 pp. 1997 reprint HC $40/$60
- *Permanent Fortifications & Sea-Coast Defences* by U.S. Congress, 1862, 544 pp. 1998 reprint HC
- $30/$45
- *American Coast Artillery Matériel* Ordnance Dept. Doc#2042 by U.S. War Dept., 1922, 528 pp., 2001 reprint HC $45/$65
- *American Seacoast Defenses: A Reference Guide* (3rd Edition) by Mark A. Berhow, (2015) 732 pp. HC $45/$80
- The Endicott & Taft Board Reports, reprint of original reports of 1886 and 1905 by U.S. Congress, 525 pp. 2007 reprint HC $45/$80
- *Artillerists and Engineers: The Beginnings of US Fortifications 1794-1815* by Col. Wade, U.S. Army. PB, 226 pp. $25/$40
- CDSG Logo Hats each $20.00 domestic and $25.00 foreign. CDSG Logo Patches each $ 4.00 domestic & foreign.
- CDSG T-Shirts (XXXL, XXL, XL, L; Red, Khaki, Navy, Black) $18.00 Domestic and $26.00 Foreign.

Send order to: CDSG Press Attn: Terry McGovern 1700 Oak Lane, McLean, VA 22101-3326
Or order via our online store at cdsg.org.

The CDSG Digital Library
The CDSG has digitized an extensive set of historic manuals, reports, records and documents on the harbor defenses of the United States Army.

- The CDSG provides back issues of the CDSG Publications (from 1985) in electronic format.
- The CDSG Documents covers a range of historical material related to seacoast defenses -- most are from the National Archives. Included are the annual reports of the chief of coast artillery and chief of engineers; several board proceedings and reports; army directories; text books; tables of organi- zation and equipment; WWII command histories; drill, field, training manuals and regulations; ordnance department documents; ordnance tables and compilations; and the ordnance gun and carriage cards.
- CDSG Documents related to specific harbor defenses. These PDF documents form the basis of the conference and special tour handouts that have been held at these locations. They include RCBs/RCWs; maps; annexes to defense projects; CD engineer notebooks; quartermaster building records; and aerial photos taken by the signal corps 1920-40. Please consult cdsg.org for more details.

Information on the CDSG ePress items can be obtained from info@cdsg.org.

www.ingramcontent.com/pod-product-compliance
Lightning Source LLC
Chambersburg PA
CBHW062019090426
42811CB00005B/897